ADVANCED OXIDATION PROCESSES FOR WASTEWATER TREATMENT

ADVANCED OXIDATION PROCESSES FOR WASTEWATER TREATMENT

Emerging Green Chemical Technology

Edited by

SURESH C. AMETA
PAHER University, Udaipur, Rajasthan, India

RAKSHIT AMETA
J. R. N. Rajasthan Vidyapeeth (Deemed-to-be University), Udaipur, Rajasthan, India

ACADEMIC PRESS
An imprint of Elsevier

Academic Press is an imprint of Elsevier
125 London Wall, London EC2Y 5AS, United Kingdom
525 B Street, Suite 1800, San Diego, CA 92101-4495, United States
50 Hampshire Street, 5th Floor, Cambridge, MA 02139, United States
The Boulevard, Langford Lane, Kidlington, Oxford OX5 1GB, United Kingdom

Notices
Knowledge and best practice in this field are constantly changing. As new research and experience broaden our
understanding, changes in research methods, professional practices, or medical treatment may become necessary.

Practitioners and researchers must always rely on their own experience and knowledge in evaluating and using
any information, methods, compounds, or experiments described herein. In using such information or methods
they should be mindful of their own safety and the safety of others, including parties for whom they have a
professional responsibility.

To the fullest extent of the law, neither the Publisher nor the authors, contributors, or editors, assume any liability
for any injury and/or damage to persons or property as a matter of products liability, negligence or otherwise, or
from any use or operation of any methods, products, instructions, or ideas contained in the material herein.

British Library Cataloguing-in-Publication Data
A catalogue record for this book is available from the British Library

Library of Congress Cataloging-in-Publication Data
A catalog record for this book is available from the Library of Congress

ISBN: 978-0-12-810499-6

For Information on all Academic Press publications
visit our website at https://www.elsevier.com/books-and-journals

Working together
to grow libraries in
developing countries

www.elsevier.com • www.bookaid.org

Publisher: Candice Janco
Acquisition Editor: Louisa Hutchins
Editorial Project Manager: Emily Thomson
Production Project Manager: Swapna Srinivasan
Cover Designer: Miles Hitchen

Typeset by MPS Limited, Chennai, India

Dedication

Dedicated to:
Mr. Panna Lal Ameta, who encouraged us at all times;
and to Mrs. Sushila Sharma, who took care of our early education.

Contents

12. Catalytic Wet Peroxide Oxidation

ALI R. TEHRANI-BAGHA AND TAREK BALCHI

List of Contributors

Rakshit Ameta J. R. N. Rajasthan Vidyapeeth (Deemed-to-be University), Udaipur, Rajasthan, India

Suresh C. Ameta PAHER University, Udaipur, Rajasthan, India

Mohammad M. Amin Isfahan University of Medical Sciences, Isfahan, Iran

Ghorban Asgari Hamadan University of Medical Sciences, Hamadan, Iran

Tarek Balchi American University of Beirut (AUB), Beirut, Lebanon

Surbhi Benjamin PAHER University, Udaipur, Rajasthan, India

Krzysztof Bobrowski Institute of Nuclear Chemistry and Technology, Warsaw, Poland; Radiation Laboratory, University of Notre Dame, Notre Dame, IN, United States

Anna Bojanowska-Czajka Institute of Nuclear Chemistry and Technology, Warsaw, Poland

Afsane Chavoshani Isfahan University of Medical Sciences, Isfahan, Iran

Anil K. Chohadia M. P. Govt. P. G. College, Chittorgarh, Rajasthan, India

Majid Hashemi Kerman University of Medical Sciences, Kerman, Iran

Keisuke Ikehata Pacific Advanced Civil Engineering, Inc., Fountain Valley, CA, United States

Abhilasha Jain St. Xavier's College, Mumbai, Maharashtra, India

Yuan Li Pacific Advanced Civil Engineering, Inc., Fountain Valley, CA, United States

Enrique J. Martínez de la Ossa University of Cádiz, Puerto Real, Spain

José C. Mierzwa University of São Paulo, São Paulo, SP, Brazil

Juan R. Portela University of Cádiz, Puerto Real, Spain

Dorian Prato-Garcia National University of Colombia, Palmira, Colombia

Pinki B. Punjabi M. L. Sukhadia University, Udaipur, Rajasthan, India

Kuravappullam V. Radha Anna University, Chennai, India

Raphael Rodrigues University of São Paulo, São Paulo, SP, Brazil

Jezabel Sánchez-Oneto University of Cádiz, Puerto Real, Spain

Abdolmotaleb Seidmohammadi Hamadan University of Medical Sciences, Hamadan, Iran

Efraim A. Serna-Galvis University of Antioquia, Medellín, Colombia

Karunamoorthy Sirisha Anna University, Chennai, India

Meenakshi S. Solanki B. N. University, Udaipur, Rajasthan, India

Tomasz Szreder Institute of Nuclear Chemistry and Technology, Warsaw, Poland

Ali R. Tehrani-Bagha American University of Beirut (AUB), Beirut, Lebanon

Antonio C.S.C. Teixeira University of São Paulo, São Paulo, SP, Brazil

Ricardo A. Torres-Palma University of Antioquia, Medellín, Colombia

Marek Trojanowicz Institute of Nuclear Chemistry and Technology, Warsaw, Poland; University of Warsaw, Warsaw, Poland

Violeta Vadillo   University of Cádiz, Puerto Real, Spain
Ruben Vasquez-Medrano   Iberoamericana University, Mexico City, Mexico
Michel Vedrenne   Technical University of Madrid (UPM), Madrid, Spain

About the Authors

Rakshit Ameta, PhD, is an Associate Professor of Chemistry, J.R.N. Rajasthan Vidyapeeth (Deemed-to-be University), Udaipur, India. He has several years of experience in teaching and research in chemistry as well as industrial chemistry and polymer science. He is presently guiding seven research students for their PhD theses, and several students have already obtained their PhDs under his supervision in green chemistry. Dr. Rakshit Ameta has received various awards and recognition in his career, including being awarded first position and the gold medal for his MSc, and receiving the Fateh Singh Award from the Maharana Mewar Foundation, Udaipur, for his meritorious performance. He has served at M.L. Sukhadia University, Udaipur; the University of Kota, Kota; and PAHER University, Udaipur. He has over 80 research publications to his credit in journals of national and international repute. He holds one patent, and two more are under way. Dr. Rakshit has organized several national conferences as Organizing Secretary at the University of Kota and PAHER University. He has delivered invited lectures and has chaired sessions in conferences held by the Indian Chemical Society and the Indian Council of Chemists. Dr. Rakshit was elected as a council member of the Indian Chemical Society, Kolkata (2011–13) and Indian Council of Chemists, Agra (2012–14) as well as Associate Editor, Physical Chemistry Section (2014–16) of Indian Chemical Society. Recently, Dr. Rakshit has been elected as Scientist-in-Charge in the Industrial and Applied Chemistry Section of Indian Chemical Society for 3 years (2014–16). He has written five degree-level books and has contributed chapters to books published by several international publishers. He has published three previous books with Apple Academic Press: *Green Chemistry: Fundamentals and Applications* (2014) and *Microwave-Assisted Organic Synthesis: A Green Chemical Approach* (2015), *Chemical Applications of Symmetry and Group Theory* (2016) and two books *Solar Energy Conversion and Storage* (2015) and *Photocatalysis: Principles and Applications* (2016) with Taylor and Francis. His research areas focus on wastewater treatment, photochemistry, green chemistry, microwave-assisted reactions, environmental chemistry, nanochemistry, solar cells, and bioactive and conducting polymers.

Suresh C. Ameta, PhD, is currently the Dean of Faculty of Science at PAHER University, Udaipur. He has served as Professor and Head of the Department of Chemistry at North Gujarat University, Patan (1994) and at M.L. Sukhadia University, Udaipur (2002–05), and as Head of the Department of Polymer Science (2005–08). He also served as Dean of Postgraduate Studies. He is presently guiding ten research students for their PhD theses, and 82 students have already obtained their PhDs under his guidance. Prof. Ameta has held the position as the President of Indian Chemical Society, Kolkata, and is now a lifelong Vice President. He was awarded a number of prestigious awards during his career, such as national prizes twice for writing chemistry books in Hindi; he received the Prof. M.N. Desai Award, the Prof. W.U. Malik Award, the National Teacher Award, the Prof. G.V. Bakore Award, Lifetime Achievement Awards by Indian Chemical Society and Indian Council of Chemists, etc. Indian Chemical Society, Kolkata has instituted an award in his honor as the Professor Suresh C. Ameta Award since 2003 and published a special issue of the journal on the occasion of his 60th birthday. With more than 350 research publications to his credit in journals of national and international repute, he is also the author of many undergraduate- and postgraduate-level books. He has published three previous books with Apple Academic Press: *Green Chemistry: Fundamentals and Applications* (2014) and *Microwave-Assisted Organic Synthesis: A Green Chemical Approach* (2015), *Chemical Applications of Symmetry and Group Theory* (2016), and two books *Solar Energy Conversion and Storage* (2015) and *Photocatalysis: Principles and Applications* (2016) with Taylor and Francis. He has also written chapters in books published by several international publishers. Prof. Ameta has delivered lectures and chaired sessions at national conferences throughout India and is a reviewer of a number of international journals. In addition, he has completed five major research projects from different funding agencies, such as DST, UGC, CSIR, and Ministry of Energy, Government of India.

Preface

The world is in a cancerous grip of ever increasing pollution of one kind or another, in which water pollution is the main culprit. Major sources of these hazardous organic wastes are industrial, domestic, hospital, municipal effluents, etc. These wastes represent one of the greatest challenges to environmental engineers. Continuous progress in industrialization, transportation, and urbanization supported by population growth and deforestation, is creating a regular pressure on the depleting freshwater resources in the majority of the world. Various methods of treatment of wastewaters are available, and used presently are associated with disadvantages. Therefore, there is an urgent need to develop an alternate method, which can deal with the problem of wastewater treatment, and in an eco-friendly manner. Here, advanced oxidation processes (AOPs) enter the scene. AOPs have been defined as "near ambient temperature and pressure water treatment processes involving the generation of hydroxyl radicals in sufficient quantity to affect water purification." Hydroxyl radical is a powerful, non-selective and green chemical oxidant, which attacks very rapidly most of the organic contaminants, even the recalcitrant molecules. Photocatalytic oxidation, sonolysis, photo-Fenton processes, catalytic wet peroxide oxidation, microwave/hydrogen peroxide, supercritical water oxidation, ozone based processes, X-ray, γ-ray, and electron beam based processes, etc. have already proved their efficiency at pilot-scale for wastewater treatment. Time is not far off when these technologies or combinations will prove their worth on a large scale also to assist us in combating against ever increasing water pollution on a global scale.

Suresh C. Ameta
Rakshit Ameta

CHAPTER

1

Introduction

Suresh C. Ameta

PAHER University, Udaipur, Rajasthan, India

1.1 ENVIRONMENT

Ever increasing and continuous exploitation of natural resources is posing a serious threat to the environment. The word environment is derived from the French word "environ", which means "surrounding." Our surrounding includes some biotic factors such as human beings, plants, animals, microbes, etc. and some abiotic factors such as light, air, water, soil, etc. Therefore, the environment has been defined by World Health Organization (WHO) as:

> The environment is all the physical, chemical, and biological factors external to the human host, and all related behaviors, but excluding those natural environments that cannot reasonably be modified.

As all the living beings are the integrated parts of environment; it is important for all. It signifies the existence of a permanent relationship between living beings and the environment. WHO also defined "health" as a state of complete physical, mental, and social well-being and not merely the absence of disease or infirmity. Presently, environmental deterioration, or the ecological crisis is the most serious problem faced by mankind. Unfortunately, an ignorant chase for industrial and technological developments has created many environmental issues like global warming, loss of global diversity, forests and energy resources, depletion of the ozone layer, acid rain, etc. At present, the condition has worsened to the extent that all living beings on Earth are now more or less exposed to some kind of pollution, e.g. water, air, soil, etc.

1.2 POLLUTION

The term pollution is derived from the Latin word "polluts", which means defiled. It has also been defined as "the release of substances into any environmental medium from

1

any process, which is capable of causing harm to man or any other living organisms supported by the environment", or "the introduction of substances or energy into the environment, which is liable to cause hazards to human health, harm to living resources and ecological systems, damage to structure, amenity, or interference with legitimate uses of the environment."

Rapid population growth, and corresponding increase in consumption and ultimately development of large scale technologies exploiting our environment is a result of the present upsetting state. Some significant types of environmental pollution are air pollution, soil pollution, water pollution, noise pollution, thermal pollution, radioactive pollution, etc.

1.3 WATER POLLUTION

Since water is basic to life, water contamination issues have been of great importance. Water is at the center point of life on Earth. It exists as one of the primary components of the human body (about 70%). Effect of water pollution on environmental and human health highlights several important water-related issues and there is an urgent need for some potential answers related to water pollution.

Theslightest change in biological, physical, or chemical properties of water is called water pollution.

Water pollution can be of two types:

- surface water and
- ground water

Surface water sources can be further divided into two sources:point sources and nonpoint sources. Any discernible, confined, and discrete conveyances from which pollutants are or may be discharged, are point sources. Nonpoint sources pollution is commonly the additive result of little amounts of contaminants gathered from a large space, which refers to diffuse contamination that doesn't originate from a single distinct supply. When pollutants present on the ground enter the water bodies under earth, then it becomes the cause of ground water pollution.

Although water covers almost three-fourths of the Earth's surface, even then only 0.002% of the water is available for human consumption. Scarcity of fresh water has become a pressing issue and requires urgent worldwide attention. It has been estimated that in 2025, about two billion people will face water scarcity and about half the world's population will be living in highly water stressed areas. Therefore, reutilization of water is becoming essential as most of areas within the world are facing or likely to face water stress problems in the near future.

On a global scale, environmental pollution and the scarcity of sufficient clean energy sources have attracted the attention of scientists to develop an ecofriendly green chemical approach for different materials and processes. Wastewaters from diverse industries are important issues and are of great concern to the Earth's ecosystem. These discharged wastes consist of organic pollutants, which are mostly toxic to various microorganisms, aquatic life, and human beings; and therefore, destroying such harmful properties of chemicals is a cause of genuine concern. Various chemicals like azo dyes, herbicides, and pesticides are actually found in rivers and lakes, and are considered harmful; some of these are endocrine disrupting chemicals also.

Textile dyes and other industrial dyestuffs incorporate one of the major groups of organic pollutants that impose an increasing environmental danger. As almost as 1%−20% of the total world production of dyes is lost during the dyeing process and is discharged in the textile effluents. This effluent is the origin of nonaesthetic pollution and eutrophication in the environment, which generates toxic byproducts as a result of oxidation, hydrolysis, or other chemical reactions taking place in the wastewater phase. Dyes have some toxic effects and these also hinder the penetration of light in contaminated waters. They color the water sources and damage living organisms by ceasing the reoxygenation ability of water, arresting sunlight, and therefore, natural growth activity of aquatic life is disturbed.

The textile industry is the largest consumer of high quality fresh water per kg of treated material. This wastewater also contains toxic, carcinogenic, and persistent chemicals such as formaldehyde, azo dyes, dioxins, and heavy metals. Dyes, pesticides, etc. are complex structures and are naturally degraded under high temperatures, alkaline conditions, ultraviolet (UV) radiation, and other radical initiators, forming toxic by-products, which may be many times more toxic to the environment than the parent compound causing perturbations to the aquatic life and food. These are resistant to microbial, chemical, thermal, and photolytic degradation as many of them are recalcitrant products. Thus, there is a need of an effective, ecofriendly and economical technique for removing these from wastewaters. Several conventional methods for treating wastewater from different industries are used to date, such as photodegradation, adsorption, filtration, coagulation, biological treatments, etc. However, some of these techniques are not completely effective and/or viable, due to the stability of pollutant molecules; thus, advanced oxidation processes (AOPs) have been reported in order to degrade such molecules, to reduce the load of organic pollutants into wastewater. The conventional techniques are not efficient in reducing heavy metals, nitrogen, phosphorous, etc. and there is no unique method to treat most of the compounds in a single step. The presence of pharmaceuticals and personal care products in the aquatic environment is also harmful. Sources of the appearance of pollutants in water and soils include agricultural run-off, hospital effluent, municipal sewage, landfill leaching, etc.

1.4 WASTEWATER TREATMENT

Municipal, industrial, and agricultural wastewaters are some of the major types. It largely comes from human and domestic wastewaters, industrial and animal wastes, groundwater infiltration, etc. The wastewater consists of 99.9% water by weight, and the remaining 0.1% is suspended or dissolved material. Mainly, three types of treatment are commonly used and these are:

- physical processes,
- chemical processes, and
- biological processes

Various methods used in these processes are given in Fig. 1.1.

Wastewater treatment, its management and disposal, has been significantly increased in modern era as it has become a topic of leading interest for human health. Conventional wastewater treatment methods are classified as physical, chemical, and biological processes. The sequential or concurrent use of these processes in a combined manner may provide an efficient method in removing or degrading pollutants, but restraints related to efficiency, execution, and cost are still main factors to limit their practical use. Biological treatments are promising for dairy and agricultural wastewater treatment. Other treatment methods have some disadvantages like pH limits, range of organic-load alterations, physico-chemical behavior of effluents, etc.

Wastewater treatment is a process, which is used to convert wastewater into a useful effluent (with no negligible health and environmental issues), which is then returned back to the water-cycle or in other words, it can be reused. The physical infrastructure used for wastewater treatment is called a wastewater treatment plant (WWTP). Such plants are classified on the basis of the types of wastewater to be treated.

In an industrialized country, a typical sewage treatment plant is comprised of primary secondary and tertiary treatments.

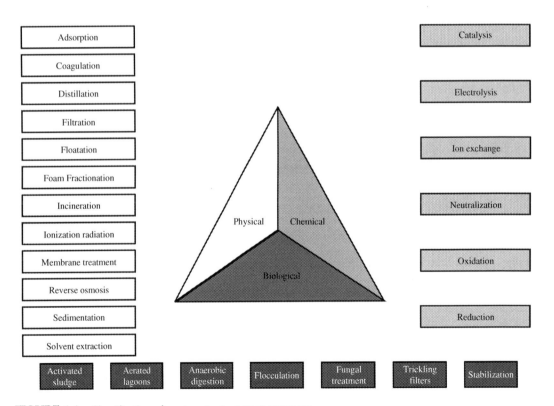

FIGURE 1.1   Classification of wastewater treatment processes.

1.4.1 Primary Treatment

1.4.1.1 *Phase Separation*

Impurities are converted into a nonaqueous phase at intermediate points in a treatment sequence to remove solids generated during oxidation or polishing in phase separation. Grease and oil can also be retrieved for fuel or saponification. Dissolved inorganic solids can be eliminated by methods like ion-exchange, reverse osmosis, distillation, etc.

1.4.1.1.1 SEDIMENTATION

Solids and nonpolar liquids are removed from wastewater by gravity due to differences in density. Gravity separation of solids in sedimentation tanks is the primary treatment of sewage. Heavier solids will accumulate at the bottom of quiescent settling basins.

1.4.1.1.2 FILTRATION

Fine solids present in colloidal suspensions can be eliminated by filtering them through fine physical barriers distinguished from coarser screens or sieves, as particles are smaller than the openings through which the water is allowed to pass.

1.4.2 Secondary Treatment

1.4.2.1 *Oxidation*

Oxidation reduces the biochemical oxygen demand (BOD) of wastewater, thus, the toxicity of some impurities. In this treatment, some impurities are converted to carbon dioxide, water, and bio-solids. Chemical oxidation is also frequently used for disinfection.

1.4.2.1.1 BIOCHEMICAL OXIDATION

Biochemical oxidation of dissolved and colloidal organic compounds, is commonly used for some agricultural and industrial wastewaters. Biological oxidation preferentially removes organic pollutants useful as a food supply for the treatment ecosystem.

1.4.2.1.2 CHEMICAL OXIDATION

Chemical oxidation generally removes some persistent organic compounds and left over concentrations after biochemical oxidation. Disinfection by using ozone, chlorine or hypochlorite in chemical oxidation kills bacteria and microbial pathogens.

1.4.2.2 *Polishing*

Treatments following the oxidation are known as polishing. Sometimes, these treatments are used independently for some industrial wastewaters. After chemical oxidation, chemical reactivity of wastewater can be reduced by chemical reduction or pH adjustment. Then, remaining contaminants will be removed by chemical absorption using activated carbon. Filtration by sand (calcium carbonate) or fabric filter is also commonly used for municipal wastewater treatment.

1.4.3 Tertiary Treatment

Tertiary treatment is also a polishing technique used after a traditional sewage treatment sequence. It is generally applied in industrialized countries by micro-filtration or synthetic membranes. Nitrates can be removed from wastewater by natural processes in wetlands but also via microbial denitrification. Ozone wastewater treatment with an ozone generator is also becoming popular; the ozone generator decontaminates the water as ozone bubbles percolate through the tank, but this treatment is energy intensive. The latest and very promising treatment technology is the use of aerobic granulation. Agricultural wastewater treatment for continuous confined animal operations (e.g. milk and egg production) may be performed in plants using mechanized treatment units. If sufficient land is available for ponds, settling basins, and facultative lagoons, then the operational cost is lower. Leachate treatment plants are used to treat leachate from landfills. Treatment options include: biological treatment, mechanical treatment by ultrafiltration, treatment with active carbon filters, and reverse osmosis using disc tube module technology.

It was observed that the majority of these approaches were only useful to transfer pollutants from aqueous to another phase; thus, generating another pollution, which requires further processing of solid-wastes and regeneration of the adsorbent, which adds more cost to the process. Some other methods like, microbiological or enzymatic decomposition, biodegradation, ozonation, and advanced oxidation processes such as Fenton and photo-Fenton catalytic reactions, H_2O_2/UV processes, etc. have also been used for the removal of various contaminants from wastewaters.

In the last few decades, photocatalytic degradation processes have been widely employed for the destruction of organic pollutants in wastewater and effluents. Advanced oxidation processes (AOPs), have been developed as an emerging destruction technology resulting in the total mineralization of most of the organic contaminants.

1.5 ADVANCED OXIDATION PROCESSES

Environmental pollution and industrialization has drawn our attention for developing advanced, hygienic and environment friendly purification technology. The most common purification processes have limitations like use of electricity and/or other energy sources.

Advanced oxidation processes are genuinely efficient for treating various toxic, organic pollutants and complete destruction of contaminants of emerging concern like, naturally occurring toxins, pesticides, dyes, and other deleterious contaminants. Different advanced oxidation processes for the degradation of various recalcitrant compounds have been reviewed from time to time (Ameta et al., 2012, 2013; Bethi et al., 2016; Boczkaj and Fernandes, 2017; Litter et al., 2017; Parsons, 2004; Poyatos et al., 2010; Wang and Xu, 2011).

Advanced oxidation processes (AOPs) refer to a set of oxidative water treatments that are used to treat toxic effluents at industrial level, hospitals, and wastewater treatment plants. AOPs include UV/O_3, UV/H_2O_2, Fenton, photo-Fenton, nonthermal plasmas, sonolysis, photocatalysis, radiolysis, supercritical water oxidation processes, etc. AOPs were first introduced by Glaze et al. (1987) as processes involving hydroxyl radicals in sufficient quantity to affect water purification. Its definition, development, and various methods to generate hydroxyl radical and other reactive oxygen species like superoxide anion radical, hydrogen peroxide, and singlet oxygen during the process have also been discussed. However, hydroxyl radical is the most efficient and effective species in AOPs.

Generally, organic pollutants interact with hydroxyl radical via addition or hydrogen abstraction pathways, resulting in a carbon-centered radical, which then reacts with molecular oxygen to form a peroxyl radical that undergoes subsequent reactions; thus, generating a host of oxidation products like ketones, aldehydes, or alcohols. Hydroxyl radicals can also form a radical cation by abstracting an electron from electron-rich substrates, which can readily hydrolyze in aqueous media giving an oxidized product. The oxidation products are often less toxic and more susceptible to bioremediation like CO_2, water, etc.

Generally, its efficiency depends profoundly upon the selected AOP method, physical and chemical properties of targeting pollutant, and operating conditions. AOPs are less applied for disinfection as these radicals have a too short half-life (of the order of microseconds), which is not sufficient for disinfection due to extremely low concentrations of radical, but some efforts are being made in this direction. AOPs are low cost to install but have high operating costs due to the input of chemicals and energy required. To limit the costs, AOPs are often used as a pre-treatment combined with a biological treatment. Advanced oxidation can also be used as a quaternary treatment or a polishing step to eliminate micro-pollutants from wastewater and for the disinfection of water. The combination of various AOPs is a quite efficient way to enhance removal of pollutants apart from reducing costs.

Many methods are classified under the broad definition of AOPs, and some of the most studied processes are given in Table 1.1. Advanced oxidation usually involves the use of strong oxidizing agents like hydrogen peroxide (H_2O_2) or ozone (O_3), catalysts (iron ions, electrodes, metal oxides), and irradiation (UV light, solar light, ultrasounds) separately or in combination under mild conditions (low temperature and pressure). Light driven AOPs are supposed to be the most attractive method for wastewater treatment due to the abundance of solar light in regions where water scarcity is high, and comparatively low costs and high efficiencies.

Different commonly used oxidants in different oxidation processes and their electrochemical oxidation potentials are summarized in Table 1.2.

TABLE 1.1   Some Common Photochemical and Nonphotochemical AOPs

Nonphotochemical AOPs	Photochemical AOPs
Ozone (O_3)	Photolysis (UV + H_2O_2)
Fenton (Fe^{2+} + H_2O_2)	Photocatalysis (Light + Catalyst)
Electrolysis (Electrodes + Current)	Photo-Fenton (Solar Light + Fenton)
Sonolysis (Ultrasound)	
Microwaves + H_2O_2	

TABLE 1.2 A Comparison of the Oxidizing Potential of Various Oxidants

Oxidizing Agent	Electrochemical Oxidation Potential (EOP) (V)	EOP Relative to Chlorine
Fluorine	3.06	2.25
Hydroxyl radical	2.80	2.05
Oxygen atomic	2.42	1.78
$TiO_2 + hv$	2.35	1.72
Ozone	2.08	1.52
Persulfate	2.01	1.48
Perbromate	1.85	1.35
Hydrogen peroxide	1.78	1.30
Perhydroxyl radical	1.70	1.25
Hypochlorite	1.49	1.10
Bromate	1.48	1.09
Chlorine	1.36	1.00
Dichromate	1.33	0.98
Chlorine dioxide	1.27	0.93
Permanganate	1.24	0.91
Oxygen (molecular)	1.23	0.90
Perchlorate	1.20	0.89
Bromine	1.09	0.80
Iodine	0.54	0.39

It is clear from these data that the hydroxyl radical is one of the most active oxidants next to fluorine. These hydroxyl radicals react with the dissolved contaminants and initiate a series of oxidation reactions until the are completely mineralized to CO_2, H_2O, and inorganic ions.

Advanced oxidation processes are superior to other treatment processes because compounds present in wastewater are degraded rather than concentrated or transferred into diffused phase, thus, preventing the generation and disposal of secondary waste material.

In AOPs, hydroxyl radical and other active oxidizing species can be generated by different methods, and some of such important AOPs for $^\bullet$OH generation are presented in Fig. 1.2.

Hydroxyl radical is the most reactive oxidizing species used in water treatment, having oxidation potential between 2.8 V (pH 0) and 1.95 V (pH 14) vs. SCE (saturated calomel electrode).

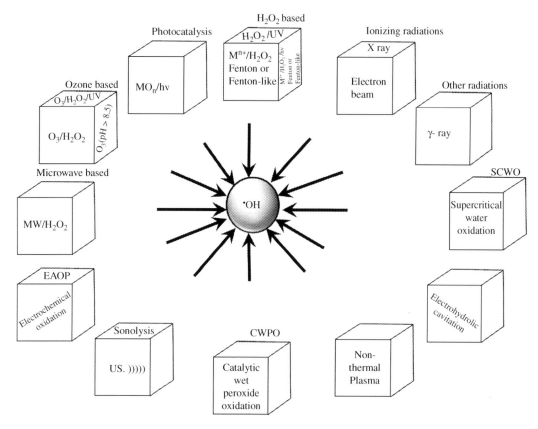

FIGURE 1.2 Different AOPs generating •OH radicals.

Advanced oxidation involves these basic steps (Fig. 1.3).

1. The first step involves the formation of strong oxidants like $^{\bullet}OH$, $HO_2^{\bullet}$, $O_2^{\bullet-}$, etc.
2. In the second step, these oxidants react with organic contaminants present in the wastewater converting them into biodegradable compounds.
3. The last step is the oxidation of these biodegradable intermediates leading to complete mineralization in water, carbon dioxide, and inorganic salts.

1.6 ADVANTAGES

AOPs have many advantages over other traditional methods for the treatment of wastewater:

- They have high reaction rates.
- They have a potential to reduce toxicity and completely mineralize organic contaminants.

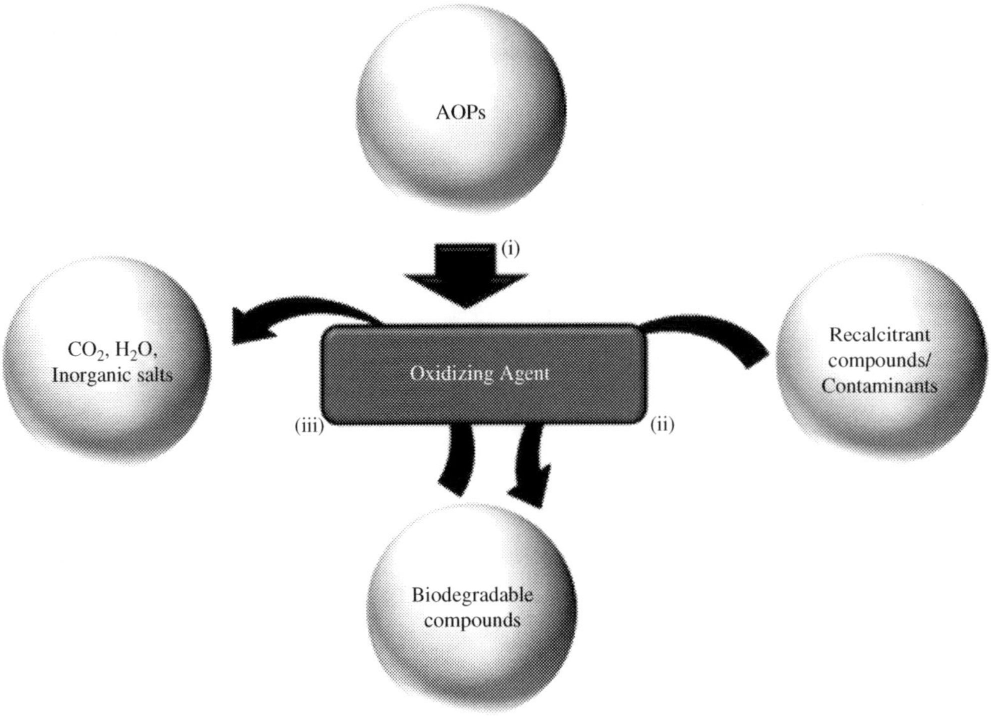

FIGURE 1.3   AOP mechanism.

- They do not concentrate waste for further treatment like methods using membranes or activated carbon absorption.
- They do not create sludge as in the case of physical, chemical, or biological processes.
- Their nonselective pathway allows for the treatment of different organics at a time.
- They are capital intensive installation, cost is relatively low as their chemistry can be tailored to have a specific application.

1.7 APPLICATIONS

Any organic contaminant, that is reactive with the hydroxyl radical, can potentially be treated through advanced oxidation processes. These include petroleum hydrocarbons, aromatic hydrocarbons (toluene, benzene, xylene, etc.), phenols, chlorinated hydrocarbons (TCE, PCE, vinyl chloride, etc.), dyes, pesticides, pharmaceuticals, explosives (TNT, RDX, and HMX), and many more.

Some other applications of AOPs are given in Fig. 1.4.

In view of the global concern about the presence of organic pollutants in waters and the necessity for the development of some innovative and efficient technologies for water remediation and purification, the AOPs enter the scene. AOPs constitute important,

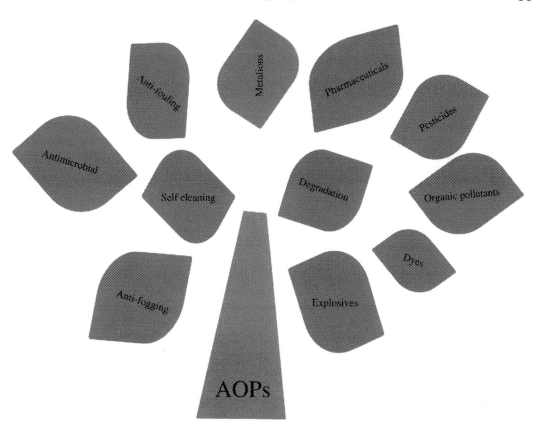

FIGURE 1.4 Applications of AOPs.

promising, efficient, and environmental-friendly methods to remove persistent organic contaminants from wastewaters. Generally, AOPs are based on the in situ generation of a powerful oxidizing agent like hydroxyl radicals (˙OH). Time is not far off, when AOPs will almost replace existing conventional treatment technologies of wastewater treatment.

References

Ameta, R., Benjamin, S., Ameta, A., Ameta, S.C., 2012. Photocatalytic degradation of organic pollutants: a review in photocatalytic materials & surfaces for environmental cleanup-II, Tayade, R. (Ed.), Mater. Sci. Forum, 734, 247–272.

Ameta, R., Kumar, A., Punjabi, P.B., Ameta, S.C., 2013. In: Rao, D.G., Senthilkumar, R., Byrne, J.A., Feroz, S. (Eds.), Advanced Oxidation Processes: Basics and Applications in Waste Water Treatment: Advanced Processes and Technologies. IWA & CRC Press, London.

Bethi, B., Sonawane, S.H., Bhanvase, B.A., Gumfekar, S.P., 2016. Nanomaterials-based advanced oxidation processes for wastewater treatment: a review. Chem. Eng. Proc. 109, 178–189.

Boczkaj, G., Fernandes, A., 2017. Wastewater treatment by means of advanced oxidation processes at basic pH conditions: a review. Chem. Eng. J. 320, 608–633.

Glaze, W.H., Kang, J.W., Chapin, D.H., 1987. The chemistry of water treatment processes involving ozone, hydrogen peroxide and UV-radiation. Ozone Sci. Eng. 9, 335–352.

Litter, M.I., Candal, R.J., Meichtry, J.M., 2017. Advanced Oxidation Technologies: Sustainable Solutions For Environmental Treatments. CRC Press, London.

Parsons, S. (Ed.), 2004. Advanced Oxidation Processes for Water and Wastewater Treatment. IWA Publishing, London.

Poyatos, J.M., Munio, M.M., Almecija, M.C., Torres, J.C., Hontoria, E., Osorio, F., 2010. Advanced oxidation processes for wastewater treatment: State of the art. Water Air Soil Pollut. 205, 187–204.

Wang, J.L., Xu, L.J., 2011. Advanced oxidation processes for wastewater treatment: formation of hydroxyl radical and application. Crit. Rev. Env. Sci. Technol. 42 (3), 251–325.

UV-Hydrogen Peroxide Processes

José C. Mierzwa, Raphael Rodrigues
and Antonio C.S.C. Teixeira
University of São Paulo, São Paulo, SP, Brazil

2.1 INTRODUCTION

With the increasing influence of human activities, synthetic organic chemicals such as pharmaceuticals, personal care products, and products from the nanotechnology industry have been used and released into the environment (Houtman, 2010). Due to increased regulations regarding drinking water quality and better understanding of the benefits of ultraviolet (UV) technology in water treatment, a growing number of utilities consider the use of UV disinfection and/or UV-based advanced oxidation processes (AOP) to degrade and mineralize those potentially harmful and recalcitrant environmental pollutants.

The UV/H_2O_2 system consists of adding hydrogen peroxide (H_2O_2) in the presence of UV light to generate hydroxyl radicals ($^{\bullet}OH$). The main advantage of using UV/H_2O_2 for water and wastewater systems resides in the fact that UV radiation can work simultaneously as a disinfectant, by physically inactivating microorganisms, and helping in the photolysis of peroxide, breaking it into the highly reactive hydroxyl radical species.

2.2 FUNDAMENTALS

The photochemical advanced oxidation process using UV light and hydrogen peroxide is a quite well established technology for water and wastewater treatment (Andreozzi et al., 1999; Comninellis et al., 2008; Legrini et al., 1993; Munter, 2001; Ribeiro et al., 2015).

This process relies on the production of an effective oxidant, the hydroxyl radical ($^\bullet$OH), through the irradiation of the hydrogen peroxide with ultraviolet radiation, according to Eq. (2.1) (USEPA, 1998).

$$H_2O_2 \xrightarrow{hv} 2\ OH \tag{2.1}$$

The photoinduced decomposition of hydrogen peroxide in pure water is rationalized by the Haber-Weiss radical chain mechanism (1932), according to which the hydroxyl radicals originated by the homolytic cleavage of the O—O bond initiate a sequence of reactions characterized by an initiation step (reaction 2.1), followed by a propagation step given by reactions (2.2) and (2.3).

$$H_2O_2 +\ OH \rightarrow H_2O + HO_2 \tag{2.2}$$

$$HO_2 + H_2O_2 \rightarrow H_2O + O_2 +\ OH \tag{2.3}$$

It is worth observing that the net reaction of reactions (2.2) and (2.3) corresponds to the decomposition of hydrogen peroxide into water and molecular oxygen (Oppenländer, 2003). Finally, different termination reactions involve radical-radical recombination (reactions 2.4—2.6):

$$OH +\ OH \rightarrow H_2O_2 \tag{2.4}$$

$$OH + HO_2 \rightarrow H_2O + O_2 \tag{2.5}$$

$$HO_2 + HO_2 \rightarrow H_2O_2 + O_2 \tag{2.6}$$

More complex mechanisms explaining H_2O_2 decomposition are found in the literature and include species like HO_2^-, $HO_2\ ^-$, and $O_2\ ^-$ in the chain decomposition mechanism (Crittenden et al., 1999; Edalatmanesh et al., 2008).

The efficiency of $^\bullet$OH radical production depends on the capability of hydrogen peroxide to absorb UV radiation, as well as on the physical and chemical characteristics of the fluid that will be submitted to the oxidation process. The absorption of UV radiation by H_2O_2 molecules depends on the molar extinction coefficient $-$, (Beers and Sizer, 1952), or on the UV absorption cross section $-$ (Molina et al., 1977); both related to the Beer-Lambert law, according to Eqs. (2.7) and (2.8) (Kaltenegger, 2011; Lide, 2006—2007) valid for H_2O_2 solution in pure water.

$$I = I_0\ e^{-\ C\ l} \tag{2.7}$$

$$I = I_0\ e^{-\ nl} \tag{2.8}$$

Where I_0 and I are the incident and transmitted intensities of UV radiation,   is the molar extinction coefficient (L/mol per cm), C is the H_2O_2 molar concentration (mol/L),   is the absorption cross section (cm^2/molecule), n is the molecular density (molecules/cm^3), and l is the optical path (cm).

Whatever definition is adopted, it could be concluded by the analysis of the data available that UV absorption by H_2O_2 increases as the wavelength decreases. In fact, the UV radiation absorption by H_2O_2 in the gas phase and in aqueous solution is significantly

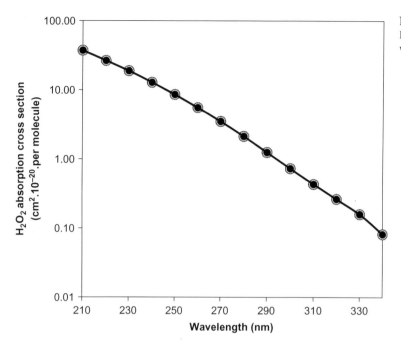

FIGURE 2.1 Variation of H_2O_2 absorption cross section with wavelength.

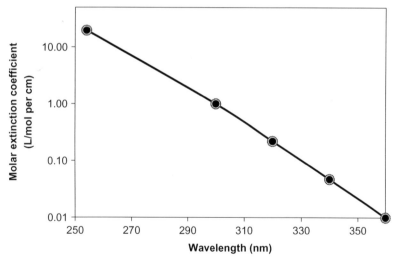

FIGURE 2.2 Variation of H_2O_2 molar extinction coefficient with wavelength.

below c.310 nm (Oppenländer, 2003). Figs. 2.1 and 2.2 prepared with the data of Vaghjiani and Ravishankara (1989), and Cataldo (2014), depict the variation of the absorption cross section, and the molar extinction coefficient for H_2O_2 with UV light wavelength.

It can be easily concluded that the smaller the UV wavelength is, the higher the energy absorption by the H_2O_2 is, increasing the potential for hydroxyl radical production. The molar absorption coefficient of H_2O_2 in the range 200–204 nm is about 180 L/mol per cm, while at 254 nm, it is approximately 19.6 L/mol per cm (Cataldo, 2014), decreasing to 0.88 L/mol per cm close to 300 nm. Therefore, according to Eq. (2.7), and as remarked by Oppenländer (2003), adequate H_2O_2 concentrations and sufficient irradiation path length in the reaction medium must be considered in order to compensate for its poor absorption in the UV region. This is considerably more critical, when the effluent to be treated absorbs UV radiation in the same irradiation wavelength, thus intensively competing for photons with hydrogen peroxide.

Though most UV/H_2O_2 treatments are carried out in neutral to acidic conditions, it is worth mentioning that under alkaline conditions, H_2O_2 deprotonates with the formation of HO_2^- according to the equilibrium $H_2O_2 \rightleftharpoons HO_2^- + H^+$ ($pK_a \sim 10.4$). The hydroperoxide anion, HO_2^-, has a significantly higher molar absorption coefficient than H_2O_2 itself (240 L/mol per cm at 254 nm) and, as a consequence, efficiently competes for UV photons and dissociates with the formation of hydroxyl radicals and oxygen anion radicals ($O^{\bullet -}$). The latter combine with water molecules, giving hydroxyl radicals and hydroxide anions (Oppenländer, 2003).

Another fundamental concept related to any photoinduced process is the quantum yield ( ), a quantitative measure of the overall efficiency of the process. By definition, the quantum yield corresponds to the number of reactant molecules (or photoproduct molecules formed) per unit time divided by the number of photons absorbed during this period (Oppenländer, 2003). According to Eq. (2.1), the photolysis of 1 mol of hydrogen peroxide should result in the generation of 2 mols of $^\bullet OH$ radicals. However, in water, the measured quantum yield of H_2O_2 decomposition [$(-H_2O_2)$] at 254 nm, equals 0.5, while the measured quantum yield of hydroxyl radicals formation [$(+^\bullet OH)$] is 0.98; $(-H_2O_2)$ in water reduces to 0.30 at 313 nm. In the gas phase, experiments indicate $(-H_2O_2) = 1.0$ (200–350 nm), $(+^\bullet OH) = 1.58$ (248 nm), and $(+^\bullet OH) = 1.22$ (193 nm) (Bolton and Cater, 1994; Legrini et al., 1993; Pehkonen et al., 1998).

It is worth mentioning that the calculated energy for the homolytic cleavage of the central O–O bond is 213 kJ/mol for H_2O_2 (Lide, 2006–2007), which corresponds to a threshold photon wavelength of 562 nm, as easily determined by the Planck equation. On the other hand, the absorption of UV radiation by hydrogen peroxide molecules in the liquid phase begins (i.e., reaches 1.0 L/mol per cm by definition) at about 310 nm. Therefore, the excess energy of the hydroxyl radicals equals 173 kJ/mol, a value that is surmounted by irradiating H_2O_2 with shorter wavelengths (<310 nm).

As a result, hydroxyl radicals formed from H_2O_2 dissociation in the gas phase at 248 nm and 193 nm have excess translational energy. On the other side, in aqueous solution, only part of the $^\bullet OH$ radicals with excess energy can escape the water (solvent) cage as "free" species; the remaining $^\bullet OH$ radicals within the cage immediately recombine (Eq. 2.4) and, as a consequence, the quantum yields of hydrogen peroxide decomposition and hydroxyl radicals formation are lower.

In addition to the UV radiation wavelength, the $^\bullet OH$ radical production is also affected by many other factors, including:

- type of UV lamp used (low pressure or medium pressure),
- UV light transmittance of the quartz sleeve,

- optical path length in the reactor medium, and
- optical properties of the effluent, such as UV radiation absorption by specific chemicals or particles.

2.2.1 UV Lamps

The type of UV lamp that should be considered for the development of an effective oxidation reactor should have a wavelength emission spectrum below 260 nm. This limitation restricts the use of medium pressure UV lamps, which have the wavelength emission spectrum spans from 200 to 700 nm (Fig. 2.3). Another issue related to the medium pressure UV lamps is its higher power input.

UV lamps useful for UV/H_2O_2 treatment processes include the low pressure mercury vapor lamps and excimer UV-vacuum lamps. The first emits virtually monochromatic radiation at a wavelength close to 254 nm. Xe_2^* excimer sources emit at about 172 nm. Table 2.1 presents examples of low pressure mercury vapor and excimer UV-vacuum lamps that can be used for the UV/H_2O_2 oxidation process. Comparing the characteristics of the different types of UV radiation sources available, it is evident that the final decision to choose a particular lamp is not so easy, mainly if this decision is to be taken based on the costs of low-pressure mercury vapor lamps and excimers, which are more expensive.

That is because the system designer will need to counterbalance the UV lamp potential for $^•OH$ radical production, effective UV radiation output, electrical power usage, lamp life span, and of course, its cost and the costs for the lamp's accessories. The main accessories for the adequate operation of a UV lamp for an oxidation process include the electrical components, such as the ballast, and the quartz sleeve. Since the electrical components are specific for each UV lamp type, no additional discussion is necessary.

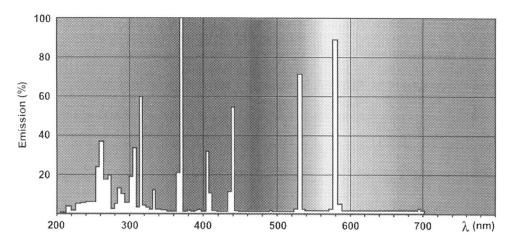

FIGURE 2.3 Medium pressure UV lamp wavelength spectrum. *From http://www.lighting.philips.com/.*

TABLE 2.1 Specification of Low Pressure Mercury Vapor and Excimer Vacuum UV Lamps From Different Manufacturers[a]

Manufacturer	Product Description	Model	Electrical Power (W)	UV Irradiance/ Output	Diameter (mm)	Radiation Length (mm)	Life Span (h)	Internet Homepage
Helios Quartz	High output	HOGL436T5L	48	120 W/cm² (254 nm)	15	360	n.s.	http://www.heliosquartz.com/
		HOGL64T5L	155	395 W/cm² (254 nm)	15	1421	n.s.	
LightTech	High output quartz	GHO436T5L	48	120 W/cm² (254 nm) 13 W (254 nm)	15	360	16,000	http://www.light-sources.com/
		GHO893T5L	95	270 W/cm² (254 nm) 30 W (254 nm)		815		
	Amalgam lamps	GPHA357T5L	42	110 W/cm² (254 nm) 11 W (254 nm)		278		
		GPHHA1000T6L	207	570 W/cm² (254 nm) 65 W (254 nm)		921		
Osram	Puritec HNS	ZMP 4021026	25	6.9 W (254 nm)	26	~438	9000	https://www.osram.com/
		ZMP 4021031	55	18.0 W (254 nm)	26	~895		
	Xeradex (Excimer)	L40/375/DB-AZ48/90	60	40 mW/cm² (172 nm) 19 W (172 nm)	40	375	2500	
		L40/910/DB-AZ48/90	150	40 mW/cm² (172 nm) 39 W (172 nm)	40	910		
Philips	TUV TL Mini	927971204099	11	2.6 W (254 nm)	19	~328	9000	http://www.lighting.philips.com/
		927972204099	23	8.0 W (254 nm)				
	TUV T5	927970204099	40	15 W (254 nm)	19.3	~1556.6		
		927971104099	145	48 W (254 nm)				
Quark Technology	Excimer VUV	QEX 1600	n.s.	70 mW/cm² (172 nm)	n.s.	1600	2500	http://www.quark-tec.com/english/product/5.html

[a] All the data presented in this table are available for download at the manufacturers internet homepages.
Note: n.s., not specified.

2.2.2 Quartz Sleeve

The quartz sleeve is a fundamental component in any UV/H_2O_2 oxidation reactor, because it will be used to isolate the electrical components from the water or wastewater being irradiated.

Considering this, the quartz sleeve transmittance for the specific UV radiation wavelength is of paramount importance. That is because the final $^\bullet$OH radical production depends on the amount of the UV radiation transmitted through the quartz sleeve wall, which depends on the glass purity, sleeve surfaces finishing, and wall thickness. The information about the optical properties of quartz sleeves is not always available in scientific papers focusing on UV/H_2O_2 oxidation processes, but one specific optical property is crucial, i.e., the UV radiation transmittance. That is because the UV intensity attenuation of the quartz sleeve will affect the final UV dose obtained. Table 2.2 presents the transmittances of some quartz sleeves available in the market, obtained at the manufacturers' internet homepage.

Most of the quartz sleeves available in the market are suitable for operating an UV oxidation process at UV-light wavelength of 254 nm. In contrast, quartz sleeves with higher UV-radiation transmittances at 172 nm are more limited.

TABLE 2.2   Transmittances of Some Quartz Sleeves Commercially Available

Manufacturer	Product Specification	Wall Thickness (mm)	UV Light Wavelength (nm)	Transmittance (%)[a]	Internet Homepage
GVB GmbH—Solutions in glass	Quartz—EN09UB	1.5	254	89.7	http://www.g-v-b.de/quartz_tubes__335.htm
			185	43.5	
Helios Quartz Group S/A	Fused quartz—NHI	n.s.	254	80–95	http://www.heliosquartz.com/prodotti/tubes-and-domed-tubes/?lang = en
			180	55–85	
Heraeus Quarzglas GmbH & Co. KG	Purasil	2	254	~90	https://www.heraeus.com/en/hqs/products_hqs/tubes/tubes_lamps/tubes_lamp_manufacturing.aspx
			172	~60	
	Purasil—XP		254	~90	
			172	85–90	
	Suprasil—130		254	~90	
			172	85–90	
	Suprasil—310		254	~90	
			172	85–90	
Technical Glass Products, Inc.	Fuse quartz—Tubing	1.0	254	85	https://www.technicalglass.com/downloads.html
			172	~15	

[a]*The values are approximated, because most of them were obtained from the graphics made available by the manufacturers.*
Note: n.s., not specified.

2.2.3 Optical Path on the UV Reactor

Another important parameter for the UV/H_2O_2 oxidation process is the optical path on the UV reactor, which is the distance that the UV-radiation needs to penetrate in the effluent in order to effectively react with H_2O_2 to produce $^{\bullet}OH$ radicals.

The Beer-Lambert law, Eqs. (2.7) and (2.8), states that the higher the optical path is, the higher the attenuation of UV-radiation will be. Attenuation will also certainly depend on the characteristics of the effluent to be treated.

Rosenfeldt et al. (2006) pointed out that the efficiency of any UV-light based oxidation process depends on the reactor geometry, and on the characteristics of the effluent being treated. In any case, an optical path higher than 1 cm is not recommended because above this value, the UV-light intensity will decrease exponentially. This attenuation can significantly increase, depending on the characteristics of the effluent to be treated.

2.2.4 Effluent Optical Properties and Characteristics

The effectiveness of the UV/H_2O_2 oxidation process relies on the production of $^{\bullet}OH$ radicals, which are responsible for the oxidation reaction. In this sense, any effluent optical property that can interfere in the production of $^{\bullet}OH$ radicals, or the presence of any chemical that can scavenge $^{\bullet}OH$ radicals should be carefully evaluated.

The optical properties of effluent are much associated with specific constituents contained on it. Dissolved chemicals can significantly affect the effluent transmittance, reducing the $^{\bullet}OH$ radical production; while certain compounds such as natural organic matter (NOM), organic acids, and some inorganic ions, can reduce the removal efficiency of specific contaminants (Gogate and Pandit, 2004; Liao et al., 2001; Stasinakis, 2008; Tokumura et al., 2016; Zhang and Li, 2014).

Measurement of UV-radiation transmittance at specific wavelength, 254 and 172 nm, is the easiest way to evaluate the effluent absorption capacity. NOM, and various aromatic compounds, strongly absorb ultraviolet radiation, making the UV-radiation absorption measurement, a surrogate indicator of such substances (APHA, AWWA, and WEF, 2012). The treatment of effluents with higher UV-radiation absorbance, which means lower transmittance at these UV-light wavelengths, will be more difficult and cost intensive.

Considering these factors, the design of a photochemical oxidation reactor for an effluent treatment is not straight forward. In fact, the final reactor design depends not only on the specific contaminant to be addressed, but also on the characteristics of the matrix in which it is contained. The characteristics regarding the UV lamp and quartz sleeve that will be used and the reactor geometry are quite relevant for any successful application.

2.3 KINETICS

The reaction kinetics of the UV/H_2O_2 oxidation process as it occurs for most of the chemical oxidation reactions are a function of the reactants involved in the reaction, basically the chemical that will be oxidized and the oxidant. The expected final products resulting from the UV/H_2O_2 oxidation processes are the carbon dioxide, water, and inorganic acids and salts (Hugül et al., 2000).

Hydroxyl radicals are the main species responsible for the oxidation process in an UV-H_2O_2 process. $^\bullet$OH radicals are extremely reactive, and nonselective transient oxidant, which quickly react with most substances present in an effluent (Deng and Zhao, 2015). Their short lifetimes result in very low steady-state concentrations ($[^\bullet OH]_{ss}$) in NOM-containing water and wastewater (Oppenländer, 2003). In fact, values of $[^\bullet OH]_{ss}$ as low as 10^{-16} mol/L have been experimentally determined (Silva et al., 2015). Procedures for the detection and quantification of hydroxyl radicals are very well described (Fernández-Castro et al., 2015).

A detailed description of the main mechanisms involved in the oxidation reaction with $^\bullet$OH radicals can be obtained (Huang et al., 1993; Legrini et al., 1993; Matileinen and Sillanpaa, 2010). A brief description of the main mechanisms involved according to Legrini et al. (1993) is presented here.

- *Hydrogen abstraction*

  This mechanism refers to a process, in which a hydrogen atom is removed from the contaminant subjected to the oxidation process, producing an organic radical. In this process, the removed hydrogen reacts with the $^\bullet$OH radical, resulting in a water molecule.

- *Hydroxyl (electrophilic) addition*

  As it suggests, this mechanism refers to the addition of a hydroxyl radical on the structure of the contaminant. This mechanism is of interest for understanding the dechlorination of chlorinated phenols, producing chloride ions.

- *Electron transfer*

  It refers to the oxidation mechanism, in which the $^\bullet$OH radical works as an electron acceptor and is reduced to the hydroxide anion. This is the basic mechanism involved in common oxidation reaction.

The attack of hydroxyl radicals on substrate molecules produce carbon-centered organic radicals that instantaneously react with dissolved oxygen molecules giving peroxyl radicals, through diffusion controlled, irreversible bimolecular reactions (Oppenländer, 2003). Aliphatic peroxyl radicals form unstable intermediary tetroxides that decompose through electrocyclic rearrangements, with the participation of water or not, or via one-electron shifts (Oppenländer, 2003), according to the so-called Russell reactions. The final picture is the formation of a myriad of oxidation products including ketones, alcohols, aldehydes, molecular oxygen, hydrogen peroxide and aliphatic peroxides. In addition, the degradation pathway of aromatic compounds by $^\bullet$OH radicals is expected to be quite more complex. It is worth mentioning that secondary oxidant species, such as hydroperoxyl radicals (HO_2) and superoxide anion radicals ($O_2{}^-$), formed during many decomposition reactions, are involved in chain oxidation reactions with remaining substrate and decomposition product molecules as well.

Moreover, $^\bullet$OH radicals and others are scavenged by carbonate and bicarbonate ions (mainly the first) through electron transfer reactions, giving radical products (HCO_3 and $CO_3{}^-$) according to the following reactions (Buxton and Elliot, 1986; Liao et al., 2001).

$$CO_3{}^{2-} + {}^\bullet OH \rightarrow CO_3{}^- + OH^- \tag{2.9}$$

$$HCO_3{}^- + {}^\bullet OH \rightarrow CO_3{}^- + H_2O \tag{2.10}$$

The CO_3^- species is considered a very selective oxidant. The scavenging by HCO_3^-/CO_3^{2-} decreases with decreasing pH of the water medium, as a consequence of the equilibrium reactions involving these species. Finally, cyanide and halide ions are also effective $^\bullet OH$ radicals scavengers, while sulfate, phosphate, and nitrate are of very little importance. In this case, chloride ions, originate chlorine atoms with the following reactions, which can add to double bonds of carbon compounds, and therefore, generate chlorinated degradation products (Liao et al., 2001).

$$Cl^- + \ ^\bullet OH \leftrightarrow HOCl^{\bullet -} \tag{2.11}$$

$$HOCl^{\bullet -} + H^+ \rightarrow Cl^\bullet + H_2O \tag{2.12}$$

Considering the non-selectivity of the $^\bullet OH$ radical, it is evident that all these mechanisms can occur simultaneously during this oxidation process.

From the practical point of view, when the application of the UV/H_2O_2 oxidation process for wastewater treatment focuses on the complete mineralization of target contaminants, the expected outcome of the oxidation reaction is the production of carbon dioxide, water, and inorganic salts as the final products.

2.3.1 UV-H$_2$O$_2$ Oxidation Kinetics

According to classical literature regarding chemical reaction kinetics, a second order rate law describes the bimolecular oxidation reaction of an $^\bullet OH$ radical with a substrate molecule because it depends on the contaminant and $^\bullet OH$ radical concentrations. Second order rate constants are usually measured by competition kinetics methods (Shemer et al., 2006; Silva et al., 2015), and most values for organic molecules $k_{HO^\bullet,M}$ lie in the range of $10^9 - 10^{10}$ L/mol per s (Oppenländer, 2003).

Considering Eq. (2.1), it is easy to verify that the $^\bullet OH$ radical production rate is directly related to the photolysis of the hydrogen peroxide by the UV radiation. It is possible to obtain the rates of H_2O_2 consumption and $^\bullet OH$ radical production through a set of equations (Table 2.3) (Crittenden et al., 1999; Edalatmanesh et al., 2008).

Considering the dissociation of hydrogen peroxide (Reaction 23 in Table 2.3), and defining the total concentration of the main active substances in equilibrium as $[C_{tot}]$ (Eq. 2.13), it is possible to calculate the concentration of each species, using the H_2O_2 equilibrium dissociation constant (K_a), Eqs. (2.14) and (2.15).

$$C_{tot} = H_2O_2 + HO_2^- \tag{2.13}$$

$$\frac{H_2O_2}{C_{tot}} = \frac{H^+}{K_A + H^+} \tag{2.14}$$

A is the effluent absorbance, with the specific H_2O_2 dosage, and $A_{H_2O_2}$ is the absorbance of H_2O_2 solution (in pure water), with the same concentration applied in the effluent, and $(-H_2O_2)$ is the quantum yield of H_2O_2. The absorbance should be measured at the same optical path that will be used in the reactor. I_0 is the UV light irradiance (UV intensity per exposition area) at a specific wavelength.

TABLE 2.3 Reaction Equations Involved on the UV/H2O2 Oxidation Process and Its Kinetic Constants

Reaction Number	Chemical Reaction	Rate Constants (L/mol per s)
1	$H_2O_2/HO_2^- \xrightarrow{h\nu} 2\ OH$	k_1 (obtained for the specific system, s^{-1}), or $r_{UV,H_2O_2} = -r_{OH} = -\ _{H_2O_2}.I_0.\frac{A_{H_2O_2}}{A}.1-e^{-A}$ [a] $_{H_2O_2} = _{HO_2} = 0.5$
2	$OH + H_2O_2 \rightarrow HO_2 + H_2O$	$k_2 = 2.7 \times 10^7$
3	$OH + HO_2^- \rightarrow HO_2 + OH^-$	$k_3 = 7.5 \times 10^9$
4	$H_2O_2 + HO_2 \rightarrow +\ OH + H_2O + O_2$	$k_4 = 3.0$
5	$H_2O_2 + O_2^- \rightarrow OH + O_2 + OH^-$	$k_5 = 0.13$
6	$OH +\ OH \rightarrow H_2O_2$	$k_6 = 5.5 \times 10^9$
7	$OH + HO_2 \rightarrow O_2 + H_2O$	$k_7 = 6.6 \times 10^9$
8	$HO_2 + HO_2 \rightarrow H_2O_2 + O_2$	$k_8 = 8.3 \times 10^5$
9	$HO_2 + O_2^- \rightarrow HO_2^- + O_2$	$k_9 = 9.7 \times 10^7$
10	$HO + O_2^- \rightarrow O_2 + OH^-$	$k_{10} = 7.0 \times 10^9$
11	$OH + CO_3^{2-} \rightarrow CO_3^- + OH^-$	$k_{11} = 3.9 \times 10^8$
12	$OH + HCO_3^- \rightarrow CO_3^- + H_2O.$	$k_{12} = 8.5 \times 10^6$
13	$OH + HPO_4^{2-} \rightarrow HPO_4^- + OH^-$	$k_{13} = 1.5 \times 10^5$
14	$OH + H_2PO_4^- \rightarrow HPO_4^- + H_2O$	$k_{14} = 2.0 \times 10^4$
15	$H_2O_2 + CO_3^{2-} \rightarrow HCO_3^- + HO_2$	$k_{15} = 4.3 \times 10^5$
16	$HO_2^- + CO_3^- \rightarrow CO_3^{2-} + HO_2$	$k_{16} = 3.0 \times 10^7$
17	$H_2O_2 + HPO_4^- \rightarrow H_2PO_4^- + HO_2$	$k_{17} = 2.7 \times 10^7$
18	$OH + CO_3^- \rightarrow ?$	$k_{18} = 3.0 \times 10^9$
19	$CO_3^- + O_2^- \rightarrow CO_3^{2-} + O_2$	$k_{19} = 6.0 \times 10^8$
20	$CO_3^- + CO_3^- \rightarrow ?$	$k_{20} = 3.0 \times 10^7$
21	$HO_2 \rightarrow O_2^- + H^+$	$k_{21} = 1.58 \times 10^5$
22	$O_2^- + H^+ \rightarrow HO_2$	$k_{22} = 1.0 \times 10^{10}$
23	$H_2O_2 \rightarrow HO_2^- + H^+$	$pK_A = 11.6$

[a]This is the simplified version of the equation presented in the work developed by Crittenden et al. (1999).

$$\frac{HO_2^-}{C_{tot}} = \frac{K_A}{K_A + H^+} \tag{2.15}$$

Using Eqs. (2.14) and (2.15), it is possible to obtain the equilibrium curves for the molar fraction of each species as a function of the solution pH. Through the examination of the curves presented in Fig. 2.4, it can be verified that the $[HO_2^-]$ concentration is negligible up to pH 9, which means that the use of reactions (3), (9), and (16) is limited for systems, which operate with higher pH values.

By using all the presented equations and performing a mass balance (molar basis), it is possible to develop a model for the oxidation process (Edalatmanesh et al., 2008).

Taking the derivatives of Eqs. (2.13), (2.14), and (2.15) by the time, and performing the mass balance for the species involved, it is possible to obtain the expressions that represent the variation of the concentration of H_2O_2 and HO_2^-.

$$\frac{d\,C_{tot}}{dt} = \frac{d\,H_2O_2}{dt} + \frac{d\,HO_2^-}{dt} \tag{2.16}$$

$$\frac{d\,H_2O_2}{dt} = \frac{d\,C_{tot}}{dt}\,\frac{H^+}{K_A + H^+} \tag{2.17}$$

$$\frac{d\,HO_2^-}{dt} = \frac{d\,C_{tot}}{dt}\,\frac{K_A}{K_A + H^+} \tag{2.18}$$

The development of Eq. (2.15), considering the data presented in Table 2.3, results in:

$$\frac{d\,C_{tot}}{dt} = r_{UV,\,H_2O_2} - H_2O_2\ \left(k_2\ OH\ + k_4\ HO_2\ + k_5\ O_2^-\right.$$

$$+\ k_{15}\ CO_3^{2-}\ + k_{17}\ HPO_4^-\ - k_3\ HO_2^-\ OH \tag{2.19}$$

$$+\ k_6\ OH^2 + k_8\ HO_2^{\ 2} + k_9\ HO_2\ O_2^-$$

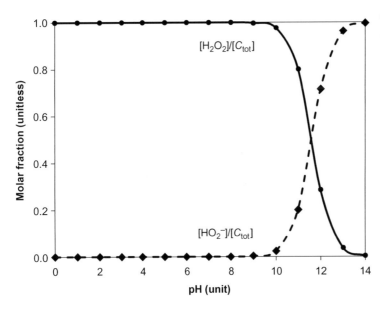

FIGURE 2.4 Equilibrium curves for the H_2O_2 dissociation.

By replacing the term r_{UV,H_2O_2} by the rate constant given in Table 2.3 (Reaction 1) in Eq. (2.19), and the resulting expression in Eqs. (2.17) and (2.18), the reactions rates for H_2O_2 and HO_2^- are represented as Eqs. (2.20) and (2.21).

$$\frac{d\,H_2O_2}{dt} = \left(\frac{H^+}{K_A + H^+}\right) \quad - \quad _{H_2O_2}I_0\frac{A_{H_2O_2}}{A}\left(1 - e^{-A}\right) - \left[H_2O_2\right]k_2\left[OH\right]$$
$$+ k_4\left[HO_2\right] + k_5\left[O_2^-\right] + k_{15}\left[CO_3^{2-}\right] + k_{17}\left[HPO_4^-\right] - k_3\left[HO_2^-\right]\left[OH\right]$$
$$+ k_6\left[OH\right]^2 + k_8\left[HO_2\right]^2 + k_9\left[HO_2\right]\left[O_2^-\right]$$

(2.20)

$$\frac{d\,HO_2^-}{dt} = \left(\frac{K_A}{K_A + H^+}\right) \quad - \quad _{H_2O_2}I_0\frac{A_{H_2O_2}}{A}\left(1 - e^{-A}\right) - \left[H_2O_2\right]k_2\left[OH\right] + k_4\left[HO_2\right]$$
$$+ k_5\left[O_2^-\right] + k_{15}\left[CO_3^{2-}\right] + k_{17}\left[HPO_4^-\right] - k_3\left[HO_2^-\right]\left[OH\right] + k_6\left[OH\right]^2$$

(2.21)

$$+ k_8\left[HO_2\right]^2 + k_9\left[HO_2\right]\left[O_2^-\right]$$

It is important to note that the adequate modeling of the UV/H_2O_2 oxidation process requires a complete set of chemical equations related to the contaminant or contaminant's reactions inside the reactor, including the photocatalytic degradation ones, to develop the reaction's rates for all the species involved in the UV/H_2O_2 oxidation process.

A close examination of the work developed by Crittenden et al. (1999), and Edalatmanesh et al. (2008) reveals the complexity involved in the use of the proposed approach for real conditions. That is because in the application of the UV/H_2O_2 oxidation for real effluents, most of the necessary information regarding its composition is not available or the costs involved in obtaining them could be prohibitive. Moreover, the equations depicted above consider a uniform radiation field inside the photochemical reactor, which may not be the case for real systems because of the inherent complexities and non-homogeneities of the reacting medium. In this case, the uncertainty involved will be quite high, meaning that a simple approach should be applied.

2.4 A SIMPLIFIED MODEL FOR PERFORMANCE EVALUATION

The modeling or design of an UV/H_2O_2 oxidation process for an effluent treatment is not easily performed using the complete theoretical approach. The main reasons for this are:

- Effluent characteristics are quite variable and complex.
- The full effluent characterization could be quite expensive.
- The mathematical model can become too complex for practical applications, requiring the use of advanced numerical solutions.
- The results obtained with a more simplified model are quite precise, considering all the uncertainties involved in the process design.

For overcoming limitations regarding the use of the comprehensive model, it is possible to use a model based on the pseudo-first order kinetics. It is possible through some simplified assumptions, which can give quite reliable results (Klavarioti et al., 2009; Qiao et al., 2005; Shu et al., 2016; Zhang et al., 2003).

The basic kinetic equation that is applied for the degradation of a specific organic contaminant using the UV/H_2O_2 oxidation, would consider the concentrations of the contaminant and the $^{\bullet}OH$ radical in the systems, which can be represented by Eq. (2.22), a second-order reaction.

$$r_C = -k\,C\quad OH \tag{2.22}$$

Considering the complexity related to the determination of the $^{\bullet}OH$ radical concentration, which is quite reactive, it is possible to simplify the Eq. (2.22), if the concentration of the $^{\bullet}OH$ radical is adopted at a steady state. It is possible, if an excess of H_2O_2 is maintained in the system, with an adequate UV irradiance. With these assumptions, two different equations can be proposed for the contaminant degradation, Eqs. (2.23) and (2.24).

$$r_C = -k'\,C \tag{2.23}$$

$$r_C = -k''\,C\,I \tag{2.24}$$

The k' and k'' are pseudo-first order degradation rate constants for the contaminant oxidation, and I is the UV irradiance applied, expressed in W/m^2. The kinetic constant k' includes the $^{\bullet}OH$ radical concentration and the UV light irradiance, while k'' only includes the $^{\bullet}OH$ radical concentration.

For practical purposes, the use of the proposed simplified model will require the development of bench-scale or pilot-plant tests for obtaining the degradation kinetic constants, which will be specific for each type of effluent studied.

Based on the results obtained with the bench-scale or pilot-plant experiments tests, it will be possible to design the treatment unit. It is worth noting that more precise results will be obtained with a more detailed evaluation, performing evaluation tests with different UV light irradiances and hydrogen peroxide concentrations. Experiments in a batch reactor are more adequate for bench-scale experiments, while for pilot-plant units, a tubular reactor is the best reactor configuration, i.e., plug-flow. Pilot-plant tests are necessary for obtaining the oxidation process design parameters.

Continuous stirred tank reactors (CSTR), could be an option for performing the UV/H_2O_2 oxidation process; however, for large scale treatment units, its intrinsic design characteristics can result in very complex equipment. In addition to this, the CSTR performance is inferior compared to tubular or batch reactors.

2.4.1 Fundamental on Reactor Design

The complete theory related to chemical reactors; fundamentals, design, and simulation can be obtained in classical literature (Froment et al., 2011; Levenspiel, 1999; Rawlings, 2002; Schmidt, 1988).

Considering the fundamental theory related to chemical reactors, a chemical oxidation reaction using the UV/H_2O_2 process can be performed in batch, continuous stirred, and tubular reactors. The main characteristics of each reactor type are presented here.

- *Batch reactor*

    It is characterized by the lack of flow through the reactor, while it is in operation. It means that the composition of the reactor content will change with the time, until the reaction equilibrium is reached, or the desired reactants conversion is obtained. At this point, the reactor content is removed, and a new batch can be started.

- *Continuously stirred tank reactor (CSTR)*

    The CSTR is characterized by the continuous flow of the reactants into and products out of the reactor, with its content maintained homogenized using a mechanical stirrer. The ideal CSTR is perfectly mixed, i.e., composition of the reaction content is homogeneous, meaning that the reactor spatial retention time should be enough for obtaining the desired reactants conversion. Its operation is only stopped, if necessary.

- *Tubular reactor*

    The tubular reactor is also characterized by a continuous flow of reactants and products. In the ideal tubular plug-flow reactor, there is no axial or back mixing of its content. The plug-flow reactor can be imagined as a set of batch reactors operating in series, with the content of each been transferred to the next one. The first reactor receives the reactant mixture and keeps it for a small period of time for the reaction, and then its content flows to the next reactor, and so on, until the mixture reaches the last reactor and then exits as a final product.

Considering ideal reactors, it is possible to simulate their behavior building a simple mathematical model derived from a mass balance. To help the development of the mass balance, it is necessary to define all the variables involved in the process and to make a flow diagram indicating all the flows entering and leaving the reactor. The flow diagrams used for performing the mass balances in the three different types of reactors are depicted in Fig. 2.5.

By performing a mass balance for each reactor configuration, it is possible to obtain the equations that correlate the variation of a specific contaminant with the chemical reaction rate, and reaction time or retention time.

- *Batch reactor* (Fig. 2.5A)

    Combining the mass balance with the rate law given by Eq. (2.23) and rearranging:

$$\frac{d\,C}{C} = k'dt \tag{2.25}$$

$$\int_{C_0}^{C} \frac{d\,C}{C} = -k' \int_{0}^{t} dt \tag{2.26}$$

Integrating the expression and rearranging,

$$C = C_0 \; e^{-k't} \tag{2.27}$$

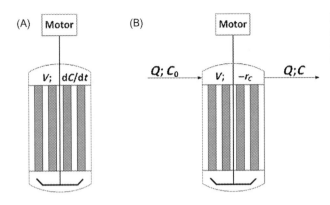

FIGURE 2.5 Flow diagrams of the three typical chemical reactors used for the UV/ H_2O_2 oxidation process. (A) Batch reactor, (B) CSTR, and (C) plug-flow reactor.

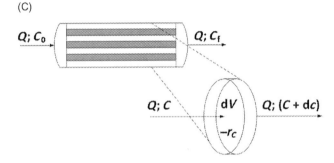

For the expression, in which the UV irradiance is disclosed, the following relation is obtained:

$$C = C_0 \; e^{-k''I \; t} \tag{2.28}$$

$$I \; t = D \tag{2.29}$$

D is defined as the UV dose, expressed in Joules per square meter.

The UV light irradiance is obtained considering the UV power applied on the irradiation area, the quartz sleeve, and effluent transmittances as well as the geometrical characteristics of the photochemical reactor-irradiation source. The UV power is obtained considering the conversion factor informed by the UV-lamp manufacturers (Table 2.1).

- *Continuous stirred tank reactor* (Fig. 2.5B)

Performing a mass balance in the reactor.

$$Q \; C_0 = Q \; C + \; -r_C \; V \tag{2.30}$$

Substituting r_C (Eq. 2.23), and considering that V/Q is the spatial detention time ( ).

$$C = \frac{C_0}{k' \; +1} \tag{2.31}$$

Disclosing the UV irradiance results:

$$C = \frac{C_0}{k' \, I + 1} \tag{2.32}$$

$$I = D \tag{2.33}$$

- *Plug-flow reactor* (Fig. 2.5C)

  Performing a mass balance on the volume element (dV).

$$Q \, C = Q \, C + dc + -r_c \, dV \tag{2.34}$$

Rearranging,

$$Q \, d C = - -r_C \, dV \tag{2.35}$$

Substituting r_C (Eq. 2.23), in the previous expression,

$$Q \, d C = - k' \, C \, dV \tag{2.36}$$

$$\frac{d C}{C} = - \frac{k'}{Q} \, dV \tag{2.37}$$

$$\int_{C_0}^{C} \frac{d C}{C} = - \frac{k'}{Q} \int_0^V dV \tag{2.38}$$

Integrating the expression and rearranging,

$$C = C_0 \, e^{-k'} \tag{2.39}$$

For the expression, in which the UV irradiance is disclosed, the following relation is obtained:

$$C = C_0 \, e^{-k'' \, I} \tag{2.40}$$

The choice of the reactor type to be used for the UV/H_2O_2 oxidation process depends on technical, operational, and economic issues. Batch reactors considering its operational characteristics are limited for the treatment of small volumes of effluents or for bench-scale evaluation tests. On the other hand, CSTR and tubular plug-flow reactors are more suitable for the treatment of large volumes of effluent continuously.

The option between the CSTR and tubular reactors will be based on the complexity of the UV/H_2O_2 configuration and the performance of each one. Considering the reactor complexity, it is evident that the CSTR reactor has a significant disadvantage, because its performance depends on effective mixing. The occurrence of a hydraulic short-circuit will reduce the reactor efficiency.

For any UV/H_2O_2 oxidation reactor configuration, the lamps need to be installed inside the reactor, which will make the installation of an appropriate mixer in the CSTR reactor difficult. In addition to this, ideal CSTR reactors present a lower performance in comparison to batch and tubular plug-flow reactors. As it can be observed, the plug-flow reactor performs better than the CSTR (as revealed by Fig. 2.6), reaching an effective degradation of the contaminant much faster and with higher efficiency.

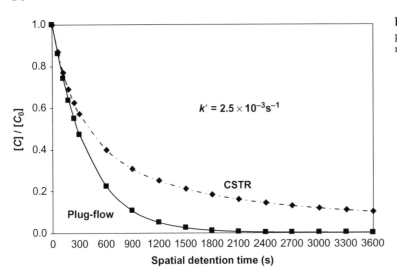

FIGURE 2.6 CSTR and plug-flow reactors performance comparison.

The UV/H$_2$O$_2$ oxidation process should preferably be performed by using the tubular plug-flow reactor configuration. As an alternative, the designer should use a series of CSTR reactors to improve the oxidation process performance.

The information presented provides the necessary base for the conception, design, and performance evaluation of new or existing effluent treatment units.

2.5 UV/H$_2$O$_2$ OXIDATION PROCESS DESIGN

From the engineering point of view, the design of any system used for wastewater treatment requires the development of bench-scale tests for evaluating the potential application of a specific process and for obtaining the necessary design parameters. The available literature related to the UV/H$_2$O$_2$ oxidation process deals with the fundamental and potential applications of the process (Halmann, 1995; Pearsons, 2004; USEPA, 1998), but does not present a straightforward procedure for the evaluation and design of treatment units, which is quite relevant for the process engineers.

Considering this limitation, a practical approach for the concept, design, and evaluation of effluent treatment units based on the UV/H$_2$O$_2$ oxidation process has been presented. It is important to note that the proposed approach was based on the fundamentals of the chemical engineering process and on the practical experience acquired over the years.

The first step for the development of any engineering project is to understand the problem that needs to be managed or solved. After that, it is necessary to identify and evaluate the possible options for dealing with the problem, in order that the designer can define the best alternative. This means that the option for using the UV/H$_2$O$_2$

oxidation process will be compared with different treatment technologies, and the decision about its application should be very well justified and understood. The procedure for the concept and design of the UV-H_2O_2 oxidation process is presented in the following sections:

2.5.1 Effluent Characteristics

It is evident that effluent matrix characteristics are of paramount relevance in the potential application of this process for an effluent treatment. This statement can be supported by studies developed for evaluating the application of the UV/H_2O_2 oxidation process for effluent and water treatment (Antonopoulou et al., 2014; Giannakis et al., 2015; James et al., 2014; Michel-Kordatou et al., 2015; Mierzwa et al., 2012; Yang et al., 2016).

It is possible to identify the most relevant effluent constituents that affect performance of the UV/H_2O_2 oxidation process, and the need for the effluent pre-treatment. These constituents can be grouped in three main categories, as follows:

- *Suspended solids*

    These represents all the effluent constituents that are insoluble and remain in suspension including sand, clay, and silt particles, as well as particulate NOM. The presence of these materials significantly affects the effluent optical properties, mainly the UV radiation transmission. The easiest way to estimate the concentration of suspended solids in an effluent is to measure its turbidity. Effluents with high turbidity values are less suitable for treatment by the UV/H_2O_2 oxidation process.

- *Dissolved organic carbon*

    It represents all the *organic* constituents contained in the effluent, including the NOM, and the variety of organic compounds used in our homes, industry, and agriculture. Some specific constituents are surely of interest including a large variety of chemical compounds, like pharmaceutical products, personal care products, pesticides, fertilizers, food additives, and others specific chemicals. These constituents can be measured directly, if they are the target of the treatment process, or by measuring the concentration of the dissolved organic carbon (DOC) or by the determination of the chemical oxygen demand (COD). In any case, the sample that is to be analyzed should be submitted to a filtration step for removing the suspended solids. It is accepted that the filtration through a 0.45 m pore size filter separates the suspended particles from the dissolved solids.

- *Dissolved inorganic chemicals*

    These represent all the inorganic chemicals that can be present in the effluent. Some of these chemicals can scavenge the $^{\bullet}OH$ radical, like bicarbonates and carbonates, and/or can absorb the UV radiation. The determination of these constituents can be specific for a constituent of interest or can be used as an indirect method like the measurement of the color or the total dissolved solids.

Though the presented parameters can significantly affect the performance of the UV/H_2O_2 oxidation process, the relevant information regarding the effluent characteristic is its

TABLE 2.4   Minimal Set of Parameters to be Analyzed for Evaluating the Potential Application of the UV/H_2O_2 Oxidation Process

Parameter	Relevance
Turbidity	Evaluation of the presence of suspended solids, which are responsible for the attenuation of UV radiation indicating the need for an effluent pre-treatment.
Dissolved organic carbon	It can affect effluent UV radiation transmission and will be responsible for the consumption of the H_2O_2 and $^\bullet OH$ radical. It will affect the intensity of UV radiation that will be necessary for the oxidation process.
Bicarbonate and carbonate alkalinity	These chemicals scavenge the $^\bullet OH$ radical, which can negatively affect the oxidation process performance.
Soluble iron and manganese	These species can be oxidized by hydrogen peroxide and precipitate on to the quartz sleeve, reducing the UV radiation. They indicate the pre-treatment needs.
UV light transmittance	Effluents with small transmittance will require a higher UV radiation intensity, or the UV/H_2O_2 oxidation process won't be an effective option for its treatment.
Specific organic contaminant	They will be used for evaluating the efficiency of the UV/H_2O_2 oxidation process.

capacity to absorb the UV radiation at the wavelength that will be used in the process. For this reason, the UV transmittance is a quite simple measurement that can be used for evaluating the potential application of the UV/H_2O_2 oxidation process. This is because effluents with higher UV transmittances will require small doses of UV radiation for producing the $^\bullet OH$ radical for effectively oxidizing the contaminant of interest.

Considering what is exposed, Table 2.4 presents the minimal set of parameters that should be considered for evaluating the potential application of the UV/H_2O_2 oxidation process for effluent treatment.

It is worth observing that other effluent quality variables can be monitored to better characterize the effluent to be submitted to the UV/H_2O_2 oxidation process, since they can be used for the design of the pre-treatment units. The choice of the effluent quality variables to be monitored will be based on the experience of the treatment system designer.

2.5.2 Bench Scale Evaluation Tests

Real world effluents have a complex composition and the design of any treatment system requires a preliminary evaluation of the potential use of any specific treatment technology for obtaining the most relevant design parameters.

Bench-scale tests should be performed using a reactor geometry that can represent, as close as possible, the reactor, which will be used in full scale treatment unit. That is relevant because most of the process parameters are affected by the reactor geometric characteristics, and if it changes, the process performance will also change. Maybe the most relevant geometric parameter is the optical path, which is related to the distance that the UV radiation will need to penetrate in the effluent and hence, the effective production of $^\bullet OH$ radicals.

Based on the Beer-Lambert law, the higher the optical path is, the lower the light transmittance will be through the solution, and this effect is more significant with the increasing concentration and absorption coefficient of the constituents contained in the effluent. So, optical paths higher than 1 cm (0.01 m) are not recommended.

With the adequate geometric configuration of the bench-scale oxidation reactors, it is necessary to perform some experiments for obtaining the kinetic degradation constant. For this purpose, different assays should be performed including the evaluation of pre-treatment methods.

The procedure for the bench-scale test is straightforward and consists of the following steps:

1. The reactor UV-lamp is turned on to stabilize the UV-photon flux.
2. The UV-reactor is filled with the effluent to be treated. It is important to use the pre-treated effluent, assuring that the best optical properties were obtained.
3. Hydrogen peroxide solution is added to obtain the desired concentration. It is important to homogenize the reactor contents.
4. Periodic samples of the reactor content are taken out to measure the concentration of the control contaminant. It is important to take a sample just before adding the hydrogen peroxide solution.
5. The assay is stopped after an adequate period, assuring the oxidation reaction has completed or stabilized.
6. The reaction constant (kinetic) is obtained by plotting the values of the natural logarithm of $[C_0]/[C]$ versus the reaction time. If a straight line is obtained, its inclination is the kinetic reaction rate. Otherwise, the H$_2$O$_2$, or the UV-radiation dosage should be increased, and the test repeated.
7. The test (steps 1−6) are repeated using different concentrations of H$_2$O$_2$ concentration, to obtain the best concentration.
8. The test (steps 1−4) are repeated using the best H$_2$O$_2$ and a different UV-radiation dosage, to obtain the UV dose.

The number of assays, as well as the duration of each one will depend on the final design aims and involved costs. It is recommended to conduct a more detailed evaluation for the design of treatment units with a high flow output.

If a tubular reactor or CSTR is used, steady state experiments should be performed using different spatial detention times, which can be done by changing the UV-reactor feed flow. An example of the graphic that should be obtained after each bench-scale test is presented in Fig. 2.7, which is based on the data used to construct Fig. 2.6. It is important to note that the results for the batch reactor will be mathematically equivalent to the ones obtained for the plug-flow reactor, considering that the spatial retention time is equivalent to the reaction time.

At the end of the bench-scale experiments, the main UV/H$_2$O$_2$ oxidation process parameters will be defined like UV light irradiation, H$_2$O$_2$ concentration, and the degradation kinetic constant. This information will be necessary for the design of a small UV/H$_2$O$_2$ oxidation treatment facility or a pilot-plant unit.

Regarding the UV light irradiation, it is quite important to specify the effective UV dose applied on the effluent, which should consider the energy loss on the quartz sleeve and on

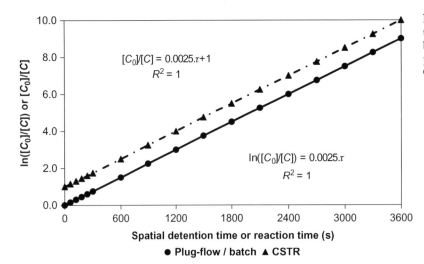

FIGURE 2.7 Plots of the results obtained in the bench scale tests using plug-flow/batch and CSTR reactors.

the effluent, because of the respective absorbance of each medium. This is done by multiplying the total applied UV irradiance by quartz sleeve and effluent transmittances (Eq. 2.41).

$$I_{effective} = \frac{I_{applied}\ \%T_{qs}\ \%T_{effluent}}{100} \qquad (2.41)$$

2.5.3 Pilot-Plant Evaluation Units

A pilot-plant evaluation unit is used to better evaluate any treatment process and identify relevant issues that can arise during continuous operation. It is important to note that its design should consider the results obtained during the bench-scale treatment tests. Additionally, it is necessary to define all the complementary units and components that will be necessary for an adequate operation. The complementary units include the pre-treatment unit, chemical dosing systems, and the necessary hydraulic, electric, and electronic components.

At this point, it is important to develop a process flow sheet of the UV/H$_2$O$_2$ oxidation treatment unit, and the complementary documents involved in its design, such as the calculation memorial and system description. This is necessary for registering the necessary information that will be needed for the evaluation of the treatment process and as guidelines for the development of the final treatment unit.

It should be emphasized that the hydraulic and geometric characteristics of the UV/H$_2$O$_2$ oxidation reactor need to be compatible with the one of the reactor that will be used in the final effluent treatment plant, mainly the optical path inside the reactor.

In addition to its importance for obtaining more precise design parameters, the pilot-plant unit can be maintained operational after the design and construction of the final treatment unit, in case of any process changes are intended, or for evaluating different treatment components, such as UV-lamps. It can be also used for training purposes.

The design of the pilot-plant unit should be performed with same criteria that would be used for the design of the final treatment unit.

- *UV/H₂O₂ reactor design*

Based on the results obtained on the bench-scale UV/H₂O₂ oxidation tests, and with the information regarding the treatment unit capacity, effluent characteristics, design, contaminant removal efficiency, and the data related to the UV-lamp and quartz sleeve, it is possible to design the UV/H₂O₂ oxidation reactor.

After the reactor has been designed, the first step is to calculate the reactor volume by Eq. (2.42), which is based on the reactor spatial retention time and the desired contaminant removal efficiency, using Eqs. (2.31) or (2.32) for the CSTR or Eqs. (2.39) or (2.40) for the plug-flow reactor. The contaminant removal efficiency is given by Eq. (2.43).

$$V = Q \tag{2.42}$$

$$\text{Removal efficiency } \% = 100 \quad 1 - \frac{C}{C_0} \tag{2.43}$$

Using the desired contaminant removal efficiency in Eq. (2.43), the value of [C]/[C₀] is obtained, which allows to obtain the reactor spatial detention time ( ), and also the reactor volume (Eq. 2.42).

With the reactor volume, the quartz sleeve diameter, and the irradiation optical path, it is possible to calculate the total length of an equivalent reactor using Fig. 2.8 as a reference.

$$d_R = d_{qs} + 2OP \tag{2.44}$$

$$V = S_{annular}L_{eq} \tag{2.45}$$

$$S_{annular} = S_{reactor} - S_{qs} \tag{2.46}$$

$$S_{annular} = \frac{d_R^2}{4} - \frac{d_{qs}^2}{4} \tag{2.47}$$

Replacing Eq. (2.44) and rearranging, Eq. (2.47) is obtained, that is then replaced in Eq. (2.45) to obtain Eq. (2.49).

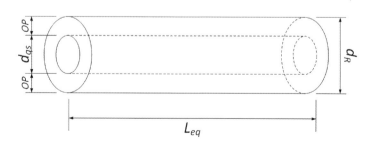

FIGURE 2.8 Total length of a reactor.

$$S_{annular} = d_{qs}OP + OP^2 \tag{2.48}$$

$$L_{eq} = \frac{V}{d_{qs}OP + OP^2} \tag{2.49}$$

For obtaining the UV power applied, it is assumed that the exposed area is the internal reactor surface area, and it can be calculated using Eq. (2.50).

$$S_{exp} = (d_{qs} + 2OP)L_{eq} \tag{2.50}$$

With the exposed area (S_{exp}), and the effective irradiance, the UV power that should be applied can be calculated, using Eq. (2.51).

$$Applied_{UV_{Power}} = I_{applied}S_{exp} \tag{2.51}$$

Rearranging Eq. (2.41) and replacing it in Eq. (2.48) and considering the UV-lamp efficiency loss (el), through its life span, the applied UV power is obtained with the Eq. (2.52).

$$Applied_{UV_{power}} = \frac{100^3 I_{effective}S_{exp}}{\%T_{qs}\%T_{effluent} \, 100 - \%UV_{el}} \tag{2.52}$$

The $\%T_{qs}$ is obtained from Table 2.2, or consulting the quartz sleeve manufacturer, while the $\%T_{effluent}$ is obtained in the laboratory. By selecting the appropriate UV-lamp model (Table 2.1), it is possible to calculate the total number of UV-lamps required for the treatment unit, as well as the number of quartz sleeves, which are the same as given in Eq. (2.53).

$$N_{UV\text{-}lamps} = N_{qs} = \frac{Applied_{UV_{power}}}{Lamp_{UV_{output}}} \tag{2.53}$$

Considering the number of quartz sleeves obtained and the net UV-lamp length ($L_{UV\text{-}net}$), the reactor's total volume is calculated by Eq. (2.54).

$$V_{total} = V + \frac{N_{qs} \, d_{out_{qs}}{}^2 L_{UV\text{-}net}}{4} \tag{2.54}$$

With the reactor's total volume, the reactor internal diameter is obtained using Eq. (2.55).

$$d_R = \sqrt{\frac{4V_{total}}{L_{UV\text{-}net}}} \tag{2.55}$$

Considering the use of a plug-flow reactor, it is important to use a $L_{UV\text{-}net}/d_R$ ratio higher or equal to three, but for lower ratios, it is recommended to use a larger number of reactors, which operates in parallel. In this case, the feed flow should be split for calculating the new $L_{UV\text{-}net}/d_R$ ratio, using the previous equations.

With the main reactor dimensions obtained, it is necessary to evenly distribute the quartz sleeves over its cross section.

TABLE 2.5 Characteristics of the Pre-treated Effluent

Parameter	Value	Unit
Flow	5	m^3/h
pH	6.5	Unitless
Transmittance	90	% for an optical path of 1 cm
Turbidity	<1	NTU
Color	<15	Color units
Bicarbonate	35	mg/L

Once the reactor (or reactors), was dimensioned, it is necessary to design the other components of the UV/H$_2$O$_2$ oxidation process, such as effluent storage tanks, pre-treatment unit, chemical dosage units, and post-treatment units, if necessary. The example below illustrates the procedure for dimensioning an UV/H$_2$O$_2$ oxidation reactor.

Example 1: A pharmaceutical industry needs to treat one of its effluents to remove the residual of an active compound used in the fabrication of a medicine used for chemotherapy. This effluent is pre-treated through a physical-chemical process, but the residual concentration of the active compound is twenty times higher than the limit defined by the Environmental Protection Agency. The engineer responsible for the effluent treatment facility performed some bench-scale tests using the UV/H$_2$O$_2$ oxidation process in a batch reactor and obtained the pseudo-first order kinetic degradation constant for the contaminant ($k' = 8.7 \times 10^{-2}$ s^{-1}), for an effective UV dose of 200 mJ/cm^2, optical path of 1 cm, and H$_2$O$_2$ dose of 10 mg/L. The effluent relevant characteristics are presented in Table 2.5. Based on the presented data, design a continuous plug-flow reactor to obtain a removal efficiency higher than 95%. Estimate the H$_2$O$_2$ consumption considering the use of a stock solution with an active H$_2$O$_2$ concentration of 37% by weight (density = 1130 g/L^1).

Resolution
Removal efficiency considered = 99%
Data conversion to the International System Units:
Flow (Q) = 5/3600 = 1.29 × 10^{-3} m^3/s
D (UV dose) = 200 × 10,000/1000 = 2000 J/m^{-2}
OP (Optical length) = 1/100 = 1.0 × 10^{-2} m
First, it is necessary to calculate the [C]/[C$_0$] ratio, using Eq. (2.43).

$$\frac{C}{C_0} = 1 - \frac{\text{Removal efficiency \%}}{100}$$

$$= 1 - \frac{99}{100} \rightarrow \frac{C}{C_0}$$

$$= 0.01$$

Now, the spatial detention time can be obtained using Eq. (2.39).

$$= -\frac{\ln\left(C / C_0\right)}{k'}$$

$$= -\frac{\ln 0.01}{8.7 \times 10^{-2}}$$

$$= 52.93 \text{ s}$$

Using the spatial detention time, it is possible to calculate the degradation kinetic constant, when the UV irradiance is disclosed. This is done using Eqs. (2.39) and (2.40).

$$e^{-k'} = e^{-k''\ D}$$

$$\text{or } k' = k''\ D$$

$$\text{or } k'' = \frac{k'}{D} = \frac{8.7 \times 10^{-2} \times 52.93}{2000}$$

$$k'' = 2.3 \times 10^{-3} \text{ m}^2/\text{W per s}$$

With the spatial detention time, the reactor volume can be calculated using Eq. (2.42).

$$V = Q \quad = 1.39 \times 10^{-3} \times 52.93$$

$$= 0.074 \text{ m}^3$$

The reactor equivalent length is obtained using Eq. (2.46) and the OP value informed. For this calculation, it is necessary to select the diameter of the quartz sleeve. From Table 2.2, the selected quartz sleeve that will be used in the project is the Suprasil−310, with an outside diameter of 3 cm (3×10^{-2} m).

$$L_{eq} = \frac{V}{\left(d_{qs}OP + OP^2\right)}$$

$$= \frac{0.074}{3.1416 \times\ 3 \times 10^{-2} \times 1 \times 10^{-2} + \left(1 \times 10^{-2}\right)^2}$$

$$= 58.9 \text{ m}$$

Now, using Eq. (2.50), it is possible to calculate the exposed area:

$$S_{exp} = L_{eq} \quad \left(d_{qs} + 2\ OP\right)$$

$$= 58.9 \times 3.1416 \times\ 3 \times 10^{-2} + 2.1 \times 10^{-2}$$

$$= 9.4 \text{ m}^2$$

The applied UV power is calculated using Eq. (2.49) and the quartz sleeve transmittance is obtained in Table 2.3 (90%). It is assumed that the UV-lamp efficiency loss is 20%, and the effective irradiance is obtained with Eq. (2.33).

$$\text{Applied}_{\text{UV}_{\text{power}}} = \frac{100^3 \times I_{\text{effective}} S_{\text{exp}}}{\%T_{\text{qs}} \times \%T_{\text{effluent}} \, 100 - \%\text{UV}_{\text{el}}}$$

$$I_{\text{effective}} = \frac{D}{-} = \frac{2000}{52.93}$$

$$= 37.8 \text{ W/m}^2$$

$$\text{Applied}_{\text{UV}_{\text{power}}} = \frac{100^3 \times 37.8 \times 9.4}{90 \times 90 \times \, 100 - 20}$$

$$= 548 \text{ W}$$

The number of UV lamps and quartz sleeves, can be obtained from Eq. (2.53). The lamp that will be used in this project (Table 2.1), is from LightTech, higher output, GHO893T5L.

UV-lamp specifications
Electrical power = 95 W
UV power (254 nm) = 30 W
Length ($L_{\text{UV-net}}$) = 0.815 m
Diameter = 15 mm

$$N_{\text{UV-lamps}} = N_{\text{qs}} = \frac{\text{Applied}_{\text{UV}_{\text{power}}}}{\text{Lamp}_{\text{UV}_{\text{output}}}} = \frac{548}{30}$$

$$= N_{\text{qs}} \cong 19$$

With the total number of quartz sleeves, it is possible to calculate the reactor total volume by Eq. (2.54).

$$V_{\text{total}} = V + \frac{N_{\text{qs}} \, d_{\text{out}_{\text{qs}}}{}^2 L_{\text{UV-net}}}{4}$$

$$= 0.074 + \frac{19 \times 3.1416 \times \, 3.0 \times 10^{-2} \, {}^2 \times 0.815}{4}$$

$$= 0.085 \text{ m}^3$$

Now, it is possible to calculate the UV-reactor diameter by Eq. (2.52).

$$d_R = \sqrt{\frac{4V_{\text{total}}}{L_{\text{UV-net}}}}$$

$$= \sqrt{\frac{4 \times 0.085}{3.1416 \times 0.815}}$$

$$= 0.36 \text{ m}$$

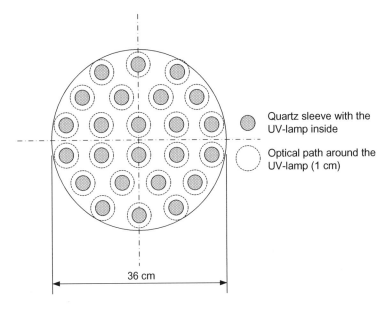

FIGURE 2.9 Final arrangement for the quartz sleeves and UV-lamps distribution.

Quartz sleeve with the UV-lamp inside

Optical path around the UV-lamp (1 cm)

36 cm

For a plug-flow reactor, a L_{UV-net}/d_R ratio higher than 3 is recommended. Considering the calculated values, the L_{UV-net}/d_R ratio is 2.3, which is lower than the recommended value. However, the design will follow to demonstrate that the reactors characteristics are not appropriate.

Quartz sleeves distribution inside the reactor:

Fig. 2.9 presents the final arrangement for the quartz sleeves and UV-lamps distribution, allowing the use of 24 quartz sleeves.

The resulting UV-lamps distribution layout will not provide an effective irradiance of the effluent, indicating that this design is not adequate (Fig. 2.9). This problem can be overcome by changing the UV-lamp model to obtain a more adequate L_{UV-net}/d_R ratio, or by using more than one reactor.

Example 2: Calculation with a different UV-lamp model:

To overcome the L_{UV-net}/d_R ratio issue, a longer UV-lamp with a lower UV-output will be evaluated. The UV-lamp selected is from Philips TUV-T5—927970204099 (Table 2.1).

UV-lamp specifications
Electrical power = 40 W
UV power (254 nm) = 15 W
Length (L_{UV-net}) = 1.5 m
Diameter = 19.3 mm

Based on the previous results related to the UV-reactor volume and required UV-power, it is possible to calculate the number of UV-lamps, the total volume and diameter of the UV-reactor and the new L_{UV-net}/d_R ratio, which are presented in Table 2.6. The resulting UV-lamps and quartz sleeves distribution layout is presented in Fig. 2.10.

TABLE 2.6 Reactor's Characteristics Considering the Selected UV-Lamp

Characteristics	Value	Unit
Number of UV-lamps and quartz sleeves calculated	40	Units
Reactor total volume	0.113	m^3
Reactor internal diameter	0.3	m
L_{UV-net}/d_R	5.0	Unitless
Optical path considered for the UV-lamps distribution	0.5	cm
Electrical power	1600	W
UV power	600	W

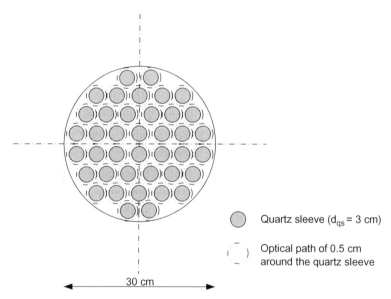

FIGURE 2.10 Final arrangement for the quartz sleeves and UV-lamps distribution considering a different UV-lamp model and quartz sleeves separation distance.

Quartz sleeve (d_{qs} = 3 cm)

Optical path of 0.5 cm around the quartz sleeve

30 cm

The resulting reactor configuration seems to be much more suitable for the oxidation process than the previous one (Fig. 2.10). This is because the distribution of the UV-light will be more uniform, which will result in a more effective irradiance. It is also important to note that the total number of lamps in this arrangement is higher than the one previously calculated. With the new distribution layout, the total number of UV-lamps is 48.

With this configuration, it is necessary to check the reactor performance for the obtained conditions.

Determination of the reactor volume (V) for oxidation (Eq. 2.54).

$$V = V_{\text{total}} - \frac{N_{qs} \; d_{\text{out}_{qs}}^2 L_{\text{UV-net}}}{4}$$

$$= \frac{3.1416 \times 0.32^2 \times 1.5}{4} - \frac{48 \times 3.1416 \times (3 \times 10^{-2})^2 \times 1.5}{4}$$

$$= 0.070 \; \text{m}^3$$

Spatial detention time (Eq. 2.42):

$$= \frac{V}{Q} = \frac{0.064}{1.39 \times 10^3}$$

$$= 46.064 \; \text{s}$$

Contaminant efficiency removal (Eqs. (2.39) and (2.43)):

$$\frac{C}{C_0} = e^{-k'}$$

$$= e^{-8.7 \times 10^{-2} \times 46.064}$$

$$= 0.018$$

$$\text{Removal efficiency \%} = \left(1 - \frac{C}{C_0}\right) 100 = 1 - 0.018 \times 100$$

$$= 98.2$$

The value obtained for the removal efficiency is lower than the one used for the reactor design, but it is higher than the removal efficiency required, i.e., 95%. This means that the obtained design is adequate.

Estimation of the apparent optical path (Eq. 2.47):

$$\left(d_{qs}OP + OP^2\right) = \frac{V}{L_{eq}}$$

$$= \frac{0.064}{40 \times 1.5 \times 3.1416}$$

$$\text{or} \quad OP^2 + 3 \times 10^{-2}OP - 3.4 \times 10^{-4} = 0$$

Solving the equation, the OP value is obtained:

$$OP = 0.9 \times 10^{-2} \; \text{m}$$

Determination of the exposed area (Eq. 2.50):

$$S_{\text{exp}} = \left(d_{qs} + 2OP\right)L_{eq}$$
$$= 3.1416 \times (3 \times 10^{-2} + 2.0.9 \times 10^{-2}) \times 40.1.5$$
$$= 9.04 \; \text{m}^2$$

Calculation of the UV irradiance (Eq. 2.49):

$$I_{effective} = \frac{Applied_{UV_{power}} \times \%T_{qs} \times \%T_{effluent}\ 100 - \%UV_{el}}{100^3 S_{exp}}$$

$$= \frac{40.15 \times 90 \times 90 \times\ 100 - 20}{100^3 \times 9.04}$$

$$= 42.96\ \text{W/m}^2$$

The value obtained is higher than the one used to develop the reactor design, meaning the new reactor configuration is adequate. Now it is possible to verify the contaminant efficiency removal using Eqs. (2.40) and (2.43) with the k'' value.

$$\text{Removal efficiency } \% = 100\ \left(1 - e^{-k''\ I}\ \right)$$

$$= 100\ \ 1 - e^{-2.3 \times 10^{-3} \times 42.96 \times 46.04}$$

$$= 98.94\%$$

Considering the higher UV irradiance, the reactor will be able to remove the contaminant with the specified efficiency.

For the design conclusion, it is necessary to obtain the H_2O_2 consumption.

H_2O_2 consumption:

The H_2O_2 consumption is obtained considering the concentration of the stock solution, effluent flow, and final concentration.

$$H_2O_{2\ effluent} = 10\ \text{mg/L or } 10\ g/m^3$$

$$Q = 5\ \text{m}^3/h$$

$$H_2O_{2\ Dosage}\ = Q\ H_2O_{2\ Effluent} = 5 \times 10$$
$$= 50\ g/h$$

2.6 PRACTICAL APPLICATIONS

The UV/H_2O_2 process is used for treatment of wastewater in some specific effluents.

2.6.1 Slaughterhouse Wastewater

Bustillo-Lecompte et al. (2016) evaluated the effect of UV/H_2O_2 in a continuous photoreactor to treat a slaughterhouse effluent. The system operated with a recycle effluent ratio, which was optimized based on the prediction made by statistical models and results collected during the design and operation of the treatment unit. The application of AOPs for effluents like this is particularly interesting because of the high rate of inactivation of microorganisms without further addition of chemicals in high concentrations (which may also lead to the formation of harmful by-products).

Based on the Beer-Lambert law, they developed a model for calculating the volumetric rate of energy absorption and used a four-factor central composite design with five levels of process variables and a response surface methodology to maximize the TOC removal and minimize the residual H_2O_2 in the effluent. As the result of this optimization, the total organic carbon removal was 81% and the minimum residual H_2O_2 was less than 2% with a flow rate 15 mL/min with 18% recycle ratio.

2.6.2 Oil-Water Emulsion

Mierzwa et al. (2012) evaluated the application of a physical-chemical clarification process and hollow fiber ultrafiltration membrane as a pre-treatment for increasing the efficiency of a UV/H_2O_2 unit treating an effluent with high organic load (827 ± 109 mg/L TOC).

The photochemical reactor was equipped with a 1600 W medium pressure UV lamp. The TOC removal by UV/H_2O_2 was around 90%, regardless the pre-treatment method applied. However, when the pre-treatment processes were applied, the rate of TOC degradation increased, because of the better light transmission in the reactor. The overall time to reach a 90% removal was cut in half by applying pre-treatment methods that remove color and turbidity.

2.6.3 Pharmaceuticals

Rosario-Ortiz et al. (2010) studied the use of low pressure UV light and H_2O_2 for the oxidation of six pharmaceuticals (meprobamate, carbamazepine, dilantin, atenolol, primidone, and trimethoprim) in three wastewater effluents under different UV and H_2O_2 dosages. The pharmaceuticals chosen have specific functionalities that make them reactive to photolysis as well as to oxidation.

The role of the effluent organic matter was evaluated utilizing an empirical model based on a second order reaction rate constant. Based on this model, the scavenging rates were estimated. In the case of the UV/H_2O_2 treatment with 500 mJ/cm^2, pharmaceutical oxidation increased compared to a UV dosage of 300 mJ/cm^2 with equivalent H_2O_2 dosages. For the dosage of 700 mJ/cm^2, the photolysis itself (without peroxide) was enough to break some of the compounds.

Kim et al. (2009) evaluated the effectiveness of UV/H_2O_2 on the removal of 41 different types of pharmaceuticals in a real wastewater effluent in Japan. UV itself was not enough to remove 29 pharmaceuticals, despite the high dosage of 2768 mJ/cm^2. The association with peroxide reduced the UV dose to 923 mJ/cm^2 and it could remove 39 components with at least 90% efficiency.

In a similar study, Afonso-Olivares et al. (2016) also evaluated the degradation of 23 pharmaceutical components present in an urban wastewater treatment plant effluent. Some anti-inflammatory compounds were removed just by photolysis, while caffeine displayed resistance to oxidation by hydroxyl radicals. UV doses of 2369 mJ/cm^2 were required to reduce caffeine and a dose nearly twice as much was needed to reduce its metabolite (paraxanthine) to one tenth of its initial value. An optimal H_2O_2 concentration of 20 mg/L was enough to remove 93% of the components with a spatial retention time of 75 min.

Wols et al. (2013) evaluated the removal of 40 different types of pharmaceuticals in water and their degradation, when exposed to low pressure and medium pressure UV lamps. Results showed that low pressure photolysis could slowly degrade most of the compounds. The medium pressure lamps could improve the degradation, and the addition of peroxide helped to degrade most of the compounds that were not degraded before. Considering that medium pressure lamps consume more energy and the association of UV lamps and peroxide were equally efficient for both the lamps, they concluded that the low pressure lamps are an effective alternative, since they consume less energy than the medium pressure lamps. The UV dosage for 90% degradation (at 10 mg/L H_2O_2) was 500 mJ/cm^2 (medium pressure) and 1000 mJ/cm^2 (low pressure).

2.6.4 Dyes

Azo dyes are hard to be biodegraded in conventional biological wastewater treatment processes. A successful approach to treat these contaminants in water is their mineralization by AOPs, mainly UV/H_2O_2.

Chang et al. (2010) proposed a kinetic model for dye decomposition, which consisted of varying peroxide concentration, dye concentration, pH and UV irradiation power. They considered pseudo-steady state for hydroxyl and hydroperoxyl free radicals and calculated reaction rates in different pathways; thus validating the model with coefficients found in literature, and using nonlinear regression. The main conclusion of this evaluation was that with increasing H_2O_2 concentration or dye concentration, the reaction rate could be increased, but it can also accelerate the consumption of hydroxyl radicals and hinder the process, if the concentrations are too high.

2.6.5 Removal of Estrogens

Estrogens are among the micropollutants not removed by conventional wastewater treatment plants. They have the potential to disrupt the endocrine system and interfere in aquatic environments. AOPs have been successfully tested to remove estrogens from water.

Cédat et al. (2016) proposed a method to evaluate the removal of estrogens by UV photolysis and UV/H_2O_2 association in a pilot unit in real conditions. Their study focused not only on the removal of the estrogens but also on the formation of toxic by-products. Their results show that UV itself is not strong enough to remove estrogens from water, but the association with peroxide was able to remove up to 80% of initial estrogenic compounds and estrogenic activity from treated wastewater. The UV dosage of 423 mJ/cm^2 with 50 mg/L of H_2O_2 and 520 mJ/cm^2 with 30 mg/L of H_2O_2 were enough to achieve the removal. No high estrogenic or toxic by-products were detected after the UV photolysis or UV/H_2O_2 process.

Advances in chemical water and wastewater treatment have led to a range of treatment processes, notably the AOPs. The correct choice of the UV lamp and quartz sleeve set is important since these components can significantly affect the final process capital and operational costs. The manufacturers provide portfolios with full specifications about the lamps' efficiency and energy consumption, as well as information regarding the quartz sleeves absorption spectrum.

For the reactor design, there is not a simple path to be followed. The choice must consider not only the effluent quantity but also its quality. In general, batch reactors are more suitable for smaller volumes and bench-scale experiments, while CSTR and plug-flow reactors are more flexible to work with larger volumes (such as pilot test units). For either case, the complexity of the effluent may interfere on the final performance. Quality variables like suspended solids, DOC and dissolved inorganic compounds may change the light transmission or scavenge $^{\bullet}OH$ radicals. Therefore, it is important to have an adequate effluent characterization before defining the design and operational conditions of the treatment unit.

Despite some of its applications are more viable than others (such as mineralization of some compounds found in water, including some residual pharmaceuticals), the overall conclusion is that the UV/H_2O_2 systems have found their place on the top of the list of the most promising technologies for water and wastewater treatment.

References

Afonso-Olivares, C., Fernández-Rodríguez, C., Ojeda-González, R.J., Sosa-Ferrera, Z., Santana-Rodríguez, J.J., Doña Rodríguez, J.M., 2016. Estimation of kinetic parameters and UV doses necessary to remove twenty-three pharmaceuticals from pre-treated urban wastewater by UV/H_2O_2. J. Photochem. Photobiol. A: Chem. 329, 130–138.

American Public Health Association (APHA), American Water Works Association (AWWA), and Water Environment Federation (WEF), 2012. Method 5910 UV-Absorbing Organic Constituents. Standard Methods for the Examination of Water and Wastewater, 22nd ed. American Public Health Association, Washington, DC.

Andreozzi, R., Caprio, V., Insola, A., Marotta, R., 1999. Advanced oxidation processes (AOP) for water purification and recovery. Catal. Today 53, 51–59.

Antonopoulou, M., Evgenidou, E., Lambropoulou, D., Konstantinou, I., 2014. A review on advanced oxidation processes for the removal of taste and odor compounds from aqueous media. Water Res. 53, 215–234.

Beers Jr., R.F, Sizer, I.W., 1952. A spectrophotometric method for measuring the breakdown of hydrogen peroxide by catalase. J. Biol. Chem. 195, 133–140.

Bolton, J.R., Cater, S.R., 1994. Homogeneous photodegradation of pollutants in contaminated water: an introduction. In: Helz, G.R. (Ed.), Aquatic and Surface Photochemistry. Lewis Publishers, Boca Raton, FL, pp. 467–490.

Bustillo-Lecompte, C.F., Ghafoori, S., Mehrvar, M., 2016. Photochemical degradation of an actual slaughterhouse wastewater by continuous UV/H_2O_2 photoreactor with recycle. J. Environ. Chem. Eng. 4, 719–732.

Buxton, G.V., Elliot, A.J., 1986. Rate constant for reaction of hydroxyl radicals with bicarbonate ions. Radiat. Phys. Chem. 27 (3), 241–242.

Cataldo, F., 2014. Hydrogenperoxide photolysis with different UV light sources including a new UV-Led light source. New Front. Chem. 23 (2), 99–110.

Cédat, B., de Brauer, C., Métivier, H., Dumont, N., Tutundjan, R., 2016. Are UV photolysis and UV/H_2O_2 process efficient to treat estrogens in waters? Chemical and biological assessment at pilot scale. Water Res. 100, 357–366.

Chang, M.W., Chung, C.C., Chern, J.M., Chen, T.S., 2010. Dye decomposition kinetics by UV/H_2O_2: initial rate analysis by effective kinetic modelling methodology. Chem. Eng. Sci. 100, 135–140.

Comninellis, C., Kapalka, A., Malato, S., Parsons, S.A., Poulios, I., Mantzavinos, D., 2008. Perspective—advanced oxidation processes for water treatment: advances and trends for R&D. J. Chem. Technol. Biotechnol. 83, 769–776.

Crittenden, J.C., Hu, S., Hand, D.W., Green, S.A., 1999. A kinetic model for H_2O_2/UV process in a completely mixed batch reactor. Water Res. 33 (10), 2315–2328.

Deng, Y., Zhao, R., 2015. Advanced oxidation processes (AOPs) in wastewater treatment. Curr. Pollut. Rep. 1, 167–176.

Edalatmanesh, M., Dhib, R., Mehrvar, M., 2008. Kinetic modeling of aqueous phenol degradation by UV/H_2O_2 process. Int. J. Chem. Kinet. 40 (1), 34–43.

Fernández-Castro, P., Vallejo, M., San Román, M.F., Ortiz, I., 2015. Insight on the fundamentals of advanced oxidation processes. Role and review of the determination methods of reactive oxygen species. J. Chem. Technol. Biotechnol. 90, 796–820.

Froment, G.F., Bischoff, K.B., De Wilde, J., 2011. Chemical Rector Analysis and Design., third ed. John Wiley & Sons, Inc., Hoboken, NJ, 860 p.

Giannakis, S., Vives, F.A.G., Grandjean, D., Magnet, A., De Alencastro, L.F., Pulgarin, C., 2015. Effect of advanced oxidation processes on the micropollutants and the effluent organic matter contained in municipal wastewater previously treated by three different secondary methods. Water Res. 84, 295–306.

Gogate, P.R., Pandit, A.B., 2004. A review of imperative technologies for wastewater treatment II: hybrid methods. Adv. Environ. Res. 8, 553–597.

Halmann, M.M., 1995. Photodegradation of Water Pollutants. CRC Press, Inc, Boca Raton, FL, 303 p.

Houtman, C.J., 2010. Emerging contaminants in surface waters and their relevance for the production of drinking water in Europe. J. Integr. Environ. Sci. 7 (4), 271–295.

Huang, C.P., Dong, C., Tang, Z., 1993. Advanced chemical oxidation: its present role and potential future in hazardous waste treatment. Waste Manage. 13, 361–377.

Hugül, M., Apak, R., Demirci, S., 2000. Modeling the kinetics of UV/hydrogen peroxide oxidation of some mono-, di-, and trichlorophenols. J. Hazard. Mater. B77, 193–208.

James, C.P., Germain, E., Judd, S., 2014. Micropollutant removal by advanced oxidation of microfiltered secondary effluent for water reuse. Sep. Purif. Technol. 127, 77–83.

Kaltenegger, L., 2011. Encyclopedia of Astrobiology. Springer, Berlin, Heidelberg, pp. 3–4, 1853.

Kim, I., Yamashita, N., Tanaka, H., 2009. Performance of UV and UV/H_2O_2 processes for the removal of pharmaceuticals detected in secondary effluent of a sewage treatment plant in Japan. J. Hazard. Mater. 166, 1134–1140.

Klavarioti, M., Mantzavinos, D., Kassinos, D., 2009. Removal of residual pharmaceuticals from aqueous systems by advanced oxidation processes. Environ. Int. 39, 402–417.

Legrini, O., Oliveros, E., Braun, A.M., 1993. Photochemical process for water treatment. Chem. Rev. 93, 671–698.

Levenspiel, O., 1999. Chemical Reaction Engineering., third ed. John Wiley & Sons, Inc, New York, 668 p.

Liao, C.-H., Kang, S.-F., Wu, F.-A., 2001. Hydroxyl radical scavenging role of chloride and bicarbonate ions in the H_2O_2/UV process. Chemosphere 44, 1193–1200.

Lide, D.R. (Ed.), 2006–2007. CRC Handbook of Chemistry and Physics. A Ready-Reference Book of Chemical and Physical Data. 87th ed. CRC Press, Boca Raton, FL.

Matilainen, A., Sillanpaa, M., 2010. Removal of natural organic matter from drinking water by advanced oxidation processes (review). Chemosphere 80, 351–365.

Michel-Kordatou, I., Michael, C., Duan, H., He, X., Dionysiou, D.D., Mills, M.A., et al., 2015. Dissolved organic matter: characteristics and potential implications in wastewater treatment and reuse applications. Water Res. 77, 213–248.

Mierzwa, J.C., Subtil, E.L.S., Hespanhol, I., 2012. UV/H_2O_2 process performance improvement by ultrafiltration and physicochemical clarification systems for industrial effluent pretreatment. Rev. Ambiente Água 7 (3), 31–40.

Molina, L.T., Schinke, S.D., Molina, M.J., 1977. Ultraviolet absorption spectrum of hydrogen peroxide vapor. Geophys. Res. Lett. 4 (12), 580–582.

Munter, R., 2001. Advanced oxidation processes – current status and prospects. Proc. Estonian Acad. Sci. Chem. 50 (2), 59–80.

Oppenländer, T., 2003. Photochemical Purification of Water and Air: Advanced Oxidation Processes (AOPs): Principles, Reaction Mechanisms, Reactor Concepts. Wiley-VCH, Weinheim, 368 p.

Pearsons, S. (Ed.), 2004. Advanced Oxidation Processes for Water and Wastewater Treatment. IWA Publishing, London, UK, 368 p.

Pehkonen, S., Pettersson, M., Lundell, J., Khriachtchev, L., Räsänen, M., 1998. Photochemical studies of hydrogen peroxide in solid rare gases: formation of the HOH... $O(^3P)$ complex. J. Phys. Chem. 102, 7643–7648.

Qiao, R.-P., Li, N., Qi, X.-H., Wang, Q.-S., Zhuang, Y.-Y., 2005. Degradation of mycrocystin-RR by UV radiation in the presence of hydrogen peroxide. Toxicon 45, 745–752.

Rawlings, J.B., 2002. Chemical Reactor Analysis and Design Fundamentals. Nob Hill Publishing, Madison, WI, 609 p.

Ribeiro, A.R., Nunes, O.C., Pereira, M.F.R., Silva, A.M.T., 2015. An overview on the advanced oxidation processes applied for the treatment of water pollutants defined in the recently launched Directive 2013/39/EU. Environ. Int. 75, 33–51.

Rosario-Ortiz, F.L., Wert, E.C., Snyder, S.A., 2010. Evaluation of UV/H_2O_2 treatment for the oxidation of pharmaceuticals in wastewater. Water Res. 44, 1440–1448.

Rosenfeldt, E.J., Linden, K.G., Canonica, S., von Gunten, U., 2006. Comparison of the efficiency of HO• radical formation during ozonation and the advanced oxidation processes O_3/H_2O_2 and UV/H_2O_2. Water Res. 40, 3695–3704.

Schmidt, L.D., 1988. The Engineering of Chemical Reactions. Oxford University Press, Inc, New York, 536 p.

Shemer, H., Sharpless, C.M., Elovitz, M.S., Linden, K.G., 2006. Relative rate constants of contaminant candidate list pesticides with hydroxyl radicals. Environ. Sci. Technol. 40, 4460–4466.

Shu, Z., Singh, A., Klamerth, N., McPhedran, K., Bolton, J.R., Belosevic, M., et al., 2016. Pilot-scale UV/H_2O_2 advanced oxidation process for municipal reuse water: assessing micropollutant degradation and estrogenic impacts on goldfish (*Carassius auratus* L.). Water Res. 101, 157–166.

Silva, M.P., Mostafa, S., Mckay, G., Rosario-Ortiz, F.L., Teixeira, A.C.S.C., 2015. Photochemical fate of amicarbazone in aqueous media: laboratory measurement and simulations. Environ. Eng. Sci. 32, 730–740.

Stasinakis, A.S., 2008. Use of selected advanced oxidation processes (AOPs) for wastewater treatment – a mini review. Global NEST J. 10 (3), 376–385.

Tokumura, M., Sugawara, A., Raknuzzaman, M., Habibullah-Al-Mamun, M.D., Masunaga, S., 2016. Comprehensive study on effects of water matrices on removal of pharmaceuticals by three different kinds of advanced oxidation processes. Chemosphere 159, 317–325.

United States Environmental Protection Agency (USEPA), 1998. Advanced Photochemical Oxidation Processes – Handbook. Office of Research and Development, Washington, DC, EPA/625/R-98/004.

Vaghjiani, G.L., Ravishankara, A.R., 1989. Absorption cross sections of CH_3OOH, H_2O_2, and D_2O vapors between 210 and 365 nm at 297 K. J. Geophys. Res. D3, 3487–3492.

Wols, B.A., Hofman-Caris, C.H.M., Harmsen, D.J.H., Beerendonk, E.F., 2013. Degradation of 40 selected pharmaceuticals by UV/H_2O_2. Water Res. 47, 5876–5888.

Yang, Y., Pignatello, J.J., Ma, J., Mitch, W.A., 2016. Effect of matrix components on UV/H_2O_2 and UV/$S_2O_8^{-2}$ advanced oxidation processes for trace organic degradation in reverse osmosis brine from municipal wastewater reuse facilities. Water Res. 89, 192–200.

Zhang, A., Li, Y., 2014. Removal of phenolic endocrine disrupting compounds from waste activated sludge using UV, H_2O_2, and UV/H_2O_2 oxidation processes: effects of reaction conditions and sludge matrix. Sci. Total Environ. 493, 307–323.

Zhang, W., Xiao, X., An, T., Song, Z., Fu, J., Sheng, G., et al., 2003. Kinetics, degradation pathway and reaction mechanism of advanced oxidation of 4-nitrophenol in water by a UV/H_2O_2 process. J. Chem. Technol. Biotechnol. 78, 788–794.

Fenton and Photo-Fenton Processes

Rakshit Ameta[1], Anil K. Chohadia[2], Abhilasha Jain[3] and Pinki B. Punjabi[4]

[1]J. R. N. Rajasthan Vidyapeeth (Deemed-to-be University), Udaipur, Rajasthan, India
[2]M. P. Govt. P. G. College, Chittorgarh, Rajasthan, India [3]St. Xavier's College, Mumbai, Maharashtra, India [4]M. L. Sukhadia University, Udaipur, Rajasthan, India

3.1 INTRODUCTION

Global economic growth is increasing exponentially in the first century of the new millennium, but at the same time, rapid urbanization and industrialization release enormous volumes of wastewater imposing various adverse effects on human health and grading the quality of the environment as a whole. It has been revealed that generation of wastewaters with complex and recalcitrant molecules is increasing day by day. The presence of these organic compounds in water poses a serious threat to public health since most of them are toxic, endocrine disrupting, mutagenic, or potentially carcinogenic to humans, animals. and aquatic life. There is a pressing demand for newer technologies for the complete mineralization of wastewaters.

Several conventional treatment methods are available such as biological, adsorption, chemical treatment, filtration, flocculation, activated charcoal and ion exchange resins for wastewater remediation. It has been frequently observed that pollutants not amenable to biological treatments may also be characterized by high chemical stability and/or by strong difficulty to be completely mineralized. In this context, oxidation processes are preferred to degrade such biorefractory substances present in wastewater. However, pollution load, process limitations, and operating conditions are the key factors to be considered during the selection of the most appropriate oxidation process for the degradation of a particular compound. Apart from high degradation efficiency, direct oxidation processes demand specified operating conditions to degrade the target compounds, which will increase the operation cost of the process.

Advanced Oxidation Processes for Wastewater Treatment
DOI: https://doi.org/10.1016/B978-0-12-810499-6.00003-6

The scope should encompass degradation as well as mineralization of organic contaminants to effectively treat recalcitrant effluent, i.e. conversion of the target molecule to its highest stable oxidation state: water, carbon dioxide, and the oxidized inorganic anions or to more easily degradable molecules, that can be removed biologically. Advanced oxidation processes (AOPs) are suggested as an alternative powerful technique for wastewater treatment processes to degrade recalcitrant organic compounds.

Glaze et al. (1987) defined AOPs as water treatment processes that produce very active radicals for degradation of pollutants at near ambient temperature and pressure. Generally, AOPs means a group of processes that produce hydroxyl radical, in which ozone and H_2O_2 act as oxidants with the assistance of light, a catalyst (e.g. Fe^{2+}, Fe^{3+}, and TiO_2), ultrasonic insertion, and/or thermal input; and there are several combinations such as Fenton (H_2O_2/Fe^{2+}), photo-Fenton ($H_2O_2/UV/Fe^{2+}$), peroxidation combined with ultraviolet light (H_2O_2/UV), peroxone (O_3/H_2O_2), peroxone combined with ultraviolet light ($O_3/H_2O_2/UV$), O_3/UV system, $O_3/TiO_2/H_2O_2$, O_3/TiO_2/electron beam irradiation, etc. AOPs typically operate with less energy requirement than direct oxidation. The classification of conventional AOPs is based on the source used for the generation of hydroxyl radicals.

Several AOPs were identified and employed for the wastewater treatment, but Fenton and photo-Fenton processes have proved to be the most powerful, effective, energetically efficient, cost effective, and less tedious method for the treatment of recalcitrant compounds, when used exclusively or coupled with conventional and biological methods. These processes do not require sophisticated equipment or costly reagents and are ecologically viable because of their relatively simpler approach, use of less hazardous chemicals, and cyclic in nature so that a less concentration of these chemicals are needed.

3.2 TYPES OF FENTON PROCESSES

Different types of Fenton processes are available such as Fenton, photo-Fenton, electro-Fenton, photo-electro-Fenton, sono-Fenton, sono-photo-Fenton, sono-electro-Fenton, homogeneous and heterogeneous Fenton, photo-Fenton, hybrid Fenton, and Fenton type processes (Goi and Trapido, 2002).

3.2.1 Fenton Processes

The Fenton reaction was first reported by H. J. Fenton (1894) and it is described as the enhanced oxidative potential of H_2O_2, when iron (Fe) is used as a catalyst under acidic conditions. The reactions involved in Fenton processes are (Haber and Willstätter, 1931):

$$Fe^{2+} + H_2O_2 \rightarrow Fe^{3+} + \ OH + OH^- \qquad (3.1)$$

$$OH + H_2O_2 \rightarrow HO_2 + H_2O \qquad (3.2)$$

$$Fe^{2+} + \ OH \rightarrow Fe^{3+} + OH^- \qquad (3.3)$$

$$Fe^{3+} + HO_2 \rightarrow Fe^{2+} + O_2 + H^+ \qquad (3.4)$$

$$OH + \ OH \rightarrow H_2O_2 \qquad (3.5)$$

$$\text{Organic pollutant} + \ OH \rightarrow \text{Degraded products} \qquad (3.6)$$

Fenton-like reactions are reactions in which other metals, such as cobalt and copper, are used at a low oxidation state (Nerud et al., 2001; Prousek, 1995).

$$Cu^+ + H_2O_2 \rightarrow Cu^{2+} + \ OH + OH^- \tag{3.7}$$

A study on the synergistic pretreatment of sugarcane bagasse (SCB) using the Fenton reaction and NaOH extraction was conducted by Zhang and Zhu (2016). Sequential pretreatments were performed in combination with the NaOH extraction and it was concluded that among all the pretreatments, the Fenton pretreatment followed by the NaOH extraction had the highest efficiency for enzymolysis and simultaneous saccharification fermentation (SSF). The analyses revealed that the Fenton pretreatment disrupts the structure of SCB to facilitate the degradation of lignin by NaOH. Nidheesh et al. (2013) reviewed the degradation of dyes from aqueous solution by Fenton processes in which the classification, characteristics, and problems of dyes were discussed. Fenton processes have been classified and represented as Fenton circle. They also included some recent studies on Fenton processes analyzing different configurations of reactors used for dye removal, efficiency, and the effects of various operating parameters of Fenton processes.

Üstün et al. (2010) observed that degradation and mineralization of 3-indole butyric acid (IBA) using Fenton and Fenton-like processes proceeded via two distinctive kinetic regimes. The initial step of the reaction was a Fenton reaction (faster step) and it was followed by a slower degradation step involving Fenton-like reactions, which was due to the lower rate of generation of hydroxyl radicals. They showed that the Fenton process may be more useful, if only IBA removal is required, but if mineralization of IBA is also needed, then the Fenton-like process is more important due to high mineralization efficiency.

Matilainen and Sillanpää (2010) employed Fenton and photo-Fenton processes for the removal of natural organic matter from drinking water. The quantum yield and electrical energy per order (E(Eo)) efficiency of azo dyes by three advanced oxidation processes (AOPs) has been assessed by Muruganandham et al. (2007). Both dyes were completely decolorized by all these processes. Among three AOPs, the Fenton process showed the high decolorization efficiencies and low electrical energy consumption. Moreover, the higher quantum yield was observed at a high initial dye concentration.

3.2.2 Photo-Fenton Processes

The combination of hydrogen peroxide and UV radiation with a Fe^{2+} or Fe^{3+} ion produces more hydroxyl radicals and in turn, it increases the rate of degradation of organic pollutants. Such a process is known as the photo-Fenton processes (Kim and Vogelpohl, 1998). Fenton reaction accumulates Fe^{3+} ions in the system and the reaction does not proceed, once all Fe^{2+} ions are consumed. The photochemical regeneration of ferrous ions by photo-reduction of ferric ions occurs in the photo-Fenton reaction (Eqs. 3.8 and 3.9). The newly generated ferrous ions react again with H_2O_2 generating hydroxyl radical and ferric ion, and in this way, the cycle continues:

$$Fe^{3+} + H_2O + h \ \rightarrow Fe^{2+} + \ OH + H^+ \tag{3.8}$$

$$Fe^{3+} + H_2O_2 + h \ \rightarrow Fe^{2+} + HO_2 + H^+ \tag{3.9}$$

The photo-Fenton process offers better performance at pH 3.0, when the hydroxy−Fe^{3+} complexes are more soluble and $Fe(OH)^{2+}$ is more photoactive. The photo-Fenton process was reported as more efficient than the Fenton treatment. In some cases, the use of sunlight instead of UV irradiation also reduced the costs; however, this offers a lower degradation rate of pollutants. Acidic conditions (about pH 3) were reported to be favorable, and this may be mainly due to the conversion of carbonate and bicarbonate species into carbonic acid, which has a low reactivity with hydroxyl radicals.

Malakootian et al. (2016) employed the photo-Fenton process in the removal of sodium dodecyl sulphate from synthetic solutions and soft drink wastewater. They removed this surfactant from real sample with efficiency of 71% in 30 min. Fenton and photo-Fenton degradation of the -blocker metoprolol (MET) at initial circumneutral pH was achieved by Romero et al. (2015), on addition of resorcinol. The effect of various parameters was examined and intermediates were identified. The Fenton and photo-Fenton degradation pathway of MET has also been proposed.

Carboxylic acids are formed during the degradation of organic pollutants in the photo-Fenton reactions and these carboxylic acids form complexes with iron. A kinetic model simulating the dynamic behaviors of the iron redox cycle and hydroxyl radical generation was developed to know contributions of acids like citric acid, formic acid, malonic acid, and oxalic acid. It was observed that citric acid and oxalic acid enhanced the rate of the iron redox cycle as well as the photo-Fenton reaction, whereas formic acid and malonic acid showed insignificant and rather adverse effects on the rate of degradation. However the effects of carboxylic acids on the Fenton-like reaction were found to be insignificant. These results in the photo-Fenton reaction have been attributed to the changes in the UV light absorbance with formed Fe^{3+}−carboxylic acid complexes (Baba et al., 2015).

Photo-Fenton degradation of rhodamine B was observed by Lagori et al. (2015) to compare the use of two different laser wavelengths for improving dental bleaching. Degradation of phenol by the solar photo-Fenton process using the compound parabolic collectors reactor was studied by Alalm et al. (2017) and observed in the rate of a degradation reaction from costs point of view. The total costs for best results with maximum degradation of phenol was found to be 2.54 €/m^3. Da Silva et al. (2015) observed the inhibitory effect of inorganic ion mixtures, such as chloride, nitrate, sulfate, carbonate, and monophosphate on the photo-Fenton mineralization of phenol by two experimental designs, i.e. fractional experimental design and central composite rotatable design (CCRD). The inhibitory effect followed the sequence:

$$H_2PO_4^- \gg Cl^- > SO_4^{2-} > NO_3^- \quad CO_3^{2-}$$

The photo-Fenton degradation of antibiotic oxytetracycline (OTC) at near neutral pH was reported by using ferricarboxylates (Pereira et al., 2014). Fe^{3+}/Oxalate/UV−Vis or Fe^{3+}/Citrate/H_2O_2/UV−Vis process was used to avoid the formation of Fe-OTC complexes. A comparative study has also been made in the presence of different inorganic anions and humic acids, and in two different real wastewater matrixes. The rice-husk-based free electron-rich mesoporous activated carbon (MAC) was precarbonized and activated by using phosphoric acid and then applied as a heterogeneous catalyst along with the Fenton reagent for the oxidation of persistent organic compounds in the tannery wastewater carrying high total dissolved solids wastewater (Sekaran et al., 2014).

Xiao et al. (2014) examined the distinct effect of oxalate vs. malonate on iron-based photocatalytic technologies like the photo-Fenton process by evaluating the degradation efficiency of rhodamine B. It was observed that photolysis of the Fe(III)-oxalato complex was favorable due to the formation of O_2^-, HO_2, and $^•OH$ while the activities of UV/Fe(III) significantly decreased because of Fe(III)-malonate complexes.

A strategy was developed by Carra et al. (2013) to remove recalcitrant pollutants from water using the solar photo-Fenton process at initial neutral pH by repeated iron additions at highly acidic pH (2.8). The effect of a UV light source such as UV-A (near-UV), UV-C (short-UV), and visible-light assisted the Fenton-like treatment and water matrix on toxicity, and transformation products were observed by Molkenthin et al. (2013). Different parameters like degradation of bisphenol A, removal of dissolved organic carbon (DOC) and rates of H_2O_2 consumption were selected to assess the treatment performances in pure water and raw, fresh water samples. Various transformation products were identified and a reaction mechanism hydroxylation, dimerization, and ring opening steps were proposed.

Rubio et al. (2013) compared the H_2O_2/UV_{254} and photo-Fenton technologies as well as the effect of bicarbonates and organic matter on the photo-Fenton reaction of *Escherichia coli*. The efficiency of the treatment was compared for Milli-Q water, Leman lake water, and artificial seawater. It was observed that the presence of bicarbonate showed a negative effect on reaction by scavenging of the $^•OH$ radicals whereas dissolved organic matter showed a positive effect by increasing the rate of inactivation. They also opined that despite the negative effect of bicarbonate ions, the AOPs, especially photo-Fenton processes are effective for disinfection treatment.

Klamerth et al. (2013) developed a modified photo-Fenton process capable of working at neutral pH for the treatment of wastewater. In this modification, ethylenediamine-N, N'-disuccinic acid (EDDS) was used. Ortega-Gómez et al. (2012) developed a control strategy to automate the hydrogen peroxide dosage to increase the efficiency of the photo-Fenton process. A model was designed using paracetamol as the model pollutant to explain the link between dynamic behavior of DO and hydrogen peroxide consumption.

De la Cruz et al. (2012) highlighted the possibility of the degradation of micropollutants such as pharmaceuticals, corrosion inhibitors, and biocides/pesticides in the presence of dissolved organic matter in much higher concentrations at natural or near neutral pH. They used UV-light (254 nm), dark Fenton ($Fe^{2+,3+}/H_2O_2$), and photo-Fenton ($Fe^{2+,3+}/H_2O_2$/light) processes. Guimarães et al. (2012b) investigated the efficiency of various AOPs for the degradation of RB-19 dye in which the photo-Fenton process was found to be most efficient method showing removal of 94.5% of dissolved organic carbon and 99.4% of color. It was also reported that by combining a biological system with the photo-Fenton process, the textile effluent can be degraded at a much higher level.

The effect of various aromatic derivatives on the advanced photo-Fenton degradation of amaranth dye was studied by Devi et al. (2011) in which the following order of the rate of reaction was reported:

Hydroquinones > Chlorophenol > Dichlorobenzene > Aromatic carboxylic acids > Anilidine > Nitrophenol

A probable degradation mechanism has been proposed on the basis on the intermediates, and it was concluded that the degradation pathway is not altered by the addition of these aromatic derivatives.

Maezono et al. (2011) examined the hydroxyl radical concentration profile in the photo-Fenton oxidation process. The hydroxyl radical concentration was measured during the decoloration of azo-dye orange II by using the scavenger probe or trapping technique. The effects of various operating parameters were studied and it was found that the hydroxyl radical generation was enhanced with increasing dosages of Fenton reagents. The dynamic behavior of the hydroxyl radical concentration was found to be strongly linked with the change in H_2O_2 concentration. Catalytic effect of inorganic additives like sodium thiosulfate and potassium bromate was observed by Jain et al. (2011) on the photo-Fenton degradation of phenol red. The effect of variation of different parameters was observed and a tentative mechanism for the reaction was also proposed.

Evaluation of two approaches namely, solar Fenton, homogeneous photocatalysis, and heterogeneous photocatalysis with titanium dioxide suspensions for the degradation of the fluoroquinolone ofloxacin in secondary treated effluents was made by Michael et al. (2010). It was concluded that solar Fenton is the more effective technique for complete degradation of this substrate, and DOC reduction of about 50% in 30 min of the photocatalytic treatment could be achieved. A parametric study was designed by Bouafia-Chergui et al. (2010) to know the effects of $[H_2O_2]$, $[Fe^{3+}]$, and $[H_2O_2]/[Fe^{3+}]$ ratio on the photo-Fenton degradation of an azo dye basic blue 41. It was reported that the concentration of H_2O_2, ferric iron, and BB41 as well as their ratio $[H_2O_2]/[Fe^{3+}]$ and the $[H_2O_2]/[BB41]$ are key factors deciding the removal of BB41. 90% of TOC removal was obtained after 8 h of irradiation time under these conditions of real wastewater. Decolorization of mordant red 73 azo dye in water using H_2O_2/UV and the photo-Fenton treatment was investigated by Elmorsi et al. (2010). The effect of various operating parameters as well as the effect of inorganic anions was also observed. It was found that the presence of chloride ion caused a large decrease in the photodegradation rate, but the effect of nitrate and carbonate ions was insignificant. The photo-Fenton treatment using Fe powder (a source of ferrous ions) was found to be highly efficient. It was also found that the photo-Fenton treatment is more efficient over the H_2O_2/UV process.

A comparison among photocatalysis, the Fenton and photo-Fenton processes for the treatment of catechol was made by Lofrano et al. (2009) in which the Fenton and photo-Fenton processes exihibited a very high efficiency in the mineralization of catechol and removal of aromaticity. Klamerth et al. (2009) identified degradation products and proposed pathway of photo—Fenton degradation of chlorfenvinphos (CFVP). It was reported that CFVP and its degradation products such as 2,4-dichlorophenol, 2,4-dichlorobenzoic acid and triethylphosphate were decomposed into organic substances like acetate, formate, maleate, and inorganic ions like chloride and phosphate. On the basis of the complete absence of chlorinated aliphatic substances and chlorinated acids, it was concluded that chlorine is removed rapidly, and that residual DOC does not correspond to any chlorinated compound.

Manzano et al. (2009) investigated the Fenton and photo-Fenton mediated degradation process of orange II in a flow photo-reactor mediated by Fe-Nafion membranes. It was reported that complete decolorization of orange II can be achieved within 1 h and dye is mineralized in the presence of ferric ion (2 ppm) and an H_2O_2/orange II ratio of 20.

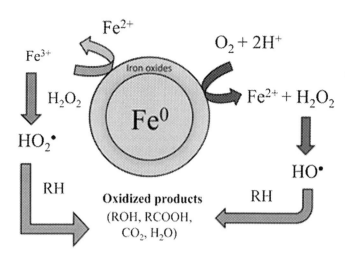

FIGURE 3.1 Simplified mechanism of heterogeneous Fenton rections with zerovalent iron. *Adapted from Litter, M. I., Slodowicz, M., 2017. J. Adv. Oxid. Technol., doi: 10.1515/jaots-2016-0164. With permission.*

Inhibitory effect of surfactants such as linear alkylbenzenesulfonates (LAS) and K-perfluoroalkylsulfonate (FT 800) was also reported for the photo-Fenton processes.

The effect of system parameters and of inorganic salts on the decolorization and degradation of procion H-exl dyes was investigated by Riga et al. (2007). Various advanced oxidation processes such as H_2O_2/UV, Fenton, UV/Fenton, TiO_2/UV, and $TiO_2/UV/H_2O_2$ were compared. They also reported adverse effects of some salts such as NaCl, Na_2CO_3, $NaHCO_3$, Na_2SO_4, $NaNO_3$, and Na_3PO_4.

Litter and Slodowicz (2017) reviewed different aspects of Fenton and Fenton-like processes for the removal of pollutants using zerovalent iron materials, including nanoparticles. The use of the iron materials together with hydrogen peroxide addition has been discussed. The use of biogenic generated iron nanoparticles for the removal of pollutants may be preferred due to their novelty and economy of synthesis (Fig. 3.1).

A kinetic study has been made by Kusic et al. (2006a) for the degradation of azo dyes using Fenton and photo-Fenton type processes, in which the highest mineralization extent was observed for $UV/Fe^0/H_2O_2$ system. The iron catalyst concentration and the iron catalyst/H_2O_2 ratio were optimized for Fenton type processes and then, Fe^{2+}/H_2O_2 and Fe^0/H_2O_2 at optimal process parameters were combined with UV radiation to enhance dye degradation. A quantum yield of acridine orange 7 was also determined.

The use of zerovalent iron (Fe^0) microspheres for the degradation of 1,4-dioxane by photo-Fenton processes was optimized in terms of pH and reagent dosage for the successful treatment of industrial wastewaters (Barndok et al., 2016). In UV photo-Fenton treatment of synthetic waters, 1,4-dioxane was completely removed in 5 min. A significant biodegradability enhancement (up to 60%) and almost total COD removal ($\geq$90%) was achieved in 10 min. and after 30 min, respectively. Sludge production and pH adjustments can be avoided in this process and it makes the Fe^0 microspheres an efficient catalyst for the treatment of relatively alkaline wastewaters containing 1,4-dioxane. They opined that the solar driven process could be an important economical alternative to the UV catalyzed process (Fig. 3.2).

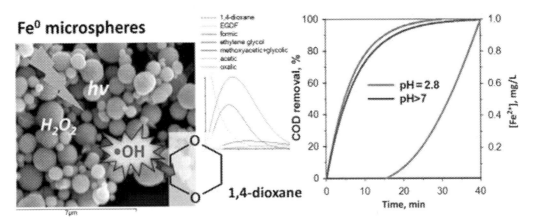

FIGURE 3.2 Heterogeneous photo-Fenton processes using zerovalent iron microspheres. *Adapted from Barndok, H., Blanco, L., Hermosilla, D., Blanco, A., 2016. Chem. Eng. J., 284, 112–121. With permission.*

Ma et al. (2005) found that the presence of dyes having a quinone unit like alizarin violet 3B can accelerate the photo-Fenton degradation of organic compounds. Complete mineralization of compounds like salicylic acid, sodium benzenesulfonate, benzyltrimethylammonium chloride, and trichloroacetic acid could be achieved by this dye catalyzed photo-Fenton process. According to the proposed mechanism, this dye acts as a co-catalyst to enhance the photo-Fenton reaction by catalyzing the cycle of Fe^{3+}/Fe^{2+} and transferring an electron from the excited dye molecule to Fe^{3+}.

The role of ferrous ion in Fenton and photo-Fenton processes for the degradation of phenol was examined by Kavitha and Palanivelu (2004). The efficiencies of Fenton, solar-Fenton, and UV-Fenton were compared, where UV-Fenton processes showed maximum mineralizing efficiency for degradation of phenol, which was used as a model compound. These processes were also applied for the abatement of effluent from the phenol resin-manufacturing unit.

The effect of substituents on the oxidation of azo dyes in the $Fe^{III}-EDTA-H_2O_2$ system was examined at neutral pH by Nam et al. (2001). Methyl, methoxy, and halo substituents on the phenolic ring of 4-(4-sulfophenylazo) phenol and 2-(4-sulfophenylazo) phenol, were chosen as model systems. It was reported that the mono- or dihalo- substituted dyes were oxidized to a greater extent than the corresponding methyl- or methoxy-substituted dyes. This effect can be attributed to the fact that halogen substituents make the phenolic moieties more acidic, which favors the phenolate anion, and could be readily attacked by hydroxyl radicals. The effect of Fe(III)-ligand properties on the effectiveness of modified photo-Fenton processes was studied by Aplin et al. (2001). A modified photo-Fenton (UV/Fe oxalate/H_2O_2) process was used for the degradation of reactive red 235. It was concluded that oxalate ligand forms Fe(III)-oxalato complexes, which could be easily photolyzed; and it is relatively unreactive with hydroxyl radicals.

The replacement of H_2O_2 by O_2 in Fenton and photo-Fenton degradation of aniline was examined by Utset et al. (2000). It was found that the presence of H_2O_2 is necessary along the reaction. It was also reported that an effective partial replacement of H_2O_2 by O_2 as an

TABLE 3.1 Degradation by the Photo-Fenton Process

Pollutant	References
Direct red 28	Ay et al. (2009)
E. coli k12	Spuhler et al. (2010)
Malachite green and direct fast light black G	Zheng et al. (2007)
Methyl orange	Devi et al. (2009)
Winery wastewater	Lucas et al. (2009)

electron acceptor takes place, when the ratio amount of aniline mineralized vs. initial aniline concentration is minimal. The photo-Fenton reaction and the TiO_2/UV process for wastewater treatment were investigated by Bauer et al. (1999) in which it was reported that the solar driven Fenton reaction is the most inexpensive method for water treatment and could be used for highly contaminated effluents. Bandara et al. (1996) studied the kinetics and quantum yield for sunlight induced photo-Fenton mineralization/decoloration of concentrated orange II solutions. It was observed that besides $^{\bullet}OH$ radicals, highly stable Fe-complexes in combination with H_2O_2 are active in the abatement of this azo dye. Degradation of some more pollutants and inactivation of bacteria is given in Table 3.1.

3.3 ELECTRO-FENTON PROCESSES

Anodic oxidation and indirect electrooxidation are employed to achieve the mineralization of toxic and biorefractory organic pollutants. In anodic oxidation, pollutants are mineralized by direct electron transfer reactions or action of radical species (hydroxyl radicals) formed on the electrode surface. Anodic oxidation is carried out in the anodic compartment of the divided cell, where the contaminated solution is treated using an anode. But it was observed that under these conditions, most of the contaminants are poorly mineralized. In electro-Fenton (EF) process, pollutants are destroyed by the action of Fenton's reagent in the bulk together with anodic oxidation at the anode surface.

The electro-Fenton (EF) process comprises of in situ electrochemical generation of H_2O_2, which readily decomposes to produce the hydroxyl radical species in an aqueous medium catalyzed by iron ions. The relevant electrochemical reaction corresponds to the two electron reduction of dissolved oxygen under slightly acidic conditions. The presence of iron ions accelerates the production of $^{\bullet}OH$ radicals. This catalytic reaction is propagated from Fe^{2+} regeneration, which mainly takes place by the reduction of Fe^{3+} ions with the H_2O_2. When the EF process is illuminated by means of UV-A light, it is named as photo-electro-Fenton (PEF) process, which can accelerate the degradation of organic compounds by the photolysis of Fe^{3+}-oxidation products complexes, and improving the ferrous ion regeneration from the photoreduction of ferric ions. The ability of EF and PEF processes to produce the $^{\bullet}OH$ radical with consequent oxidation of organic compounds make them suitable for wastewater treatment.

TABLE 3.2   Degradation by Electro-Fenton Processes

Pollutant	References
4-Amino-3-hydroxy-2-p-tolylazo-naphthalene-1-sulfonic acid	Labiadh et al. (2015)
Humic acids	Trellu et al. (2016)
Medical waste sterilization plant wastewater and for the treatment of landfill Leachate	Gökkuş and Yildiz (2016)
	Suresh et al. (2016)
Orange G	Nurhayati et al. (2016)
Tyrosol	Ammar et al. (2015)
Yellow-52	Peralta-Hernández et al. (2006)

Le et al. (2015) developed a new cathode for the electro-Fenton process by electrochemical deposition of reduced graphene oxide (RGO) on the surface of carbon felt (CF), which showed very high efficiency in the removal of dyes and good stability. An enhancement in the production of hydrogen peroxide was also confirmed.

A comparative study was made by Olvera-Vargas et al. (2015) for the degradation of pharmaceutical ranitidine (RNTD) by electro-Fenton and solar photoelectro-Fenton (SPEF) processes using a pre-pilot flow plant with a Pt/air-diffusion cell. In SPEF, Pt anode and air-diffusion cathode were used and the cell was coupled to a flat solar photoreactor to irradiate the solution with sunlight. The higher oxidation ability of SPEF compared to EF was attributed to the potent combined action of hydroxyl radicals and photolysis. Malic, pyruvic, acetic, oxalic, oxamic, and formic acids were detected as final products. Electro-Fenton process has also been utilized in degradation of other pollutants (Table 3.2).

3.4 SONO-FENTON AND SONO-PHOTO-FENTON PROCESSES

Now-a-days, ultrasound has also been used for the degradation of chemical contaminants. Ultrasound is a sound wave with a frequency greater than the upper limit of human hearing, i.e. approximately 20 kHz. The application of an ultrasound wave creates expansion and compression cycles. The expansion cycle can result in acoustic cavitation. Before implosion, the cavities oscillate in size, following the expansion and compression cycles of the ultrasound wave. When these cavitation bubbles implode, several hundred of atmosphere pressure and several thousand Kelvin temperature can be achieved (Suslick et al., 1990). Under these conditions, organic compounds are decomposed either by direct pyrolytic cleavage, or the hydroxyl radicals formed by pyrolysis.

Fenton and sonolysis can be combined together. These methods exploit the advantages of both ultrasound and Fenton's reagent; which can enhance the degradation of organic pollutants (Liang et al., 2007). Hydrogen peroxide is again produced by the recombination of hydroxyl radicals, but by the application of UV lightthe amount of hydroxyl radicals is

TABLE 3.3 Degradation by Sono-Fenton and Sono-Photo-Fenton Processes

Pollutant	References
Azure-B	Vaishnave et al. (2014)
Nitrobenzene	ElShafei et al. (2014)
Non-volatile organic compounds, dyes carbofuran	Ammar (2016)
Reactive blue	Siddique et al. (2014)
Reactive blue 69	Khataee et al. (2016a,b)
Wastewater from the production of coke	

again increased. The intermediate complex formed due to the reaction of Fe^{3+} with H_2O_2 during the Fenton reaction could be reduced to Fe^{2+} by sonolysis and photolysis. It was observed that the amount of ferrous salt required under SPF condition is very small as compared to the Fenton condition. SPF process also reduces the amount of ferrous ions present in the treated water, and this is very important from an industrial point of view.

The plasma-modified clinoptilolite (PMC) nanorods were prepared from natural clinoptilolite (NC) by Khataee et al. (2016a,b) using environmental benign corona discharge plasma. The catalytic performance of the PMC in the heterogeneous sono-Fenton-like process was observed and compared with NC. The greater efficiency of PMC catalyst was reported for treatment of phenazopyridine (PhP). Similarlily, PMC nanorods were prepared from oxygen, nitrogen nonthermal plasma, and argon glow discharge plasma to achieve expedient degradation of reactive blue 69 (RB69) degradation and reactive red 84, respectively. Patil and Gogate (2015) studied different combined treatment processes such as ultrasound (US)/TiO_2, solar/TiO_2, US/solar/TiO_2, US/H_2O_2, US/Fenton, and US/ozone for the degradation of dichlorvos pesticide. A high efficacy for removal of dichlorvos was reported for TiO_2/solar (78.42%) and US/Fenton (81.19%) processes. Sono-Fenton and sono-photo-Fenton systems were used for degradation of more pollutants (Table 3.3).

3.5 HETEROGENEOUS FENTON AND PHOTO-FENTON PROCESSES

Major disadvantages associated with homogeneous Fenton processes include formation of high amounts of metal-containing sludges at the end of the process, which imposes a detrimental effect on the environment and a large amount of the catalytic metals are lost; thus, leading to an increase in the costs associated. Due to these difficulties, several efforts have been made to use solid supports for the active iron species. These materials are required to exhibit high catalytic activity and stability; and metal should not be lost by leaching. Attempts have also been directed toward the immobilization of iron species on different solid supports or the use of insoluble iron oxides such as goethite, magnetite, and hematite in order to simplify iron separation. This process is named as the heterogeneous Fenton reaction. The heterogeneous Fenton process is conceptually attractive and practical, because it does not require the sludge separation step; thus it reduces cost of operation. Various supports for the Fe species such as silica, clay, activated carbon, resins, nafion, zeolite etc. have been used to fabricate heterogeneous Fenton catalysts. A novel technique using peroxymonosulfate (PMS) and peroxydisulfate (PDS) in the presence of zerovalent metallic iron (Fe^0) in the photo-Fenton process has been developed by Devi et al. (2016). PMS was found to be a better oxidant due to its dipolar unsymmetrical structure, higher oxidation potential, and lower LUMO energy. The following order was observed in the rate of degradation reaction for various oxidation processes:

Fe^0/PMS/UV　　Fe^0/HP/UV > Fe^0/PDS/UV > HP/UV > PDS/UV > PMS/UV > Fe^0/PMS/ dark > Fe^0/HP/dark > Fe^0/PDS/dark > Fe^0/UV > Fe^0/dark.

Degradation of 4-nitrophenol has been investigated on iron-enriched hybrid montmorillonite-alginate beads (Fe-MABs) by Barreca et al. (2014) using the solar photo-Fenton process. These beads were prepared by the ion-gelation method from alginate and montmorillonite clay suspension dropped in a calcium chloride solution whereas iron-enriched beads (Fe-MABs) were prepared using iron-exchanged montmorillonite. A good removal efficiency of these hybrid beads was also reported for the first three cycles. Wang et al. (2014) prepared Fe(II)Fe(III)-LDHs by coprecipitation method, and studied its catalytic effect in heterogeneous-Fenton degradation of methylene blue.

It is of great importance to synthesize easily separable and eco-friendly efficient catalysts for photocatalytic and photo-Fenton degradation for environment remediation application. Ammonia-modified graphene (AG) sheets decorated with Fe_3O_4 nanoparticles (AG/Fe_3O_4) as a magnetically recoverable photocatalyst was synthesized by Boruah et al. (2017). They functionalized graphene oxide (GO) sheets by an amide functional group followed by doping Fe_3O_4 nanoparticles (NPs) onto the functionalized GO surface. As-prepared AG/Fe_3O_4 nanocomposite showed higher photocatalytic activity toward degradation of phenol (92.43%), 2-nitrophenol (2-NP) (98%), and 2-chlorophenol (2-CP) (97.15%) within 70−120 min whereas 93.56% phenol, 98.76% 2-NP, and 98.06% of 2-CP degradation were achieved within 50−80 min. using photo-Fenton degradation under sunlight irradiation. It was proposed that the synergistic effect between amide functionalized graphene and Fe_3O_4 nanoparticles enhances the photocatalytic activity by preventing the recombination rate of the electron-hole-pair in these NPs. In addition, AG/Fe_3O_4 nanocomposite can be reused up to ten cycles (Fig. 3.3).

A ternary nanocomposite with TiO_2 nanoparticles anchored on reduced graphene oxide (rGO)-encapsulated Fe_3O_4 spheres (Fe_3O_4@rGO@TiO_2) was synthesized by depositing TiO_2 nanoparticles on the surface of the Fe_3O_4 spheres wrapped by graphene oxide (GO) and used as a high efficient heterogeneous catalyst for photo-Fenton degradation of recalcitrant pollutants under neutral pH (Yang et al., 2015a,b). The as-prepared catalyst reflected good ferromagnetism and superior stability, making its separation convenient so that it can be recycled. A swift reduction of Fe^{3+} to Fe^{2+} can be achieved due to the synergic effects between the different components composing the catalyst. This catalyst exhibited enhanced activity for the degradation of azo-dyes as compared with Fe_3O_4, Fe_3O_4@SiO_2@TiO_2, or SiO_2@rGO@TiO_2. It was concluded that the composite catalyst possesses a great potential for visible-light driven destruction of organic pollutants (Fig. 3.4).

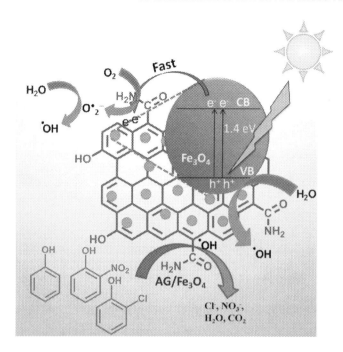

FIGURE 3.3 Ammonia-modified graphene sheets decorated magnetic Fe_3O_4 nanoparticles for degradation of phenolic compounds. *Adapted from Boruah, P.K., Sharma, B., Karbhal, I., Shelke, M.V., Das, M.R., 2017. J. Hazard. Mater., 325, 90–100. With permission.*

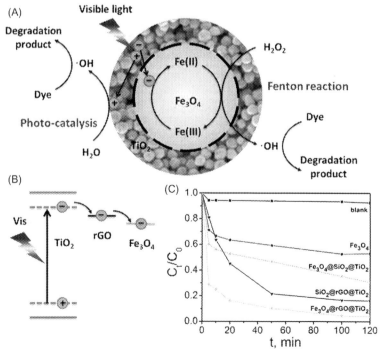

FIGURE 3.4 Degradation of methylene blue in a novel heterogeneous $Fe_3O_4@rGO@TiO_2$ catalyzed photo-Fenton system. (A) Degradation of methylene blue (B) Electron from TiO_2 to Fe_3O_4 via rGO (C) Kinetics of reaction with different combinations. *Adapted from Yang, X., Chen, W., Huang, J., Zhou, Y., Zhu, Y., Li, C., 2015. Sci. Rep. doi:10.1038/srep10632. With permission.*

Pradhan et al. (2013) prepared the mesoporous iron modified Al_2O_3 pillared montmorillonite nanocomposite by using sodium exchanged montmorillonite by cation-exchange, gallery-templated synthesis, and impregnation method. This could be used as a catalyst in the photo-Fenton process for the quick removal of organic dyes such as acid blue and reactive blue, in which 100% degradation was reported within 30 min with a high concentration of dye. The efficacy of this catalyst can be attributed to small particle sizes, mesoporosity, and fast reduction of Fe^{3+}.

Zhang et al. (2013) examined the novel reactor for the degradation of organic pollutants by coupling the heterogeneous photo-Fenton like oxidation with membrane separation. This reactor consisted of a Xe lamp, a submerged membrane module, and $FeVO_4$ (catalyst) with high activity. It was found that the catalyst was successfully left in the reactor, which can be continuously reused.

Zhao et al. (2012) developed bifunctional iron modified rectorite as an efficient adsorbent and photo-Fenton catalyst for removal of organic contaminants. It was reported that this modification has increased the layer-to-layer spacing and the surface area of rectorite, which in turn increased the adsorption of rhodamine B on rectorite. This modification also enabled rectorite to show efficient visible light photocatalytic ability by catalyzing the decomposition of H_2O_2 into hydroxyl radicals.

A novel catalyst was prepared by immobilizing Fe(III) onto collagen fiber; and its high activity as a catalyst for the Fenton degradation of dyes under irradiation of UVC was reported by Tang et al. (2005). A laponite clay-based Fe nanocomposite (Fe-Lap-RD) was also employed for the photo-Fenton degradation of reactive red HE-3B by Feng et al. (2003). A high specific surface area and high total pore volume was reported for this nanocomposite, which led to a high catalytic performance in the degradation process. Degradation of rhodamine B and 2,4-dichlorophenol by the photo-Fenton-like system with hematite was reported by Zhang et al. (2012). A number of other contaminants have been successfully degraded by heterogeneous Fenton and photo-Fenton processes (Table 3.4).

3.6 COMBINED (HYBRID) FENTON AND PHOTO-FENTON PROCESSES

Solar photo-Fenton treatment combined with nanofiltration to remove microcontaminants in real municipal effluents was investigated in order to achieve low cost strategy by Miralles-Cuevas et al. (2017). The photo-Fenton process was operated at circumneutral pH using (S,S)-ethylenediamine-N,N -disuccinic acid trisodium salt (EDDS) as the iron complexing agent. The effect of homogeneous and heterogeneous Fenton type reagents (using Fe containing ZSM-5 zeolite) on the removal of orange II has been investigated together with the effect of UV irradiation in combination with ultrasound both alone, and with the homogeneous Fenton-type reagent (Dükkanci et al., 2014). Chen et al. (2014) deigned a novel method by combining the heterogeneous Fenton-like reaction and photocatalysis using Co-TiO_2 nanocatalyst for activation of $KHSO_4$ with visible light irradiation at ambient conditions. Excellent catalytic stability and reusability have been observed with almost no dissolution of Co^{2+}. The effects of the calcination temperature and cobalt concentration

TABLE 3.4 Degradation by Heterogeneous Fenton and Photo-Fenton Processes

Pollutant	References
Acid blue 29	Soon and Hameed (2013)
Acid red 14	Kasiri et al. (2010)
Azo dye	Khataee et al. (2015)
Indigo blue	Almazán-Sánchez et al. (2016)
Methyl orange	Hassan et al. (2016)
Methyl orange	Xu et al. (2013)
Methylene blue	Yang et al. (2015a,b)
Methylene blue and Congo red	Zhang and Nan (2014)
Microcystin-LR	Fang et al. (2011)
Orange II	Feng et al. (2006)
Orange II	Gong et al. (2010)
Organic compound	Tireli et al. (2015)
Phenol	Zhang et al. (2016)
Reactive black B	Shih et al. (2013)
Reactive brilliant orange X-GN	Chen et al. (2009, 2010)
Rhodamine B	Liu et al. (2006)
Tetrabromobisphenol A	Zhong et al. (2012)

on the synergistic performance were examined and a possible mechanism for the synergistic system has also been proposed.

A magnetic $ZnFe_2O_4$-reduced graphene oxide hybrid was successfully developed and employed as the heterogeneous catalyst for photo-Fenton-like decolorization of various dyes using peroxymonosulfate (PMS) by Yao et al. (2014). It was found that this combination of $ZnFe_2O_4$ NPs with graphene sheets leads to a much higher catalytic activity and stability than pure $ZnFe_2O_4$.

Kim et al. (2012) assessed the synergistic effects of TiO_2 photocatalysis in combination with Fenton-like reactions on oxidation of organic compounds at circumneutral pH. It was concluded that the synergistic effects result from dual roles of iron as an electron acceptor to facilitate charge separation in a TiO_2 photocatalyst and as a Fenton reagent. It was also suggested that the adsorption of iron onto the photoexcited TiO_2 surface modifies electron transfer properties of iron toward H_2O_2 at neutral pH to convert the resultant reactive oxidant from Fe(IV) into a stronger form, likely $^{\bullet}OH$.

The degradation of bisphenol A by an advanced oxidation process that combines sonolysis, Fe^{2+}, and TiO_2 in a photoassisted process was reported by Torres-Palma et al. (2010). Lee and Sedlak (2009) reported that the addition of phosphotungstate ($PW_{12}O_{40}^{3-}$), a polyoxometalate, extends the working pH range of the $Fe(III)/H_2O_2$ system up to pH 8.5 by forming a soluble complex with iron that converts H_2O_2 into oxidants. The coordination of Fe(II) by $PW_{12}O_{40}^{3-}$ also alters the mechanism of the reaction of Fe(II) with H_2O_2 at neutral pH, resulting in formation of an oxidant capable of oxidizing aromatic compounds.

3.7 APPLICATIONS

3.7.1 Dyes

Photocatalytic decolorization of dyes using AOP is a recent concern among researchers since it offers an attractive method for decolorization of dyes and breaks them into simple mineral form. The oxidation using Fenton reagent has been found to be a promising treatment method for the effective decolorization and degradation of dyes. An investigation of photodegradation of basic orange 12 using H_2O_2/Fe^{3+} has been carried out (Kumar and Ameta, 2013).

Dark- and photo-Fenton type processes, Fe^{2+}/H_2O_2, Fe^{3+}/H_2O_2, Fe^0/H_2O_2, $UV/Fe^{2+}/H_2O_2$, $UV/Fe^{3+}/H_2O_2$, and $UV/Fe^0/H_2O_2$, were applied for the treatment of model colored wastewater containing two reactive dyes, reactive blue 49 and reactive blue 137; and degradation kinetics have been compared by Kusic et al. (2009) Reactive azo dye procion red H-E7B solutions have been submitted to solar-assisted photo-Fenton degradation. Heteroatom oxidation products like NH_4^+, NO_3^-, Cl^-, SO_4^{2-} and short chain carboxylic acid were quantified (García-Montaño et al., 2008a,b). Zheng et al. (2007) investigated the oxidation of acidic dye eosin Y with the Fenton process and photo-Fenton process (solar light or artificial light source), and also proved that oxalic acid could improve the photocatalytic efficiency in the solar-Fenton process.

The UV−vis spectral changes of amido black 10B in aqueous solution during the Fenton treatment process at low H_2O_2 and Fe^{2+} concentrations were studied by Sun et al. (2007). It was observed that it is easier to destruct the azo linkage group than to destruct the aromatic rings of amido black 10B by Fenton oxidation. The experimental results showed that the Fenton oxidation process was an effective process for the degradation of azo dye amido black 10B.

Daneshvar and Khataee (2006) compared the removal of azo dye, acid red 14, a common textile, and leather dye, from contaminated water using Fenton, UV/H_2O_2, $UV/H_2O_2/Fe(II)$, $UV/H_2O_2/Fe(III)$, and $UV/H_2O_2/Fe(III)/oxalate$ processes, and the results showed that its decolorization efficiency at the reaction time of 2 min follows the order:

$$UV/H_2O_2/Fe\ III\ /oxalate > UV/H_2O_2/Fe\ III\ > UV/H_2O_2/Fe\ II\ > UV/H_2O_2$$

Zheng and Xiang (2004) used photo-Fenton oxidation processes for the degradation of rhodamine B and COD, which can be removed effectively at the same time. The photodegradation of three commercially available dyestuffs (reactive black 5, direct yellow 12, and direct red 28) by UV, UV/H_2O_2, and $UV/H_2O_2/Fe^{2+}$ processes was investigated in a laboratory scale batch photoreactor equipped with a 16 W immersed-type low-pressure mercury vapor lamp (2004).

Neamtu et al. (2003) evaluated the degradation of two azo reactive dyes, reactive yellow 84 and reactive red 120 by photo-Fenton and Fenton-like oxidation, and the results show that the color removal of reactive yellow 84 after 15 min reaction time follows the decreasing order:

$$Solar/Fe^{2+}/H_2O_2 > UV/Fe^{2+}/H_2O_2 > UV/Cu^{2+}/Fe^{3+}/H_2O_2 > UV/Fe^{3+}oxalate/H_2O_2 > UV/Fe^{3+}/H_2O_2 > Dark/Fe^{2+}/H_2O_2 > Solar/Fe^{3+}oxalate/H_2O_2 > UV/H_2O_2 > UV/Fe^{2+} = UV$$

During the same reaction period, the relative order for the reactive red 120 removal rate was slightly different:

$$Solar/Fe^{2+}/H_2O_2 > UV/Fe^{2+}/H_2O_2 > UV/Fe^{3+}/H_2O_2 = UV/Cu^{2+}/Fe^{3+}/H_2O_2 > UV/Fe^{3+}oxalate/H_2O_2 = UV/H_2O_2 > UV$$

The homogeneous photocatalytic degradation of cationic acridine orange monohydrochloride and anionic alizarin violet 3B in the photo-Fenton system under visible irradiation has been studied by Xie et al. (2000). The generated intermediates during degradation processes of acridine orange were found to be N,N-dimethyl formaldehyde and N,N-dimethyl acetamide. Ammonium ions were also formed during the degradation of acridine orange through the ion-selective electrode analysis.

Wu et al. (1999) studied the photodegradation of malachite green under visible light irradiation ($\lambda > 470$ nm) in the presence of Fe^{3+}/H_2O_2 or Fe^{2+}/H_2O_2 in comparison with the dark reaction. A probable degradation mechanism of malachite green by Fe^{3+}/H_2O_2 under visible irradiation has been discussed, which involves an electron transfer from the excited dye by visible light into Fe^{3+} in the initial photo-Fenton degradation. Conversion of Fe^{3+} to Fe^{2+} was detected in the degradation process of malachite green. More dyes were degraded by these processes and are summarized in Table 3.5.

3.7.2 Agrochemicals

The occurrence of significant amounts of biocidal finishing agents in the environment as a consequence of intensive textile finishing activities has become a subject of major public health concern and scientific interest.

Chlorpyrifos, lambda-cyhalothrin, and diazinon were the major contaminants found in the wastewater. Therefore, solar photo-Fenton and solar TiO_2 photocatalysis processes were used by Alalm et al. (2015b) for degradation of these from real industrial wastewater. The degradation of 2-chloro-4,6-diamino-1,3,5-triazine in aqueous solutions by the photo-Fenton process has been investigated by Dbira et al. (2014). The results of total organic carbon (TOC) and total Kjeldahl nitrogen (TKN) analyses have shown that no carbon dioxide (CO_2) and ammonia (NH_3) are formed during photo-Fenton treatment of aqueous solutions of triazine. These results indicate that only substituent groups of the triazine ring are released; however, nitrogen atoms of the triazine ring remain unaffected. The degradation of 2-chloro-4,6-diamino-1,3,5-triazine starts by a rapid release of chlorine atoms as chloride ions to form 2-hydroxy-4,6-diamino-1,3,5-triazine. The amino groups of 2-hydroxy-4,6-diamino-1,3,5-triazine are oxidized into nitro groups by hydroxyl radicals to form 2-hydroxy-4,6-dinitro-1,3,5-triazine, which undergoes a slow release of nitro groups,

TABLE 3.5 Degradation of Some Dyes

Dyes	References
Acid blue 161	Trovó et al. (2016)
Acid orange 24	Chacón et al. (2006)
Acid orange 7	Scheeren et al. (2002)
Basic Blue 99	Tavares et al. (2016)
Dyes in polyelectrolyte microshells	Tao et al. (2005)
Great green SF	Zheng et al. (2006)
H-exl dyes	Ntampegliotis et al. (2006)
Monuron	Bobu et al. (2006)
Mordant red 73	Elmorsi et al. (2010)
Reactive black 5	Lucas and Peres (2007)
Reactive black 5	Meriç et al. (2004)
Reactive black 5	Verma et al. (2015)
Reactive blue 181	Basturk and Karatas (2014)
Reactive brilliant red X3B	Hu and Xu (2004)
Reactive dyestuff	Kang et al. (2000)
Reactive orange 113	Gutowska et al. (2007)
Reactive orange 4	Muruganandham and Swaminathan (2004)
Reactive red 198 dye	Dehghani et al. (2015)
Reactive red 2	Emilia et al. (2016)
Reactive red 45	Peternel et al. (2007)
Reactive yellow 86	Katsumata et al. (2010)
Vat green 3, Reactive black 5, Acid orange 7, Food red 17, and Food yellow 3	Luna et al. (2012)

and their substitution with hydroxyl groups to form cyanuric acid and nitrate ions. The degradation of cyanuric acid by the photo-Fenton process has also been investigated. Sánchez et al. (2013) reported the economic evaluation of a combined photo-Fenton/MBR process using pesticides as the model pollutant.

Vilar et al. (2012) investigated the enhancement in biodegradability of a pesticide-containing wastewater from phytopharmaceutical plastic containers washing. They used a solar photo-Fenton treatment step followed by a biological oxidation process. Transformation of 2,4-dichlorophenol by H_2O_2/UV-C, Fenton, and photo-Fenton oxidation products and toxicity has been evaluated by Karci et al. (2012). Hydroquinone and formic acid were the common oxidation products and 3,5-dichloro-2-hydroxybenzaldehyde, phenol, 4-chlorophenol, and 2,5-dichlorohydroquinone were identified as the additional H_2O_2/UV-C oxidation products of 2,4-dichlorophenol.

Farré et al. (2005) degraded some biorecalcitrant pesticides (alachlor, atrazine, chlorfenvinphos, diuron, isoproturon, and pentachlorophenol) by homogeneous photo-Fenton/ozone (PhFO) and heterogeneous TiO_2-photocatalysis/ozone (PhCO) coupled systems; while Derbalah et al. (2004) investigated the photocatalytic removal kinetics of fenitrothion at a concentration of 0.5 mg/L in pure and natural waters in Fe^{3+}/H_2O_2/UV−Vis, Fe^{3+}/UV−Vis and H_2O_2/UV−Vis oxidation systems, with respect to decrease in fenitrothion concentrations with irradiation time using a solar simulator. Other agrochemicals degraded by these processes are tabulated in Table 3.6.

3.7.3 Pharmaceuticals

The pharmaceutical residues in municipal, surface, ground, and potable water create water pollution as these have adverse impacts on human health, even if present in low concentrations. These drugs are released into the wastewater during manufacture and improper disposal of unused or expired medicines. The risks associated with pharmaceutical contamination of the aquatic environment had been an emerging issue. Drugs are designed to have a particular biological effect and, therefore, some adverse environmental risks are expected from the exposure to these drugs. A number of pharmaceuticals pose undesired impacts on humans, animals, and aquatic organisms.

Orbeci et al. (2014) evaluated a modified photo-Fenton procedure as a suitable advanced oxidative process to degrade antibiotics in aqueous solutions. The classical photo-Fenton procedure was modified using a photocatalytic reactor with continuous recirculation, which was equipped with a high pressure cylindrically shaped mercury lamp centrally and coaxially positioned and surrounded by iron mesh. Photo-Fenton degradation of the sulfonamide antibiotics, sulfadiazine and sulfathiazole, mediated by Fe^{3+}-oxalate was studied by Batista and Nogueira (2012). Characterization of the degradation performance of the sulfamethazine antibiotic by the photo-Fenton process has been reported by Pérez-Moya et al. (2010). Characterization of intermediate products of solar photocatalyzed degradation by a titanium dioxide semiconductor and Fenton reagent (Fe^{2+}/H_2O_2) of ranitidine at pilot-scale

TABLE 3.6 Degradation of Some Agrochemicals

Pollutant	References
Atrazine	Benzaquén et al. (2015)
Cymoxanil, methomyl, oxamyl, dimethoate, pyrimethanil and telone	Oller et al. (2006)
2,4-Dichlorophenoxyacetic acid and 2,4-dichlorophenol	Luna et al. (2012)
Diuron, imidacloprid	Malato et al. (2001, 2003)
Thiabendazole Textar 60 T Imazalil sulphate Fruitgard IS 7.5	Santiago et al. (2011)

has been studied by Radjenoviæ et al. (2010). The degradation of flumequine and nalidixic acid in distilled water by two solar photocatalytic processes, TiO_2 and photo-Fenton, was also evaluated.

Dipyrone is an analgesic drug, which is very quickly hydrolyzed to 4-methylaminoantipyrine (4-MAA) in an acidic solution. A batch study was conducted by Giri and Golder (2014). to compare the performance of different oxidation processes such as Fenton (FP), photo-Fenton (PFP), UV/H_2O_2 photolysis (UVP), and UV/TiO_2 photocatalysis (UVPC) for removal of 4-MAA from an aqueous solution. They evaluated degradation efficiency in terms of total organic carbon (TOC) reduction and enhancement of biodegradability. Maximum 4-MAA removals of 94.1%, 96.4%, 74.4%, and 71.2% were achieved in FP, PFP, UCP, and UVPC, respectively, against mineralization of 49.3%, 58.2%, 47%, and 24.6%, respectively. The proposed mechanisms suggest that the cleavage of three methyl moieties followed by pyrazolinone ring breakage led to formation of various intermediates. These intermediates were hydroxylated and carboxylic derivatives.

3-Aminopyridine (3AP) is used in the manufacture of anti-inflammatory drugs. It is also as a plant growth regulator, which is one of the emerging pollutants, because of its toxic nature, carcinogenic potential, and hazardous effects on the natural environment. Its degradation by advanced oxidation technologies like Fenton and Photo-Fenton oxidation has been reported by Karale et al. (2014). The effect of various operating parameters like pH, hydrogen peroxide, iron salts (both ferrous and iron extracted from laterite soil) and reaction time were observed. Optimum ratio of $[H_2O_2]:[Fe^{2+}]$: 24–40:1 showed up to (90%–77%) removal efficiency for Fenton's oxidation at pH 3. Photo-Fenton oxidation studies showed 100% removal for concentration range (10–30 mg/L) under both iron salts. Fenton's oxidation required reaction time of 5 h for 10–30 mg/L to 7 h for 40–60 mg/L, and finally to 8.5 h for 70–80 mg/L of 3AP while photo-Fenton oxidation required much less reaction time equal to 1.5 h for 10–30 mg/L to 2 h for 40–60 mg/L, and 4 h for 70–80 mg/L.

The solar photo-Fenton process for degradation of four types of pharmaceuticals (amoxicillin, ampicillin, diclofenac, and paracetamol) was studied by Alalm et al. (2015a,b). Oxidation of each pharmaceutical was investigated individually using 100 mg/L of a synthetic solution. Effect of irradiation time, initial pH value, and dosage of the Fenton reagent were investigated. Paracetamol and amoxicillin were completely removed after 60 and 90 min of irradiation, respectively while complete degradation of ampicillin and diclofenac required more than 120 min. Complete removal of all pharmaceuticals was achieved under strongly acidic conditions (pH = 3). Optimum dosage of H_2O_2, $FeSO_4$ and, $7H_2O$ were 1.5 and 0.5 g/L, respectively.

Elmolla and Chaudhuri (2010) examined photo-Fenton treatment of an antibiotic wastewater containing amoxicillin and cloxacillin. The effect of operating conditions (H_2O_2:COD molar ratio and H_2O_2:Fe^{2+} molar ratio) on biodegradability (BOD5:COD ratio) improvement and mineralization was evaluated. Optimum operating conditions for treatment of the antibiotic wastewater were found to be H_2O_2:COD molar ratio 2.5 and H_2O_2:Fe^{2+} molar ratio 20 (COD:H_2O_2:Fe^{2+} molar ratio 1:2.5:0.125) at pH = 3 and reaction time being 30 min. Under these conditions, complete degradation of amoxicillin and cloxacillin was observed in 1 min, biodegradability improved from 0.09 to 0.50 ± 0.01 in 30 min, and COD and DOC removal were found to be $67 \pm 1\%$ and $51 \pm 2\%$, respectively in 30 min.

Degradation rates and removal efficiencies of propranolol using UV-C direct photolysis, H_2O_2, UV-C/H_2O_2, UV-C/Fe^{2+}, H_2O_2/Fe^{2+}, and UV-C/$H2O_2$/Fe^{2+} were studied by Baydum et al. (2012). The results showed that the oxidation rate was increased on increasing the hydrogen peroxide concentrations during the UV-C/H_2O_2 treatment. Addition of more ferrous ions enhanced the oxidation rate for the H_2O_2/Fe^{2+} and UV-C/H_2O_2/Fe^{2+} processes. It was possible to remove more than 80% from the propranolol concentration of 20 mg/L in 5 min using the photo-Fenton process.

Characterization and treatment of a real pharmaceutical wastewater containing 775 mg dissolved organic carbon per liter were studied by Sirtori et al. (2009) using a solar photo-Fenton/biotreatment, This wastewater contained nalidixic acid (45 mg/L), an antibiotic belonging to the quinolone group. Thephoto-Fenton process enhanced the biodegradability of nalidixic acid followed by a biological treatment. Nalidixic acid completely disappeared after 190 min. The photo-Fenton treatment time (190 min) and H_2O_2 dose (66 mM) were found necessary for adequate biodegradability of the wastewater. Overall degradation efficiency of the combined photo-Fenton and biological treatment was over 95%, out of which, 33% corresponds to the solar photochemical process and remaining 62% to the biological treatment.

3.7.4 Petroleum Refinery Effluents

Water produced in oil fields is one of the main sources of wastewater generated in the industry. It contains several organic compounds such as, benzene, toluene, ethyl benzene, and xylene (BTEX). Presently, the rise in the generation of extremely toxic and refractory wastewaters(as petroleum refinery plant effluents), is demanding increasingly efficient treatment technologies, where photo-Fenton advanced oxidation processes have entered the scene due to their rapid and effective degradation of petroleum-derived pollutants. Petrochemical wastewater treatment using the photo-Fenton process has been suggested by Rubio-Clemente et al. (2015), while Cho et al. (2006) treated the petroleum gas station's groundwater samples contaminated by BTEX and total petroleum hydrocarbons (TPHs) with advanced oxidation processes (AOPs), such as TiO_2 photocatalysis and Fe^{2+}/H_2O_2 exposed to solar light.

Detoxification of synthetic and real groundwater contaminated with gasoline and diesel using H_2O_2, H_2O_2 + UV, Fenton, photo-Fenton, and biofilters was made by Torres et al. (2016). A comparative study of solar photo-Fenton, and solar photocatalysis of TiO_2 combined with the Fenton process to treat petroleum wastewater was also made by Aljubourya et al. (2016).

Rocha et al. (2013) investigated the treatment of petroleum-extraction wastewater by means of a photo-Fenton like process using sunlight as an irradiation source. A reduction of the wastewater UV absorption spectral intensity indicates that polycyclic aromatic hydrocarbons and aromaticity removal of approximately 92.7% and 96.2%, respectively were obtained after 7 h of sunlight exposure. Da Silva et al. (2012) reported a treatment of polluted water by integrating two processes, i.e., induced air flotation (IAF) and photo-Fenton for treatment of residual waters contaminated with xylene. Guimarães et al. (2012a) assessed the application of AOPs in the treatment of stripped sour water from the Petrobras Replan oil refinery. Oxidative mineralization of petroleum refinery effluent using a Fenton-like process has been described by Hasan et al. (2012). Tony et al. (2012) also treated the oil refinery wastewater, contaminated with hydrocarbon oil, using physico-chemical, Fenton and photo-Fenton oxidation processes.

3.7.5 Surfactants

Surfactants are harmful for fish and human beings; and they are the main ingredients of synthetic detergents. Surfactants are widely used in soap, toothpaste, cleaning detergents, shampoo, etc. They also have numerous industrial applications in cosmetic and medicinal products, textiles, foodstuffs, paper, oil recovery, paints, polymers, pesticides, mining, etc. They are discharged to surface water and groundwater through household and industrial wastewaters. They also cause foam in rivers and effluent treatment plants, and reduce water quality; surfactants also cause short and long term damage to the environment. Combined Fenton oxidation and aerobic biological processes for treating a surfactant wastewater containing abundant sulfate have been investigated by Wang et al. (2008).

Pagano et al. (2008) carried out oxidation of 10 nonionic surfactants (6 alcohol ethoxylates and 4 alkylphenol ethoxylates) by Fenton and H_2O_2/UV processes in synthetic (deionized water) and real aqueous matrices. Mixtures of the 10 surfactants were treated in both; synthetic and real matrices. Fenton treatment of municipal secondary effluents containing the surfactants mixture led to its complete removal on using a molar ratio [Total surfactants]:$[H_2O_2]$:$[Fe^{2+}]$ equal to 1:17:12 was used.

Anionic alkylbenzene sulfonate (ABS) and linear alkylbenzene sulfonate (LAS), are being widely used in household and industrial detergents; and these were degraded by Lin et al. (1999) using the Fenton treatment. The effects of pH, amounts of ferrous sulfate and hydrogen peroxide, and temperature on the surfactant removal were studied. It was observed that chemical coagulation improved the turbidity and removal of dissolved Fe after the Fenton process. A first-order kinetic model was adopted to represent the Fenton oxidation of surfactant wastewater.

The oxidative degradation of two dyes and a nonionic surfactant by iron(II) sulphate and hydrogen peroxide was studied in the presence and absence of UV irradiation (Fenton and photo-Fenton processes) (Ferrero, 2000). The degradation kinetics were followed by measuring the residual total organic carbon (TOC) value as a function of time at different initial TOC concentrations and temperatures. It was observed that the residual TOC values attained a stable level after a given treatment time; even in the case of some dyehouse effluent samples.

Miranzadeh et al. (2016) investigated the effectiveness of Fenton and photo-Fenton processes for the treatment of anionic surfactants from aqueous solutions. Linear alkyl benzene sulfonate (LAS) was selected as a model system. It was revealed that the mean removal efficiency of LAS in Fenton and photo-Fenton in 20 min reaction time with constant concentration of hydrogen peroxide (100 mg/L) and 20 mg/L ferrous iron (20 mg/L) were 20.16% and 22.47%, respectively. LAS removal efficiency was increased to 69.38% and 86.66% in 80 min reaction time for hydrogen peroxide (800 mg/L) and ferrous ion (120 mg/L), respectively.

Solar photodegradation of two commercial surfactants, sodium dodecyl sulfate (SDS) and dodecyl benzene sulfonate (DBS), has been studied by Amat et al. (2004). It was observed that photo-Fenton reaction was the most effective method. In the photo-Fenton process, one can also use Cr(III). This could be interesting in the case of wastewaters containing chromium from a leather related industry. It was found easier to degrade DBS than SDS. Degradation yields higher than 80% were obtained in less than 3 h of exposure to sunlight; however, the best results were obtained again using iron salts as photocatalysts (photo-Fenton).

3.7.6 Leachates

Landfilling is the most common method of solid waste management in the country. Leachate management is a major problem concerned with landfill. Leachate can pollute surface and subsurface water sources. Advanced oxidation processes, such as photo-Fenton is an effective treatment method, where hydroxyl radicals are generated and used in pollutant degradation. Landfill leachate is one kind of the most harmful types of high concentration organic wastewaters. The effective processing method is necessary to improve the treating efficiency of landfill leachate. Aqueous oxidation treatment of COD and color in landfill leachate with the combination of hydrogen peroxide and ferrous ion (Fenton reagent) has been studied by Chen et al. (2014) while Ding et al. (2012) treated the landfill leachate by the microwave-Fenton oxidation process catalyzed by Fe^{2+} loaded on granular activated carbon. Optimization of Fenton and photo-Fenton-based advanced oxidation processes for post treatment of composting leachate of municipal solid waste by an activated sludge process has been suggested by Mahdad et al. (2016).

Advanced oxidation processes like Fenton and photo-Fenton have been effectively applied to oxidize the persistent organic compounds in solid waste leachate and convert them to harmless products. Zazouli et al. (2012) studied the effect of different variables on the treatment of municipal landfill leachate by Fenton, Fenton-like, and photo-Fenton processes. The results showed that the optimum concentration of H_2O_2 required was equal to 5 g/L for the Fenton-like process and 3 g/L for the Fenton and photo-Fenton processes. The optimum ratio of H_2O_2: Fe^{2+}/Fe^{3+} were 8:1 in all processes. At optimum conditions, the amount of COD removal was found to be 69.6%, 65.9%, and 83.2% in Fenton, Fenton-like, and photo-Fenton processes, respectively. Biodegradability (BOD5/COD ratio) of the treated leachate was increased after all processes as compared to that of the raw leachate. Highest increase in BOD5/COD ratio was observed in the photo-Fenton process (Fig. 3.5).

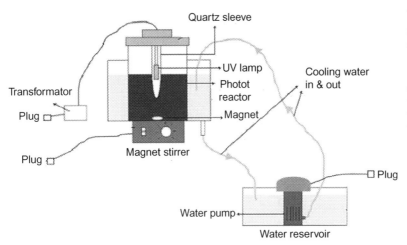

FIGURE 3.5 Municipal solid waste landfill leachate treatment by Fenton, photo-Fenton, and Fenton-like processes. *Adapted from Zazouli, M.A., Yousefi, Z., Eslami, A., Ardebilian, M.B., Iran. J. Environ. Health Sci. Eng., 9(1), doi: 10.1186/1735-2746-9-3, 2012. With permission.*

The solar photo-Fenton reaction parameters in the oxidation of a sanitary landfill leachate at lab-scale have been given by Tireli et al. (2015). They also tried to elucidate the role of ferric hydroxides, ferric sulphate, and ferric chloride species, by taking advantage of ferric speciation diagrams, in the efficiency of the photo-Fenton reaction, when applied to leachate oxidation. The molar fraction of the most photoactive ferric species, $FeOH^{2+}$, was linearly correlated with the photo-Fenton pseudo-first order kinetic constants obtained at different solution pH and temperature values. Ferric ion speciation diagrams showed that the presence of high amounts of chloride ions adversely affected the photo-Fenton reaction, due to the decrease of the solubility of ferric ions and scavenging of hydroxyl radicals for chlorine radical formation. The increment of the photo-Fenton reaction rates with temperature was mainly associated with the increase of the molar fraction of $FeOH^{2+}$.

3.7.7 Other Pollutants

Wide use of methyl tert-butyl ether (MTBE) as fuel oxygenates led to worldwide environment contamination, basically due to fuel leaks from storage or pipelines. Therefore, Levchuk et al. (2014) provided an overview of recent developments in technologies for MTBE removal from water. Bisphenol-A is ubiquitous in the environment and it is a controversial endocrine disruptor. Its degradation using the heterogeneous photo-Fenton with the Fe-Y catalyst system was demonstrated by Jiang et al. (2014).

Phenolic compounds (vanillin, protocatechuic acid, syringic acid, *p*-coumaric acid, gallic acid, and l-tyrosine) are present in high concentrations in various types of agro-industrial wastes. As these are highly biorecalcitrant, the possibility of treatment by photo-Fenton reaction under artificial light in laboratory experiments and in pilot-plant experiments under sunlight has been investigated by Gernjak et al. (2003) (Table 3.7).

TABLE 3.7 Degradation of Pollutants

Pollutant	References
5-Amino-6-methyl-2-benzimidazolone	Sarria et al. (2001)
Benzothiazole	Andreozzi et al. (2000)
Cellulose bleaching effluents	Torrades et al. (2003)
4-Chlorophenol, dichloroacetic acid	Nogueira et al. (2004)
Cork boiling and bleaching wastewaters	Vilar et al. (2009)
Cork manufacturing wastewater	Silva et al. (2004)
Cosmetic	Marcinowski et al. (2014)
Dichloroethane, dichloromethane and trichloromethane	Rodríguez et al. (2005)
Domoic acid	Bandala et al. (2009)
E10 sunset yellow FCF	Gosetti et al. (2005)
Escherichia coli and *Enterococcus faecalis*	Rodríguez-Chueca et al. (2014)
Formaldehyde	Kajitvichyanukul et al. (2008)
Gabapentin, metformin, metoprolol, atenolol, clarithromycin, primidone, methylbenzotriazole, and benzotriazole	Neamțu et al. (2014)
Kraft pulp mill effluent	Rabelo et al. (2014)
4-Mercaptobenzoic acid	Kumar (2015)
Metol, *N*-methyl-*p*-aminophenol	Lunar et al. (2000)
5-Nitro-1,2,4-triazol-3-one	Campion et al. (1999)
p-Nitroaniline	Sun et al. (2008)
p-Nitrotoluene-o-sulfonic acid	Parra et al. (2001)
Paper mill effluents	Muñoz et al. (2006)
Paper pulp treatment effluents	Pérez et al. (2002)
Phenol	Huang et al. (2010)
Polyphenols	Sabaikai et al. (2014)
Phenol	Kusic et al. (2006b)
Salmonella typhimurium	Rengifo-Herrera et al. (2013)
2,4,6-Trinitrophenol, ammonium picronitrate, 2,4-dinitrotoluene, methyl-2,4,6-trinitrophenylnitramine and 2,4,6-trinitrotoluene, hexahydro-1,3,5-trinitro-1,3,5-triazine, octahydro-1,3,5,7-tetranitro-1,3,5,7-tetrazocine	Liou et al. (2003)

3.8 RECENT DEVELOPMENTS

In order to increase the enzymatic digestibility of rice straw for saccharification, Fenton's reagent, ultrasound, and the combination of Fenton's reagent and ultrasound were used by Xiong et al. (2017). It was revealed that the rice straw pretreated by ultrasound-assisted Fenton's reagent exhibited the largest specific surface area and the lowest degree of polymerization value.

The expedient degradation of rhodamine B, acid red G, and metronidazole by the iron sludge-graphene catalyst was reported by Guo e al. (2017), in which iron sludge existed as FeOOH particles that were mainly entrapped inside the graphene sheet. This catalyst exhibited wide pH operating range, excellent stability, and reusability. It was concluded that the mesoporous structure increases the adsorption ability and the promoted the Fe^{3+}/ Fe^{2+} cycle. Trapido et al. (2017) employed aerobic biological treatment (BIO) combined with the Fenton process. This elaborated BIO-Fenton-BIO approach was reported to be cost-effective for the total organic load reduction and for the selected specific pollutants removal.

Degradation of various pollutants have been reported in recent years (Table 3.8).

Lonfat et al. (2016) reported the application of the Fe-citrate-based photo-Fenton process for the inactivation of *E. coli* and obtained promising results of bacterial inactivation at near-neutral and alkaline pH conditions using very low iron concentration and avoiding precipitation of ferric hydroxides. The effects of the solution pH and Fe-citrate complex concentration on *E. coli* inactivation were also observed. The efficiency of the homogeneous photo-Fenton process using Fe-citrate complex (as a source of iron) enhanced bacterial inactivation as compared with the heterogeneous photo-Fenton treatment using $FeSO_4$ and goethite as sources of iron, at near-neutral pH. It was observed that the bacterial inactivation rate increased in the order:

$$\text{Goethite} < FeSO_4 < \text{Fe-citrate}$$

Encouraging results were also obtained, when this treatment was tried for bacterial inactivation in natural water samples from Lake Geneva (Switzerland) at pH 8.5. No bacterial reactivation and/or growth were observed after the photo-Fenton treatment (Fig. 3.6).

It can be concluded that the different types of Fenton and photo-Fenton processes are highly efficient, less tedious, and act as an eco-friendly technique with the capability of degrading a wide spectrum of organic pollutants. In the last few decades, various attempts have been made to achieve a high rate of decomposition and mineralization of recalcitrant complex organic compounds at neutral pH, and minimize the sludge production. Various improvements have been suggested and incorporated in order to achieve low operation cost. Exploitation of solar energy in order to accomplish the expedient degradation of various pollutants will fulfill the need of today and years to come. This field opens the avenue for new researches in which cost reduction, less energy consumption, and practical applicability of these processes at larger scale and natural conditions could be aimed.

TABLE 3.8 Photo-Fenton degradation of various pollutants

Pollutant	References
Acid Red 26, malachite green, reactive blue 4	Yanchao Jin et al. (2017)
Azo dye	Li et al. (2017)
Cibacron red	García-Montaño et al. (2006)
Cibacron red FN-R	García-Montaño et al. (2008a,b)
Ciprofloxacin, sulfamethoxazole, trimethoprim	Lima et al. (2017)
N,N-dimethylaniline and 2,4-dichlorophenol	Wang et al. (2007)
Direct blue 71	Ertugay and Acar (2017)
Gallic acid	Quici and Litter (2009)
Geosmin and 2-methylisoborneol	Park et al. (2017)
Ibuprofen	Méndez-Arriaga et al. (2009)
Indole and its derivatives	Kaczmarek and Staninski (2017)
Methyl orange, methylene blue, p-nitrophenol	Ferroudj et al. (2017)
Mordant yellow 10	He et al. (2002)
Municipal landfill leachate	Jung et al. (2017)
Nitrilotriacetic acid	Zhang et al. (2017)
Orange G	Cai et al. (2017)
Orange II	Tokumura et al. (2006)
Organic pollutants	Zhao et al. (2017)
Paper and pulp mill effluents	Brink et al. (2017)
Phloroglucinol	Wang et al. (2017)
Promazine, promethazine, chlorpromazin, and thioridazine	Wilde et al. (2017)
Reactive red HE-3B	Feng et al. (2003)
Reactive yellow 81	Acisli et al. (2017)
Rhodamine B and 4-nitrophenol	Zhang et al. (2010)
Sodium alginate	Zhou et al. (2017)
Tebuthiuron, ametryn	Gozzi et al. (2017)
p-Toluenesulfonic acid, benzoic acid, p-nitrobenzoic acid, acetaminophen, caffeic p-nitrobenzoic acid	Santos-Juanes et al. (2017)

(Continued)

TABLE 3.8 (Continued)

Pollutant	References
o-toluidine	Garcia-Segura et al. (2017)
Triphenyltinchloride	Yong et al. (2017)
Venlafaxine	Giannakis et al. (2017)
Vinasse from sugar cane ethanol distillery	Rodríguez et al. (2017)
Wastewater	Villegas-Guzman et al. (2017)
Winery wastewater	Díez et al. (2017)

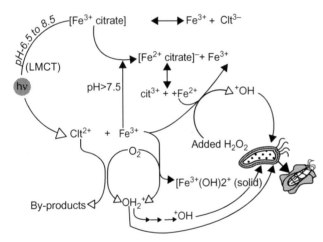

FIGURE 3.6 Bacterial inactivation during Fe-citrate-based photo-Fenton process. *Lonfat, C.R., Barona, J.F., Sienkiewicz, A., Velaza, J., Benitez, L.N., Pulgarin, C., 2016. Appl. Catal. B Environ. 180, 379–390. With permission.*

References

Acisli, O., Khataee, A., Darvishi, R., Soltani, C., Karaca, S., 2017. Ultrasound-assisted Fenton process using siderite nanoparticles prepared via planetary ball milling for removal of reactive yellow 81 in aqueous phase. Ultrasonics Sonochem. 35 A, 210–218.

Alalm, M.G., Tawfik, A., Ookawara, S., 2015a. Degradation of four pharmaceuticals by solar photo-Fenton process: kinetics and costs estimation. J. Environ. Chem. Eng. 3 (1), 46–51.

Alalm, M.G., Tawfik, A., Ookawara, S., 2015b. Comparison of solar TiO$_2$ photocatalysis and solar photo-Fenton for treatment of pesticides industry wastewater: operational conditions, kinetics, and costs. J. Water Proc. Eng. 8, 55–63.

Alalm, M.G., Tawfik, A., Ookawara, S., 2017. Investigation of optimum conditions and costs estimation for degradation of phenol by solar photo-Fenton process. Appl. Water Sci. 7 (1), 375–382.

Aljubourya, D.A., Palaniandy, P., Aziz, H.B.A., Feroz, S., 2016. Comparative study to the solar photo-Fenton, solar photocatalyst of TiO_2 and solar photocatalyst of TiO_2 combined with Fenton process to treat petroleum wastewater by RSM. J. Pet. Environ. Biotechnol. 7, 260. Available from: http://dx.doi.org/10.4172/2157-7463.1000260.

Almazán-Sánchez, P.T., Solache-Ríos, M.J., Linares-Hernández, I., Martínez-Miranda, V., 2016. Adsorption-regeneration by heterogeneous Fenton process using modified carbon and clay materials for removal of indigo blue. Environ. Technol. 37 (14), 1843−1856.

Amat, A.M., Arques, A., Miranda, M.A., Seguí, S., 2004. Photo-Fenton reaction for the abatement of commercial surfactants in a solar pilot plant. Solar Energy 77 (5), 559−566.

Ammar, H.B., 2016. Sono-Fenton process for metronidazole degradation in aqueous solution: Effect of acoustic cavitation and peroxydisulfate anion. Ultrasonics Sonochem. 33, 164−169.

Ammar, S., Oturan, M.A., Labiadh, L., Guersalli, A., Abdelhedi, R., Oturan, N., et al., 2015. Degradation of tyrosol by a novel electro-Fenton process using pyrite as heterogeneous source of iron catalyst. Water Res. 74, 77−87.

Andreozzi, R., D'Apuzzo, A., Marotta, R., 2000. A kinetic model for the degradation of benzothiazole by Fe^{3+}-photo-assisted Fenton process in a completely mixed batch reactor. J. Hazard. Mater. 80 (1−3), 241−257.

Aplin, R., Feitz, A.J., Waite, T.D., 2001. Effect of Fe(III)-ligand properties on effectiveness of modified photo-Fenton processes. Water Sci. Technol. 44 (5), 23−30.

Ay, F., Catalkaya, E.C., Kargi, F., 2009. A statistical experiment design approach for advanced oxidation of direct red azo-dye by photo-Fenton treatment. J. Hazard. Mater. 162 (1), 230−236.

Baba, Y., Yatagai, T., Harada, T., Kawase, Y., 2015. Hydroxyl radical generation in the photo-Fenton process: Effects of carboxylic acids on iron redox cycling. Chem. Eng. J. 277, 229−241.

Bandala, E.R., Brito, L., Pelaez, M., 2009. Degradation of domoic acid toxin by UV-promoted Fenton-like processes in seawater. Desalination 245 (1−3), 135−145.

Bandara, J., Morrison, C., Kiwi, J., Pulgarin, C., Peringer, P., 1996. Degradation/decoloration of concentrated solutions of orange II. Kinetics and quantum yield for sunlight induced reactions via Fenton type reagents. J. Photochem. Photobiol. A Chem. 99 (1), 57−66.

Barndok, H., Blanco, L., Hermosilla, D., Blanco, A., 2016. Heterogeneous photo-Fenton processes using zerovalent iron microspheres for the treatment of wastewaters contaminated with 1,4-dioxane. Chem. Eng. J. 284, 112−121.

Barreca, S., Velez, J.J., Pace, A., Orecchio, S., Pulgarin, C., 2014. Neutral solar photo-Fenton degradation of 4-nitro-phenol on iron-enriched hybrid montmorillonite-alginate beads (Fe-MABs). J. Photochem. Photobiol. A Chem. 282, 33−40.

Basturk, E., Karatas, M., 2014. Advanced oxidation of reactive blue 181 solution: A comparison between Fenton and sono-Fenton process. Ultrasonics Sonochem. 21 (5), 1881−1885.

Batista, A.P.S., Nogueira, R.F.P., 2012. Parameters affecting sulfonamide photo-Fenton degradation-iron complexation and substituent group. J. Photochem. Photobiol. A Chem. 232 (15), 8−13.

Bauer, R., Waldner, G., Fallmann, H., Hager, S., Klare, M., Krutzler, T., et al., 1999. The photo-Fenton reaction and the TiO_2/UV process for waste water treatment − Novel developments. Catal. Today 53 (1), 131−144.

Baydum, V.P.A., Dantas, R.F., Teixeira, A., Pacheco, J.G.A., Silva, V.L., 2012. Pre-treatment of propranolol effluent by advanced oxidation processes. Afinidad. LXIX 211−216.

Benzaquén, T.B., Isla, M.A., Alfano, O.M., 2015. Fenton and photo-Fenton processes for the degradation of atrazine: a kinetic study. J. Chem. Technol. Biotechnol. 90, 459−467.

Bobu, M., Wilson, S., Greibrokk, T., Lundanes, E., Siminiceanu, I., 2006. Comparison of advanced oxidation processes and identification of monuron photodegradation products in aqueous solution. Chemosphere 63 (10), 1718−1727.

Boruah, P.K., Sharma, B., Karbhal, I., Shelke, M.V., Das, M.R., 2017. Ammonia-modified graphene sheets decorated magnetic Fe_3O_4 nanoparticles for photocatalytic and photo-Fenton degradation of phenolic compounds under sunlight irradiation. J. Hazard. Mater. 325, 90−100.

Bouafia-Chergui, S., Oturan, N., Khalaf, H., Oturan, M.A., 2010. Parametric study on the effect of the ratios $[H_2O_2]/[Fe^{3+}]$ and $[H_2O_2]/[substrate]$ on the photo-Fenton degradation of cationic azo dye basic blue 41. J. Environ. Sci. Health A 45, 622−629.

Brink, A., Sheridan, C.M., Harding, K.G., 2017. The Fenton oxidation of biologically treated paper and pulp mill effluents: a performance and kinetic study. Proc. Saf. Environ. Prot. 107, 206−215.

Cai, M.-Q., Zhu, Y.-Z., Wei, Z.-S., Hu, J.-Q., Pan, S.-D., Xiao, R.-Y., et al., 2017. Rapid decolorization of dye orange G by microwave enhanced Fenton -like reaction with delafossite-type $CuFeO_2$. Sci. Total Environ. 580, 966–973.

Campion, L.L., Giannotti, C., Ouazzani, J., 1999. Photocatalytic degradation of 5-nitro-1,2,4-triazol-3-one NTO in aqueous suspension of TiO_2, Comparison with Fenton oxidation. Chemosphere 38 (7), 1561–1570.

Carra, I., Casas, J.L., Santos-Juanes, L., Malato, S., Sánchez, J.A., 2013. Iron dosage as a strategy to operate the photo-Fenton process at initial neutral pH. Chem. Eng. J. 224, 67–74.

Chacón, J.M., Leal, M.T., Sánchez, M., Bandala, E.R., 2006. Solar photocatalytic degradation of azo-dyes by photo-Fenton process. Dyes Pigments 69 (3), 144–150.

Chen, Q., Wu, P., Li, Y., Zhu, N., Dang, Z., 2009. Heterogeneous photo-Fenton photodegradation of reactive brilliant orange X-GN over iron-pillared montmorillonite under visible irradiation. J. Hazard. Mater. 168 (2-3), 901–908.

Chen, Q., Wu, P., Dang, Z., Zhu, N., Li, P., Wu, J., et al., 2010. Iron pillared vermiculite as a heterogeneous photo-Fenton catalyst for photocatalytic degradation of azo dye reactive brilliant orange X-GN. Separ. Purif. Technol. 71 (3), 315–323.

Chen, Y., Liu, C., Nie, J., Wu, S., Wang, D., 2014. Removal of COD and decolorizing from landfill leachate by Fenton's reagent advanced oxidation. Clean Technol. Environ. Policy 16 (1), 189–193.

Cho, I.H., Kim, Y.G., Yang, J.K., Lee, N.H., Lee, S.M., 2006. Solar-chemical treatment of groundwater contaminated with petroleum at gas station sites: ex situ remediation using solar/TiO_2 photocatalysis and Solar Photo-Fenton. J. Environ. Sci. Health A Toxicol. Hazard. Subst. Environ. Eng. 41 (3), 457–473.

Clément, T., Yoan, P., Nihal, O., Emmanuel, M., David, H., Eric, D., et al., 2016. Comparative study on the removal of humic acids from drinking water by anodic oxidation and electro-Fenton processes: mineralization efficiency and modeling. Appl. Catal. B: Environ 194, 32–41.

Da Silva, S.S., Chiavone-Filho, O., de Barros Neto, E.L., Foletto, E.L., 2015. Oil removal from produced water by conjugation of flotation and photo-Fenton processes. J. Environ. Manage. 147, 257–263.

Da Silva, S.S., Chiavone-Filho, O., De Barros, E.L., Nascimento, C.A., 2012. Integration of processes induced air flotation and photo-Fenton for treatment of residual waters contaminated with xylene. J. Hazard. Mater. 199–200, 151–157.

Daneshvar, N., Khataee, A.R., 2006. Removal of azo dye C.I. acid red 14 from contaminated water using Fenton, UV/H_2O_2, UV/H_2O_2/Fe(II), UV/H_2O_2/Fe(III) and UV/H_2O_2/Fe(III)/oxalate processes: A comparative study. J. Environ. Sci. Health A Toxicol. Hazard. Subst. Environ. Eng. 41 (3), 315–328.

Dbira, S., Bedoui, A., Bensalah, N., 2014. Investigations on the degradation of triazine herbicides in water by photo-Fenton process. Am. J. Anal. Chem. 5 (8), 500–517.

De la Cruz, N., Giménez, J., Esplugas, S., Grandjean, D., de Alencastro, L.F., Pulgarín, C., 2012. Degradation of 32 emergent contaminants by UV and neutral photo-Fenton in domestic wastewater effluent previously treated by activated sludge. Water Res. 46 (6), 1947–1957.

Dehghani, M., Taghizadeh, M.M., Gholami, T., Ghadami, M., Keshtgar, L., Elhameyan, Z., et al., 2015. Optimization of the parameters influencing the photo-Fenton process for the decolorization of reactive red 198 (RR198). Jundishapur. J. Health. Sci. 7 (2), 38–43.

Derbalah, A.S., Nakatani, N., Sakugawa, H., 2004. Photocatalytic removal of fenitrothion in pure and natural waters by photo-Fenton reaction. Chemosphere 57 (7), 635–644.

Devi, L.G., Kumar, S.G., Mohan Reddy, K., Munikrishnappa, C., 2009. Photodegradation of methyl orange an azo dye by advanced Fenton process using zero valent metallic iron: Influence of various reaction parameters and its degradation mechanism. J. Hazard. Mater. 164 (2–3), 459–467.

Devi, L.G., Rajashekhar, K.E., Raju, K.S.A., Kumar, S.G., 2011. Influence of various aromatic derivatives on the advanced photo-Fenton degradation of amaranth dye. Desalination 270 (1–3), 31–39.

Devi, L.G., Srinivas, M., Kumari, M.L.A., 2016. Heterogeneous advanced photo- Fenton process using peroxymonosulfate and peroxydisulfate in presence of zero valent metallic iron: A comparative study with hydrogen peroxide photo-Fenton process. J. Water Proc. Eng. 13, 117–126.

Díez, A.M., Rosales, E., Sanromán, M.A., Pazos, M., 2017. Assessment of LED-assisted electro- Fenton reactor for the treatment of winery wastewater. Chem. Eng. J. 310 (2), 399–406.

Ding, X., Ai, Z., Zhang, L., 2012. Design of a visible light driven photo-electrochemical/electro-Fenton coupling oxidation system for wastewater treatment. J. Hazard. Mater. 239-240, 233–240.

Dükkanci, M., Vinatoru, M., Mason, T.J., 2014. The sonochemical decolourisation of textile azo dye orange II: effects of Fenton type reagents and UV light. Ultrasonics Sonochem. 21 (2), 846–853.

Elmolla, E.S., Chaudhuri, M., 2010. Photo-Fenton treatment of antibiotic wastewater. Nat. Environ. Pollut. Technol. 9 (2), 365–370.

Elmorsi, T.M., Riyad, Y.M., Mohamed, Z.H., Abd El Bary, H.M., 2010. Decolorization of mordant red 73 azo dye in water using H_2O_2/UV and photo-Fenton treatment. J. Hazard. Mater. 174 (1-3), 352–358.

ElShafei, G.M.S., Yehia, F.Z., Dimitry, O.I.H., Badawi, A.M., Eshaq, G., 2014. Ultrasonic assisted-Fenton-like degradation of nitrobenzene at neutral pH using nanosized oxides of Fe and Cu. Ultrasonics Sonochem. 21, 1358–1365.

Emilia, A.T., Wijaya, A.Y., Mermaliandi, F., 2016. Degradation of reactive red 2 by Fenton and photo-Fenton oxidation processes. J. Eng. Appl. Sci. 11 (8), 2006–2016.

Ertugay, N., Acar, F.N., 2017. Removal of COD and color from direct blue 71 azo dye wastewater by Fenton's oxidation: kinetic study. Arab. J. Chem. 10 (Suppl. 1), 1158–1163.

Fang, Y.-F., Chen, D.-X., Huang, Y.-P., Yang, J., Chen, G.-W., 2011. Heterogeneous Fenton photodegradation of Microcystin-LR with visible light irradiation. Chin. J. Anal. Chem. 39 (4), 540–543.

Farré, M.J., Franch, M.I., Malato, S., Ayllón, J.A., Peral, J., Doménech, X., 2005. Degradation of some biorecalcitrant pesticides by homogeneous and heterogeneous photocatalytic ozonation. Chemosphere 58 (8), 1127–1133.

Feng, J., Hu, X., Yue, P.L., Zhu, H.Y., Lu, G.Q., 2003. Discoloration and mineralization of reactive red HE-3B by heterogeneous photo-Fenton reaction. Water Res. 37 (15), 3776–3784.

Feng, J., Hu, X., Yue, P.L., 2006. Effect of initial solution pH on the degradation of orange II using clay-based Fe nanocomposites as heterogeneous photo-Fenton catalyst. Water Res. 40 (4), 641–646.

Fenton, H., 1894. Oxidation of tartaric acid in the presence of iron. J. Chem. Soc. 65, 899–910.

Ferrero, F., 2000. Oxidative degradation of dyes and surfactant in the Fenton and photo-Fenton treatment of dyehouse effluents. Color. Technol. 116 (5–6), 148–153.

Ferroudj, N., Talbot, D., Michel, A., Davidson, A., Abramson, S., 2017. Increasing the efficiency of magnetic heterogeneous Fenton catalysts with a simple halogen visible lamp. J. Photochem. Photobiol. A: Chem. 338, 85–95.

García-Montaño, J., Torrades, F., García-Hortal, J.A., Domènech, X., Peral, J., 2006. Combining photo-Fenton process with aerobic sequencing batch reactor for commercial hetero-bireactive dye removal. Appl. Catal. B Environ. 67 (1–2), 86–92.

García-Montaño, J., Domènech, X., García-Hortal, J.A., Torrades, F., Peral, J., 2008a. The testing of several biological and chemical coupled treatments for Cibacron Red FN-R azo dye removal. J. Hazard. Mater. 154 (1–3), 484–490.

García-Montaño, J., Torrades, F., Perez-Estrada, L.A., Oller, I., Malato, S., Maldonado, M.I., et al., 2008b. Degradation pathways of the commercial reactive azo dye procion red H-E7B under solar-assisted photo-Fenton reaction. Environ. Sci. Technol. 42 (17), 6663–6670.

Garcia-Segura, S., Anotai, J., Singhadech, S., Lu, M.-C., 2017. Enhancement of biodegradability of o-toluidine effluents by electro-assisted photo-Fenton treatment. Proc. Saf. Environ. Prot. 106, 60–67.

Gernjak, W., Krutzler, T., Glaser, A., Malato, S., Caceres, J., Bauer, R., et al., 2003. Photo-Fenton treatment of water containing natural phenolic pollutants. Chemosphere 50 (1), 71–78.

Giannakis, S., Hendaoui, I., Rtimi, S., Fürbringer, J.-M., Pulgarin, C., 2017. Modeling and treatment optimization of pharmaceutically active compounds by the photo-Fenton process: The case of the antidepressant venlafaxine. J. Environ. Chem. Eng. 5 (1), 818–828.

Giri, A.S., Golder, A.K., 2014. Fenton, photo-Fenton, H_2O_2 photolysis, and TiO_2 photocatalysis for dipyrone oxidation: drug removal, mineralization, biodegradability, and degradation mechanism. Ind. Eng. Chem. Res. 53, 1351–1358.

Glaze, W.H., Kang, J.W., Chapin, D.H., 1987. The chemistry of water treatment involving ozone, hydrogen peroxide and ultraviolet radiation. Ozone Sci. Eng 9, 335–352.

Goi, A., Trapido, M., 2002. Hydrogen peroxide photolysis, Fenton reagent and photo-Fenton for the degradation of nitrophenols: a comparative study. Chemosphere 46 (6), 913–922.

Gökkuş, Ö., Yıldız, Y.Ş., 2016. Application of electro-Fenton process for medical waste sterilization plant wastewater. Desalin. Water Treat. 57, 24934–24945.

Gong, Y.H., Zhang, H., Li, Y.L., Xiang, L.J., Royer, S., Valange, S., et al., 2010. Evaluation of heterogeneous photo-Fenton oxidation of orange II using response surface methodology. Water Sci. Technol. 62 (6), 1320–1326.

Gosetti, F., Gianotti, V., Polati, S., Gennaro, M.C., 2005. HPLC-MS degradation study of E10 sunset yellow FCF in a commercial beverage. J. Chromatogr. A 1090 (1–2), 107–115.

Gozzi, F., Sires, I., Thiam, A., de Oliveira, S.C., Brilla, E., 2017. Treatment of single and mixed pesticide formulations by solar photoelectro-Fenton using a flow plant. Chem. Eng. J. 310 (2), 503–513.

Guimarães, J.R., Gasparini, M.C., Maniero, M.G., Mendes, C.G.N., 2012a. Stripped sour water treatment by advanced oxidation processes. J. Braz. Chem. Soc. 23 (9), 1680–1687.

Guimarães, J.R., Maniero, M.G., de Araújo, R.N., 2012b. A comparative study on the degradation of RB-19 dye in an aqueous medium by advanced oxidation processes. J. Environ. Manage. 110, 33–39.

Guo, S., Yuan, N., Zhang, G., Yu, J.C., 2017. Graphene modified iron sludge derived from homogeneous Fenton process as an efficient heterogeneous Fenton catalyst for degradation of organic pollutants. Microp. Mesop. Mater. 238, 62–68.

Gutowska, A., Kałużna-Czaplińska, J., Jóźwiak, W.K., 2007. Degradation mechanism of reactive orange 113 dye by H_2O_2/Fe^{2+} and ozone in aqueous solution. Dyes Pigments 74 (1), 41–46.

Haber, F., Willstätter, R., 1931. Unparrigkit und radikalatten im reaktionsmechanismus organischer und enzymatischer vorgage. Ber. Dtsch. Chem. Ges. 64, 2844–2856.

Hasan, D.B., Abdul Aziz, A.R., Daud, W.M., 2012. Using D-optimal experimental design to optimise remazol black B mineralisation by Fenton-like peroxidation. Environ. Technol. 33 (10–12), 1111–1121.

Hassan, M.E., Chen, Y., Liu, G., Zhu, D., Cai, J., 2016. Heterogeneous photo-Fenton degradation of methyl orange by Fe_2O_3/TiO_2 nanoparticles under visible light. J. Water Proc. Eng. 12, 52–57.

He, J., Ma, W., He, J., Zhao, J., Yu, J.C., 2002. Photooxidation of azo dye in aqueous dispersions of $H_2O_2/$-FeOOH. Appl. Catal. B Environ. 39 (3), 211–220.

Hu, M., Xu, Y., 2004. Photocatalytic degradation of textile dye X3B by heteropolyoxometalate acids. Chemosphere 54 (3), 431–434.

Huang, Y.-H., Huang, Y.-J., Tsai, H.-C., Chen, H.-T., 2010. Degradation of phenol using low concentration of ferric ions by the photo-Fenton process. J. Taiwan Inst. Chem. Eng. 41 (6), 699–704.

Jain, A., Vaya, D., Sharma, V.K., Ameta, S.C., 2011. Photo-Fenton degradation of phenol red catalyzed by inorganic additives: a technique for wastewater treatment. Kinet. Catal. 52 (1), 40–47.

Jiang, C., Xu, Z., Guo, Q., Zhuo, Q., 2014. Degradation of bisphenol A in water by the heterogeneous photo-Fenton. Environ. Technol. 35 (5–8), 966–972.

Jin, Y., Wang, Y., Huang, Q., Zhu, L., Cui, Y., Lin, C., 2017. The experimental study of a hybrid solar phot-Fenton and photovoltaic system for water purification. Energy Convers. Manage. 135, 178–187.

Jung, C., Deng, Y., Zhao, R., Torrens, K., 2017. Chemical oxidation for mitigation of UV-quenching substances (UVQS) from municipal landfill leachate: Fenton process versus ozonation. Water Res. 108, 260–270.

Kaczmarek, M., Staninski, K., 2017. Terbium(III) ions as sensitizers of oxidation of indole and its derivatives in Fenton system. J. Lumin. 183, 470–477.

Kajitvichyanukul, P., Lu, M.-C., Jamroensan, A., 2008. Formaldehyde degradation in the presence of methanol by photo-Fenton process. J. Environ. Manage. 86 (3), 545–553.

Kang, S.-F., Liao, C.-H., Po, S.-T., 2000. Decolorization of textile wastewater by photo-Fenton oxidation technology. Chemosphere 41 (8), 1287–1294.

Karale, R., Manu, S.B., Shrihari, S., 2014. Fenton and photo-Fenton oxidation processes for degradation of 3-aminopyridine from water. APCBEE Proc. 9, 25–29.

Karci, A., Arslan-Alaton, I., Olmez-Hanci, T., Bekbölet, M., 2012. Transformation of 2,4-dichlorophenol by $H_2O_2/$UV-C, Fenton and photo-Fenton processes: oxidation products and toxicity evolution. J. Photochem. Photobiol. A Chem. 230 (1), 65–73.

Kasiri, M.B., Aleboyeh, A., Aleboyeh, H., 2010. Investigation of the solution initial pH effects on the performance of UV/Fe-ZSM5/H_2O_2 process. Water Sci. Technol. 61 (8), 2143–2149.

Katsumata, H., Koike, S., Kaneco, S., Suzuki, T., Ohta, K., 2010. Degradation of reactive yellow 86 with photo-Fenton process driven by solar light. J. Environ. Sci. (China) 22 (9), 1455–1461.

Kavitha, V., Palanivelu, K., 2004. The role of ferrous ion in Fenton and photo-Fenton processes for the degradation of phenol. Chemosphere 55 (9), 1235–1243.

Khataee, A., Salahpour, F., Fathinia, M., Seyyedi, B., Vahid, B., 2015. Iron rich laterite soil with mesoporous structure for heterogeneous Fenton-like degradation of an azo dye under visible light. J. Ind. Eng. Chem. 26, 129–135.

Khataee, A., Gholami, P., Vahid, B., Joo, S.W., 2016a. Heterogeneous sono-Fenton process using pyrite nanorods prepared by non-thermal plasma for degradation of an anthraquinone dye. Ultrasonics Sonochem. 32, 357–370.

Khataee, A., Rad, T.S., Vahid, B., Khorram, S., 2016b. Preparation of zeolite nanorods by corona discharge plasma for degradation of phenazopyridine by heterogeneous sono-Fenton-like process. Ultrasonics Sonochem. 33, 37–44.

Kim, H.E., Lee, J., Lee, H., Lee, C., 2012. Synergistic effects of TiO_2 photocatalysis in combination with Fenton-like reactions on oxidation of organic compounds at circumneutral pH. Appl. Catal. B Environ. 115, 219–224.

Kim, S.-M., Vogelpohl, A., 1998. Degradation of organic pollutants by the Ppoto-Fenton-Ppocess. Chem. Eng. Technol. 21, 187–191.

Klamerth, N., Gernjak, W., Malato, S., Agüera, A., Lendl, B., 2009. Photo-Fenton decomposition of chlorfenvinphos: determination of reaction pathway. Water Res. 43 (2), 441–449.

Klamerth, N., Malato, S., Agüera, A., Fernández-Alba, A., 2013. Photo-Fenton and modified photo-Fenton at neutral pH for the treatment of emerging contaminants in wastewater treatment plant effluents: a comparison. Water Res. 47 (2), 833–840.

Kumar, D., 2015. Photo-Fenton and Fenton processes for the degradation of 4-mercaptobenzoic acid. Acta Chim. Pharm. Indica 5 (2), 88–92.

Kumar, D., Ameta, R., 2013. Use of photo-Fenton reagent for the degradation of basic orange 2 in aqueous medium. J. Chem. Pharm. Res. 5 (1), 68–74.

Kusic, H., Koprivanac, N., Srsan, L., 2006a. Azo dye degradation using Fenton type processes assisted by UV irradiation: a kinetic study. J. Photochem. Photobiol. A Chem. 181 (2–3), 195–202.

Kusic, H., Koprivanac, N., Bozic, A.L., Selanec, I., 2006b. Photo-assisted Fenton type processes for the degradation of phenol: a kinetic study. J. Hazard. Mater. 136 (3), 632–644.

Kusic, H., Koprivanac, N., Horvat, S., Bakija, S., Bozic, A.L., 2009. Modeling dye degradation kinetic using dark- and photo-Fenton type processes. Chem. Eng. J. 155 (1–2), 144–154.

Labiadh, L., Oturan, M.A., Panizza, M., Ben Hamadi, N., Ammar, S., 2015. Complete removal of AHPS synthetic dye from water using new electro-Fenton oxidation catalyzed by natural pyrite as heterogeneous catalyst. J. Hazard. Mater. 297, 34–41.

Lagori, G., Rocca, J.P., Brulat, N., Merigo, E., Vescovi, P., Fornaini, C., 2015. Comparison of two different laser wavelengths' dental bleaching results by photo-Fenton reaction: in vitro study. Lasers Med. Sci. 30 (3), 1001–1006.

Le, T.X.H., Bechelany, M., Lacour, S., Oturan, N., Oturan, M.A., Cretin, M., 2015. High removal efficiency of dye pollutants by electron-Fenton process using a graphene based cathode. Carbon 94, 1003–1011.

Lee, C., Sedlak, D.L., 2009. A novel homogeneous Fenton-like system with Fe(III)–phosphotungstate for oxidation of organic compounds at neutral pH values. J. Mol. Catal. A Chem. 311 (1–2), 1–6.

Levchuk, I., Bhatnagar, A., Sillanpää, M., 2014. Overview of technologies for removal of methyl tert-butyl ether (MTBE) from water. Sci. Total Environ. 476–477, 415–433.

Li, X., Jin, X., Zhao, N., Angelidaki, I., Zhang, Y., 2017. Novel bio-electro-Fenton technology for azo dye wastewater treatment using microbial reverse-electrodialysis electrolysis cell. Biores. Technol. 228, 322–329.

Liang, J., Komarov, S., Hayashi, N., Eiki, K., 2007. Recent trends in the decomposition of chlorinated aromatic hydrocarbons by ultrasound irradiation and Fenton's reagent. J. Mater. Cycles Waste Manage. 9, 47–55.

Lima, M.J., Silva, C.G., Silva, A.M.T., Lopes, J.C., Dias, M.M., Faria, J.L., 2017. Homogeneous and heterogeneous photo-Fenton degradation of antibiotics using an innovative static mixer photoreactor. Chem. Eng. J. 310 (2), 342–351.

Lin, S.H., Lin, C.M., Leu, H.G., 1999. Operating characteristics and kinetic studies of surfactant wastewater treatment by Fenton oxidation. Water Res. 33 (7), 1735–1741.

Liou, M.-J., Lu, M.-C., Chen, J.-N., 2003. Oxidation of explosives by Fenton and photo-Fenton processes. Water Res. 37 (13), 3172–3179.

Litter, M.I., Slodowicz, M.J., 2017. An overview on heterogeneous Fenton and photo-Fenton reactions using zerovalent iron materials. Adv. Oxid. Technol. 20 (1). Available from: http://dx.doi.org/10.1515/jaots-2016-0164.

Liu, Y., Li, Y.M., Wen, L.H., Hou, K.Y., Li, H.Y., Pu, G., et al., 2006. Synthesis, characterization and photo degradation application for dye-rhodamine B of nano-iron oxide/bentonite. Guang pu 26 (10), 1939–1942.

Lofrano, G., Rizzo, L., Grassi, M., Belgiorno, V., 2009. Advanced oxidation of catechol: A comparison among photocatalysis, Fenton and photo-Fenton processes. Desalination 249 (2), 878–883.

Lonfat, C.R., Barona, J.F., Sienkiewicz, A., Velaza, J., Benitez, L.N., Pulgarin, C., 2016. Bacterial inactivation with iron-citrate complex: a new sourcs of dissolved iron in solar photo-Fenton process at near-neutral and alkalike pH. Appl. Catal. B Environ. 180, 379–390.

Lucas, M.S., Peres, J.A., 2007. Degradation of reactive black 5 by Fenton/UV-C and ferrioxalate/H_2O_2/solar light processes. Dyes Pigments 74 (3), 622–629.

Lucas, M.S., Mosteo, R., Maldonado, M.I., Malato, S., Peres, J.A., 2009. Solar photochemical treatment of winery wastewater in a CPC reactor. J. Agric. Food Chem. 57 (23), 11242–112428.

Lunar, L., Sicilia, D., Rubio, S., Pérez-Bendito, D., Nickel, U., 2000. Degradation of photographic developers by Fenton's reagent: condition optimization and kinetics for metol oxidation. Water Res. 34 (6), 1791–1802.

Luna, A.J., Chiavone-Filho, O., Machulek, A., de Moraes, J.E.F., Nascimento, C.A.O., 2012. Photo-Fenton oxidation of phenol and organochlorides (2,4-DCP and 2,4-D) in aqueous alkaline medium with high chloride concentration. J. Environ. Manage. 111, 10–17.

Ma, J., Song, W., Chen, C., Ma, W., Zhao, J., Tang, Y., 2005. Fenton degradation of organic compounds promoted by dyes under visible irradiation. Environ. Sci. Technol. 39 (15), 5810–5815.

Maezono, T., Tokumura, M., Sekine, M., Kawase, Y., 2011. Hydroxyl radical concentration profile in photo-Fenton oxidation process: Generation and consumption of hydroxyl radicals during the discoloration of azo-dye orange II. Chemosphere 82 (10), 1422–1430.

Mahdad, F., Younesi, H., Bahramifar, N., Hadavifar, M., 2016. Optimization of Fenton and photo-Fenton-based advanced oxidation processes for post-treatment of composting leachate of municipal solid waste by an activated sludge process. KSCE J. Civ. Eng. 20, 2177–2188.

Malakootian, M., Jaafarzadeh, N., Dehdarirad, A., 2016. Efficiency investigation of photo-Fenton process in removal of sodium dodecyl sulphate from aqueous solutions. Desalin. Water Treat. 57 (51), 24444–24449.

Malato, S., Caceres, J., Agüera, A., Mezcua, M., Hernando, D., Vial, J., et al., 2001. Degradation of imidacloprid in water by photo-Fenton and TiO_2 photocatalysis at a solar pilot plant: a comparative study. Environ. Sci. Technol. 35 (21), 4359–4366.

Malato, S., Cáceres, J., Fernández-Alba, A.R., Piedra, L., Hernando, M.D., Agüera, A., et al., 2003. Photocatalytic treatment of diuron by solar photocatalysis: evaluation of main intermediates and toxicity. Environ. Sci. Technol. 37 (11), 2516–2524.

Manzano, M.A., Riaza, A., Quiroga, J.M., Kiwi, J., 2009. Optimization of the solution parameters during the degradation of orange II in a photo-reactor mediated by Fe-Nafion membranes. Water Sci. Technol. 60 (4), 833–840.

Marcinowski, P.P., Bogacki, J.P., Naumczyk, J.H., 2014. Cosmetic wastewater treatment using the Fenton, Photo-Fenton and H_2O_2/UV processes. J. Environ. Sci. Health, Part A. 49 (13), 1531–1541.

Matilainen, A., Sillanpää, M., 2010. Removal of natural organic matter from drinking water by advanced oxidation processes. Chemosphere 80 (4), 351–365.

Méndez-Arriaga, F., Torres-Palma, R.A., Pétrier, C., Esplugas, S., Gimenez, J., Pulgarin, C., 2009. Mineralization enhancement of a recalcitrant pharmaceutical pollutant in water by advanced oxidation hybrid processes. Water Res. 43 (16), 3984–3991.

Meriç, S., Kaptan, D., Ölmez, T., 2004. Color and COD removal from wastewater containing reactive black 5 using Fenton's oxidation process. Chemosphere 54 (3), 435–441.

Michael, I., Hapeshi, E., Michael, C., Fatta-Kassinos, D., 2010. Solar Fenton and solar TiO_2 catalytic treatment of ofloxacin in secondary treated effluents: evaluation of operational and kinetic parameters. Water Res. 44 (18), 5450–5462.

Miralles-Cuevas, S., Oller, I., Agüera, A., Malato, S., 2017. Strategies for reducing cost by using solar photo-Fenton treatment combined with nanofiltration to remove microcontaminants in real municipal effluents: toxicity and economic assessment. Chem. Eng. J. 318, 161–170.

Miranzadeh, M.B., Zarjam, R., Dehghani, R., Haghighi, M., Badi, H.Z., Marzaleh, M.A., et al., 2016. Comparison of Fenton and photo-Fenton processes for removal of linear alkyle benzene sulfonate (LAS) from aqueous solutions. Pol. J. Environ. Stud. 25 (4), 1639–1648.

Molkenthin, M., Olmez-Hanci, T., Jekel, M.R., Arslan-Alaton, I., 2013. Photo-Fenton-like treatment of BPA: effect of UV light source and water matrix on toxicity and transformation products. Water Res. 47 (14), 5052–5064.

Muñoz, I., Rieradevall, J., Torrades, F., Peral, J., Domènech, X., 2006. Environmental assessment of different advanced oxidation processes applied to a bleaching Kraft mill effluent. Chemosphere 62 (1), 9–16.

Muruganandham, M., Swaminathan, M., 2004. Decolourisation of Reactive Orange 4 by Fenton and photo-Fenton oxidation technology. Dyes Pigments 63 (3), 315–321.

Muruganandham, M., Selvam, K., Swaminathan, M., 2007. A comparative study of quantum yield and electrical energy per order (E(Eo)) for advanced oxidative decolourisation of reactive azo dyes by UV light. J. Hazard. Mater. 144 (1–2), 316–322.

Nam, S., Renganathan, V., Tratnyek, P.G., 2001. Substituent effects on azo dye oxidation by the FeIII–EDTA–H_2O_2 system. Chemosphere 45 (1), 59–65.

Neamțu, M., Grandjean, D., Sienkiewicz, A., Le Faucheur, S., Slaveykova, V., Velez, J.J., et al., 2014. Degradation of eight relevant micropollutants in different water matrices by neutral photo-Fenton process under UV254 and simulated solar light irradiation—a comparative study. Appl. Catal. B Environ. 158-159, 30–37.

Neamtu, M., Yediler, A., Siminiceanu, I., Kettrup, A., 2003. Oxidation of commercial reactive azo dye aqueous solutions by the photo-Fenton and Fenton-like processes. J. Photochem. Photobiol. A Chem. 161 (1), 87–93.

Nerud, F., Baldrian, P., Gabriel, J., Ogbeifun, D., 2001. Decolourization of synthetic dyes by the Fenton reagent and the Cu/pyridine/H_2O_2 system. Chemosphere 44 (5), 957–961.

Nidheesh, P.V., Gandhimathi, R., Ramesh, S.T., 2013. Degradation of dyes from aqueous solution by Fenton processes: a review. Environ. Sci. Pollut. Res. Int. 20 (4), 2099–2132.

Nogueira, R.F., Trovó, A.G., Paterlini, W.C., 2004. Evaluation of the combined solar TiO_2/photo-Fenton process using multivariate analysis. Water Sci. Technol. 49 (4), 195–200.

Ntampegliotis, K., Riga, A., Karayannis, V., Bontozoglou, V., Papapolymerou, G., 2006. Decolorization kinetics of Procion H-exl dyes from textile dyeing using Fenton-like reactions. J. Hazard. Mater. 136 (1), 75–84.

Nurhayati, E., Yang, H., Chen, C., Liu, C., Juang, Y., Huang, C., et al., 2016. Electro-photocatalytic Fenton decolorization of orange G using mesoporous TiO_2/stainless steel mesh photo-electrode prepared by the sol-gel dip-coating method. Int. J. Electrochem. Sci. 11, 3615–3632.

Oller, I., Gernjak, W., Maldonado, M.I., Pérez-Estrada, L.A., Sánchez-Pérez, J.A., Malato, S., 2006. Solar photocatalytic degradation of some hazardous water-soluble pesticides at pilot-plant scale. J. Hazard. Mater. 138 (3), 507–517.

Olvera-Vargas, H., Oturan, N., Oturan, M.A., Brillas, E., 2015. A pre-pilot flow plant scale for the electro-Fenton and solar photoelectro-Fenton treatments of the pharmaceutical ranitidine. Sep. Purif. Technol. 146, 127–137.

Orbeci, C., Untea, I., Nechifor, G., Segneanu, A.E., Craciun, M.E., 2014. Effect of a modified photo-Fenton procedure on the oxidative degradation of antibiotics in aqueous solutions. Sep. Purif. Technol. 122, 290–296.

Ortega-Gómez, E., Moreno, J.C., Álvarez, J.D., Casas, J.L., Santos-Juanes, L., Sánchez, J.A., 2012. Automatic dosage of hydrogen peroxide in solar photo-Fenton plants: Development of a control strategy for efficiency enhancement. J. Hazard. Mater. 237-238, 223–230.

Pagano, M., Lopez, A., Volpe, A., Mascolo, G., Ciannarella, R., 2008. Oxidation of nonionic surfactants by Fenton and H_2O_2/UV processes. Environ. Technol. 29 (4), 423–433.

Park, J.-A., Nam, H.-L., Choi, J.-W., Ha, J., Lee, S.-H., 2017. Oxidation of geosmin and 2-methylisoborneol by the photo-Fenton process: Kinetics, degradation intermediates, and the removal of microcystin-LR and trihalomethane from Nak-Dong River water, South Korea. Chem. Eng. J. 313 (1), 345–354.

Parra, S., Malato, S., Blanco, J., Péringer, P., Pulgari, C., 2001. Concentrating versus non-concentrating reactors for solar photocatalytic degradation of p-nitrotoluene-o-sulfonic acid. Water Sci. Technol. 44 (5), 219–227.

Patil, P.N., Gogate, P.R., 2015. Degradation of dichlorvos using hybrid advanced oxidation processes based on ultrasound. J. Water Process Eng. 8, e58–e65.

Peralta-Hernández, J.M., Meas-Vong, Y., Rodríguez, F.J., Chapman, T.W., Maldonado, M.I., Godínez, L.A., 2006. In situ electrochemical and photo-electrochemical generation of the Fenton reagent: A potentially important new water treatment technology. Water Res. 40 (9), 1754–1762.

Pereira, J.H.O.S., Queirós, D.B., Reis, A.C., Nunes, O.C., Borges, M.T., Boaventura, R.A.R., et al., 2014. Process enhancement at near neutral pH of a homogeneous photo-Fenton reaction using ferricarboxylate complexes: application to oxytetracycline degradation. Chem. Eng. J. 253, 217–228.

Pérez, M., Torrades, F., García-Hortal, J.A., Domènech, X., Peral, J., 2002. Removal of organic contaminants in paper pulp treatment effluents under Fenton and photo-Fenton conditions. Appl. Catal. B Environ. 36 (1), 63–74.

Pérez-Moya, M., Graells, M., Castells, G., Amigó, J., Ortega, E., Buhigas, G., et al., 2010. Characterization of the degradation performance of the sulfamethazine antibiotic by photo-Fenton process. Water Res. 44 (8), 2533–2540.

Peternel, I.T., Koprivanac, N., Bozić, A.M., Kusić, H.M., 2007. Comparative study of UV/TiO_2, UV/ZnO and photo-Fenton processes for the organic reactive dye degradation in aqueous solution. J. Hazard. Mater. 148 (1–2), 477–484.

Pradhan, A.C., Varadwaj, G.B., Parida, K.M., 2013. Facile fabrication of mesoporous iron modified Al_2O_3 nanoparticles pillared montmorillonite nanocomposite: A smart photo-Fenton catalyst for quick removal of organic dyes. Dalton Trans 42 (42), 15139–15149.

Prousek, J., 1995. Fenton reaction after a century. Chem. Listy 89, 11–21.

Quici, N., Litter, M.I., 2009. Heterogeneous photocatalytic degradation of gallic acid under different experimental conditions. Photochem. Photobiol. Sci. 8 (7), 975–984.

Rabelo, M.D., Bellato, C.R., Silva, C.M., Ruy, R.B., da Silva, C.A.B., Nunes, W.G., 2014. Application of photo-Fenton process for the treatment of Kraft pulp mill effluent. Adv. Chem. Eng. Sci. 4 (4), 483–490.

Radjenović, J., Sirtori, C., Petrović, M., Barceló, D., Malato, S., 2010. Characterization of intermediate products of solar photocatalytic degradation of ranitidine at pilot-scale. Chemosphere 79 (4), 368–376.

Rengifo-Herrera, J.A., Castaño, O.L., Sanabria, I.J., 2013. Culturability and viability of *Salmonella Typhimurium* during photo-Fenton process at pH 5.5 under solar simulated irradiation. J. Water Resour. Prot. 5 (8), 21–27.

Riga, A., Soutsas, K., Ntampegliotis, K., Karayannis, V., Papapolymerou, G., 2007. Effect of system parameters and of inorganic salts on the decolorization and degradation of procion H-exl dyes, comparison of H_2O_2/UV, Fenton, $UV/Fenton$, TiO_2/UV and $TiO_2/UV/H_2O_2$ processes. Desalination 211 (13), 72–86.

Rocha, O.R.S.D., Dantas, R.F., Duarte, M.M.M.B., Duarte, M.M.L., Silva, V.L., 2013. Solar photo-Fenton treatment of petroleum extraction wastewater. Desalin. Water Treat. 51 (28–30), 5785–5791.

Rodríguez, S.M., Gálvez, J.B., Rubio, M.I., Ibáñez, P.F., Gernjak, W., Alberola, I.O., 2005. Treatment of chlorinated solvents by TiO_2 photocatalysis and photo-Fenton: influence of operating conditions in a solar pilot plant. Chemosphere 58 (4), 391–398.

Rodrigues, C.S.D., Neto, A.R., Duda, R.M., de Oliveira, R.A., Madeira, L.M., 2017. Combination of chemical coagulation, photo-Fenton oxidation and biodegradation for the treatment of vinasse from sugar cane ethanol distillery. J. Clean. Prod. 142 (4), 3634–3644.

Rodríguez-Chueca, J., Polo-López, M.I., Mosteo, R., Ormad, M.P., Fernández-Ibáñez, P., 2014. Disinfection of real and simulated urban wastewater effluents using a mild solar photo-Fenton. Appl. Catal. B Environ. 150–151, 619–629.

Romero, V., Acevedo, S., Marco, P., Giménez, J., Esplugas, S., 2015. Enhancement of Fenton and photo-Fenton processes at initial circumneutral pH for the degradation of the -blocker metoprolol. Water Res. 88, 449–457.

Rubio-Clemente, A., Chica, E., Peñuela, G.A., 2015. Petrochemical wastewater treatment by photo-Fenton process. Water Air Soil Pollut 226 (62), 61–79.

Rubio, D., Nebot, E., Casanueva, J.F., Pulgarin, C., 2013. Comparative effect of simulated solar light, UV, UV/H_2O_2 and photo-Fenton treatment ($UV-vis/H_2O_2/Fe^{2+,3+}$) in the *Escherichia coli* inactivation in artificial seawater. Water Res. 47, 6367–6379.

Sabaikai, W., Sekine, M., Tokumura, M., Kawase, Y., 2014. UV light photo-Fenton degradation of polyphenols in oolong tea manufacturing wastewater. J. Environ. Sci. Health Part A. 49 (2), 193–202.

Sánchez, J.A., Román, I.M., Carra, I., Cabrera, A., Casas, J.L., Malato, S., 2013. Economic evaluation of a combined photo-Fenton/MBR process using pesticides as model pollutant, Factors affecting costs. J. Hazard. Mater. 244–245, 195–203.

Santiago, D.E., Melián, E.P., Rodríguez, C.F., Méndez, J.A.O., Pérez-Báez, S.O., Doña-Rodríguez, J.M., 2011. Degradation and detoxification of banana post-harvest treatment water using advanced oxidation techniques. Green Sustain. Chem. 1 (3), 39–46.

Santos-Juanes, L., García Einschlag, F.S., Amat, A.M., Arques, A., 2017. Combining ZVI reduction with photo-Fenton process for the removal of persistent pollutants. Chem. Eng. J. 310 (2), 484–490.

Sarria, V., Parra, S., Invernizzi, M., Peringer, P., Pulgarin, C., 2001. Photochemical-biological treatment of a real industrial biorecalcitrant wastewater containing 5-amino-6-methyl-2-benzimidazolone. Water Sci. Technol. 44 (5), 93–101.

Scheeren, C.W., Paniz, J.N., Martins, A.F., 2002. Comparison of advanced processes on the oxidation of acid orange 7 dye. J. Environ. Sci. Health A. 37 (7), 1253–1261.

Sekaran, G., Karthikeyan, S., Boopathy, R., Maharaja, P., Gupta, V.K., Anandan, C., 2014. Response surface modeling for optimization heterocatalytic Fenton oxidation of persistence organic pollution in high total dissolved solid containing wastewater. Environ. Sci. Pollut. Res. 21 (2), 1489–1502.

Shih, Y.J., Ho, C.H., Huang, Y.H., 2013. Photo-Fenton oxidation of azo dye reactive black B using an immobilized iron oxide as heterogeneous catalyst. Water Environ. Res. 85 (4), 340–345.

Siddique, M., Farooq, R., Price, G.J., 2014. Synergistic effects of combining ultrasound with the Fenton process in the degradation of reactive blue. Ultrasonics Sonochem. 21 (3), 1206–1212.

Silva, C.A., Madeira, L.M., Boaventura, R.A., Costa, C.A., 2004. Photo-oxidation of cork manufacturing wastewater. Chemosphere 55 (1), 19–26.

Sirtori, C., Zapata, A., Oller, I., Gernjak, W., Agüera, A., Malato, S., 2009. Decontamination industrial pharmaceutical wastewater by combining solar photo-Fenton and biological treatment. Water Res. 43 (3), 661–668.

Soon, A.N., Hameed, B.H., 2013. Degradation of acid blue 29 in visible light radiation using iron modified mesoporous silica as heterogeneous Photo-Fenton catalyst. Appl. Catal. A Gen. 450, 96–105.

Spuhler, D., Rengifo-Herrera, J.A., Pulgarin, C., 2010. The effect of Fe^{2+}, Fe^{3+}, H_2O_2 and the photo-Fenton reagent at near neutral pH on the solar disinfection (SODIS) at low temperatures of water containing *Escherichia coli* K12. Appl. Catal. B Environ. 96 (1–2), 126–141.

Sun, J.-H., Sun, S.-P., Wang, G.-L., Qiao, L.-P., 2007. Degradation of azo dye amido black 10B in aqueous solution by Fenton oxidation process. Dyes Pigments 74 (3), 647–652.

Sun, J.-H., Sun, S.-P., Fan, M.-H., Guo, H.-Q., Lee, Y.-F., Sun, R.-X., 2008. Oxidative decomposition of p-nitroaniline in water by solar photo-Fenton advanced oxidation process. J. Hazard. Mater. 153 (1–2), 187–193.

Suresh, A., Minimol Pieus, T., Soloman, P.A., 2016. Treatment of landfill leachate by membrane bioreactor and electro Fenton process. Int. J. Eng. Sci. Res. Technol. 5 (8), 689–697.

Suslick, K.S., Doktycz, S.J., Flint, E.B., 1990. On the origin of sonoluminescence and sonochemistry. Ultrasonics 28, 280–290.

Tang, R., Liao, X.P., Liu, X., Shi, B., 2005. Collagen fiber immobilized Fe(III): A novel catalyst for photo-assisted degradation of dyes. Chem. Commun. 47, 5882–5884.

Tao, X., Su, J., Chen, J., Zhao, J., 2005. A novel route for waste water treatment: Photo-assisted Fenton degradation of dye pollutants accumulated in natural polyelectrolyte microshells. Chem. Commun. 36, 4607–4609.

Tavares, M.G., Santos, D.H., Torres, S.J., Pimentel, W.R., Tonholo, J., Zanta, C.L., 2016. Efficiency and toxicity: comparison between the Fenton and electrochemical processes. Water Sci. Technol. 74 (5), 1143–1154.

Tireli, A.A., Guimarães, Ido, R., Terra, J.C., da Silva, R.R., Guerreiro, M.C., 2015. Fenton-like processes and adsorption using iron oxide-pillared clay with magnetic properties for organic compound mitigation. Environ. Sci. Pollut. Res. Int. 22 (2), 870–881.

Tokumura, M., Ohta, A., Znad, H.T., Kawase, Y., 2006. UV light assisted decolorization of dark brown colored coffee effluent by photo-Fenton reaction. Water Res. 40 (20), 3775–3784.

Tony, M.A., Purcell, P.J., Zhao, Y.Q., 2012. Oil refinery wastewater treatment using physicochemical, Fenton and Photo-Fenton oxidation processes. J. Environ. Sci. Health Part A. 47 (3), 435–440.

Torrades, F., Pérez, M., Mansilla, H.D., Peral, J., 2003. Experimental design of Fenton and photo-Fenton reactions for the treatment of cellulose bleaching effluents. Chemosphere 53 (10), 1211–1220.

Torres, G.F., Méndez, J.A.O., Tinoco, D.L., Marin, E.D., Araña, J., Herrera-Melián, J.A., et al., 2016. Detoxification of synthetic and real groundwater contaminated with gasoline and diesel using Fenton, photo-Fenton, and biofilters. Desalin. Water Treat. 57 (50), 23760–23769.

Torres-Palma, R.A., Nieto, J.I., Combet, E., Pétrier, C., Pulgarin, C., 2010. An innovative ultrasound, Fe^{2+} and TiO_2 photoassisted process for bisphenol A mineralization. Water Res. 44 (7), 2245–2252.

Trapido, M., Tenno, T., Goi, A., Dulovaa, N., Kattela, E., Klausona, D., et al., 2017. Bio-recalcitrant pollutants removal from wastewater with combination of the Fenton treatment and biological oxidation. J. Water Proc. Eng. 16, 277−282.

Trovó, A.G., Hassan, A.K., Mika, S., Walter, Z.T., 2016. Degradation of acid blue 161 by Fenton and photo-Fenton processes. Int. J. Environ. Sci. Technol. 13, 147−158.

Üstün, G.E., Solmaz, S.K.A., Morsünbül, T., Azak, H.S., 2010. Advanced oxidation and mineralization of 3-indole butyric acid (IBA) by Fenton and Fenton-like processes. J. Hazard. Mater. 180 (1−3), 508−513.

Utset, B., Garcia, J., Casado, J., Domènech, X., Peral, J., 2000. Replacement of H2O2 by O2 in Fenton and photo-Fenton reactions. Chemosphere 41 (8), 1187−1192.

Vaishnave, P., Kumar, A., Ameta, R., Punjabi, P.B., Ameta, S.C., 2014. Photo oxidative degradation of azure-B by sono-photo-Fenton and photo-Fenton reagents. Arab. J. Chem. 7 (6), 981−985.

Verma, A., Hura, A.K., Dixit, D., 2015. Sequential photo-Fenton and sono-photo-Fenton degradation studies of reactive black 5 (RB5). Desalin. Water Treat. 56 (3), 677−683.

Vilar, V.J., Maldonado, M.I., Oller, I., Malato, S., Boaventura, R.A., 2009. Solar treatment of cork boiling and bleaching wastewaters in a pilot plant. Water Res. 43 (16), 4050−4062.

Vilar, V.J.P., Moreira, F.C., Ferreira, A.C.C., Sousa, M.A., Goncalves, C., Alpendurada, M.F., et al., 2012. Biodegradability enhancement of a pesticide-containing bio-treated wastewater using a solar photo-Fenton treatment step followed by a biological oxidation process. Water Res. 46 (15), 4599−4613.

Villegas-Guzman, P., Giannakis, S., Torres-Palma, R.A., Pulgarin, C., 2017. Remarkable enhancement of bacterial inactivation in wastewater through promotion of solar photo-Fenton at near-neutral pH by natural organic acids. Appl. Catal. B Environ. 205, 219−227.

Wang, Y., Ma, W., Chen, C., Hu, X., Zhao, J., Yu, J.C., 2007. Fe^{3+}/Fe^{2+} cycling promoted by Ta_3N_5 under visible irradiation in Fenton degradation of organic pollutants. Appl. Catal. B Environ. 75 (3−4), 256−263.

Wang, X.-J., Song, Y., Mai, J.-S., 2008. Combined Fenton oxidation and aerobic biological processes for treating a surfactant wastewater containing abundant sulfate. J. Hazard. Mater. 160 (2), 344−348.

Wang, Q., Tian, S., Long, J., Ning, P., 2014. Use of Fe(II)Fe(III)-LDHs prepared by co-precipitation method in a heterogeneous-Fenton process for degradation of methylene blue. Catal. Today 224, 41−48.

Wang, Y., Lin, X., Shao, Z., Shan, D., Irini, A., 2017. Comparison of Fenton, UV-Fenton and nano-Fe_3O_4 catalyzed UV-Fenton in degradation of phloroglucinol under neutral and alkaline conditions: role of complexation of Fe^{3+} with hydroxyl group in phloroglucinol. Chem. Eng. J. 313, 938−945.

Wilde, M.L., Schneider, M., Kummerer, K., 2017. Fenton process on single and mixture components of phenothiazine pharmaceuticals: assessment of intermediaries, fate, and preliminary ecotoxicity. Sci. Total Environ. 583, 36−52.

Wu, K., Xie, Y., Zhao, J., Hidaka, H., 1999. Photo-Fenton degradation of a dye under visible light irradiation. J. Mol. Catal. A: Chem. 144 (1), 77−84.

Xiao, D., Guo, Y., Lou, X., Fang, C., Wang, Z., Liu, J., 2014. Distinct effects of oxalate versus malonate on the iron redox chemistry: Implications for the photo-Fenton reaction. Chemosphere. 103, 354−358.

Xie, Y., Chen, F., He, J., Zhao, J., Wang, H., 2000. Photoassisted degradation of dyes in the presence of Fe^{3+} and H_2O_2 under visible irradiation. J. Photochem. Photobiol. A: Chem. 136 (3), 235−240.

Xiong, Z.-Y., Qin, Y.H., Ma, J.-Y., Yang, L., Wu, Z.-K., Wang, T.-L., et al., 2017. Pretreatment of rice straw by ultrasound-assisted Fenton process. Bioresource Technol. 227, 408−411.

Xu, Z., Zhang, M., Wu, J., Liang, J., Zhou, L., Lu, B., 2013. Visible light-degradation of azo dye methyl orange using $TiO_2/$ -FeOOH as a heterogeneousphoto-Fenton-like catalyst. Water Sci. Technol. 68 (10), 2178−2185.

Yang, B., Tian, Z., Zhang, L., Guo, Y., Yan, S., 2015a. Enhanced heterogeneous Fenton degradation of methylene blue by nanoscale zero valent iron (nZVI) assembled on magnetic Fe_3O_4/reduced graphene oxide. J. Water Proc. Eng. 5, 101−111.

Yang, X., Chen, W., Huang, J., Zhou, Y., Zhu, Y., Li, C., 2015b. Rapid degradation of methylene blue in a novel heterogeneous Fe_3O_4@rGO@TiO_2 catalyzed photo-Fenton system. Sci. Rep. 5. Available from: http://dx.doi.org/10.1038/srep10632.

Yao, Y., Qin, J., Cai, Y., Wei, F., Lu, F., Wang, S., 2014. Facile synthesis of magnetic $ZnFe_2O_4$-reduced graphene oxide hybrid and its photo-Fenton-like behavior under visible irradiation. Environ. Sci. Pollut. Res. Int. 21 (12), 7296−7306.

Yong, X.-Y., Gu, D.-Y., Wu, Y.-D., Yan, Z.-Y., Zhou, J., Wu, X.-Y., et al., 2017. Bio-electron- Fenton (BEF) process driven by microbial fuel cells for triphenyltin chloride (TPTC) degradation. J. Hazard. Mater. 324 B, 178–183.

Zazouli, M.A., Yousefi, Z., Eslami, A., Ardebilian, M.B., 2012. Municipal solid waste landfill leachate treatment by Fenton, photo-Fenton and Fenton-like processes: Effect of some variables. Iran. J. Environ. Health Sci. Eng. 9 (1). Available from: http://dx.doi.org/10.1186/1735-2746-9-3.

Zhang, T., Nan, Z.-R., 2014. Decolorization of methylene blue and Congo red by attapulgite-based heterogeneous Fenton catalyst. Desalin.Water Treat. 57 (10), 4633–4640.

Zhang, T., Zhu, M.J., 2016. Enhancing enzymolysis and fermentation efficiency of sugarcane bagasse by synergistic pretreatment of Fenton reaction and sodium hydroxide extraction. Bioresour. Technol. 214, 769–777.

Zhang, G., Gao, Y., Zhang, Y., Guo, Y., 2010. Fe_2O_3-pillared rectorite as an efficient and stable Fenton-like heterogeneous catalyst for photodegradation of organic contaminants. Environ. Sci. Technol. 44 (16), 6384–6389.

Zhang, Y., Gu, Y., Yang, H., He, Y., Li, R.P., Huang, Y.P., et al., 2012. Degradation of organic pollutants by photo-Fenton-like system with hematite. Huan Jing Ke Xue. 33 (4), 1247–1251.

Zhang, Y., Xiong, Y., Tang, Y., Wang, Y., 2013. Degradation of organic pollutants by an integrated photo-Fenton-like catalysis/immersed membrane separation system. J. Hazard. Mater. 244–245, 758–764.

Zhang, J., Zhang, X., Wang, Y., 2016. Degradation of phenol by a heterogeneous photo-Fenton process using Fe/Cu/Al catalysts. RSC Adv. 6, 13168–13176.

Zhang, Y., Klamerth, N., Chelme-Ayala, P., El-Din, M.G., 2017. Comparison of classical Fenton, nitrilotriacetic acid (NTA)-Fenton, UV-Fenton, UV photolysis of Fe-NTA, UV-NTA-Fenton, and UV-H_2O_2 for the degradation of cyclohexanoic acid. Chemosphere 175, 178–185.

Zhao, X., Zhu, L., Zhang, Y., Yan, J., Lu, X., Huang, Y., et al., 2012. Removing organic contaminants with bifunctional iron modified rectorite as efficient adsorbent and visible lightphoto-Fenton catalyst. J. Hazard. Mater. 215–216, 57–64.

Zhao, K., Zeng, Q., Bai, J., Li, J., Xia, L., Chen, S., et al., 2017. Enhanced organic pollutants degradation and electricity production simultaneously via strengthening the radicals reaction in a novel Fenton-photocatalytic fuel cell system. Water Res. 108, 293–300.

Zheng, H.L., Xiang, X.Y., 2004. Photo-Fenton oxidation processes used in the degradation of rhodamine B. Guang pu 24 (6), 726–729.

Zheng, H., Pan, Y., Xiang, X., 2007. Oxidation of acidic dye eosin Y by the solar photo-Fenton processes. J. Hazard. Mater. 141 (3), 457–464.

Zheng, H.L., Zhang, Z.M., Tang, M.F., Yi, Q., Chen, C.Y., Peng, Z.C., 2006. The oxidation degradation of great green SF by Fenton reagent. Guang pu 26 (4), 768–771.

Zhong, Y.H., Liang, X.L., Zhong, Y., Zhu, J., Zhu, S., Yuan, P., et al., 2012. Heterogeneous UV/Fenton degradation of TBBPA catalyzed by titanomagnetite: Catalyst characterization, performance and degradation products. Water Res. 46 (15), 4633–4644.

Zhou, Q., Liu, Y., Yu, G., He, F., Chen, K., Xiao, D., et al., 2017. Degradation kinetics of sodium alginate via sono-Fenton, photo-Fenton and sono-photo-Fenton methods in the presence of TiO_2 nanoparticles. Polym. Degrad. Stab. 135, 111–120.

CHAPTER

4

Ferrioxalate-Mediated Processes

Ruben Vasquez-Medrano[1], Dorian Prato-Garcia[2] and Michel Vedrenne[3]

[1]Iberoamericana University, Mexico City, Mexico [2]National University of Colombia, Palmira, Colombia [3]Technical University of Madrid (UPM), Madrid, Spain

4.1 INTRODUCTION

The industrial and urban wastewaters contain a great variety of pollutants such as nutrients, heavy metals, salts, pathogens, soluble, and particulate organic matter (Howe et al., 2012; Tchobanoglous et al., 2003). Over the years, different phenomena such as economic development, changes in land use, improvement of life standards, changes in agricultural practices, population growth, and urbanization have significantly changed the characteristics of wastewaters (Grandclément et al., 2017).

The conventional treatment of wastewaters focuses on the abatement of organic matter, nutrients, and other compounds through the combination of mechanical, biological, physical, and chemical operations. The use of a biological treatment is essential for the elimination of solubilized organic matter, producing decontamination through the metabolic action of microorganisms (Malato et al., 2002). While this chain of operations is appropriate for most wastewaters of domestic origin, treating industrial effluents is usually more complicated due to their high load of organic matter, salinity, non-neutral pHs, color, or the presence of synthetic chemical compounds with high persistence and low biodegradability. These compounds can frequently bring about more stable and toxic species through the biotic and abiotic transformations that normally take place in wastewater treatment plants or pass unchanged through the process and exert their toxicity on the aquatic biocoenosis (Binelli et al., 2015). As a consequence, specific technological options need to be implemented aimed at eliminating chemical compounds that are difficult to remove by the application of conventional treatments, or decreasing the overall pollution levels of the effluent.

Essentially, the recalcitrant character of an effluent can be determined by the presence of one or several compounds whose properties are directly related to the processes that produced such effluent as waste. Often, these compounds exhibit xenophore substituents (e.g., halogens), highly condensed aromatic hydrocarbons, unusual bonds or bond sequences, or an excessive molecular size (Ibáñez et al., 2007). Examples of such compounds can be pesticides, fertilizers, pigments and dyes, chemical precursors, etc. In numerous cases, treatability problems are caused by the presence of micropollutants, which include substances such as pharmaceuticals, personal care products, steroid hormones, and other emerging compounds (Luo et al., 2014; Martínez-Alcalá et al., 2017).

A great variety of recalcitrant compounds and their concurrence with other pollutants calls for tailoring specific treatment processes to the particularities of the effluents. Thus, a significant number of treatment options have been reported to date as effective in the degradation of pollutants through physicochemical pathways. While the application of a specific technique will be a function of the effluent's composition as well as other variables (e.g., operation conditions, costs, space, environmental legislation), the best-known options are based on the chemical oxidation of the recalcitrant compounds.

Advanced oxidation processes (AOPs) are a group of techniques that rely on the use of radicals to destroy complex organic compounds into simpler end products. The logic behind these processes is to reduce the recalcitrant character of an effluent to a point that it can be passed on to conventional wastewater treatment processes. The use of AOPs has been widely supported over the past two decades, mainly because of the non-selective character of the degradation processes of organic and inorganic species, as well as microorganisms (Giannakis et al., 2016a,b). Most of these processes rely on the formation of hydroxyl radicals ($^{\bullet}OH$), which are considered the most powerful oxidizing agents in aqueous phase (E : 2.8 V/SHE, Standard Hydrogen Electrode). Moreover, hydroxyl radicals exhibit higher reaction rates than other conventional oxidizing agents such as $KMnO_4$, Cl_2, O_2, O_3, or hydrogen peroxide (H_2O_2) (Bautista et al., 2008).

There are several technical alternatives available for the generation of hydroxyl radicals, such as electrochemical processes (Zhou et al., 2005), sonochemical processes (Mantzavinos et al., 2004), through the exposure to artificial or natural radiation (Bremner et al., 2006; Esplugas et al., 2002), through homogeneous and heterogeneous catalysis (Esplugas et al., 2005; Lelario et al., 2016; Martínez et al., 2005), and in acid or basic conditions (Kavitha and Palanivelu, 2004; Neyens and Baeyens, 2003). Within the overall chain of treatments in a wastewater treatment facility, the use of AOPs can be found upstream and downstream a conventional biological process, depending on whether the biodegradability of an effluent needs to be conditioned or to clean it before discharge to any receiving water bodies (Blanco and Malato, 2003).

4.2 THE FENTON AND PHOTO-FENTON REACTIONS

The Fenton reaction was described for the first time by Fenton, who observed the oxidation of tartaric acid by hydrogen peroxide in the presence of ferrous ions (Fenton, 1894). However, it was not until the decade of 1960, when the Fenton reaction was first applied to the destruction of hazardous organic compounds (Barbuśinski, 2009; Neyens and Baeyens, 2003). The combination of iron (Fe^{2+}) and hydrogen peroxide (H_2O_2) generates hydroxyl radicals ($^{\bullet}OH$, HO_2) through Eqs. 4.1–4.2 (Pignatello et al., 2006).

$$Fe^{2+} + H_2O_2 \rightarrow Fe^{3+} + OH + OH^- k = 40 - 80 L/mol per s \tag{4.1}$$

$$Fe^{3+} + H_2O_2 \rightarrow Fe^{2+} + HO_2 + H^+ k = 9 \times 10^{-7} L/mol per s \tag{4.2}$$

Since the reaction of H_2O_2 with Fe^{3+} is significantly slower as compared to Fe^{2+}, the catalyst regeneration phase is considered the most critical stage of the process as it reduces the radical production rate and could end up depleting the iron available for the reaction (Eqs. 4.3–4.4) (Malato et al., 2009).

$$Fe^{3+} + HO_2 \rightarrow Fe^{2+} + O_2 + H^+ k = 0.3 - 2 \times 10^6 L/mol per s \tag{4.3}$$

$$Fe^{3+} \leftrightarrow FeOH^{2+} \leftrightarrow Fe OH_2^{+} \leftrightarrow Fe OH_2^{4+} \leftrightarrow Fe_2O_3nH_2O \tag{4.4}$$

The Fenton reaction requires a specific set of conditions to achieve maximal reduction of organic matter. In this respect, the efficiency of the process is determined by factors such as temperature, pH, as well as hydrogen peroxide, and iron concentrations. In the case of pH, the chemistry of iron requires that the Fenton reaction must be carried out in acidic conditions (pH = 3–4) to avoid its precipitation as hydroxide. The reaction presents its maximum catalytic activity at pH 2.8–3.0, after which its efficiency decreases due to the precipitation of iron as $Fe(OH)_3$ and the degradation of H_2O_2 into O_2 and H_2O, respectively (Barbuśinski, 2009; Pignatello et al., 2006). Regarding the temperature of the reaction, higher temperatures result in increased rates; however, it may also result in an increased decomposition of H_2O_2 into O_2 and H_2O. H_2O_2, being a costly reagent to generate or supply, its thermal decomposition generally results in reaction temperatures being close to room temperature.

The dosage of reagents (H_2O_2 and Fe^{2+}) is perhaps the most important consideration in the design of the treatment process, as it will not only condition the overall efficiency of the process but its cost also. The concentration of H_2O_2 that must be supplied for the adequate degradation of pollutants is a function of the chemical oxygen demand (COD) of the effluent (Pignatello et al., 2006). Similarly, increasing the concentration of Fe^{2+} leads to higher reaction rates until it reaches a certain concentration above which all rate increases appear to be marginal (Rivas et al., 2001). Furthermore, an enormous increase in the concentration of Fe^{2+} may contribute to higher concentrations of total dissolved solids, which may be problematic for further treatment steps or discharge (Babuponnusami and Muthukumar, 2014). It is generally recommended that an optimal dose of reagents is tailored to meet the desired removal of pollutants for a specific effluent using experimental trials and mathematical optimization techniques (Bautista et al., 2008; Szpyrkowicz et al., 2001; Vedrenne et al., 2012a,b).

Under normal conditions (the "dark Fenton"), an accumulation of Fe^{3+} and its complexation with carboxylate species from the degradation of organic matter generally reduces the degradation rates of the Fenton reaction (Pignatello et al., 2006). However, its rate can be substantially increased in the presence of radiation (UV, visible) due to the photochemical properties of the iron (Fe^{3+}) complexes that undergo a ligand-to-metal charge transfer excitation, and the formation of complexes with diverse carboxylates and polycarboxylates that exhibit photoactive properties as in Eqs. 4.5–4.6 (Malato et al., 2009; Pignatello et al., 2006):

$$Fe OH^{2+} + h\upsilon \rightarrow Fe^{2+} + OH \tag{4.5}$$

$$Fe OOCR^{2+} + h\upsilon \rightarrow Fe^{2+} + CO_2 + R \tag{4.6}$$

The Fenton reaction has several advantages compared to other AOPs. First of all, iron and hydrogen peroxide (H_2O_2) are abundant and safe from an environmental point of view, if optimally dosed. Furthermore, the process follows relatively simple operating principles and the absence of mass transfer limitations (Mirzaei et al., 2017). In the case of the photo-Fenton reaction, the exposure to sunlight is one of the main advantages as it avoids the use of UV lamps, with associated energy consumptions and potential hazards. It also drastically reduces iron sludge waste and enhances the decomposition of H_2O_2 (Bokare and Choi, 2014).

Even though the Fenton and the photo-Fenton reactions have proved to be effective in the removal of complex organic matter, the treatment exhibits a number of drawbacks, which have limited its upscaling beyond pilot plant (Garcia-Segura et al., 2016; Spasiano et al., 2015). The acidic conditions that need to be kept to avoid the precipitation of iron and its further reconditioning before other treatments downstream require the use of acidifying and alkalinizing agents, which necessarily increase the cost of the treatment. Additional difficulties can arise due to corrosion of steel equipment. In consequence, substantial research has been carried out to allow achieving similar efficiency levels in neutral or circumneutral conditions (Carra et al., 2014).

In this context, the use of organic acids has been described as an effective way for avoiding acid pHs, since these form complexes with iron, which are more soluble as acidity decreases. Examples of organic ligands used to enhance the Fenton reaction are oxalic acid, citric acid, and oxalic acid, which allow for forming strong, stable, and soluble complexes with iron, avoiding the sequestration of the catalyst by other organic and inorganic compounds in the effluent. Moreover, these complexes have a much higher quantum yield than ferric iron-water complexes and can use a higher fraction of the solar radiation spectrum (580 nm) (Manenti et al., 2015; Prato-García et al., 2009; Prato-García and Buitrón, 2011).

Here, an alternative to the traditional Fenton/photo-Fenton process for the oxidation of organic matter using ferrioxalate complexes enters the scene.

4.3 THE FERRIOXALATE-MEDIATED FENTON REACTION

4.3.1 Influence of pH

The pH of a residual effluent is a critical parameter that conditions the overall performance of the Fenton and photo-Fenton processes because it controls the catalyst's speciation and determines the occurrence of undesirable parallel reactions forming less-oxidant radicals (Prato-García and Buitrón, 2012; Pouran et al., 2015). At acidic conditions (pH $= 1.0-2.5$), it has been observed that $^{\bullet}OH$ radicals are consumed by protons (H^+) (Eq. 4.7).

$$OH + H^+ + e^- \rightarrow H_2O \qquad (4.7)$$

It is worth mentioning that the predominant form of the catalyst $[Fe(H_2O)_6]^{3+}$ reacts slowly with the oxidant reducing the efficiency of the process even more. When the reaction medium is highly acidic (pH < 1.0), the production rate of the $^{\bullet}OH$ radicals is limited due to the electrophilic character of $H_3O_2^+$ ions formed from H_2O_2 (Mirzaei et al., 2017).

At an alkaline pH (7.0−14.0), the hydrolysis-precipitation processes, and the presence of parallel reactions severely limit the availability of the catalyst for the generation of $^{\bullet}OH$ radicals. In addition, the self-decomposition of H_2O_2 further decreases the efficiency of the process (Eq. 4.8). Proof of this influence is the fact that an increase in pH from 0 to 14 reduces the oxidation potential of $^{\bullet}OH$ radicals from 2.59 V to 1.64 V.

$$2H_2O_2 \rightarrow 2H_2O + O_2 \tag{4.8}$$

Iron is highly soluble in an aqueous phase in the pH range between 3.0 and 4.0, as it is being presented as species such as Fe^{2+} or $[Fe\,H_2O\,_6^{2+}]$. In the absence of complexing agents and in acidic conditions, $[Fe\,H_2O\,_6^{2+}]$ hydrolyzes into $Fe(H_2O)_5(OH)^+$ and $Fe(H_2O)_4(OH)_3$ (or $FeOH^+$ and $Fe(OH)_2$, respectively). This last species is crucial in Fenton processes because it is significantly more reactive than Fe^{2+}. Although an acidic pH facilitates the generation of $^{\bullet}OH$ radicals, the observed reaction rate constants for the regeneration of the catalyst through the consumption of oxidants (Eq. 4.2) or through the photocatalytic pathway (Eq. 4.6) are relatively low. This provokes the accumulation of Fe^{3+} in the reaction medium as well as reduction in the degradation rate of pollutants. It is important to remember that in the absence of reagents, the photoreduction of the catalyst is responsible for the production of aqueous $^{\bullet}OH$ radicals. The quantum yield for the production of Fe^{2+} through Eq. 4.2 depends on the wavelength of the incident radiation; in general, such a yield shows a decrease inversely proportion to the wavelength of radiation, e.g., the quantum yield of Eq. 4.6 is reported to be within the range 0.13−0.19 at 313 nm and 0.017 at 360 nm. This in turn translates directly in an important reduction of the catalyst regeneration speed. It should be highlighted that in many cases, such a reaction does not play an important role unless Fe^{2+} can be quickly oxidized by any other species present in the medium (O_2, $HO_2^{\bullet}$, H_2O_2) (Pignatello et al., 2006).

4.3.2 Iron Complexes with Organic and Inorganic Substances

Perhaps one of the most notable characteristics of Fe^{3+} is its ability to form complexes with organic and inorganic substances. In the absence of solar radiation, Fe^{3+} and Fe^{2+} form stable complexes with chloride and sulfate ions (Eqs. 4.9−4.13), which hinder catalyst regeneration (Devi et al., 2013; Grebel et al., 2010; Machulek et al., 2007; Micó et al., 2013).

$$Fe^{2+} + Cl^- \leftrightarrow Fe\,Cl\,^+ \tag{4.9}$$

$$Fe^{3+} + Cl^- \leftrightarrow Fe\,Cl\,^{2+} \tag{4.10}$$

$$Fe^{2+} + SO_4^{2-} \rightarrow FeSO_4 \tag{4.11}$$

$$Fe^{3+} + SO_4^{2-} \rightarrow Fe\,SO_4\,^+ \tag{4.12}$$

$$Fe^{3+} + 2SO_4^{2-} \rightarrow Fe\,SO_4\,^{2-} \tag{4.13}$$

When the process occurs at an acidic pH (<3.5) in the presence of solar radiation, an adverse effect is caused by the presence of chloride and sulfate ions, which photoreduces Fe^{3+} and generates additional oxidant radicals such as $Cl^{\bullet}$ ($E = 2.41$ V vs NHE) (Eqs. 4.14−4.15) (De Laat et al., 2004).

$$FeSO_4^+ + h\upsilon \rightarrow Fe^{2+} + SO_4^- \tag{4.14}$$

$$FeCl^+ + h\upsilon \rightarrow Fe^{2+} + CO_2 + Cl \tag{4.15}$$

On the other hand, the carboxylate-Fe^{3+} and polycarboxylate-Fe^{3+} present greater molar absorption coefficients and allowed the use of wavelengths of the visible spectrum for catalyst regeneration. The formation of this type of complexes is interesting as it makes operating at circumneutral conditions possible (pH = 5.5–7.5), increases $^\bullet OH$ radicals production rate, the regeneration of the iron catalyst, and the formation of additional radicals such as CO_2^- ($E = 1.90\,V$ vs NHE) (Eqs. 4.16–4.19).

$$Fe\,C_2O_4 + H_2O_2 \rightarrow Fe\,C_2O_4^+ + \,^\bullet OH + OH^- \tag{4.16}$$

$$Fe\,C_2O_{4\,3}^{\,3-} + h\upsilon \rightarrow Fe^{2+} + 2C_2O_4^{2-} + C_2O_4^- \tag{4.17}$$

$$Fe\,C_2O_{4\,3}^{\,3-} + C_2O_4^- \rightarrow Fe^{2+} + 3C_2O_4^{2-} + 2CO_2 \tag{4.18}$$

$$C_2O_4^- \rightarrow CO_2^- + CO_2 \tag{4.19}$$

In recent years, a wide range of organic and inorganic substances capable of forming complexes with Fe^{3+} have been reported. The efficiency of the chelating agent depends on its concentration, the type of pollutant present in the effluent, the pH of operation and the iron-to-chelating agent ratio. Any iron-chelating agent system should present characteristics such as low cost, high catalytic activity toward the pollutant of interest, environmentally-friendly, and high quantum yield. Additionally, it should be resistant to oxidation and capable of generating additional chemical species that increase the oxidative capacity of the reaction medium (Jeong and Yoon, 2004; Sun and Pignatello, 1992).

Some of the byproducts generated during the treatment of aromatic compounds with the Fenton reaction generate agents that catalyze the destruction of the organic matter due to their oxidation-reduction potential (Eq. 4.19). Additionally, substances such as quinone-hydroquinone undergo alternative pathways for the regeneration of the catalyst (Eqs. 4.20–4.21), which allows overcoming, albeit transitorily, the iron regeneration bottleneck (Eq. 4.2). Finally, the continuous hydroxylation of organic substances originates simpler compounds such as organic acids, which in turn form very stable iron complexes that impede catalyst regeneration. This condition and the reduction of the concentration of H_2O_2 explain the decrease in the oxidative capacity of the Fenton processes.

$$C_6H_4\,OH_2 + Fe^{3+} \rightarrow C_6H_4\,OH\,O + Fe^{2+} + H^+ \tag{4.20}$$

$$C_6H_4\,OH\,O + Fe^{3+} \rightarrow C_6H_4\,O_2 + Fe^{2+} + H^+ \tag{4.21}$$

4.3.3 Reaction Mechanisms

The iron complexes with carboxylates and polycarboxylates have increased molar absorption coefficients in the near UV (300–400 nm) and visible spectra (Xiao et al., 2014). The possibility of operating in conditions close to neutrality and a higher quantum yield make these species an interesting alternative for improving the performance of conventional Fenton processes. The specific quantum yield of the complexes will be defined by variables such as pH and the ratio of iron-to-oxalate (Abrahamson et al., 1994; Wang et al., 2014). When the iron-to-oxalate ratio is 1, $Fe(C_2O_4)^+$ is the dominant species, but at higher concentration, oxalate produces species such as $Fe C_2O_4{}_2^-$ and $Fe C_2O_4{}_3^{3-}$, which are more photoactive than $Fe(C_2O_4)^+$ and $FeOH^{2+}$ (Wang et al., 2014).

Although the advantages of using organic complexes of iron to intensify the Fenton reaction are evident, the reaction mechanisms and the intermediate products remain controversial. In general, the photolysis of complexes leads to the oxidation of ligands and the reduction of the catalyst to Fe^{2+}. The degradation products of the complexes are key as they generate additional reactive species (HO_2, O_2^-, OH, CO_2^-, $C_2O_4^-$) and originate the required oxidant. Pignatello et al. (2006) have identified that under high concentrations of Fe^{3+}, the oxalyl radical ($C_2O_4^-$) decomposes itself into carbon dioxide and carboxylate radicals (CO_2^-), which are transformed into superoxide radicals (O_2^-) in the presence of oxygen (Eq. 4.22). On the other hand, at high concentrations of Fe^{3+}, the oxalyl radical reduces the catalyst to Fe^{2+}.

$$O_2 + CO_2^- \rightarrow O_2^- + CO_2 \qquad (4.22)$$

The majority of scientists defend two models of the likely reaction pathways of ferrioxalate complexes (Pozdnyakov et al., 2008). In the first model, the complex becomes dissolved after its excitation without any electron transfer occurring between the ligand and the metal (Eq. 4.23). The second model considers a process of charge transfer between the ligand and the catalyst (Eq. 4.24). In each case, the destruction of the pollutant is favored due to the photoreduction of the catalyst and the formation of additional oxidizing and reducing agents.

$$Fe C_2O_4{}_3{}^{3-} + h\upsilon \rightarrow Fe C_2O_4{}_2{}^- + 2CO_2^- \qquad (4.23)$$

$$Fe C_2O_4{}_3{}^{3-} + h\upsilon \rightarrow Fe C_2O_4{}_2{}^{2-} + C_2O_4^- \qquad (4.24)$$

4.3.4 Optimization

The cost of the Fenton reagent (Fe^{2+}, H_2O_2) and the most common reactor designs are thought to pose the greatest limitations on the use of the process at an industrial scale. Additionally, the acidification and neutralization stages further increase the operating costs and have the associated risk of the formation of sludge due to the catalyst precipitation. Therefore, the majority of the processes are to be optimized in respect to the use of reagents and operating conditions close to neutrality. The operating factors examined for an optimal use of the process are the type of organic ligand, the iron-to-ligand, ratio the oxidation state of the catalyst (Fe^{3+} or Fe^{2+}), the dose of Fenton reagent, temperature, radiation source (solar radiation vs lamps), and accumulated energy in the reactor (Table 4.1).

Due to technical and economic considerations, the preferred optimization approach for the selection of the best operative conditions in Fenton and photo-Fenton processes has changed in the last two decades (Callao, 2014; Sakkas et al., 2010). The study of individual variables (e.g., pH, temperature, H_2O_2, Fe, etc.) is a relatively simple task from an

TABLE 4.1 Complexing Agents Applied in the Intensification of the Ferrioxalate-Mediated Process

Complexing Agent	Pollutant	Operative Conditions	Comments	Source
Oxalic acid	2,4-Dichlorophenoxyacetic acid (2,4-D)	Photo-Fenton (pH = 5.0, Fe^{3+} = 3 mg/L, Fe^{3+}/oxalate = 10; H_2O_2 = 10 mg/L, T = 50 C)	The presence of radiation reduces the reaction time from 180 to 120 min. The removal efficiency of TOC increases from 20% to 40%	Conte et al. (2016)
Oxalic acid	Trimethoprim (TMP) and sulfamethoxazole (SM)	Photo-Fenton (pH = 5.0, Fe^{2+} = 5 mg/L, oxalate = 15 mg/L; H_2O_2 = 89 mg/L (TMP), and 102 mg/L (SM)	The formation of iron complex byproducts limits the photoreduction of the catalyst. The introduction of ferric oxalates improves the quantic yield of the process and the degradation rate at pH = 5	Dias et al. (2014)
Oxalic acid	Synthetic textile wastewater	Photo-Fenton (pH = 4.0, Fe^{3+} = 40 mg/L, Fe^{3+}/oxalate = 3; H_2O_2 = 50−100 mg/L)	At acidic conditions (2.8−3.0), TOC removal is associated with the precipitation of iron-pollutant complexes. The use of ligands reduces the precipitation of organic matter and achieves a decolorization of 98% while improving the effluent's biodegradability	Doumic et al. (2015)
Oxalic acid	Protocatechuic acid (PA)	Photo-Fenton (pH = 4−5, Fe^{2+} = 2 mg/L, oxalate = 60 mg/L; H_2O_2 = 100 mg/L)	The study includes a multivariable strategy for the selection of the best operative conditions (pH, temperature, radiation, air flow, and concentrations of iron, peroxide, and oxalic acid). The study also considers the use of natural and artificial radiation. An increase to pH 4 improves the efficiency of TOC removal (97%) and PA (100%)	Monteagudo et al. (2010)
Oxalic acid	Phenol	Photo-Fenton (pH = 5.5−6.5, oxalate/phenol = 1.5, oxalate/Fe^{3+} = 15, and H_2O_2/phenol > 5.0). Evaluated phenol concentrations = 100−800 mg/L	The use of oxalate allowed operating efficiently at a pH 5.5−6.5. Additionally, the capacity of iron-oxalate complexes for the generation of H_2O_2 in circumneutral conditions has been demonstrated. The efficiencies obtained at an acidic pH (2.8−3.0) allowed removing phenol (100%) and COD (> 85%) in 45−60 min	Prato-García et al. (2009)
Oxalic acid	2,4-Dichlorophenoxyacetic acid (2,4-D)	High concentration FeOx system (Fe^{3+} = 56 mg/L, oxalate = 432 mg/L, H_2O_2 = 340 mg/L, pH = 2.8−3.0). Low concentration FeOx system (Fe^{3+} = 0.56 mg/L, oxalate = 432 mg/L, H_2O_2 = 0 y 340 mg/L, pH = 2.8−3.0)	High concentrations of ferrioxalate allow the reaction to occur without further addition of oxidant, and that a removal of 60% of 2,4-D is achieved. Low concentrations of ferrioxalate favor the photoreduction of the catalyst and increase the removal of 2,4-D (60%−90%). Reaction times were between 20 and 50 min	Jeong and Yoon, 2004

Oxalic acid	Winery wastewaters	(H_2O_2 = 260 mg/L, Fe^{2+} = 0.0 mg/L, oxalic acid = 80 mg/L, T = 22−30 C irradiation = 448 W/h)	Multivariate study to identify the best operative conditions (concentration of H_2O_2, Fe^{2+}, and oxalic as well as temperature and radiation). After 360 min of treatment a 61% reduction of TOC was achieved. A native 6 mg/L of Fe^+ was present in the effluent	Monteagudo et al. (2012)
Oxalic acid	Indigo blue	Photo-Fenton (Indigo = 6.67−33.33 mg/L, oxalic/Fe^{2+} = 257, H_2O_2 = 35, and 1280 mg/L)	The use of oxalate accelerates the regeneration of the catalyst, the formation of additional species, and the operation at conditions close to neutrality (pH = 5−6). The treatment achieved a biodegradable effluent	Vedrenne et al. (2012a,b)
Oxalic and malonic acids	Rhodamine B, RHB	Photo-Fenton (Fe^{3+} = 0.1 mg/L, oxalate = 1 mg/L, malate = 1 mg/L, H_2O_2 = 0.0 mg/L, pH = 3.0). Photo-Fenton (Fe^{3+} = 0.1 mg/L, oxalate = 1 mg/L, malate = 1 mg/L, H_2O_2 = 10 mg/L, pH = 3)	In the absence of H_2O_2, the Fe^{3+}/UV system allows a decoloring of 55% while the Fe^{3+}/UV/ oxalate system allows decoloring the medium in 99%. The ferrioxalate complexes generate more powerful radicals ($O_2^{\bullet-}$, $HO_2^{\bullet}$, $^{\bullet}OH$) while the Fe-malate is less effective for the photoreduction of the catalyst and the generation of H_2O_2. In the presence of H_2O_2 the maximum decolorization was achieved with oxalate (99%), followed by malate (55%)	Xiao et al. (2014)
Oxalic, malic, tartaric, and citric acids	Bisphenol A (BPA)	Photo-Fenton (pH = 3−5, Iron oxide = 150 mg/L (Fe_3O_4 and Fe_2O_3), oxalate = 25 mg/L; H_2O_2 = 17 mg/L)	The lixiviation capacity and the formation of complexes decrease in the following order: oxalic > citric > tartaric > malic. The use of oxalic acid allows removing 100% of BPA in 120 for the studied catalysts.	Rodriguez et al. (2009)
Citrate	Atrazine	Photo-Fenton (pH = 3−5, Fe^{2+} = 3.4 mg/L, citrate = 126 mg/L; H_2O_2 = 0.0 mg/L)	The use of citrate allows extending the pH of operation from 3.5 to 5.4 without reducing the efficiency of the process. An increase in radiation (15−50 W/cm²) and the concentration of citrate (126−1260 mg/L) favor the reaction due to the generation of an oxidant	Ou et al. (2008)
EDDS and citrate	Pharmaceuticals (carbamazepine, flumequine, ibuprofen, ofloxacin, and sulfamethoxazole)	Photo-Fenton (pH = 6−7, Fe^{3+} = 5.6 mg/L, EDDS/Fe^{3+} = 2; citrate/Fe^{3+} = 5; Fe^{3+}/ EDDS = 3; H_2O_2 = 51 mg/L)	A removal of 90% (citrate) and 96% (EDDS) of the present substances was achieved. The use of complexes guaranteed the operation at neutral pH and eliminated the need for any further conditioning	Miralles-Cuevas et al. (2014)

(Continued)

TABLE 4.1 (Continued)

Complexing Agent	Pollutant	Operative Conditions	Comments	Source
Oxalic acid, citric acid, and EDDS-ethylenediamine-N,N'-disuccinic acid)	Acrylic-textile dyeing wastewater	Photo-Fenton (pH = 4.0, Fe^{3+} = 40 mg/L, Fe^{3+}/oxalate = 3; Fe^{3+}/citrate = 1; Fe^{3+}/EDDS = 3 H_2O_2 = 100−200 mg/L)	At an acidic pH, a very low mineralization is observed along with a marginal consumption of H_2O_2. The precipitation of iron and the formation of complexes generate a slow reaction induction phase. The most effective ligand was oxalate, followed by citrate and EDDS	Soares et al. (2015)
Oxalic acid, citric acid, and EDDS-ethylenediamine-N,N'-disuccinic acid)	Real textile wastewater	Photo-Fenton (pH = 2.8, Fe^{3+} = 100 mg/L, Fe^{3+}/oxalate = 3; Fe^{3+}/citrate = 1; Fe^{3+}/EDDS = 3; H_2O_2 = 1088 mg/L, T = 30 C)	All complexing agents contributed to an improvement in the decoloring efficiency (100%) and mineralization (65%). In the absence of ligands and at an acidic pH, the precipitation of the catalyst is observed. TOC removal is not related to the mineralization of the effluent. The most effective ligand was oxalate, followed by citrate and EDDS	Manenti et al. (2015)
Formate, citrate, maleate, oxalate, and ethylenediaminetetra-acetic acid (EDTA)	2,4-Dichlorophenoxyacetic acid (2,4-D)	Fenton (Fe^{2+} = 0.1 mM, ligand = 1.2 mM (formate, citrate, maleate, oxalate, EDTA, H_2O_2 = 1 mM, pH = 2.8). Photo-Fenton (Fe^{2+} = 1 mM, ligand = 1.2 mM, H_2O_2 = 1 mM, pH = 2.8)	In the absence of radiation and ligand, the process removed 60% of 2,4-D. The use of ligands in Fenton processes reduces the efficiency due to a rupture of the catalytic cycle of iron. In the presence of UV radiation, EDTA (40%) and maleate (60%) presented the least removal efficiencies. The greatest efficiencies were produced by acetate (95%), citrate (90%), and formate (90%)	Kwan and Chu, 2007
Aminopolycarboxylates, polycarboxylates, hydroxamates, N-heterocyclic carboxylates, polyhydroxy aromatics, porphyrins, sulfur compounds, phosphates	2,4-dichlorophenoxyacetic acid (2,4-D)	Fenton (2,4-D = 22 mg/L, H_2O_2 = 340 mg/L, pH = 6)	The study evaluated the capacity of 50 organic ligands for the elimination of the pesticide 2,4-D at pH = 6. At least 20 compounds solubilized iron and mineralized 2,4-D in variable times (20−300 min). The most active ligands achieved mineralizations of 80%−92%	Sun and Pignatello, 1992
Resorcinol	Metoprolol (MeTo)	Fenton (pH = 2.8, Fe^{2+} = 10 mg/L, H_2O_2 = 150 mg/L) removal of 67% of MeTo and 8% TOC. Fenton (pH = 6.2, Fe^{2+} = 10 mg/L, H_2O_2 = 150 mg/L) removal of 100% of MeTo and 17% TOC. Photo-Fenton (pH = 6.2, Fe^{2+} = 10 mg/L, H_2O_2 = 150 mg/L) removal of 100% of MeTo and 76% TOC	The addition of resorcinol allows operating at pH 6.2 and obtaining higher efficiencies at acidic conditions.	Romero et al. (2016)

operative and economic point of view; however, it overlooks any synergistic or antagonistic effects that may occur during the treatment process. The use of this type of strategies usually results in selecting extreme operative conditions (an excess of oxidant, catalyst, ligand, and accumulated energy) that favor undesirable parallel reactions (Eqs. 4.25–4.27) and reduce the technical feasibility and economic attractiveness of the process (Prato-García and Buitrón, 2012).

$$OH + H_2O_2 \rightarrow HO_2 + H_2O \tag{4.25}$$

$$OH + H_2O_2 \rightarrow O_2^- + H_2O + H^+ \tag{4.26}$$

$$Fe^{2+} + OH \rightarrow Fe^{3+} + OH^- \tag{4.27}$$

In general, an increase in the concentration of organic ligands has a positive effect on the efficiency of the process since it facilitates the regeneration of the catalyst. In some other cases, an increase in the concentration of ligand does not increase the degradation rate; this may be related with the specific characteristics of the substance that is being subjected to treatment and the applied experimental strategy (Salem et al., 2009). An excess of ligand increases the absorbance of the solution, which in turn reduces the number of photons that are available in the reaction medium for catalyst regeneration. Similarly, the excess of ligand delays the decomposition of the transient species generated after the photolysis of the iron-ligand compounds. This explains the negative effect in treatment efficiency that has been reported in the decoloring of azo dyes (Dong et al., 2008; Selvam et al., 2005). The use of elevated doses of ligand, catalyst, and oxidant have been associated with high organic loads or to the treatment of pigmented effluents. This operative restriction could derive in a high consumption of reagents and radicals for the generation of species with a lower oxidizing power. Recently, the use of dosing strategies for reagents has been promoted as an alternative to reduce the occurrence of undesirable reactions, allowing a better consumption of reagents (Doumic et al., 2015; Prato-García and Buitrón, 2012).

The use of chemometric techniques allows exploring in a rigorous way the incidence of the operating variables in the efficiency of the process (Schenone et al., 2015). The first stage of the experimental strategy usually involves the study of variables through factorial, fractioned factorial, and Placket-Burman designs with the objective of reducing the number of experiments (Callao, 2014; Myers et al., 2009; Sakkas et al., 2010). When the amount of factors increases, the use of a fractioned factorial design allows a lower number of experimental executions and has as an additional advantage of the fact that the least relevant factors are not considered (e.g., interactions between three or more factors). The Plackett–Burman design is an attractive alternative, when the objective is to determine the variables of interest. This type of design is oriented to the study of the principal effects and does not consider the interactions of two factors, some of which are essential in AOPs such as the Fenton and photo-Fenton processes (Callao, 2014). A central composite design requires a low number of executions and it is adequate in the construction of second-order response surfaces. The number of experiments (N) can be determined from Eq. 4.28 (Myers et al., 2009).

$$N = 2^k + 2k + C_0 \tag{4.28}$$

where k is the number of factors, 2^k is the required executions for the maximum and minimum levels, $2k$ is the executions in the external points of the design, and C_0 is the number of executed central points. The Box–Behnken design consists of three levels for each variable, and the number of experiments can be estimated from Eq. (4.29); its main advantage is that it avoids the use of extreme operative conditions (Callao, 2014). This in turn, will reduce the problems associated with undesirable reactions.

$$N = 2k \quad k - 1 \ + C_0 \tag{4.29}$$

It should be mentioned that the use of response surfaces allows localizing the operative optimal and the nature of the stationary point (Sakkas et al., 2010). The desirability function (D) shown in Eq. (4.30), allows achieving a specific objective through a weight-adjustment process using the factors under study (e.g., oxidant concentration, catalyst, ligand, etc.). In Fenton and photo-Fenton processes, the objectives are variable: decoloring, mineralization, or removal of an active compound, while the most common restrictions were associated to the reduction in the consumption of reagents (Monteagudo et al., 2010).

$$D = \ \prod_{i=1} d_i \ ^{1/N} \tag{4.30}$$

This strategy has been particularly useful for determining a series of optimal conditions for the various applications of the Fenton reaction, achieving reductions in the consumption of reagents within the range between 50% and 70%. The experimental strategy evidences the capacity to detoxify effluents with a fraction of the reagents theoretically required.

4.4 APPLICATIONS

The Fenton and photo-Fenton reactions have been successfully applied as a treatment for a wide array of wastewaters of different origins. Bautista et al. (2008) and Pouran et al. (2015) presented a comprehensive review of these applications, which includes wastewaters from the production of pesticides (Barbuśinski and Filipek, 2001; Zapata et al., 2009), chemicals (Collivignarelli et al., 1997), petroleum extraction and refining (Gao et al., 2004), leather tanning (Vidal et al., 2004), production of cosmetics (Bautista et al., 2014), pharmaceuticals (Michael et al., 2012), hormonal residuals (Frontistis et al., 2011), plastic containers washing (Vilar et al., 2012), dyes (Arslan-Alaton et al., 2009; García-Montaño et al., 2008; Prato-García and Buitrón, 2012), landfill leachates (Silva et al., 2013; Vedrenne et al., 2012a,b), cork boiling (De Torres-Socías et al., 2013), pulp milling (Fernandes et al., 2014), olive mill effluents (Papaphilippou et al., 2013), beverage production (Expósito et al., 2016), and wine production (Ioannou and Fatta-Kassinos, 2013), etc. It has also been used to remove yeast toxic compounds present in lignocellulosic prehydrolyzates prior to the fermentation in the production of biofuels (Vedrenne et al., 2015) as well as in several disinfection applications (Giannakis et al., 2016a,b; Ndounla et al., 2014; Polo-López et al., 2012).

Although not as vast as in the case of the ordinary photo-Fenton, numerous studies have described the application of the photo-Fenton reaction mediated with ferrioxalate complexes to different types of wastewaters. The applications of the ferrioxalate-assisted photo-Fenton process for the treatment of wastewaters from the textile, pesticides, pharmaceutical, chemical, and petroleum industries, or the treatment and disinfection of effluents have been described briefly.

4.4.1 Textile Industry

Several studies are available dealing with the degradation of dyes and pigments from the textile industries with ferrioxalate-mediated photo-Fenton processes. Arslan et al. (2000) described the degradation of a mixture of various aminochlorotriazine reactive dyes (Procion Blue HERD, Procion crimson HEXL, Procion yellow HE4R, Procion navy HEXL, and Procion yellow HEXL), along with several conditioning, anti-creasing, and sequestering agents. Under full photo-Fenton conditions, the effluent becomes free of total organic carbon (TOC) after 20 min of treatment (Table 4.2). The study was carried out at an acidic pH so as to recognize the need of further spending resources to bring it to neutrality for release.

Monteagudo et al. (2009) investigated the efficiency of the ferrioxalate photo-Fenton system in the detoxification of a simulated effluent carrying Orange II (20 mg/L). The selected operative conditions of the treatment allowed a 100% decolorization, along with a 95% removal of TOC using continuous addition of H_2O_2. When H_2O_2 was added at the beginning of the treatment, TOC removal was 80%. Monteagudo et al. (2010) also assessed the photodegradation of reactive blue 4 solutions using the ferrioxalate systems with continuous addition of H_2O_2 and air injection (Table 4.2). Five levels of operative conditions were assessed. Under the optimal conditions selected in this work; ($[Fe^{2+}] = 4$ ppm, $[C_2O_4^{2-}] = 19$ ppm, pH $= 2$ and no air bubbling, initial dose of H_2O_2), a 61% removal of TOC was achieved. The efficiency of the system was increased to 82%, when the addition of H_2O_2 was made continuously.

The degradation of methyl orange with ferrioxalate complexes was examined by Fadzil et al. (2012) and Azami et al. (2012) using different operation conditions (Table 4.2). Fadzil et al. (2012) applied the process to simulated effluents with a concentration of 20 mg/L, achieving a 67% decolorization after 90 min of treatment, while Azami et al. (2012) used solutions with a concentration of 0.025 mM of dye and achieved between 92% and 100% reduction of color after 65 min of treatment under acidic conditions (Chakma et al., 2015).

Vedrenne et al. (2012a,b) focused on the treatment of Indigo, analyzing the degradation of three different concentrations of dye (6.66, 16.67, and 33.33 mg/L, see Table 4.2). They found that after 180 min of treatment, 73% of the TOC was removed from the effluent using the lowest hydrogen peroxide dose (257 mg/L). It was also observed that after 20 kJ/L irradiation, the biodegradability of the effluent notably increased as confirmed by the presence of short chain carboxylic acids as intermediate products and by the mineralization of organic nitrogen into nitrate. The toxicity of the initial and resulting effluent was assessed for brine shrimp (*Artemia salina*), showing an overall decrease.

Chakma et al. (2015) examined the degradation of methylene blue and acid red B with an initial concentration of 20 ppm (0.063 mM) using a number of AOPs, including ferrioxalate-mediated photo-Fenton reaction. They found that after 60 min of treatment, the solution of acid red B was decolorized in 66% while the solution of methylene blue was decolorized only 45% (Table 4.2).

TABLE 4.2 Operative Conditions of Different Dye Treatments with a Ferrioxalate-Mediated Fenton Process

Dye	[Dye]$_0$	[Fe^{3+}]$_0$/[Fe^{2+}]$_0$	[H$_2$O$_2$]$_0$	[C$_2$O$_4^{2-}$]$_0$	pH	Reactor	Source
Procion Blue HERD	6.38 mg/L						
Procion Crimson HEXL	40.6 mg/L						
Procion Yellow HE4R	15.0 mg/L	0.4 mM	50 mM	1.2 mM	2.8	UV Batch Photoreactor	Arslan et al. (2000)
Procion Navy HEXL	86.3 mg/L						
Procion Yellow HEXL	33.2 mg/L						
Orange II OR	20.0 mg/L	2.0 mg/L	75 mg/L	60.0 mg/L	3.0	Continuous Flow CPC	Fadzil et al. (2012)
Methyl Orange	20.0 mg/L	2.0 mg/L	5 mM	6.0 mg/L	6.0	UV Batch Photoreactor	Azami et al. (2012)
	0.025 mM	0.3 mM	11 mM	6.5 mM	3.5	UV Batch Photoreactor	Monteagudo et al. (2010)
Reactive Blue 4	20 mg/L	0.0–15.0 mg/L	50 mg/L	0.0–60.0 mg/L	2.0–6.0	Continuous Flow CPC	Vedrenne et al. (2012a,b)
Indigo Blue	6.66–33.33 mg/L	1.87 mg/L	257–1280 mg/L	65.67 mg/L	5.0–6.0	Batch Solar CPC	Chakma et al. (2015)
Acid Red B	0.040 mM	0.2 mM	5.0 mM	0.6 mM	3.0	UV Batch Photoreactor	
Methylene Blue	0.063 mM						
Synthetic cotton-textile dyeing effluent	346 mg/L (COD)	40.0 mg/L	50–300 mg/L	1:3 ratio Fe	4.0	Solar CPC Pilot Plant	Doumic et al. (2015)

Doumic et al. (2015) analyzed the effectiveness of the photo-Fenton treatment with ferrioxalates for the degradation of a synthetic cotton-textile dyeing wastewater, which consisted of a very complex mixture of chemicals. This mixture contained several dyes (Procion yellow HEXL, Procion deep red HEXL gran), anti-oil and anti-crease agents, enzymes, water correctors, bleaching agents, electrolytes, and detergents. Under the operative conditions, decolorization of approximately 100% was achieved after 4 kJ/L of irradiation as well as a reduction of 77.7% in the total COD of the effluent (Table 4.2).

The treatment of real textile wastewater was carried out by Manenti et al. (2015) using different organic ligands for the complexation of iron. This treatment was carried out in laboratory and pilot-plant scale reactors irradiated by artificial and natural solar radiation using a wastewater sample collected from an industrial textile company located in the north of Portugal. The textile effluent had an alkaline pH, a high content of organic matter (COD = 1239 mg/L), and a moderate biodegradability. They observed that the treatment using organic ligands (e.g., oxalates) enhanced the photo-Fenton reaction, avoiding the formation of iron-organic pollutant complexes and increasing the overall catalytic activity. The optimal conditions for the treatment of this effluent were $[Fe^{3+}] = 100$ mgL and pH = 2.8, which allowed achieving a 69% of mineralization after 8.8 kJ/L of radiation. The main contribution of this study was the fact that it used samples from real wastewater from a textile factory, and not simulated effluents as done by most of the workers.

4.4.2 Chemical Industry and Pesticides

The photo-Fenton reaction assisted with ferrioxalate complexes has been applied to treat effluents from the chemical industry contaminated with different compounds. Prato-García et al. (2009) focused on the treatment of phenolic wastewaters in two compound parabolic collectors (CPC) under batch and closed flow conditions. Phenol transformation efficiencies of 100% and total COD reduction percentages of 85% were achieved after 1 h of treatment. The reagents were dosed following mass ratios of oxalate/phenol of 1.5, oxalate/Fe^{3+} of 15, and H_2O_2/phenol greater than 5. The treatment was undertaken at pH conditions close to neutrality.

Key operative considerations in the degradation of para-chlorophenol in wastewaters was investigated in Kusic et al. (2011). They evaluated the interactions between process parameters applying experimental design techniques. The optimal conditions for the maximal degradation of para-chlorophenol are: pH = 4.4, concentration of $[Fe^{3+}] = 0.1$ mM, and $[C_2O_4^{2-}] = 3.4$ mM, when UV-C radiation was used. When the effluent is exposed to UV-A radiation, the optimal operative conditions are pH = 5.6, $[Fe^{3+}] = 1.38$ mM, and $[C_2O_4^{2-}] = 8.16$ mM.

Monteagudo et al. (2011) presented a study on the mineralization of a mixture of phenolic compounds (gallic, p-coumaric, and protocatechuic acids) in wastewaters using the photo-Fenton process with ferrioxalates using a pilot-plant scale CPC. An optimization study was performed to investigate the influence of pH, temperature, solar power, air flow, and dosage of H_2O_2, Fe^{2+}, and oxalic acid. The optimal conditions in these experiments, allowed a complete removal of phenolic compounds and 94% removal of TOC after 5 and 194 min of treatment, respectively. The variables having a higher degree of influence in the efficiency of the treatment are pH and the initial concentrations of reagents. The optimal conditions of this study were $[H_2O_2] = 400\ mg/L$, $[Fe^{2+}] = 20$, $[C_2O_4^{2-}] = 60$, and pH = 4.

A novel application of the photo-Fenton reaction assisted with ferrioxalates for the degradation of 4-nitrophenol was examined in Ayodele and Hameed (2013), using solid-phase catalysts (copper pillared bentonite). This treatment's main advantage is that it avoids pH adjustments. The optimal operation conditions found in this study were 20% excess of H_2O_2, 2.0 of solid-phase catalyst, and a temperature of 40 C. A similar study by Ayodele (2013) studied the effects of phosphoric acid treatment with the ferrioxalate-assisted photo-Fenton reaction using kaolinite as a catalyst support. These two studies stress on the benefits of the heterogeneous photo-Fenton in minimizing catalyst leaching.

Schenone et al. (2015) reported the application of the ferrioxalate photo-Fenton process to the degradation of the herbicide dicholophenoxyacetic acid (2,4-D) in natural conditions (pH = 5). The influence of temperature and concentration ratios was also investigated through the use of the response surface methodology. Under optimal conditions, the treatment achieved degradation levels of 91.4% and 95.9% for low and high radiation conditions. They echo the general understanding about photo-Fenton processes, whose effectiveness are conditioned by the dose of hydrogen peroxide.

Estrada-Arriaga et al. (2016) studied the treatment of a real oil refinery effluent with high concentrations of phenols through a combined ferrioxalate photo-Fenton and membrane ultrafiltration. The effluent treated in this work had a concentration of 200 mg/L of phenols, which was obtained directly from a petroleum refinery in Mexico. The highest removals of COD (84%) and phenol (100%) were obtained with concentrations of 200 mg/L of oxalate, 20 mg/L, 500 mg/L of H_2O_2, and a pH of 5 after 120 min of treatment. The ultrafiltration process reduced COD to 22 mg/L, achieving a total removal of 94%.

4.4.3 Pharmaceutical Industry

The ferrioxalate-mediated photo-Fenton reaction has also been applied in numerous studies in the treatment of effluents from the pharmaceutical industries. These effluents are characterized by the presence of compounds such as antibiotics, hormones, and endocrine disruptors, which have a direct detrimental effect on aquatic life.

The degradation of an effluent containing the antibiotic amoxicillin with ferrioxalate photo-Fenton as well as its toxicity was observed by Trovó et al. (2011) before and after treatment. The degradation of the antibiotic was carried out in the presence of a potassium ferrioxalate complex that proved to be more effective than the sole addition of $FeSO_4$. The study identified the presence of sixteen intermediate compounds that originate during the degradation of the antibiotic: Toxicity to *Daphnia magna* was observed throughout the treatment, which was associated to the presence of intermediates and oxalate. The toxicity of the effluent was reduced to 45% after 240 min of irradiation. At the end of the treatment, the presence of short chain (and biodegradable) carboxylic acids was widely observed. They investigated the efficiency of several treatments in the degradation of an effluent containing the analgesic paracetamol, in which the ferrioxalate-mediated system exhibited a worse performance compared to the use of $FeSO_4$ only (Trovó et al., 2012).

This treatment was applied by Pereira et al. (2014) to the decontamination of a wastewater containing the antibiotic oxytetracycline. It was observed that the addition of oxalates and citrates avoided the formation of an oxytetracycline-iron complex, which was lost in filtering operations. The treatment was carried out in a CPC pilot-scale plant using an iron/oxalate molar ratio of 1:3 with an initial $[Fe^{2+}]$ of 2 mg/L and pH 5. These conditions allowed a removal of the dissolved organic carbon of approximately 51%, accompanied by the presence of short chain carboxylic acids.

Souza et al. (2014) tested the application of the photo-Fenton reaction with ferrioxalates for the removal of the nonsteroidal anti-inflammatory drug diclofenac in water. The degradation of this drug in the aqueous solution is very slow due to precipitation-redissolution mechanisms and therefore, the use of ferrioxalate complexes has an advantage since it avoids precipitation. This study evaluated three different iron/oxalate molar ratios (1:3, 1:6, and 1:9, with an initial $[Fe^{3+}]$ of 2 mg/L under natural conditions (pH = 5−6) and artificial radiation. The 1:9 ratio achieved a complete degradation of diclofenac and 63% mineralization in 90 min at pH = 6. The 1:3 ratio achieved comparable reductions but at the slightly acidic pH of 5. These last operative conditions are deemed the most suitable due to the reagent economy.

Dias et al. (2014) studied the degradation of the antibiotics trimethoprim and sulfamethoxazole in aqueous solutions by the application of the photo-Fenton reaction enhanced with ferrioxalate complexes. The addition of ferrioxalates showed down the complexation of iron by organic compounds originated in the decomposition of the antibiotics. The effluents treated were successfully detoxified using a low iron concentration (5 mg/L) achieving a total disappearance of trimethoprim and sulfamethoxazole on 2.00 kJL and 1.25 kJ/L of radiation, respectively.

Davididou et al. (2017) reported that the degradation of the analgesic antipyrine using the ferrioxalate-intensified photo-Fenton reaction was triggered by artificial radiation using a laboratory-scale reactor. The optimization of the dose of reagents was carried out using artificial neural networks. The conditions were $[H_2O_2] = 100$ mg/L, $[Fe^{2+}] = 20$ mg/L, $[H_2C_2O_4] = 100$ mg/L, and pH = 2.8, which allowed the complete degradation of antipyrine and 90% of TOC.

4.4.4 Food and Beverage Industry

The ferrioxalate-mediated photo-Fenton reaction has also been applied for the treatment of winery wastewaters by Monteagudo et al. (2012). This study reported the mineralization of wastewater from an actual winery using a pilot plant CPC along with a physicochemical pre-treatment consisting of coagulation-flocculation. The operating conditions were optimized through the application of a multivariate experimental design and neuronal networks. The resulting optimal conditions were $[H_2O_2] = 260$ mg/L (supplied in two doses), $[H_2C_2O_4] = 80$ mg/L, pH = 3.5, $[Fe^{2+}] = 30$ mg/L achieving an overall reduction of 61% in the TOC contents. Numerous papers have been focused on the degradation of the winery wastewaters, but (Ioannou et al., 2015) the application of ferrioxalate-mediated photo-Fenton process is one of the few studies on winery wastewater.

4.4.5 Water Disinfection

The use of ferrioxalates in water disinfection using the photo-Fenton reaction has been limited, mostly due to the fact that it requires the addition of compounds that are not normally found in water and may exhibit toxic effects. However, the ferrioxalate-mediated photo-Fenton reaction has been used in the disinfection of waters with microorganisms by Cho et al. (2004) and Cho and Yoon (2008). Cho et al. (2004) demonstrated that although the inactivation of *Escherichia coli* is possible with photo-ferrioxalates; of course, its action is slower than that of conventional disinfectants. Cho and Yoon (2008) reported that the use of ferrioxalates was between 10^4 and 10^7 fold more effective for the inactivation of *Cryptosporidium parvum* than typical chemical disinfectants such as ozone, chlorine dioxide, and free chlorine. The photo-Fenton reaction has also been tested with other organic ligands (e.g., citrate) in the bacterial inactivation of *E. coli* (Ruales-Lonfat et al., 2016).

4.5 FUTURE TRENDS

The application of the photo-Fenton reaction enhanced with ferrioxalates to the treatment of effluents from various origins had been a subject of intense study. The process is well characterized, as its kinetics and critical variables are understood. Most of this understanding stems from the fact that it is a variation of the traditional photo-Fenton process, which has already been studied much more. Despite the fact that the photo-Fenton (and ferrioxalate-mediated photo-Fenton) can be applied to virtually any effluent contaminated with organic matter, there are still several gaps that need to be addressed.

The majority of the studies to date about the application of the photo-Fenton system with ferrioxalates have used simulated effluents. While the use of simulated effluents is necessary at the very first stages of experimentation to understand the performance of the treatment under controlled conditions, the application of those findings cannot be immediately extrapolated to a real industrial wastewater (Bilińska et al., 2016). Real wastewaters are characterized by the presence of many concomitant compounds used in different unit operations of the industrial process. Moreover, the composition of wastewaters is site specific and has a variable character depending on the operation patterns of the industrial facility. Therefore, the effectiveness of the photo-Fenton treatment with ferrioxalates should be investigated with real wastewaters as much as possible, taking into consideration the findings and lessons learned during the treatment of simulated effluents. Further knowledge is required identifying variations in the effectiveness of treatment due to the presence of concomitant compounds and to quantify their degree of influence. In this respect, Klamerth et al. (2013) and Manenti et al. (2015) have already investigated the treatment's efficiency with real wastewaters. Following the same logic, other studies have investigated native compounds in the effluent that can act as performance enhancers, for example, in the formation of organic complexes (Papoutsakis et al., 2016).

The influence of the upscaling of the ferrioxalate-mediated photo-Fenton reaction from laboratory scale to pilot plant, pre-industrial and industrial scales still need to be properly described. Most of the studies deal with the photo-ferrioxalate system at either laboratory or pilot-plant scale and therefore, information about operational and material obstacles in the upscaling process or the overall efficiency of the treatment are generally scarce. Furthermore, greater emphasis on the role of the ferrioxalate-mediated system as a unitary process in the greater chain of treatments within a wastewater treatment facility should be made (e.g., either as a pre-treatment or as a final polishing), as opposed to its isolated study. Presently, work is being done on comprehensively describing the role of AOPs in general within the wider wastewater treatment processes, yet specific information for ferrioxalate-mediated processes is yet to be developed (Oller et al., 2011; Papaphilippou et al., 2013).

On the same lines, the cost and benefits of implementing a ferrioxalate system still need to be identified as compared to other AOPs or other treatments, particularly regarding the use of oxalate as an additional reagent as well as its associated impacts (e.g., in terms of its biodegradability or toxicity). So far, work has been carried out in the identification of factors affecting operating costs for the photo-Fenton reaction and for combined treatments with biological reactors (Carra et al., 2013; Pérez et al., 2013). Finally, efforts are still to be made to characterize the process in terms of its environmental risks and general sustainable character, usually through a life cycle assessment (Giménez et al., 2015).

The ferrioxalate-mediated photo-Fenton reaction is a variation of the basic photo-Fenton reaction in the sense that the addition of oxalate complexes in the dissolved iron catalyst allow this process to occur at less acidic conditions and to increase the photocatalytic production of hydroxyl radicals and therefore, increasing its efficiency. As with the regular photo-Fenton reaction, the ferrioxalate system also has the degradation of recalcitrant compounds that result in a decreased biodegradability due to their complex chemical structure.

The efficiency of the treatment is directly correlated with a number of operation variables such as the dose of reagents (iron catalyst, organic ligands, and hydrogen peroxide), the pH of operation, the level of irradiation, and the presence of external influencing factors such as air or salinity. The findings of numerous studies examining the effectiveness of the treatment in the detoxification of a number of effluents indicate that the most critical ones are the dose of reagents, the ratios of peroxide to iron and oxalate to iron, the pH of operation, and the type of radiation source. Although the complex composition of the effluents calls for obtaining specific and optimal operation conditions for each type of them, the interplay of these variables often conditions the efficiency of the process.

Although not as extensively described as the ordinary photo-Fenton reaction, the process mediated with ferrioxalates has been applied to wastewaters with a wide range of origins showing promising results such as effluents from the chemical, textile, pharmaceutical, food and beverage industries. The process has also been used to disinfect waters for human consumption although its use remains limited due to the toxicity of oxalates. The majority of the studies published on the ferrioxalate-mediated photo-Fenton reaction use simulated or synthetic wastewaters with laboratory and pilot-plant scale equipment. As a consequence, further investigation is needed to fill current knowledge gaps concerning the use of real wastewaters, the pre-industrial and industrial upscaling of treatments, the role of the process as a treatment itself within a wider wastewater treatment facility, as well as the costs and benefits of implementing it compared to more mature alternatives.

References

Abrahamson, H.B., Rezvani, A.B., Brushmiller, J.G., 1994. Photochemical and spectroscopic studies of complexes of iron (III) with citric acid and other carboxylic acids. Inorg. Chim. Acta 226, 117−127.

Arslan, I., Balcioğlu, I.A., Bahnemann, D.W., 2000. Advanced chemical oxidation of reactive dyes in simulated dyehouse effluents by ferrioxalate-Fenton/UV-A and TiO_2/UV-A processes. Dyes Pigments 47, 207−218.

Arslan-Alaton, I., Tureli, G., Olmez-Hanci, T., 2009. Treatment of azo dye production wastewaters using Photo-Fenton-like advanced oxidation processes: Optimization by response surface methodology. J. Photochem. Photobiol. A 202, 142−153.

Ayodele, O.B., 2013. Effect of phosphoric acid treatment on kaolinite supported ferrioxalate catalyst for the degradation of amoxicillin in batch photo-Fenton process. Appl. Clay Sci. 72, 74−83.

Ayodele, O.B., Hameed, B.H., 2013. Synthesis of copper pillared bentonite ferrioxalate catalyst for degradation of 4-nitrophenol in visible light assisted Fenton process. J. Ind. Eng. Chem. 19, 966−974.

Azami, M., Bahram, M., Nouri, S., Naseri, A., 2012. A central composite design for the optimization of the removal of the azo dye, methyl orange, from waste water using the Fenton reaction. J. Serb. Chem. Soc. 77, 235−246.

Babuponnusami, A., Muthukumar, K., 2014. A review on Fenton and improvements to the Fenton process for wastewater treatment. J. Env. Chem. Eng. 2, 557−572.

Barbuśinski, K., 2009. Fenton reaction − Controversy concerning the chemistry. Ecol. Chem. Eng. 16, 347−358.

Barbuśinski, K., Filipek, K., 2001. Use of Fenton's reagent for removal of pesticides from industrial wastewater. Pol. J. Environ. Stud. 10, 207−212.

Bautista, P., Mohedano, A.F., Casas, J.A., Zazo, J.A., Rodríguez, J.J., 2008. An overview of the application of Fenton oxidation to industrial wastewaters treatment. J. Chem. Technol. Biotechnol. 83, 1323−1338.

Bautista, P., Casas, J.A., Zazo, J.A., Rodríguez, J.J., Mohedano, A.F., 2014. Comparison of Fenton and Fenton-like oxidation for the treatment of cosmetic wastewater. Water Sci. Technol. 70, 472−478.

Bilińska, L., Gmurek, M., Ledakowicz, S., 2016. Comparison between industrial and simulated textile wastewater treatment by AOPs − Biodegradability, toxicity and cost-assessment. Chem. Eng. J. 306, 550−559.

Binelli, A., Magni, S., Della Torre, C., Parolini, M., 2015. Toxicity decrease in urban wastewaters treated by a new biofiltration process. Sci. Total Environ. 537, 235−242.

Blanco, J., Malato, S., 2003. Solar Detoxification. Plataforma Solar de Almería. United Nations Educational, Scientific and Cultural Organization (UNESCO), Almería, Spain.

Bokare, A.D., Choi, W., 2014. Review of iron-free Fenton-like systems for activating H_2O_2 in advanced oxidation processes. J. Hazard. Mater. 275, 121−135.

Bremner, D.H., Burgess, A.E., Houllemare, D., Namkung, K., 2006. Phenol degradation using hydroxyl radicals generated from zero-valent iron and hydrogen peroxide. Appl. Catal. B: Environ. 63, 15−19.

Callao, M.P., 2014. Multivariate experimental design in environmental analysis. Trends Anal. Chem. 62, 86−92.

Carra, I., Ortega-Gómez, E., Santos-Juanes, L., López, J.L.C., Sánchez-Pérez, J.A., 2013. Cost analysis of different hydrogen peroxide supply strategies in the solar photo-Fenton process. Chem. Eng. J. 224, 75−81.

Carra, I., Malato, S., Jiménez, M., Maldonado, M.I., Pérez, J.A.S., 2014. Microcontaminant removal by solar photo-Fenton at natural pH with sequential and continuous iron additions. Chem. Eng. J. 235, 132−140.

Chakma, S., Das, L., Moholkar, V.S., 2015. Dye decolorization with hybrid advanced oxidation processes comprising sonolysis/Fenton-like/photo-ferrioxalate systems: A mechanistic investigation. Sep. Purif. Technol. 156, 596−607.

Cho, M., Yoon, J., 2008. Measurement of OH radical CT for inactivating Cryptosporidium parvum using photo/ferrioxalate and photo/TiO_2 systems. J. Appl. Microbiol. 104, 759−766.

Cho, M., Lee, Y., Chung, H., Yoon, J., 2004. Inactivation of Escherichia coli by photochemical reaction of ferrioxalate at slightly acidic and near-neutral pHs. Appl. Environ. Microbiol. 70, 1129−1134.

Collivignarelli, C., Riganti, V., Teruggi, S., Montemagno, F., 1997. Treatment of industrial wastewater with Fenton's reagent. Part II. Ing. Ambient 26, 409−418.

Conte, L.O., Schenone, A.V., Alfano, O.M., 2016. Photo-Fenton degradation of the herbicide 2,4-D in aqueous medium at pH conditions close to neutrality. J. Environ. Manage. 170, 60−69.

Davididou, K., Monteagudo, J.M., Chatzisymeon, E., Durán, A., Expósito, A.J., 2017. Degradation and mineralization of antipyrine by UV-A LED photo-Fenton reaction intensified by ferrioxalate with addition of persulfate. Sep. Purif. Technol. 172, 227−235.

De Laat, J., Truong, G., Legube, B., 2004. A comparative study of the effects of chloride, sulfate and nitrate ions on the rates of decomposition of H_2O_2 and organic compounds by $Fe(II)/H_2O_2$ and $Fe(III)/H_2O_2$. Chemosphere 55, 715–723.

De Torres-Socías, E., Fernández-Calderero, I., Oller, I., Trinidad-Lozano, M.J., Yuste, F.J., Malato, S., 2013. Cork boiling wastewater treatment at pilot plant scale: Comparison of solar photo-Fenton and ozone (O_3, O_3/H_2O_2). Toxicity and biodegradability assessment. Chem. Eng. J. 234, 232–239.

Devi, L.G., Munikrishnappa, C., Nagaraj, K., Rajashekhar, E., 2013. Effect of chloride and sulfate ions on the advanced photo Fenton and modified photo Fenton degradation process of alizarin red S. J. Appl. Mol. Catal. A: Chem. 374-375, 125–131.

Dias, I.N., Souza, B.S., Pereira, J.H.O.S., Moreira, F.C., Dezotti, M., Boaventura, R.A.R., et al., 2014. Enhancement of the photo-Fenton reaction at near neutral pH through the use of ferrioxalate complexes: A case study on trimethoprim and sulfamethoxazole antibiotics removal from aqueous solutions. Chem. Eng. J. 247, 302–313.

Dong, Y., He, L., Yang, M., 2008. Solar degradation of two azo dyes by photocatalysis using Fe(III)-oxalate complexes/H_2O_2 under different weather conditions. Dyes Pigments 77, 343–350.

Doumic, L.I., Soares, P.A., Ayude, M.A., Cassanello, M., Boaventura, R.A.R., Vilar, V.J.P., 2015. Enhancement of the solar photo-Fenton reaction by using ferrioxalate complexes for the treatment of a synthetic cotton-textile dyeing wastewater. Chem. Eng. J. 277, 86–96.

Esplugas, S., Giménez, J., Contreras, S., Pascual, E., Rodríguez, M., 2002. Comparison of different advanced oxidation processes for phenol degradation. Water Res. 36, 1034–1042.

Esplugas, S., Rodríguez, M., Malato, S., Pulgarin, C., Contreras, S., Curcó, D., et al., 2005. Optimizing the solar photo Fenton process in the treatment of contaminated water. Determination of intrinsic kinetic constants for scale-up. Sol. Energy 79, 360–368.

Estrada-Arriaga, E.B., Zepeda-Aviles, J.A., García-Sánchez, L., 2016. Post-treatment of real oil refinery effluent with high concentrations of phenols using photo-ferrioxalate and Fenton's reactions with membrane process step. Chem. Eng. J. 285, 508–516.

Expósito, A.J., Monteagudo, J.M., Díaz, I., Durán, A., 2016. Photo-Fenton degradation of a beverage industrial effluent: Intensification with persulfate and the study of radicals. Chem. Eng. J. 306, 1203–1211.

Fadzil, N.A.M., Zainal, Z., Abdullah, A.H., 2012. Ozone-assisted decolorization of methyl orange via homogeneous and heterogeneous photocatalysis. Int. J. Electrochem. Sci. 7, 11993–12003.

Fenton, H.J.H., 1894. Oxidation of tartaric acid in presence of iron. J. Chem. Soc., Trans. 65, 899–910.

Fernandes, L., Lucas, M.S., Maldonado, M.I., Oller, I., Sampaio, A., 2014. Treatment of pulp mill wastewater by *Cryptococcus podzolicus* and solar photo-Fenton: A case study. Chem. Eng. J. 245, 158–165.

Frontistis, Z., Xekoukoulotakis, N.P., Hapeshi, E., Venieri, D., Fatta-Kassinos, D., Mantzavinos, D., 2011. Fast degradation of estrogen hormones in environmental matrices by photo-Fenton oxidation under simulated solar radiation. Chem. Eng. J. 178, 175–182.

Gao, Y., Yang, M., Hu, J., Zhang, Y., 2004. Fenton's process for simultaneous removal of TOC and Fe^{2+} from acidic waste liquor. Desalination 160, 123–130.

García-Montaño, J., Pérez-Estrada, L., Oller, I., Maldonado, M.I., Torrades, F., Peral, J., 2008. Pilot plant scale reactive dyes degradation by solar photo-Fenton and biological processes. J. Photochem. Photobiol. A 195, 205–214.

Garcia-Segura, S., Bellotindos, L.M., Huang, Y.H., Brillas, E., Lu, M.C., 2016. Fluidized-bed Fenton process as alternative wastewater treatment technology-A review. J. Taiwan. Inst. Chem. Eng. 67, 211–225.

Giannakis, S., López, P.M.I., Spuhler, D.S., Pérez, J.A., Ibáñez, P.F., Pulgarín, C., 2016a. Solar disinfection is an augmentable, in situ-generated photo-Fenton reaction – Part 2: A review of the applications for drinking water and wastewater disinfection. Appl. Catal. B - Environ. 198, 431–446.

Giannakis, S., Voumard, M., Grandjean, D., Magnet, A., De Alencastro, L.F., Pulgarin, C., 2016b. Micropollutant degradation, bacterial inactivation and regrowth risk in wastewater effluents: Influence of the secondary (pre) treatment on the efficiency of advanced oxidation processes. Water Res. 102, 505–516.

Giménez, J., Bayarri, B., González, O., Malato, S., Peral, J., Esplugas, S., 2015. A comparison of the environmental impact of different AOPs: Risk indexes. Molecules 20, 503–518.

Grandclément, C., Seyssiecq, I., Piram, A., Chung, P.W.-W., Vanot, G., Tiliacos, N., et al., 2017. From the conventional biological wastewater treatment to hybrid processes, the evaluation of organic micropollutant removal: A review. Water Res. 111, 297–317.

Grebel, J.E., Pignatello, J.J., Mitch, W.A., 2010. Effect of halide ions and carbonates on organic contaminant degradation by hydroxyl radical-based advanced oxidation processes in saline waters. Environ. Sci. Technol. 44, 6822–6828.

Howe, K.J., Hand, D.W., Crittenden, J.C., Rhodes Trussell, R., Tchobanoglous, G., 2012. Principles of water treatment, first ed John Wiley & Sons, Hoboken.

Ibáñez, J.G., Hernández-Esparza, M., Doria-Serrano, C., Fregoso-Infante, A., Singh, M.M., 2007. Environmental Chemistry, Fundamentals, first ed Springer, New York.

Ioannou, L.A., Fatta-Kassinos, D., 2013. Solar photo-Fenton oxidation against the bioresistant fractions of winery wastewater. J. Environ. Chem. Eng. 1, 703–712.

Ioannou, L.A., Puma, G.L., Fatta-Kassinos, D., 2015. Treatment of winery wastewater by physicochemical, biological and advanced process: A review. J. Hazard. Mater. 286, 343–368.

Jeong, J., Yoon, J., 2004. Dual roles of $CO_2^{\bullet-}$ for degrading synthetic organic chemicals in the photo/ferrioxalate system. Water Res. 38, 3531–3540.

Kavitha, V., Palanivelu, K., 2004. The role of ferrous ion in Fenton and photo-Fenton processes for the degradation of phenol. Chemosphere 55, 1235–1243.

Klamerth, N., Malato, S., Agüera, A., Fernández-Alba, A., 2013. Photo-Fenton and modified photo-Fenton at neutral pH for the treatment of emerging contaminants in wastewater treatment plant effluents: A comparison. Water Res. 47, 833–840.

Kusic, H., Koprivanac, N., Bozic, A.L., 2011. Treatment of chlorophenols in water matrix by UV/ferrioxalate system: Part I. Key process parameter evaluation by response surface methodology. Desalination 279, 258–268.

Kwan, C.Y., Chu, W., 2007. The role of organic ligands in ferrous-induced photochemical degradation of 2,4-dichlorophenoxyacetic acid. Chemosphere 67, 1601–1611.

Lelario, F., Brienza, M., Bufo, S.A., Scrano, L., 2016. Effectiveness of different advanced oxidation processes (AOPs) on the abatement of the model compound mepanipyrim in water. J. Photochem. Photobiol. A 321, 187–201.

Luo, Y., Guo, W., Ngo, H.H., Nghiem, L.D., Hai, F.I., Zhang, J., et al., 2014. A review on the occurrence of micropollutants in the aquatic environment and their fate and removal during wastewater treatment. Sci. Total Environ. 473-474, 619–641.

Machulek, A., Moraes, J.E.F., Vautier, C., Silverio, C.A., Friedrich, L.C., Nascimento, C.A.O., et al., 2007. Abatement of the inhibitory effect of chloride anions on the photo-Fenton process. Env. Sci. Technol. 41, 8459–8463.

Malato, S., Blanco, J., Vidal, A., Richter, C., 2002. Photocatalysis with solar energy at a pilot-plant scale: An overview. Appl. Catal. B - Environ. 37, 1–15.

Malato, S., Fernández, P., Maldonado, M.I., Blanco, J., Gernjak, W., 2009. Decontamination and disinfection of water by solar photo-catalysis: Recent overview and trends. Catal. Today 147, 1–59.

Manenti, D.R., Soares, P.A., Módenes, A.N., Espinoza-Quiñones, F.R., Boaventura, R.A.R., Bergamasco, R., et al., 2015. Insights into solar photo-Fenton process using iron(III)–organic ligand complexes applied to real textile wastewater treatment. Chem. Eng. J. 266, 203–212.

Mantzavinos, D., Vassilakis, C., Pantidou, A., Psillakis, E., Kalogerakis, N., 2004. Sonolysis of natural phenolic compounds in aqueous solutions: Degradation pathways and biodegradability. Water Res. 38, 3110–3118.

Martínez, F., Calleja, G., Melero, J.A., Molina, R., 2005. Heterogeneous photo-Fenton degradation of phenolic aqueous solutions over iron-containing SBA-15 catalyst. Appl. Catal. B - Environ. 60, 181–190.

Martínez-Alcalá, I., Guillén-Navarro, J.M., Fernández-López, C., Martínez-Alcalá, I., Guillén-Navarro, J.M., Fernández-López, C., 2017. Pharmaceutical biological degradation, sorption and mass balance determination in a conventional activated-sludge wastewater treatment plant from Murcia, Spain. Chem. Eng. J. 316, 332–340.

Michael, I., Hapeshi, E., Michael, C., Varela, A.R., Kyriakou, S., Manaia, C.M., et al., 2012. Solar photo-Fenton process on the abatement of antibiotics at a pilot scale: Degradation kinetics, ecotoxicity and phytotoxicity assessment and removal of antibiotic resistant enterococci. Water Res. 46, 5621–5634.

Micó, M.M., Bacardit, J., Malfeito, J., Sans, C., 2013. Enhancement of pesticide photo-Fenton at high salinities. App. Catal. B: Environ. 132–133, 162–169.

Miralles-Cuevas, S., Oller, I., Sánchez, J.A., Malato, S., 2014. Removal of pharmaceuticals from MWTP effluent by nanofiltration and solar photo-Fenton using two different iron complexes at neutral pH. Water Res. 64, 23–31.

Mirzaei, A., Chen, Z., Haghighat, F., Yerushalmi, L., 2017. Removal of pharmaceuticals from water by homo/heterogeneous Fenton-type processes. A review. Chemosphere 174, 685–688.

Monteagudo, J.M., Durán, A., San Martín, I., Aguirre, M., 2009. Effect of continuous addition of H_2O_2 and air injection on ferrioxalate-assisted solar photo-Fenton degradation of orange II. Appl. Catal. B-Environ. 89, 510–518.

Monteagudo, J.M., Durán, A., San Martín, I., Aguirre, M., 2010. Photodegradation of reactive blue 4 solutions under ferrioxalate-assisted UV/solar photo-Fenton system with continuous addition of H_2O_2 and air injection. Chem. Eng. J. 162, 702–709.

Monteagudo, J.M., Durán, A., Aguirre, M., San Martín, I., 2011. Optimization of the mineralization of a mixture of phenolic pollutants under a ferrioxalate-induced solar photo-Fenton process. J. Hazard. Mater. 185, 131–139.

Monteagudo, J.M., Durán, A., Corral, J.M., Carnicer, A., Frades, J.M., Alonso, M.A., 2012. Ferrioxalate-induced solar photo-Fenton system for the treatment of winery wastewaters. Chem. Eng. J. 181-182, 281–288.

Myers, R.H., Montgomery, D.C., Anderson, C.M., 2009. Response surface methodology, Process and product optimization using designed experiments, third ed John Wiley, Hoboken.

Ndounla, J., Kenfack, S., Wéthé, J., Pulgarín, C., 2014. Relevant impact of irradiance (vs. dose) and evolution of pH and mineral nitrogen compounds during natural water disinfection by photo-Fenton in solar CPC reactor. Appl. Catal. B-Environ. 148–149, 144–153.

Neyens, E., Baeyens, J., 2003. A review of classic Fenton's peroxidation as an advanced oxidation technique. J. Hazard. Mater. 98, 33–50.

Oller, I., Malato, S., Sánchez-Pérez, J.A., 2011. Combination of advanced oxidation processes and biological treatments for wastewater decontamination — A review. Sci. Total Environ. 409, 4141–4166.

Ou, X., Quan, X., Chen, S., Zhang, F., Zhao, Y., 2008. Photocatalytic reaction by Fe(III)–citrate complex and its effect on the photodegradation of atrazine in aqueous solution. J. Photochem. Photobiol. A 197, 382–388.

Papaphilippou, P.C., Yiannapas, C., Politi, M., Daskalaki, V.M., Michael, C., Kalogerakis, N., et al., 2013. Sequential coagulation–flocculation, solvent extraction and photo-Fenton oxidation for the valorization and treatment of olive mill effluent. Chem. Eng. J. 224, 82–88.

Papoutsakis, S., Pulgarín, C., Oller, I., Sánchez-Moreno, R., Malato, S., 2016. Enhancement of the Fenton and photo-Fenton processes by components found in wastewater from the industrial processing of natural products: The possibilities of cork boiling wastewater reuse. Chem. Eng. J. 304, 890–896.

Pereira, J.H.O.S., Queirós, D.B., Reis, A.C., Nunes, O.C., Borges, M.T., Boaventura, R.A.R., et al., 2014. Process enhancement at near neutral pH of a homogeneous photo-Fenton reaction using ferricarboxylate complexes: Application to oxytetracycline degradation. Chem. Eng. J. 253, 217–228.

Pérez, S.J.A., Sánchez, I.M.R., Carra, I., Reina, A.C., López, J.L.C., Malato, S., 2013. Economic evaluation of a combined photo-Fenton/MBR process using pesticides as model pollutant. Factors affecting costs. J. Hazard. Mater. 244-245, 195–203.

Pignatello, J.J., Oliveros, E., MacKay, A., 2006. Advanced oxidation process for organic contaminant destruction based on the Fenton reaction. Crit. Rev. Environ. Sci. Technol. 36, 1–84.

Polo-López, M.I., García-Fernández, I., Velegraki, T., Katsoni, A., Oller, I., Mantzavinos, D., et al., 2012. Mild solar photo-Fenton: An effective tool for the removal of Fusarium from simulated municipal effluents. Appl. Catal. B-Environ. 111-112, 545–554.

Pouran, S.R., Abdul Aziz, A.R., Daud, W.M.A.W., 2015. Review on the main advances in photo-Fenton oxidation system for recalcitrant wastewaters. J. Ind. Eng. Chem. 21, 53–69.

Pozdnyakov, I.P., Kel, O.K., Plyusnin, V.F., Grivin, V.P., Bazhin, N.M., 2008. New insights into the photochemistry of ferrioxalate. J. Phys. Chem. A. 112, 8316–8322.

Prato-García, D., Buitrón, G., 2011. Degradation of azo mixtures through sequential hybrid systems: Evaluation of three advanced oxidation processes. J. Photochem. Photobiol. A 223, 103–110.

Prato-García, D., Buitrón, G., 2012. Evaluation of three reagent dosing strategies in a photo-Fenton process for the decolorization of azo dye mixtures. J. Hazard. Mater. 217-218, 293–300.

Prato-García, D., Vasquez-Medrano, R., Hernández-Esparza, M., 2009. Solar photoassisted advanced oxidation of synthetic phenolic wastewaters using ferrioxalate complexes. Sol. Energy 83, 306–315.

Rivas, F.J., Beltran, F.J., Frades, J., Buxeda, P., 2001. Oxidation of p-hydroxybenzoic acid by Fenton's reagent. Water Res. 35, 387–396.

Rodríguez, E., Fernández, G., Ledesma, B., Álvarez, P., Beltrán, F.J., 2009. Photocatalytic degradation of organics in water in the presence of iron oxides: Influence of carboxylic acids. Appl. Catal. B-Environ. 92, 240–249.

Romero, V., Acevedo, S., Marco, P., Giménez, J., Esplugas, S., 2016. Enhancement of Fenton and photo-Fenton processes at initial circumneutral pH for the degradation of the b-blocker metoprolol. Water Res. 88, 449–457.

Ruales-Lonfat, C., Barona, J.F., Sienkiewicz, A., Vélez, J., Benítez, L.N., Pulgarín, C., 2016. Bacterial inactivation with iron citrate complex: A new source of dissolved iron in solar photo-Fenton process at near-neutral and alkaline pH. Appl. Catal. B-Environ. 180, 379–390.

Sakkas, V.A., Islam, A., Stalikas, C., Albanis, T.A., 2010. Photocatalytic degradation using design of experiments: a review and example of the Congo red degradation. J. Hazard. Mater. 175, 33–44.

Salem, M.A., Abdel-Halim, S.T., El-Sawy, A.E.H., Zaki, A.B., 2009. Kinetics of degradation of allura red, ponceau 4R and carmosine dyes with potassium ferrioxalate complex in the presence of H_2O_2. Chemosphere 76, 1088–1093.

Schenone, A.V., Conte, L.O., Botta, M.A., Alfano, O.M., 2015. Modeling and optimization of photo-Fenton degradation of 2,4-D using ferrioxalate complex and response surface methodology (RSM). J. Environ. Manage. 155, 177–183.

Selvam, K., Muruganandham, M., Swaminathan, M., 2005. Enhanced heterogeneous ferrioxalate photo-Fenton degradation of reactive orange 4 by solar light. Sol. Energy Mat. Sol. Cells 8, 61–74.

Silva, T.F.C.V., Fonseca, A., Saraiva, I., Vilar, V.J.P., Boaventura, R.A.R., 2013. Biodegradabtility enhancement of a leachate after biological lagooning using a solar driven photo-Fenton reaction with further combination with an activated sludge biological process, at pre-industrial scale. Water Res. 47, 3543–3557.

Soares, P.A., Batalha, M., Guello, S., Boaventura, R.A.R., Vilar, V.J.P., 2015. Enhancement of a solar photo-Fenton reaction with ferric-organic ligands for the treatment of acrylic-textile dyeing wastewater. J. Environ. Manage. 152, 120–131.

Souza, B.M., Dezotti, M.W.C., Boaventura, R.A.R., Vilar, V.J.P., 2014. Intensification of a solar photo-Fenton reaction at near neutral pH with ferrioxalate complexes: A case study on diclofenac removal from aqueous solutions. Chem. Eng. J. 256, 448–457.

Spasiano, D., Marotta, R., Malato, S., Fernández, P., Di Somma, I., 2015. Solar photocatalysis: Materials, reactors, some commercial, pre-industrialized applications. A comprehensive approach. Appl. Catal. B: Environ. 170-171, 90–123.

Sun, Y., Pignatello, J.J., 1992. Chemical treatment of pesticide wastes. Evaluation of Fe(III) chelates for catalytic hydrogen peroxide oxidation of 2,4-D at circumneutral pH. J. Agric. Food Chem. 40, 322–327.

Szpyrkowicz, L., Juzzolino, C., Kaul, S.N., 2001. A comparative study on oxidation of disperse dyes by electrochemical process, ozone, hypochlorite and Fenton reagent. Water Res. 35, 2129–2136.

Tchobanoglous, G., Burton, F.L., Stensel, H.D., 2003. Wastewater Engineering Treatment and Reuse - Metcalf & Eddy Inc, fourth Ed McGraw-Hill, New York.

Trovó, A.G., Pupo Nogueira, R.F., Agüera, A., Fernández-Alba, A.R., Malato, S., 2011. Degradation of the antibiotic amoxicillin by photo-Fenton process – Chemical and toxicological assessment. Water Res. 45, 1394–1402.

Trovó, A.G., Pupo Nogueira, R.F., Agüera, A., Fernández-Alba, A.R., Malato, S., 2012. Paracetamol degradation intermediates and toxicity during photo-Fenton treatment using different iron species. Water Res. 46, 5374–5380.

Vedrenne, M., Vasquez-Medrano, R., Prato-García, D., Fontana-Uribe, B.A., Ibáñez, J.G., 2012a. Characterization and detoxification of a mature landfill leachate using a combined coagulation–flocculation/photo Fenton treatment. J. Hazard. Mater. 205–206, 208–215.

Vedrenne, M., Vasquez-Medrano, R., Prato-García, D., Frontana-Uribe, B.A., Hernández-Esparza, M., 2012b. A ferrous oxalate mediated photo-Fenton system: Toward an increased biodegradability of indigo dyed wastewaters. J. Hazard. Mater. 243, 292–301.

Vedrenne, M., Vasquez-Medrano, R., Pedraza-Segura, L., Toribio-Cuaya, H., Ortiz-Estrada, C.H., 2015. Reducing furfural-toxicity of a corncob lignocellulosic prehydrolyzate liquid for *Saccharomyces cerevisiae* with the photo-Fenton reaction. J. Biobased Mater. Bio. 9, 476–485.

Vidal, G., Nieto, J., Mansilla, H.D., Bornhardt, C., 2004. Combined oxidative and biological treatment of separated streams of tannery wastewater. Water Sci. Technol. 49, 287–292.

Vilar, V.J.P., Moreira, F.C., Ferreira, A.C.C., Sousa, M.A., Gonçalves, C., Alpendurada, A.F., et al., 2012. Biodegradability enhancement of a pesticide-containing bio-treated wastewater using a solar photo-Fenton treatment step followed by a biological oxidation process. Water Res. 46, 4599–4613.

Wang, Z., Xiao, D., Liu, J., 2014. Diverse redox chemistry of photo/ferrioxalate system. RSC Adv. 4, 44654–44658.

Xiao, D., Guo, Y., Luo, X., Fang, C., Wang, Z., Liu, J., 2014. Distinct effects of oxalate versus malonate on the iron redox chemistry: Implications for the photo-Fenton reaction. Chemosphere 103, 354–358.

Zapata, A., Velegraki, T., Sánchez-Pérez, J.A., Mantzavinos, D., Maldonado, M.I., Malato, S., 2009. Solar photo-Fenton treatment of pesticides in water: Effect of iron concentration on degradation and assessment of ecotoxicity and biodegradability. Appl. Catal. B: Environ. 88, 448–454.

Zhou, M., Dai, Q., Lei, L., Ma, C., Wang, D., 2005. Long life modified lead dioxide anode for organic wastewater treatment: electrochemical characteristics and degradation mechanism. Environ. Sci. Technol. 39, 363–370.

Further Reading

Ballesteros Martín, M.M., Sánchez Pérez, J.A., García Sánchez, J.L., Casas López, J.L., Malato Rodríguez, S., 2009. Effect of pesticide concentration on the degradation process by combined solar photo-Fenton and biological treatment. Water Res. 43, 3838–3848.

Ozone-Based Processes

Keisuke Ikehata and Yuan Li

Pacific Advanced Civil Engineering, Inc., Fountain Valley, CA, United States

5.1 INTRODUCTION

Ozone (O_3) is a powerful oxidant that has been widely used in a number of industrial processes, such as municipal and industrial wastewater treatment, drinking water disinfection, chemical synthesis, food and beverage, agriculture, air pollution control, medical and dental applications (Loeb, 2011; Loeb et al., 2012). Ozone, in particular, has been used for more than 100 years in drinking water treatment in the United States and throughout the world (Loeb et al., 2012; Rakness, 2005; Thompson and Drago, 2015). Ozone treatment (ozonation) has also been used in wastewater treatment for effluent disinfection, odor control, color removal, oxidation of inorganic and organic contaminants (Rice, 1996; Robson and Rice, 1991). Ozone-based treatment is particularly useful in advanced water reclamation and potable reuse applications (Gerrity et al., 2014; Gerrity and Snyder, 2011).

Ozone is an unstable gas that can be generated from gaseous oxygen molecules using electrical energy, such as electric discharge and ultraviolet (UV) irradiation. In water, ozone undergoes a series of reactions to decompose into various oxidative species, including hydroxyl radical ($^\bullet$OH), which is an even stronger oxidant than the parent molecular ozone. The powerful oxidative power of ozonation is partly owing to the generation of hydroxyl radicals. There are several ways to enhance the generation of hydroxyl radicals in ozone-based water and wastewater treatment, including the addition of hydrogen peroxide (H_2O_2), UV irradiation, and metal catalysts (Andreozzi et al., 1999; Beltrán, 2003). These processes are collectively called advanced oxidation processes (AOPs).

Ozonation and ozone-based AOPs are capable of oxidizing numerous organic compounds in water and wastewater, including pharmaceuticals and personal care products (Ikehata et al., 2006), solvents (Hoigné and Bader, 1983), pesticides (Ikehata and Gamal El-Din, 2005a, 2005b), and surfactants (Ikehata and Gamal El-Din, 2004). Here, the basics of ozonation and ozone-based AOPs in wastewater treatment, such as different ozone reactions, including direct and indirect reactions, byproduct formation, and other considerations, are described, followed by historical and more recent ozone-based AOP research in wastewater treatment and several notable ozone wastewater treatment and water reuse projects in the United States.

5.2 OZONE-BASED AOPs

Ozone is a highly reactive gas with a limited solubility in water. Therefore, the ozone reactions are highly complicated in the aqueous system, which involves gas-liquid mass transfer, self-decomposition, reactions with dissolved and suspended inorganic and organic constituents. Once dissolved in water, ozone acts as an oxidant owing to its high standard redox potential (E^0) of 2.07 V (Beltrán, 2003). Ozone reacts with dissolved inorganic and organic constituents in different pathways, namely molecular ozone (direct) and hydroxyl radical (indirect). It is well known that the direct reactions are more selective than the indirect reactions. These two groups of reactions are briefly summarized here. More comprehensive explanation of these ozone reactions in water and wastewater treatment has been discussed by Beltrán (2003).

Owing to its unique chemical structure that involves four possible resonance forms with three oxygen atoms, ozone possesses both electrophilic and nucleophilic characters. According to Beltrán (2003), the aqueous molecular ozone reactions can be classified into three categories:

- oxidation—reduction reactions
- dipolar cycloaddition reactions
- electrophilic substitution reactions

Due to its high redox potential, ozone has a capacity to oxidize numerous compounds, such as iron, manganese, nitrite, sulfides, and bromide, via the oxidation—reduction reactions. In many cases, the ozone oxidation reactions can be described as oxygen transfer from ozone to the reactant. The main half-reactions of ozone in water are:

$$O_3 + 2H^+ + 2e^- \rightarrow O_2 + H_2O \tag{5.1}$$

$$O_3 + H_2O + e^- \rightarrow O_2 + 2OH^- \quad (E^0 = 1.24V) \tag{5.2}$$

Ozone is known to attack $C = C$ double bonds in organic molecules such as olefins and aromatics to form a cyclic intermediate called ozonide, which subsequently undergoes a series of reactions such as abnormal ozonolysis and produces two smaller molecules such as ketones, aldehydes, or carboxylic acids. In electrophilic substitution reactions, an ozone molecule reacts with an organic molecule, such as a substituted aromatic compound, at one of the nucleophilic positions. Ortho- and para-hydroxylation of aromatic compounds such as phenol is a typical example of electrophilic substitution reactions. Hydroxylated phenols are further decomposed by ozone into smaller organic acids, ketones, and aldehydes via a series of reactions, including abnormal ozonolysis.

In addition to these direct ozone reactions that involve ozone molecules and other compounds (contaminants in the case of water and wastewater treatment), ozone molecules decompose into various reactive oxygen species (i.e., free radicals) that react with the contaminants in water. This is called indirect reactions, or hydroxyl radical pathway, because the main reactive species in an indirect reaction is hydroxyl radical. Hydroxyl radicals are generated by a series of radical chain reactions, which involve initiation, propagation, and termination reactions. There are two main initiation reactions depending on the pH:

Acidic to neutral pH:

$$O_3 + OH^- \rightarrow HO_2 + O_2^- \qquad k_{i1} = 70 \text{ M/s} \qquad (5.3)$$

Alkaline pH:

$$O_3 + OH^- \rightarrow HO_2^- + O_2 \qquad k_{i2} = 40 \text{ M/s} \qquad (5.4)$$

$$O_3 + HO_2^- \rightarrow HO_2 + O_3^- \qquad k_{i3} = 2.2 \times 10^6 \text{ M/s} \qquad (5.5)$$

Usually, these initiation reactions become the limited step because of the relatively slow reactions of the ozone molecule and hydroxyl ion (OH$^-$) (Eqs. 5.3 and 5.4). However, under alkaline conditions the hydroperoxide ion $\left(HO_2^-\right)$ reacts with another ozone molecule much faster than the former reactions and generates a hydroperoxyl radical (HO$_2^\bullet$), which decomposes into another free radical, and the chain reactions are propagated, such as:

$$HO_2 \rightarrow O_2^- + H^+ \qquad k_1 = 7.9 \times 10^5 \text{ per } s \qquad (5.6)$$

$$O_2^- + H^+ \rightarrow HO_2 \qquad k_1 = 5 \times 10^{10} \text{ M/s} \qquad (5.7)$$

$$O_3 + O_2^- \rightarrow O_3^- + O_2 \qquad k_2 = 1.6 \times 10^9 \text{ M/s} \qquad (5.8)$$

$$O_3^- + H^+ \rightarrow HO_3 \qquad k_3 = 5.2 \times 10^{10} \text{ M/s} \qquad (5.9)$$

$$HO_3 \rightarrow O_3^- + H^+ \qquad k_4 = 3.3 \times 10^2 \text{ per s} \qquad (5.10)$$

$$HO_3 \rightarrow OH + O_2 \qquad k_5 = 1.1 \times 10^5 \text{per s} \qquad (5.11)$$

$$O_3 + OH \rightarrow HO_4 \qquad k_6 = 2 \times 10^9 \text{ M/s} \qquad (5.12)$$

$$HO_4 \rightarrow HO_2 + O_2 \qquad k_7 = 2.8 \times 10^4 \text{ per } s \qquad (5.13)$$

Under alkaline conditions, the generation of hydrogen peroxide (H$_2$O$_2$) is also suggested by the following reactions:

$$HO_2^- + H^+ \rightarrow H_2O_2 \qquad k_8 = 5 \times 10^{10} \text{ M/s} \qquad (5.14)$$

Actually, hydrogen peroxide is an initiator and promoter of ozone decomposition, which is the basis of the ozone/hydrogen peroxide (O$_3$/H$_2$O$_2$) AOPs.

The hydroxyl radical has a higher oxidation potential (2.8 V) than the molecular ozone and can attack organic and inorganic compounds non-selectively with very high reaction rates (Andreozzi et al., 1999). The second order reaction rate constants between the ozone and the reactant (k_{OH}), and can be in the order of 10^8 to 10^{10} M/s. Therefore, the indirect ozone reactions are often responsible for the destruction of many recalcitrant organic compounds in water and wastewater (Ikehata and Gamal El-Din, 2004, 2005a, 2005b; Ikehata et al., 2006). A number of reactions may occur, when organic and inorganic compounds are exposed to hydroxyl radicals, namely hydrogen abstraction, radical–radical reactions, electrophilic addition, and electron transfer reactions (Oppenländer, 2003). In addition to hydroxyl radicals, other reactive oxygen species, such as superoxide radical anions (O_2^-), hydroperoxyl radicals (HO_2), triplet oxygen $(^3O_2)$, and organic peroxyl radicals $(ROO^\bullet)$, also participate in the indirect reactions of ozonation and ozone-based AOP.

There are a number of reactions that can terminate the indirect ozone reactions. Carbonate (CO_3^{2-}), bicarbonate (HCO_3^-), *tert*-butanol, *p*-chlorobenzoate, and humic substances are known inhibitors of ozone decomposition. These substances are also called hydroxyl radical scavengers. A high concentration of hydrogen peroxide also acts as an inhibitor of ozone decomposition.

5.2.1 Ozone/Hydrogen Peroxide

The O_3/H_2O_2 AOP is probably the best-studied and best-implemented ozone-based AOP for water and wastewater treatment. The O_3/H_2O_2 process was extensively studied by Staehelin and Hoigné (1982). At a low concentration of hydrogen peroxide (10^{-5} to 10^{-4} M), ozone decomposition in water is accelerated and as a result, the hydroxyl radical concentration increases. It was found that the only ionic form of hydrogen peroxide (hydroperoxide ion, HO_2^-) reacted with ozone according to Eq. (5.5) to initiate the free radical reactions. Since the reaction rate constant for this reaction is very large (in the order of 10^6 M/s), ozone is decomposed very rapidly in the presence of hydrogen peroxide, and the contaminants liable to hydroxyl radical reactions can be degraded in the O_3/H_2O_2 AOP very effectively. It is known that the ratio of 2 mol of ozone per mol of hydrogen peroxide is the optimum stoichiometry for this process (Beltrán, 2003). In drinking water treatment, the ratios of 0.2 to 0.5 mg H_2O_2 per mg O_3 are generally used (Rakness, 2005). In the O_3/H_2O_2 AOP, unless the reactivity of the target contaminant against the molecular ozone is very high (i.e., the second order rate constant $k > 10^6$ M/s), much of contaminant degradation occurs via indirect reactions with hydroxyl radicals.

5.2.2 Ozone/UV

The ozone/UV (O_3/UV) AOP is another well-studied ozone-based AOP. Dissolved ozone molecules absorb UV light with a peak absorbance at 260 nm and a molar absorptivity of 3292 ± 70 M/cm (Hart et al., 1983). Upon the irradiation of UV, the dissolved ozone molecules undergo photolysis reactions to yield hydrogen peroxide (Beltrán, 2003):

$$O_3 + H_2O + h \rightarrow H_2O_2 \qquad (5.15)$$

This hydrogen peroxide can initiate the ozone decomposition. Alternatively, a hydrogen peroxide molecule may undergo another photolysis reaction to form two hydroxyl radicals.

$$H_2O_2 + h \rightarrow 2 OH \qquad (5.16)$$

In addition to the molecular ozone and hydroxyl radical reactions, the target contaminant may be photolyzed, if the contaminant absorbs UV light (around 254 nm, when low-pressure mercury vapor UV lamps are used) with a significant molar absorptivity (Oppenländer, 2003).

5.2.3 Catalytic Ozonation

Various homogeneous and heterogeneous catalytic ozonation processes have been proposed and tested in water and wastewater treatment (Beltrán, 2003). Many transition metal ions such as cobalt(II), nickel(II), copper(II), iron(II), manganese(II), and zinc(II) were found to enhance the contaminant degradation in water. Hill (1948) first reported the ozone decomposition in the presence of a homogeneous catalyst cobalt(II) in an acidic medium.

$$O_3 + Co^{2+} + H_2O \rightarrow OH + CoOH^{2+} + O_2 \quad k = 37M/min \quad (5.17)$$

The generation of hydroxyl radical through the direct oxidation of cobalt(II) ion with ozone suggested that the catalytic ozonation is an AOP. In addition to dissolved metal ions, various metal oxides such as copper(II) oxide (CuO), manganese dioxide (MnO_2), titanium dioxide (TiO_2), and iron(III) oxide (Fe_2O_3), as well as palladium and activated carbon, were also found to act as heterogenous catalysts of ozone decomposition. Addition of UV irradiation and hydrogen peroxide to the homogenous/heterogeneous catalytic ozonation has also been actively studied. This type of ozone-based AOP appeared to be useful in wastewater treatment, especially for high-strength industrial wastewaters and landfill leachates that contain high concentrations of recalcitrant organic contaminants. Beltrán (2003) discussed this topic in details.

5.3 OZONATION BY-PRODUCTS

The formation of smaller organic molecules during ozonation of water and wastewater treatment have the following implications:

- generation of organic acids
  - impact on pH
- generation of more biologically degradable organics
  - impact on biological stability (mostly drinking water treatment)
  - better performance of subsequent biological treatment (mostly wastewater treatment)
- generation of aldehydes
  - aldehydes such as formaldehyde are regulated, genotoxic compounds

In addition, ozonation also produces two types of disinfection byproducts, namely bromate (BrO_3^-) and N-nitrosodimethylamine (NDMA) (Najim and Trussell, 2001, Von Gunten and Hoigné, 1994). Both ozonation byproducts are probable carcinogens (Rakness, 2005; WHO, 2005, 2008) and are of public health concern. There are numerous published research papers and review articles available on this subject (Kransner et al.,

2013; Mitch et al., 2004; Ozekin et al., 1998; Wert et al., 2007). The formation of NDMA and its control/removal is one of the key issues in advanced water reclamation for potable reuse.

5.4 WASTEWATER OZONATION AND OZONE-BASED AOPs

The use of ozone in wastewater treatment can be dated back to the 1970s as both, municipal and industrial wastewater treatment plants started using ozone at that time (Rice, 1996; Robson and Rice, 1991). The major objective of ozonation in wastewater treatment is disinfection and pathogen inactivation after biological secondary treatment, especially in municipal wastewater treatment. Ozone-based treatment has also been tested and used for pre-oxidation of biorefractory compounds (Lin et al., 2001), improvement of physicochemical treatment (Jekel, 1994), odor mitigation (Kerc and Olmez, 2010), and sludge treatment (Nagare et al., 2008) in many parts of the world.

5.4.1 Municipal Wastewater Treatment

A majority of ozone uses in municipal wastewater treatment are for effluent disinfection. The uses of ozone-based AOPs in municipal wastewater treatment is limited because the effective ozone exposure for disinfection, which is measured by the product of residual concentration and time (Ct), will be lost upon the addition of hydrogen peroxide and/or UV because these agents induce the ozone decomposition. However, due to the concerns over trace organic contaminants, such as pesticides, endocrine disruptors, pharmaceuticals, and personal care products, in the wastewater and treated effluent, many researchers have studied the effectiveness of ozone-based AOPs to destroy these organic compounds in wastewater (Borikar et al., 2015; Ikehata and Gamal El-Din, 2005a, 2005b; Ikehata et al., 2006; Snyder et al., 2006; Tsuno et al., 2010).

5.4.2 Industrial Wastewater Treatment

Rice (1996) and Beltrán (2003) reviewed a number of earlier ozone applications in industrial wastewater treatment, including:

- aquaculture
- aquarium
- electroplating
- electronic chip manufacturing
- textile industry
- petroleum refineries
- oil shale wastewater
- chemical manufacturing
- pulp and paper
- food and beverage industry
- tanneries

- pharmaceutical industry
- plastic and resins
- hospital wastewater
- landfill leachate

Those industrial wastewater treatment processes typically involve the degradation of toxic organics, such as phenolics and polyaromatic hydrocarbons, removal of color, and oxidation of inorganics, such as iron, manganese, chromium, sulfides, and cyanides (Beltrán, 2003; Rice, 1996).

In general, wastewater contains high levels of organic and inorganic compounds that react with ozone very quickly, such as phenols, synthetic dyes, and other substituted aromatics. In this case, the ozone reactions are often mass transfer controlled and direct ozone reaction dominated (Beltrán, 2003). In addition, hydroxyl radical scavengers, such as carbonate and bicarbonate, are often present at high concentrations, which inhibits the indirect ozone reactions. However, there are many reports investigating the ozone-based AOPs, including O_3/H_2O_2 and/or O_3/UV AOPs, for the treatment of industrial wastewater, such as oil shale wastewater (Munter et al., 1993), landfill leachate (Haapea et al., 2002; Leitzke, 1993), pulp mill effluent (Munter et al., 1993; Murphy et al., 1993), textile wastewater (Ledakowicz and Solecka, 2001), pharmaceutical wastewater (Höfl et al., 1997), and table olive wastewater (Beltrán et al., 1999). These reports suggest a great promise of ozone-based AOP for industrial wastewater treatment.

5.5 RECENT STUDIES

Different studies on wastewater treatment using ozone-based AOPs have been reported (Table 5.1). This list only includes works dealing with actual wastewater or effluent samples. Many of them investigated several different AOPs, including ozone-based processes, as well as non-ozone-based processes, such as UV/H_2O_2, TiO_2 photocatalysis, and ultrasound-based technologies, concurrently. As can be seen in this table, ozone-based processes were tested for the treatment of a wide variety of wastewater, including landfill leachate, pulp and paper effluent, shaft furnace gas cleaning plant effluent, oil refinery effluent, industrial laundry wastewater, ballast water, swine manure, domestic sewage effluent, and hospital wastewater. The applications of unconventional AOPs, such as ultrasonic-based technologies (Babu et al., 2016; Ibanez et al., 2013), O_3/persulfate (Abu Amr et al., 2013), and electro-peroxone (Yao et al., 2016), in wastewater treatment have also been reported.

5.5.1 Landfill Leachate Treatment

Landfill leachate, especially from mature and stabilized landfills, contains high concentrations of organic and inorganic compounds that inhibit the performance of conventional biological treatment. Different types of chemical oxidation, including ozonation, have been attempted to improve the treatability of landfill leachate (Beltrán, 2003). Recently, Abu Amr et al. (2013) compared the effects of persulfate ($S_2O_8^{2-}$) and ozone, as well as the

TABLE 5.1 Ozone-Based AOPs for Wastewater Treatment

Type of Water	Type of AOP	Note	References
Landfill leachate	O_3/H_2O_2	•OH radical scavenging effect of organic matters	Ghazi et al. (2014)
	O_3, persulfate, O_3/persulfate	COD, color, and ammonia removal	Abu Amr et al. (2013)
	O_3, O_3/H_2O_2	Aged leachate treatment, 17 - estradiol, tris-(2-chloroethyl) phosphate, tris-(butoxyethyl)-phosphate degradation	Qiao et al. (2014)
Industrial wastewater	Sono-ozone process, sonophotocatalysis, sono-Fenton systems, and sonophoto-Fenton, ultrasound	Review article on ultrasound-based AOPs	Babu et al. (2016)
Pulp and paper effluent	O_3, UV, UV/H_2O_2, O_3/H_2O_2, $O_3/UV/H_2O_2$, Fenton, etc.	Review article on AOP treatment of pulp and paper effluent, reuse	Hermosilla et al. (2015)
	O_3, TiO_2 photocatalysis	Ozonation combined with MBR could remove up to 90% of COD	Merayo et al. (2013)
Shaft furnace gas cleaning plant wastewater	O_3/UV	Organics degradation in the presence of heavy metals, such as arsenic compounds, lead, zinc	Czaplicka et al. (2013)
Oil refinery wastewater	O_3, UV, H_2O_2, O_3/UV, UV/H_2O_2	Recalcitrant organic oxidation, reuse	Souza et al. (2016)
Industrial laundry wastewater	O_3/UV	Combination of biological treatment, in situ O_3 generation followed by UF-NF, reuse	Mozia et al. (2016)
Ballast water	UV/Ag-TiO_2/O_3	*Amphidinium* sp. inactivation	Wu et al. (2011)
Swine manure	O_3, H_2O_2, O_3/H_2O_2, Fenton	COD and color removal from biologically treated swine manure	Riano et al. (2014)
Domestic/municipal wastewater	O_3, UV, O_3/UV	Small scale on-site reuse	Bustos et al. (2010)
	O_3, H_2O_2/UV	Cytostatic drug removal	Tuerk et al. (2010)
	$O_3/UV/Fe^{3+}$ or Fe_3O_4	Treatment of primary effluent, subsequent biological treatment, pharmaceuticals removal	Espejo et al. (2014)
	O_3/ultrasound	Trace organic removal	Ibanez et al. (2013)
	O_3/H_2O_2, UV/H_2O_2	Energy requirement for trace organic removal	Katsoyiannis et al. (2011)
	O_3/H_2O_2, UV/H_2O_2	Trace organic removal, NDMA formation, and bromate formation, reuse	Lee et al. (2016)

(Continued)

TABLE 5.1 (Continued)

Type of Water	Type of AOP	Note	References
	O_3, UV/H_2O_2	Trace organic removal, toxicity assessment	Richard et al. (2014)
	O_3/H_2O_2, UV, Xe, O_3/Xe, Ce-TiO$_2$, Ce-TiO$_2/O_3/Xe$	Degradation of synthetic musk compounds from effluent	Santiago-Morales et al. (2012)
	O_3, H_2O_2, UV, O_3/UV, UV/H_2O_2	Degradation of estrone	Sarkar et al. (2014)
	Electro-peroxone (O_3/H_2O_2)	Degradation of pharmaceuticals	Yao et al. (2016)
Hospital wastewater	O_3, UV, UV/H_2O_2, O_3/H_2O_2, $O_3/UV/H_2O_2$	Anticancer drug (cyclophosphamide and ifosfamide) removal	Cesen et al. (2015)
	O_3, UV, O_3/UV	Glutaraldehyde degradation	Kist et al. (2013)
	O_3, O_3/H_2O_2	Trace organic removal from MBR effluent	Lee et al. (2014)

AOP, advanced oxidation process; *COD*, chemical oxygen demand; O_3, ozone; H_2O_2, hydrogen peroxide; *UV*, ultraviolet; *TiO$_2$*, titanium dioxide; *UF*, ultrafiltration; *NF*, nanofiltration; *MBR*, membrane bioreactor; *NDMA*, N-nitrosodimethylamine.

combination of two agents on the removal of chemical oxygen demand (COD), color, and ammonia-N from a leachate sample from a semi-aerobic stabilized landfill. They optimized the ozone:COD ratio for COD removal (0.60 kg O_3/kg COD) and found that the combined method ($O_3/S_2O_8^{2-}$) achieved the highest removal efficiency as compared with other methods, including the O_3/Fenton process. Similarly, Qiao et al. (2014) investigated the treatment of leachate from a stabilized landfill using the O_3/H_2O_2 AOP to improve the removal of specific organic compounds, namely 17 -estradiol and two phosphate-based flame retardants [tris-(2-chloroethyl) phosphate and tris-(butoxyethyl) phosphate] and anaerobic biodegradation of organics. Direct oxidation of compounds was possible with the O_3/H_2O_2 AOP at relatively low ozone doses (up to 7.5 mg/L). Ghazi et al. (2014) studied the impact of landfill ages on the hydroxyl radical scavenging effect of landfill leachate during the O_3/H_2O_2 AOP. They found that the dissolved organic matter in the leachate from a younger landfill exhibited about one order of magnitude faster radical scavenging effect than the leachate from a more mature landfill.

5.5.2 Industrial Wastewater Treatment

Several different types of industrial wastwaters were tested for their treatability with several ozone-based AOPs in recent years (Table 5.1). Many of the reports investigated the applicability of the AOP technology for internal reuse. Pulp and paper mills are often required to reduce their freshwater usage and achieve zero liquid discharge by internal water reuse. Hermosilla et al. (2015) reviewed the recent reports on the treatment of pulp and paper wastewater using various AOPs as one of the zero-liquid discharge technologies.

The ozone-based AOPs have been extensively studied and have been successfully implemented at an industrial scale. Merayo et al. (2013) studied the combination treatment using ozone or TiO_2 photocatalysis followed by the membrane bioreactor (MBR) to remove COD from pulp and paper effluents. Ozonation alone could reduce COD by 35%−60% and could further improve biodegradability of effluents in the subsequent biological treatment.

The O_3/UV AOP has been tested for the treatment of shaft furnace gases' cleaning water (Czaplicka et al., 2013), oil refinery wastewater (Souza et al., 2016), and industrial laundry wastewater (Mozia et al., 2016). The latter two intended to use the AOP to degrade organic matter and prevent fouling in the subsequent membrane processes, such as nanofiltration (NF) and reverse osmosis (RO). Riano et al. (2014) compared ozonation and O_3/H_2O_2 AOP with the Fenton process for the removal of color and COD from biologically treated swine manure. They found that the Fenton process was more suitable than ozone-based processes for this type of wastewater treatment. Ozone-based AOPs other than the O_3/H_2O_2 and O_3/UV processes have been studied for industrial wastewater treatment. Wu et al. (2011) examined the combination of silver-TiO_2 photocatalysis and ozone for the inactivation of marine dinoflagellate alga *Amphidinium* sp. in ballast water. Babu et al. (2016) reviewed the applications of ultrasound-based AOPs, including sono-O_3 processes, for wastewater treatment.

5.5.3 Domestic/Municipal Wastewater Treatment

Ozone-based AOPs have been studied extensively for the treatment of domestic/municipal wastewater. Most of the studies concerned the removal of trace organic contaminants, such as endocrine disruptors, pharmaceuticals, and personal care products, from biologically treated wastewater (Espejo et al., 2014; Ibanez et al., 2013; Katsoyiannis et al., 2011; Lee et al., 2016; Richard et al., 2014; Santiago-Morales et al., 2012; Sarkar et al., 2014; Tuerk et al., 2010; Yao et al., 2016). Espejo et al. (2014) investigated the use of ozonation combined with UV-A and iron(III) or iron(II,III) oxide (magnetite, Fe_3O_4) catalysts as a pretreatment for the secondary biological treatment of municipal wastewater. The effective removal of nine pharmaceuticals was demonstrated by the AOP. Richard et al. (2014) studied the degradation of bisphenol A, ciprofloxacin, metoprolol, and sulfamethoxazole in surface water and wastewater by ozonation and the UV/H_2O_2 process. Similarly, Turek et al. (2010) studied the oxidation of cytostatic drug, cyclophosphamide by ozonation, and UV/H_2O_2 process. Degradation of two synthetic musk compounds, namely galaxolide and tonalide, was studied by Santiago-Morales et al. (2012).

The potential of ozone-based AOPs in water reclamation and reuse, including potable reuse, was also actively discussed in recent years (Bustos et al., 2010; Lee et al., 2016). Bustos et al. (2010) reported the use of the O_3/UV AOP for the disinfection of effluent from a small sewage treatment plant at Universidad Autónoma Metropolitana Azcapotzalco in Mexico to achieve on-site reuse. Lee et al. (2016) suggested the use of sequential AOPs, the O_3/H_2O_2 followed by UV/H_2O_2, might be beneficial to reduce the oxidation byproducts namely bromate and NDMA during the advanced water reclamation. Formation/degradation of NDMA in surface water and wastewater by O_3/H_2O_2 and UV/H_2O_2 was also studied by Katsoyiannis et al. (2011).

5.5.4 Hospital Wastewater Treatment

Hospital wastewater contains high concentrations of pharmaceuticals and therefore, their degradation and removal by AOPs including ozone-based processes are important. Cesen et al. (2015) reported the improved biodegradation of two cytostatic drugs, cyclophosphamide, and ifosfamide, from a hospital's wastewater using the $O_3/H_2O_2/UV$ process. Similarly, Kist et al. (2013) studied the glutaraldehyde removal from secondary effluent samples from a regional hospital in southwestern Brazil by the O_3/UV process. Lee et al. (2014) investigated the degradation of 67 trace organics in the effluent from a pilot-scale MBR at a hospital using ozonation and O_3/H_2O_2 AOP. They found that a majority of the trace organics could be eliminated by the ozone-based treatment processes.

5.5.5 Ozone-Based Municipal Wastewater Treatment and Water Reuse in the United States

In the United States, ozone was first introduced to municipal water and wastewater treatment in the early 20th century (Rice, 1999). About 300 water treatment plants in the United States currently use ozone for disinfection, taste and color control, and/or contaminant oxidation (Thompson and Drago, 2015). Ozone has been used in municipal wastewater treatment since the 1970s (Robson and Rice, 1991) and gained its popularity in the 1980s, when up to 45 wastewater treatment plants were using ozone treatment in the United States. The primary treatment objective of those earlier installations was disinfection of the primary and secondary effluent. In addition, odor control, sludge oxidation, organics oxidation, and suspended solids removal were also the drivers of ozone applications in the wastewater treatment plants. However, the number of ozone applications in wastewater treatment declined significantly during the 1990s and 2000s because of the high cost of operations and maintenance, as well as the high capital cost requirement of aged equipment renewal. A few ozone facilities retired in the 2010s, including the ones in Hagerstown, Maryland and Indianapolis, Indiana. As of today, there are only nine operational ozone-based wastewater treatment facilities in the United States (Table 5.2).

The treatment capacities of the existing municipal ozone wastewater treatment facilities range from about $1000-8000 \text{ m}^3/\text{h}$, while the ozone generation capacities range from $10-100 \text{ kg/h}$. In many cases, liquid oxygen (LOX) is used as an oxygen source. Ozone gas is typically injected by a side-stream method using multiple venturi injectors and flash reactors (Fig. 5.1).

One of the major goals of ozone treatment at the wastewater treatment facilities is water reuse (Table 5.2). In fact, due to the recent drought in the Southwestern United States, more and more utilities are considering adapting advanced water reclamation trains to produce potable water from secondary or tertiary treated wastewater. Ozone can provide a number of benefits in water reuse, including odor and color removal, virus and protozoa inactivation, and contaminant oxidation (Gerrity and Snyder, 2011). As a result, the use of ozone in wastewater treatment is again becoming an active topic in the United States. Here, several state-of-the-art municipal wastewater ozonation projects in the United States are described.

TABLE 5.2   Municipal Wastewater Treatment Plants Currently Using Ozone in the United States

Municipality	State	Year (Renewed)	ADWF (m³/h)	Oxygen Source	O₃ Generation Capacity (kg/h)	O₃ Dose (mg/L)	O₃ Generator Manufacturer	Objective
Springfield	Missouri	1978 (2000, 2012)	5500	Cryogenic oxygen	106	3.5	Mitsubishi Electric	Disinfection
Frankfort	Kentucky	1980 (2007)	6300	LOX	19	4–8	Ozonia	Disinfection
El Paso	Texas	1985 (2008)	1600	LOX	17	5.4	Wedeco	Disinfection, reuse
Trion	Georgia	1997	1300	LOX	34	27	Trailigaz	Color removal, disinfection
Gwinnett County	Georgia	2003 (2006)	7900	LOX	89	2.7 + 1.3	Ozonia	Disinfection, reuse
El Segundo	California	2013	4725	LOX	38	8	Ozonia	Oxidation, reuse
Scottsdale	Arizona	2013	3150	LOX	26	8	Metawater/ Fuji Electric	Reuse, taste, and odor
Fayetteville	Arkansas	2016	1100	VSA	11	4.3	Pinnacle Ozone	Disinfection, trace organics
Clark County	Nevada	2017[a]	4700	LOX	38	3–8	Ozonia	Disinfection, reuse

[a]*Trial operation*
ADWF, average dry weather flow; *O₃*, ozone; *LOX*, liquid oxygen; *VSA*, vacuum swing absorption.
Based on Thompson, C.M., 2016. California, Nevada - Epicenter of ozone use in North America. Proceedings of the International Ozone Association-Pan American Group Annual Conference. Las Vegas, NV (Thompson, 2016) and updated by the authors.

5.6 EXISTING OZONE-BASED ADVANCED WATER RECLAMATION FACILITIES

In the United States, most of the advanced water reclamation facilities are constructed as separate facilities from the conventional wastewater treatment plants. In particular, the advanced treatment facilities designed to achieve very high reclaimed water quality for such uses as potable reuse, include a series of membrane processes such as microfiltration (MF), ultrafiltration (UF), NF, and RO, as well as UV-based AOPs (Fig. 5.2).

The MF/UF process removes suspended solids prior to RO, which removes a majority (up to 99%) of dissolved constituents including both inorganic and organic compounds. The UV/H₂O₂ AOP is added for NDMA and 1,4-dioxane destruction. There are at least ten operational advanced water reclamation facilities in the Southwestern United States, most notably in Southern California, including the Groundwater Replenishment System,

FIGURE 5.1 Venturi ozone injection system (left) and pipeline flash reactor (right) at the Scottsdale water campus AWTF (Scottsdale, AZ).

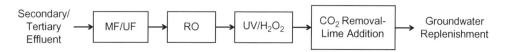

FIGURE 5.2 Typical advanced water reclamation process scheme for indirect potable reuse.

Orange County Water District (Fountain Valley, CA), the Edward C. Little Water Reclamation Facility (WRF), West Basin Municipal Water District (El Segundo, CA), and the Leo J. Vander Lans Advanced Water Treatment Facility, Water Replenishment District of Southern California (Long Beach, CA).

There are opportunities for the ozone-based treatment to complement or replace the unit processes in this advanced water reclamation process scheme, such as:

- add ozone or ozone-biological filtration prior to MF/UF
- replace the UV/H_2O_2 AOP with an ozone-based AOP
- use NF instead of RO and add an ozone-based AOP

In all cases, ozone is expected to act as a barrier against both pathogenic microorganisms (viruses and protozoa) and trace organic compounds. In addition, many studies have been reported that the addition of preozonation reduced the fouling of subsequent MF or UF (Pisarenko et al., 2012, Van Geluwe et al., 2011). Because the NDMA removal capacity of ozone-based AOPs is limited (Lee et al., 2016), UV-based treatment is still needed for the photolysis of NDMA. The ozone-based AOP can supplement the trace organic removal capacity of NF in the last case. The use of NF is beneficial, whereas brine disposal is very costly and higher water recovery (>90%) is desired.

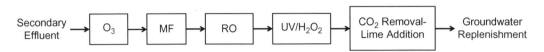

FIGURE 5.3   Current process scheme of the Edward C. Little WRF in El Segundo, CA.

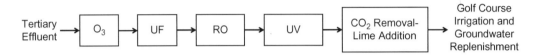

FIGURE 5.4   Process scheme of the Scottsdale Water Campus AWTF in Scottsdale, AZ.

FIGURE 5.5   Scottsdale Water Campus AWTF (Scottsdale, AZ), (Left: Ozone generators, Right: RO facility).

Two existing advanced water reclamation facilities, namely the Edward C. Little WRF and Scottsdale Water Campus Advanced Water Treatment Facility (AWTF), implemented the preozonation approach. The Edward C. Little WRF was built as one of the first major advanced water reclamation facilities using RO in 1992. In 2013, ozone was added to the process scheme anticipating the reduction of MF fouling as shown in Fig. 5.3. However, the addition of ozone induced several unexpected side effects, including elevated levels of NDMA, aldehydes, and total organic carbon in the final product water (Daniel and Oelker, 2016). NDMA and aldehydes are known ozonation by-products (Najim and Trussell, 2001, Rakness, 2005). Those side effects could be mitigated by lowering the ozone dose, which ironically sacrificed the intended effect of ozone on MF fouling reduction. NDMA and aldehydes, and other biodegradable organics such as ketones could be removed by biological filtration such as biological activated carbon (BAC) filtration.

The Scottsdale Water Campus AWTF have a very similar process scheme as the Edward C. Little WRF (Figs. 5.4 and 5.5). The differences are the absence of UV AOP and the feed water quality (tertiary vs. secondary), as well as UF vs. MF. To date, no byproduct issues have been reported at the Scottsdale plant, probably because the feed water received nitrification and denitrification in the upstream water reclamation plant.

According to the AWTF operators, the objective of ozonation is to reduce the UF fouling and to degrade trace organics in the tertiary effluent.

5.7 PLANNED OZONE-BASED ADVANCED WATER RECLAMATION PROJECTS

In the United States, the number of ozone-based advanced water reclamation projects is currently increasing. At least six active projects in several arid/semi-arid states are recognized (Table 5.3). Two notable projects are described here. Pure Water San Diego (San Diego, CA) is a well-known indirect/direct potable reuse project that employs the standard MF/UF-RO-UV AOP scheme with ozone-BAC filtration as a pretreatment (Fig. 5.6). The ozone-BAC component was actually added as an additional treatment barrier during the demonstration study (Fig. 5.7) due to the increasing need and public acceptance towards direct potable reuse in the area (Pearce et al., 2016). Phase 1 of the full-scale water purification facility will be completed by 2021 and highly purified reclaimed water will be sent to one of the source water reservoirs (Miramar Reservoir) for drinking water production (City of San Diego, 2017).

The second example is the planned advanced treatment at Rio Rancho's Cabezon Water Reclamation Plant. This is a non-RO-based potable reuse project using MBR followed by the O_3/H_2O_2 AOP and BAC (Marley, 2010). Once completed, this is probably one of the first non-RO potable reuse projects that will use an ozone-based AOP for contaminant oxidation and disinfection in the United States. (Fig. 5.8). According to Marley (2010), the

TABLE 5.3 Planned Ozone-Based Advanced Water Reclamation Projects in the United States

City	State	Commissioning	Treatment Capacity (m³/h)	Process Scheme	Biofiltration	Type of Reuse
San Diego	California	2021 (Phase I)	4725 (Phase I)	O_3-BAC-MF/UF-RO-UV/H_2O_2 or Cl_2 AOP	Yes	Indirect or direct potable reuse (surface water augmentation)
Monterey	California	2018	1100	O_3-MF-RO-UV/H_2O_2 AOP	No	Indirect potable reuse (groundwater)
Rio Rancho	New Mexico	2018	160	MBR-O_3/H_2O_2 AOP-BAC	Yes	Indirect potable reuse (groundwater)
El Paso	Texas	Unknown	1600	O_3-MF/UF-RO-UV H_2O_2 AOP-GAC-Cl_2	No	Direct potable reuse
Los Angeles	California	Unknown	12,600?	To be decided	Yes/No	Indirect potable reuse (groundwater)
Oklahoma City	Oklahoma	Unknown	1800?	To be decided	Yes	Indirect or direct potable reuse (surface water augmentation)

O_3, ozone; *BAC*, biological activated carbon; *MF*, microfiltration; *UF*, ultrafiltration; *RO*, reverse osmosis; H_2O_2, hydrogen peroxide; Cl_2, chlorine; *UV*, ultraviolet; *AOP*, advanced oxidation process; *GAC*, granular activated carbon.

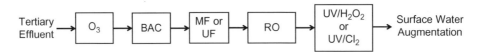

FIGURE 5.6   Current proposed process scheme of the Pure Water San Diego Demonstration Facility in San Diego, CA.

FIGURE 5.7   Ozone generator (Left) and GAC filters (Right) at the Pure Water San Diego Demonstration Facility (San Diego, CA).

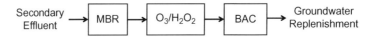

FIGURE 5.8   Proposed process scheme of the Cabezon Water Reclamation Plant in Rio Rancho, NM.

main objective of O_3/H_2O_2 AOP is the destruction of endocrine disrupting compounds, pharmaceuticals, and personal care products. The purpose of adding hydrogen peroxide, is for bromate formation control. NDMA formation and control is also a key issue in ozone-based advanced water reclamation schemes, especially where RO is not used as a barrier (Fig. 5.8).

5.8 CONCLUDING REMARKS

Owing to the strong disinfection and oxidative powers, ozone has been used for the treatment of both municipal and industrial wastewaters on an industrial scale for many decades. Ozone-based treatment is particularly useful for the treatment of wastewater that has color originating from natural (e.g., lignin) and synthetic (e.g., dyes) aromatic compounds. These aromatic compounds are highly resistant against biological oxidation. Ozone can alter the chemical structures of those recalcitrant organics into more biodegradable forms. Ozonation is also highly effective in degrading trace organic compounds in municipal and hospital wastewater, and is becoming a critical component of advanced water reclamation schemes for potable reuse. Despite their slow adaptation to full-scale treatment facilities, ozone-based AOPs have been actively studied for their potential uses in wastewater treatment for many years. Since ozone itself is an expensive and highly-technically involved process, the combination of ozone and additional chemical/physical agents such as hydrogen peroxide and UV requires significantly higher capital and operational costs and may not be feasible in many wastewater treatment projects. However, as the population growth continues and water scarcity becomes more and more severe in densely populated areas, the demand for more advanced wastewater treatment technologies including ozone-based AOPs is anticipated to increase. It is probably desirable to develop more cost-effective and more operator-friendly ozone-based AOPs for smaller wastewater treatment and water reclamation projects.

References

Abu Amr, S.S., Aziz, H.A., Adlan, M.N., 2013. Optimization of stabilized leachate treatment using ozone/persulfate in the advanced oxidation process. Waste Manage. 33, 1434–1441.

Andreozzi, R., Caprio, V., Insola, A., Marotta, R., 1999. Advanced oxidation by ozone to remove recalcitrance from waste waters – a review. Environ. Technol. 22, 51–59.

Babu, S.G., Ashokkumar, M., Neppolian, B., 2016. The role of ultrasound on advanced oxidation processes. Topics Curr. Chem. 374, 75.

Beltrán, F.J., 2003. Ozone Reaction Kinetics for Water and Waste water Systems. Lewis Publishers, Boca Raton, FL.

Beltrán, F.J., García-Araya, F., Frades, J., Álavez, P., Gimeno, O., 1999. Effects of single and combined ozonation with hydrogen peroxide or UV radiation on the chemical degradation and biodegradability of debittering table olive industrial waste waters. Water Res. 33, 723–732.

Borikar, D., Mohseni, M., Jasim, S., 2015. Evaluation and comparison of conventional and advanced oxidation processes for the removal of PPCPs and EDCs and their effect on THM-formation potential. Ozone Sci. Eng. 37, 154–169.

Bustos, Y.A., Vaca, M., Lopez, R., Torres, L.G., 2010. Disinfection of a waste water flow treated by advanced primary treatment using O_3, UV and O_3/UV combinations. J. Environ. Sci. Health Part A 45, 1715–1719.

Cesen, M., Kosjek, T., Laimou-Geraniou, M., Kompare, B., Sirok, B., Lambropolou, D., et al., 2015. Occurrence of cyclophosphamide and ifosfamide in aqueous environment and their removal by biological and abiotic waste water treatment processes. Sci. Total Environ. 527-528, 465–473.

City of San Diego, 2017. Pure Water San Diego Phase 1 Project Fact Sheet.

Czaplicka, M., Kurowski, R., Jaworek, K., Bratek, L., 2013. Application of advanced oxidation processes for cleaning of industrial water generated in wet dedusting of shaft furnace gases. Environ. Technol. 34, 1455–1462.

Daniel, U., Oelker, G., 2016. Solution pollution: ozone as pretreatment for recycled water and the side effects that can cause regulatory headaches. Proc. International Ozone Association, Pan American Group Annual Conference Las Vegas, NV.

Espejo, A., Aguinaco, A., Garcia-Araya, J.F., Beltran, F.J., 2014. Sequential ozone advanced oxidation and biological oxidation processes to remove selected pharmaceutical contaminants from an urban waste water. J. Environ. Sci. Health Part A 49, 1015–1022.

Gerrity, D., Owens-Bennett, E., Venezia, T., Stanford, B.D., Plumlee, M.H., Debroux, J., et al., 2014. Applicability of ozone and biological activated carbon for potable reuse. Ozone Sci. Eng. 36, 123–137.

Gerrity, D., Snyder, S., 2011. Review of ozone for water reuse applications: Toxicity, regulations, and trace organic contaminant oxidation. Ozone Sci. Eng. 33, 253−266.

Ghazi, N.M., Lastra, A.A., Watts, M.J., 2014. Hydroxyl radical (OH) scavenging in young and mature landfill leachates. Water Res. 56, 148−155.

Haapea, P., Korhonen, S., Tuhkanen, T., 2002. Treatment of industrial landfill leachates by chemical and biological methods: ozonation, ozonation + hydrogen peroxide, hydrogen peroxide, and biological post-treatment for ozonated water. Ozone Sci. Eng 24, 369−378.

Hart, E.J., Sehested, K., Holoman, J., 1983. Molar absorptivities of ultraviolet and visible bands of ozone in aqueous solutions. Anal. Chem. 55, 46−49.

Hermosilla, D., Merayo, N., Gasco, A., Blanco, A., 2015. The application of advanced oxidation technologies to the treatment of effluents from the pulp and paper industry: a review. Environ. Sci. Pollut. Res. 22, 168−191.

Hill, G.R., 1948. Kinetics, mechanisms, and activation energy of the cobaltous ion catalyzed decomposition of ozone. J. Am. Chem. Soc. 70, 1306−1307.

Höfl, C., Sigl, G., Specht, O., Wurdack, I., Wabner, D., 1997. Oxidative degradation of AOX and COD by different advanced oxidation processes: a comparative study with two samples of a pharmaceutical waste water. Water Sci. Technol. 35, 257−264.

Hoigné, J., Bader, H., 1983. Rate constants of reactions of ozone with organic and inorganic compounds in water − I: Non-dissociating organic compounds. Water Res. 16, 173−183.

Ibanez, M., Gracia-Lor, E., Bijlsma, L., Morales, E., Pastor, L., Hernandez, F., 2013. Removal of emerging contaminants in sewage water subjected to advanced oxidation with ozone. J. Hazard. Mater. 260, 389−398.

Ikehata, K., Gamal El-Din, M., 2004. Degradation of recalcitrant surfactants in waste water by ozonation and advanced oxidation processes: a review. Ozone Sci. Eng. 26, 327−343.

Ikehata, K., Gamal El-Din, M., 2005a. Aqueous pesticide degradation by ozonation and ozone-based advanced oxidation processes: a review (Part I). Ozone Sci. Eng. 27, 83−114.

Ikehata, K., Gamal El-Din, M., 2005b. Aqueous pesticide degradation by ozonation and ozone-based advanced oxidation processes: a review (Part II). Ozone Sci. Eng. 27, 173−202.

Ikehata, K., Naghashkar, N.J., Gamal El-Din, M., 2006. Degradation of aqueous pharmaceuticals by ozonation and advanced oxidation processes: a review. Ozone Sci. Eng. 28, 353−414.

Jekel, M.R., 1994. Flocculation effects of ozone. Ozone Sci. Eng. 16, 55−66.

Katsoyiannis, I.A., Canonica, S., von Gunten, U., 2011. Efficiency and energy requirements for the transformation of organic micropollutants by ozone, O_3/H_2O_2 and UV/H_2O_2. Water Res. 45, 3811−3822.

Kerc, A., Olmez, S.S., 2010. Ozonation of odorous air in waste water treatment plants. Ozone: Sci. Eng. 32, 199−203.

Kist, L.T., Rosa, E.C., Machado, E.L., Camargo, M.E., Moro, C.C., 2013. Glutaraldehyde degradation in hospital waste water by photoozonation. Environ. Technol. 34, 2579−2586.

Kransner, S.W., Mitch, W.A., McCurry, D.L., Hanigan, D., Westerhoff, P., 2013. Formation, precursors, control, and occurrence of nitrosamines in drinking water: a review. Water Res. 47, 4433−4450.

Ledakowicz, S., Solecka, M., 2001. Influence of ozone and advanced oxidation processes on biological treatment of textile waste water. Ozone Sci. Eng. 23, 327−332.

Lee, Y., Gerrity, D., Lee, M., Gamage, S., Pisarenko, A., Trenholm, R.A., et al., 2016. Organic contaminant abatement in reclaimed water by UV/H_2O_2 and a combined process consisting of O_3/H_2O_2 followed by UV/H_2O_2: prediction of abatement efficiency, energy consumption, and byproduct formation. Environ. Sci. Technol. 50, 3809−3819.

Lee, Y., Kovalova, L., McArdell, C.S., von Gunten, U., 2014. Prediction of micropollutant elimination during ozonation of a hospital waste water effluent. Environ. Sci. Technol. 64, 134−148.

Leitzke, O., 1993. Chemical oxidation with ozone and UV-light under a pressure of 5 bar abs. and at temperature between 10 and 60 C. In: Proceedings of the International Symposium on Ozone-Oxidation Methods for Water and Waste water Treatment. Paris, France, International Ozone Association, European-African Group.

Lin, C.K., Tsai, T.Y., Liu, J.C., Chen, M.C., 2001. Enhanced biodegradation of petrochemical waste water using ozonation and BAC advanced treatment system. Water Res. 35, 699−704.

Loeb, B.L., 2011. Ozone: Science Eng.: thirty-three years and growing. Ozone Sci. Eng. 33, 329−342.

Loeb, B.L., Thompson, C.M., Drago, J., Takahara, H., Baig, S., 2012. Worldwide ozone capacity of treatment of drinking water and waste water: a review. Ozone Sci. Eng. 34, 64−77.

Marley, R., 2010. Reclaimed water aquifer storage, recovery (ASR. Rio Rancho, New Mexico. Proceedings of the WESTCAS 2010 Winter Conference. Albuquerque, NM, February 17–19, 2010.

Merayo, N., Hermosilla, D., Blanco, L., Cortijo, L., Blanco, A., 2013. Assessing the application of advanced oxidation processes, and their combination with biological treatment, to effluents from pulp and paper industry. J. Hazard. Mater. 262, 420–427.

Mitch, W.A., Sharp, J.O., Trussell, R.R., Valentine, R.L., Alvarez-Cohen, L., Sadlak, D.L., 2004. N-Nitrosodimethylamine (NDMA) as a drinking water contaminant: a review. Environ. Eng. Sci. 20, 389–404.

Mozia, S., Janus, M., Brozek, P., Bering, S., Tarnowski, K., Mazur, J., et al., 2016. A system coupling hybrid biological method with UV/O_3 oxidation and membrane separation for treatment and reuse of industrial laundry waste water. Environ. Sci. Poll. Res. 23, 19145–19155.

Munter, R., Kallas, J., Pikkov, L., Preis, S., Kamenev, S., 1993. Ozonation and AOP treatment of the waste water from oil shale and pulp and paper industries. In: Ozone in Water and Waste water Treatment, Vol. 2, Proceedings of the 11th Ozone World Congress. San Francisco, CA, International Ozone Association, Pan American Group.

Murphy, J.K., Hulsey, R.A., Long, B.W., Amaranath, R.K., 1993. Use of ozone and advanced oxidation processes to remove color from pulp and paper mill effluents. In: Ozone in Water and Waste water Treatment, Vol. 2, Proceedings of the 11th Ozone World Congress. San Francisco, CA, International Ozone Association, Pan American Group.

Nagare, H., Tsuno, H., Saktaywin, W., Somiya, T., 2008. Sludge ozonation and its application to a new advanced waste water treatment process with sludge disintegration. Ozone Sci. Eng. 30, 136–144.

Najim, I., Trussell, R.R., 2001. NDMA formation in water and waste water. J. Am. Water Works Assoc. 93, 92–99.

Oppenländer, T., 2003. Photochemical Purification of Water and Air. Wiley-VCH Verlag, Weinheim, Germany.

Ozekin, K., Westerhoff, P., Amy, G.L., Siddiqui, M., 1998. Molecular ozone and radical pathways of bromate formation during ozonation. J. Environ. Eng. 124, 456–462.

Pearce, B., Pisarenko, A., Chen, E., Trussell, R.S., Howe, E., 2016. Evaluating and comparing 4 on-line dissolved ozone residual meters. In: Proceedings of the International Ozone Association-Pan American Group Annual Conference. Las Vegas, NV, International Ozone Association-Pan American Group.

Pisarenko, A.N., Stanford, B.D., Yan, D., Gerrity, D., Snyder, S.A., 2012. Effects of ozone and ozone/peroxide on trace organic contaminants and NDMA in drinking water and water reuse applications. Water Res. 46, 316–326.

Qiao, Y., Do, A., Yeh, D., Watt, M.J., 2014. A bench-scale assessment of ozone pre-treatments for landfill leachates. Environ. Technol. 35, 145–153.

Rakness, K.L., 2005. Ozone in Drinking Water Treatment – Process Design, Operation, and Optimization, Denver, CO, American Water Works Association.

Riano, B., Coca, M., Garcia-Gonzalez, M.C., 2014. Evaluation of Fenton method and ozone-based processes for colour and organic matter removal from biologically pre-treated swine manure. Chemosphere 117, 193–199.

Rice, R.G., 1996. Applications of ozone for industrial waste water treatment – a review. Ozone Sci. Eng. 18, 477–515.

Rice, R.G., 1999. Ozone in the United States of America - State-of-the-art. Ozone Sci. Eng. 21, 99–118.

Richard, J., Boergers, A., Vom Eyser, C., Bester, K., Tuerk, J., 2014. Toxicity of the micropollutants bisphenol A, ciprofloxacin, metoprolol and sulfamethoxazole in water samples before and after the oxidative treatment. Int. J. Hygiene Environ. Health 217, 506–514.

Robson, C.M., Rice, R.G., 1991. Waste water ozonation in the U.S.A. – history and current status. Ozone Sci. Eng. 13, 23–40.

Santiago-Morales, J., Gomez, M.J., Herrera, S., Fernandez-Alba, A.R., Garcia-Calvo, E., Rosal, R., 2012. Oxidative and photochemical processes for the removal of galaxolide and tonalide from waste water. Water Res. 46, 4435–4447.

Sarkar, S., Ali, S., Rehmann, L., Nakhla, G., Ray, M.B., 2014. Degradation of estrone in water and waste water by various advanced oxidation processes. J. Hazard. Mater. 278, 16–24.

Snyder, S.A., Wert, E.C., Rexing, D.J., Zegers, R.E., Drury, D.D., 2006. Ozone oxidation of endocrine disruptors and pharmaceuticals in surface water and waste water. Ozone Sci. Eng. 28, 445–460.

Souza, B.M., Souza, B.S., Guimaraes, T.M., Ribeiro, T.F., Cerqueira, A.C., Sant'Anna Jr., G.L., et al., 2016. Removal of recalcitrant organic matter content in waste water by means of AOPs aiming industrial water reuse. Environ. Sci. Pollut. Res. 23, 22947–22956.

Staehelin, S., Hoigné, J., 1982. Decomposition of ozone in water: Rate of initiation by hydroxide ions and hydrogen peroxide. Environ. Sci. Technol. 16, 666–681.

Thompson, C.M., 2016. California, Nevada – Epicenter of ozone use in North America. Proceedings of the International Ozone Association-Pan American Group Annual Conference. Las Vegas, NV.

Thompson, C.M., Drago, J.A., 2015. North American installed water treatment ozone systems. J. Am. Water Works Assoc. 107, 45–55.

Tsuno, H., Arakawa, K., Kato, Y., Nagare, H., 2010. Advanced sewage treatment with ozone under excess sludge reduction, disinfection and removal of EDCs. Ozone Sci. Eng. 30, 238–245.

Tuerk, J., Sayder, B., Boergers, A., Vitz, H., Kiffmeyer, T.K., Kabasci, S., 2010. Efficiency, costs and benefits of AOPs for removal of pharmaceuticals from the water cycle. Water Sci. Technol. 61, 985–993.

Van Geluwe, S., Braeken, L., Van der Bruggen, B., 2011. Ozone oxidation for the alleviation of membrane fouling by natural organic matter: a review. Water Res. 45, 3551–3570.

Von Gunten, U., Hoigné, J., 1994. Bromate formation during ozonization of bromide-containing waters: interaction of ozone and hydroxyl radical reactions. Environ. Sci. Technol. 28, 1234–1242.

Wert, E.C., Rosario-Ortiz, F.L., Drury, D.D., Snyder, S.A., 2007. Formation of oxidation byproducts from ozonation of waste water. Water Res. 41, 1481–1490.

WHO, 2005. Bromate in Drinking-water. World Health Organization, Geneva, Switzerland.

WHO, 2008. N-Nitrosodimethylamine in Drinking-water. World Health Organization, Geneva, Switzerland.

Wu, D., You, H., Zhang, R., Chen, C., Lee, D.J., 2011. Inactivation of Amphidinium sp. in ballast waters using UV/Ag-TiO$_2$ + O$_3$ advanced oxidation treatment. Bioresour. Technol. 102, 9838–9842.

Yao, W., Wang, X., Yang, H., Yu, G., Deng, S., Huang, J., et al., 2016. Removal of pharmaceuticals from secondary effluents by an electro-peroxone process. Water Res. 88, 826–835.

6

Photocatalysis

*Rakshit Ameta[1], Meenakshi S. Solanki[2], Surbhi Benjamin[3]
and Suresh C. Ameta[3]*

[1]J. R. N. Rajasthan Vidyapeeth (Deemed-to-be University), Udaipur, Rajasthan, India
[2]B. N. University, Udaipur, Rajasthan, India [3]PAHER University, Udaipur, Rajasthan, India

6.1 INTRODUCTION

Industrialization, technologies, and consumption of nonrenewable sources are increasing at a rapid rate since the last few decades, because of regular continuous increasing demands of materials related to textile, dye, fertilizer, domestic, plastic, etc. As a consequence, environmental pollution and energy crises have already reached an alarming stage. The industrial wastes are more toxic, and nonbiodegradable as compared to municipal wastes because these consist of fats, oil, grease, heavy metals, phenols, and ammonia, etc. (Mutamim et al., 2012). The agriculture and pharmaceutical effluents release pesticides and other chemicals that are responsible for some chronic diseases, harmful for human endocrine, making water not useful for drinking and other end uses (Das, 2014). There is an urgent need to develop some newer technologies that are eco-friendly in nature and lead to degradation or complete elimination of environmental pollutants and thus, prove to be an alternative clean strategy. In other words, there should be a sustainable solution to the problem.

Advanced oxidation processes (AOPs) are an environmentally friendly technique for the removal of almost all types of pollutants: air pollutants, water pollutants such as aromatics, petroleum based contents, petroleum hydrocarbons, chlorinated hydrocarbons, pesticides, insecticides, volatile organic compounds (VOC), dyes, and other organic materials (Zhang et al., 2014). AOPs are based on the generation of reactive oxygen species like hydroxyl radicals with one unpaired electron, and because of this, they possess short life times. Therefore, they actively and readily react with a series of chemical species, which are otherwise very difficult to degrade. AOPs are comparatively better than other conventional methods because they generate thermodynamically stable oxidation products such as carbon dioxide, water, and biodegradable organics. AOPs include photocatalysis process, which plays an important role in harvesting sunlight by a photocatalyst. Then, these photocatalysts have been used effectively to solve the problems related to environment pollution and energy crises in the presence of different ranges of solar spectrum (Sang et al., 2015, Solanki et al., 2015a).

6.2 PHOTCATALYSIS

The term photocatalyst is a combination of two words: photo related to photon and catalyst, which is a substance altering the reaction rate in its presence. Therefore, photocatalysts are materials that change the rate of a chemical reaction on exposure to light. This phenomenon is known as photocatalysis. Photocatalysis includes reactions that take place by utilizing light and a semiconductor. The substrate that absorbs light and acts as a catalyst for chemical reactions is known as a photocatalyst. All the photocatalysts are basically semiconductors. Photocatalysis is a phenomenon, in which an electron-hole pair is generated on exposure of a semiconducting material to light.

The photocatalytic reactions can be categorized into two types on the basis of appearance of the physical state of reactants.

- *Homogeneous photocatalysis*: When both the semiconductor and reactant are in the same phase, i.e. gas, solid, or liquid, such photocatalytic reactions are termed as homogeneous photocatalysis.
- *Heterogeneous photocatalysis*: When both the semiconductor and reactant are in different phases, such photocatalytic reactions are classified as heterogeneous photocatalysis.

The energy difference between the valence band (HOMO) and the conduction band (LUMO) are known as the band gap (Eg). On the basis of band gap, the materials are classified into three basic categories (Fig. 6.1):

- metal or conductor: Eg < 1.0 eV
- semiconductor: Eg < 1.5–3.0 eV
- insulator: Eg > 5.0 eV

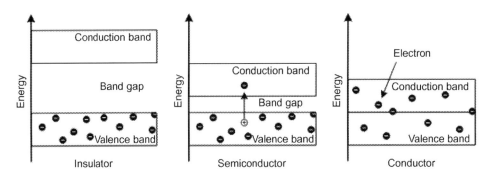

FIGURE 6.1   Different types of materials.

Semiconductors are capable of conducting electricity even at room temperature in the presence of light and hence, work as photocatalysts. When a photocatalyst is exposed to light of the desired wavelength (sufficient energy), the energy of photons is absorbed by an electron (e^-) of valence band and it is excited to conduction band. In this process a hole (h^+) is created in valence band. This process, leads to the formation of photo-excitation state, and e^- and h^+ pair is generated. This excited electron is used for reducing an acceptor in which a hole is used for oxidation of donor molecules. The importance of photocatalysis lies in the fact that a photocatalyst provides both oxidation as well as a reduction environment and that too, simultaneously.

The fate of the excited electron and hole is decided by the relative positions of the conduction and valence bands of the semiconductor and the redox levels of substrate.

There are four ways in which the semiconductor and the substrate interact with each other depending upon the relative positions of the valence and conduction bands and the redox levels. The four different combinations are (Fig. 6.2):

1. Reduction of substrate takes place, when the redox level of substrate is lower than the conduction band of the semiconductor.
2. Oxidation of substrate takes place, when the redox level of the substrate is higher than the valence band of the semiconductor.
3. Neither oxidation nor reduction is possible, when the redox level of the substrate is higher than the conduction band and lower than the valence band of the semiconductor.
4. Both reduction and oxidation of the substrate take place, when the redox level of the substrate is lower than the conduction band and higher than the valence band.

Photocatalysts may be used for antifouling, antifogging, conservation and storage of energy, deodorization, sterilization, self-cleaning, air purification, wastewater treatment, etc. Semicoductors act as sensitizers for photoredox processes due to their electronic structure. Some semiconductors are able to photocatalyze the complete mineralization of many organic pollutants like aromatics, halo hydrocarbons, insecticides, pesticides, dyes, and surfactants (Fig. 6.3).

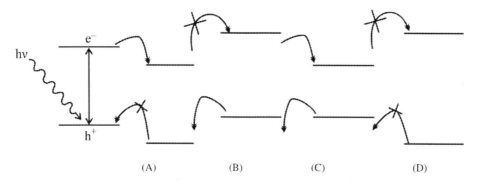

FIGURE 6.2 Different possibilities of reactions. (A) Reduction. (B) Oxidation. (C) Redox reaction. (D) No reaction.

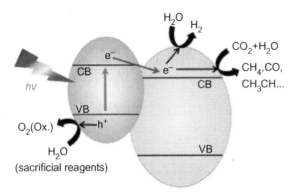

FIGURE 6.3   Degradation of organic pollutants.

Metal oxides have a wide range of application to solve environmental problems and in electronics due to their capacity to form charge carriers when they are exposed to light. Metal oxides have properties like:

- required electronic structure
- light absorption properties
- charge transport characteristics

Semiconductor-mediated photocatalysis has gained huge attention as it helps to overcome the problem related to fast charge recombination (Wang et al., 2014).

6.2.1 Binary Oxides

Chan et al. (2011) categorized photocatalyst into three categories:

1. titanium dioxide,
2. zinc oxide, and
3. other metal oxides like molybdenum oxide, vanadium oxide, indium oxide, tungsten oxide, and cerium oxide.

Since the last few decades, heterogeneous binary metal oxides photocatalysts like TiO_2, V_2O_5, ZnO, Fe_2O_3, CdO, CdS, and Al_2O_3 (Karunakaran and Senthilvelan, 2005) have been studied extensively for the removal of organic colored pollutants such as azo dye, acid orange 7 (Vinodgopal et al., 1996a,b), methylene blue (Fabiyi and Skelton, 2000), alizarin S, methyl red, Congo red, orange G (Sivalingam et al., 2003). The mineralization of organic pollutants into CO_2, NH_4^+, NO_3^-, and SO_4^{2-}, respectively (Houas et al., 2001) takes place by formation of O_2^-, HOO , or OH through a multistep reduction which can be detected by ESR spin-trapping technique (Liu et al., 2000).

TiO_2 was used for the full oxidation, detoxification and complete mineralization of dyes like, alizarin S, methyl red, Congo red, methylene blue, crocein orange G (Lachheb et al., 2002) dyes in aqueous solutions under irradiation of UV light. Curri et al. (2003) prepared nanoparticles of ZnO and TiO_2 by a nonhydrolytic process and compared its performance for elimination of methylene blue with commercial TiO_2 and ZnO powders. The enhanced efficiency observed was due to crystallinity, nano-size structure, a large amount of surface hydroxyl species and reduced band-gap (Nagaveni et al., 2004). The hierarchical hetero-structure along with morphology, specific composition, and functionalities is one of the important factors, which determine the applicability of a sample (Liu et al., 2017). On the base of TiO_2 nanosheets the MIL-100(Fe) was fabricated by a self-assembly method into hierarchical sandwich-like heterostructures for degradation of methylene blue dye (MB) under visible light ($\lambda \geq 420$ nm). Such a structure increases the absorption ability and superior separation of photogenerated electron-hole pairs. As a result, photocatalytic performance also enhances (Fig. 6.4).

Fluroxypyr (FLX) found in the surface or ground water has been removed at basic pH, i.e., 8−10, by using Degussa P-25 Titania as a photocatalyst. It followed half order kinetics Aramendía et al. (2005). First order elimination of nitrobenzene and substituted nitrobenzenes was carried out with combustion synthesized and commercial TiO_2 under illumination of UV light by Priya and Madras (2006).

The photooxidation of aniline to azobenzene in protic solvent ethanol in the presence of Fe_2O_3 photocatalyst takes place at a higher rate at 254 nm than that at 365 nm. Karunakaran and Senthilvelan (2006) observed that photooxidation under sunlight or UV light depends on the type of protic and aprotic solvents. Furthermore, Karunakaran and Dhanalakshmi (2008) reported that not all the photocatalysts have efficiency or performance. For instance, TiO_2 (anatase), Fe_2O_3, CuO, ZnO, ZnS, ZrO_2, PbO, PbO_2, CdO, HgO degrade phenol under natural sunlight whereas MoO_3, Co_3O_4, CdS, SnO_2, Sb_2O_3, La_2O_3, Pb_2O_3, Bi_2O_3, CeO_2, Sm_2O_3, and Eu_2O_3 do not degrade phenol. Tungsten oxide nanoparticles synthesized by sol-gel method exhibited complete mineralization of dyes under laser irradiation within few minutes (Qamar et al., 2009).

Polisetti et al. (2011) studied the effect of different parameters such as pH, inorganic salts, and H_2O_2 on degradation of anionic dyes such as orange G, amido black, remazol brilliant blue R, and alizarin cyanine green with help of ZrO_2. The physicochemical of the prepared sample has been characterized by FT-Raman spectroscopy, X-ray photoelectron spectroscopy, FT-infrared spectroscopy, UV−vis spectroscopy, BET surface area analysis, and zero

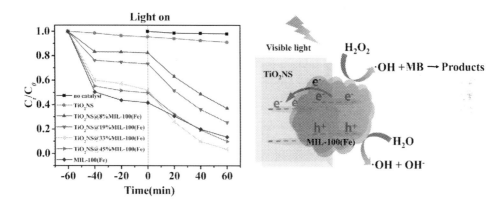

FIGURE 6.4 Hierarchical sandwich-like heterostructures for degradation of methylene blue. *Adapted from Liu X., Dang, R., Dong, W., Huang, X., Tang, J., Wang, G., Appl. Catal. B, 209, 506−513. With permission.*

point charge pH measurement. Elamin and Elsanousi (2013) applied photooxidation for removal of methyl orange by nanostructure zinc oxide (ZnO). Solution combustion method was used for the preparation of nanocrystalline -Fe$_2$O$_3$, which was tested by photocatalytic removal of rhodamine B under irradiation of ultraviolet (UV) light.

Gondal et al. (2009) studied disinfecting *Escherichia coli* microorganism in water by nano-WO$_3$ prepared by sol gel method in the presence of laser light. Binary metal oxides namely, ferric oxides, manganese oxides, aluminum oxides, titanium oxides, magnesium oxides, and cerium oxides are also efficient in removal of heavy metal due to high surface area and specific affinity for heavy metal adsorption from solution (Hua et al., 2012).

Zhang et al. (2017) studied degradation of organic pollutants like phenol, hydrochloride, and rhodamine B dye by a direct Z-scheme carbon nanodots/WO$_3$ nanorods composite in presence of visible light. The photocatalyst decorated by carbon on its surface, leads to synergistic effect and enhances performance three times in comparison to simple WO$_3$ nanorods (Fig. 6.5). The degradation percentages by various samples were:

Carbon Dots/WO$_3$ >	Prepared WO$_3$ nanorods >	Commercial WO$_3$ nanoparticles
(97.1%, 99.1%, 61.2%)	(66.6%, 69.1%, 22.4%)	(22.1%, 11.6%, ~0%)

Binary metal oxides combination alters the solar radiation harvesting capability, which directly affects its performance (Li, 2013). Composites show higher efficiency than its individual semiconductor (Vinodgopal et al., 1996a,b). Wu et al. (2006) synthesized couples CdS/TiO$_2$ through microemulsion-mediated solvothermal method at low temperature. They showed that the CdS semiconductor acts as a sensitizer for titania therefore, enhances the oxidation of methylene blue in water or nitric oxide in air under visible light irradiation. CdS/TiO$_2$ powder fabricated via a sol gel method by varying content of CdS from 5% to 50% (w/w). The increased photocatalytic degradation of an anionic azo-dye, i.e. orange II was observed for composite under a light of wavelength more than 400 nm. The composite with lower content of CdS performed better than semiconductors in pure

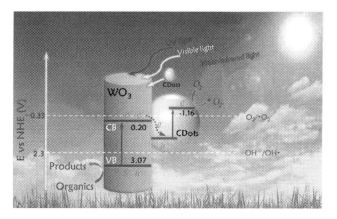

FIGURE 6.5　Elimination of organic pollutants via carbon nanodots/WO$_3$ nanorods composite. *Adapted from Zhang J., Ma, Y., Du, Y., Jiang, H., Zhou, D., Dong, S., 2017. Appl. Catal. B: Environ. 209, 253–264. With permission.*

form (Bessekhouad et al., 2006). Various composites such as $CuO-SnO_2$, ZrO_2-TiO_2, SnO_2/TiO_2, TiO_2/WO_3, Fe_2O_3/SnO_2, TiO_2-ZrO_2, $ZnO-SiO_2$, and $CNTs-CoO-TiO_2$ (Dlamini et al., 2011) have been used for treatment of acid blue 62 dye, 4-chlorophenol, rhodamine-B, methylene blue, acid blue 62, phenol, and Hg (II) reduction (Xia et al., 2007; Neppolian et al., 2007; Zhou et al., 2008; Akurati et al., 2008; Xia et al., 2008; Kambur et al., 2012; Mohamed and Aazama, 2012).

TiO_2/WO_3 and TiO_2/SnO_2 systems degrade 4-chlorophenol at a slow rate of 369 nm, whereas pure component shows no degradation at this wavelength. But at 435 nm, the degradation rate of 4-chlorophenol increases using the same composites (Lin et al., 2008). The impregnation method was used for fabrication of WO_3-TiO_2 photocatalysts. The heterojunction between monoclinic amorphous WO_3 over TiO_2 helps in trapping an electron which accelerates the degradation of bisphenol A (Zerjav et al., 2017).

Nanocomposites are formed not only by suitable combination of binary metal oxides, but it can also be formed by use of appropriate metal along with a semiconductor. Fu and Zhang (2010) treated TiO_2 with negatively charged Au nanoparticles for modification. Au get absorbed on the surface of a semiconductor at lower pH close to isolectric point of titania by electrostatic force of attraction. The degradation of bisphenol A by Au decorated TiO_2 nanocomposite film was accelerated by ~ 2.5 folds due to the ultra-fine size, uniform dispersion, physical separation, and high loading of Au nanoparticles. The Ag-coated TiO_2 composites prepared via an ultrasonication method reported to have higher activity because of the excellent distribution and interaction of Ag nanoparticles with the TiO_2 nanofiber support (Reddy et al., 2011).

TiO_2 activity has been increased by coating it on silica substrate embedded with Ag nanoparticles (TiO_2-SiO_2-Ag composites) with the help of ion implantation technique (Xu et al., 2013). The composites have potential application for removal of toxic and carcinogenic motor vehicle exhaust and solvent utilization. TiO_2-SiO_2 based photocatalysts examined for the elimination of VOCs toluene (Zou et al., 2006). Magdalane et al. (2016) carried out chemical precipitation and hydrothermal process for preparation of CeO_2-CdO nanocomposites by using cerium nitrate and cadmium nitrate. Band gap of coupled metal oxides was 3.15 eV at 393 nm and 2.90 eV at 427 nm. Introduction of transition metal oxide enhances the crystallinity, purity and stability of nanocomposites. The nanocomposite has cubic structure with embedded Cd^{2+} ions in crystal lattice. This composite has better efficiency to inhibit growth of bacteria $P.$ $aeruginosa$ (Fig. 6.6).

The addition of $Ti_{(x)}Zr_{(1-x)}O_2$ (e.g. TiO_2-ZrO_2 system) in polyethersulfone developed the ultrafiltration membrane. In this the ZrO_2 reduces the crystalline growth of titania during the process. It was observed that as Zr content increases the band gap, i.e., the blue shift also increases. The role of metal oxides is to alter the surface of the membrane by affecting the hydrophilicity and porosity of modified membranes (Sotto et al., 2014).

The use of metal free and co-catalyst positively influences the efficiency. The $Mn_{0.25}Cd_{0.75}S/MoS_2$ composites synthesized by single step hydrothermal method reported for hydrogen production from water splitting under visible-light irradiation. It shows better efficiency than $Mn_{0.25}Cd_{0.75}S$ because the interfacial layer between $Mn_{0.25}Cd_{0.75}S$ and MoS_2 accelerates the electron transfer process from $Mn_{0.25}Cd_{0.75}S$ to MoS_2, which directly helps in charge carrier separation (Huang et al., 2017) (Fig. 6.7).

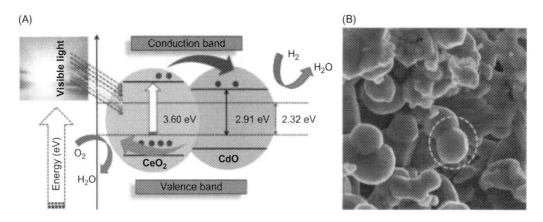

FIGURE 6.6   (A) Mechanism followed by CeO$_2$-CdO nanocomposites and (B) SEM image of CeO$_2$-CdO nanocomposites. *Adapted from Magdalane C.M., Kaviyarasu, K., Judith Vijaya, J., Siddhardha, B., Jeyaraj, B., 2016. J. Photochem. Photobiol. B: Biol. 163, 77–86. With permission.*

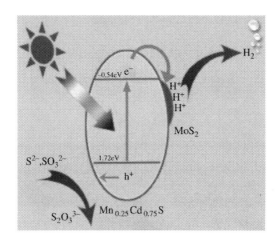

FIGURE 6.7   Scheme of hydrogen production by using Mn$_{0.25}$Cd$_{0.75}$S/MoS$_2$ composites. *Adapted from Huang Q.Z., Xiong, Y., Zhang, Q., Yao, H.C., Li, Z.J., 2017. Appl. Catal. B Environ. 209, 514–522. With permission.*

6.2.2 Ternary and Quaternary Oxides

Spins shape ZnFe$_2$O$_4$ (band gap 1.90 eV at 653 nm), Sr$_7$Fe$_{10}$O$_{22}$, and SrFe$_{12}$O$_{19}$ were synthesized by polymeric method, and solid state method and applied for photodecomposition of isopropyl alcohol (Jang et al., 2009, Doma et al., 2012). Semiconductor Ba(In$_{1/3}$Sn$_{1/3}$M′$_{1/3}$)O$_3$ (M represents the metal Sn, Pb, Nb, or Ta) has band gap in the range of 2.97–3.30 eV. The replacement of electronegative non-transition metal cations provides more sensitivity towards a broad range of solar spectrum. The change the band structure enhances the removal of 4-chlorophenol (Hur et al., 2005).

Quaternary metal oxides like Bi$_4$Nb$_x$Ta$_{1-x}$O$_8$I (Hu et al., 2012), MnCoTiO$_2$ (Kiriakidis and Binas, 2014), and Pb$_2$FeNbO$_6$ (Vijayasankara et al., 2013) photocatalysts synthesized

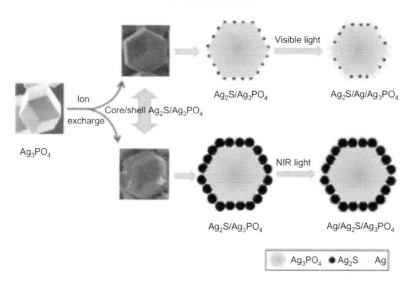

FIGURE 6.8 Preparation scheme of core shell Ag_2S/Ag_3PO_4 composites. *Adapted from Tian J., Yan, T., Qiao, Z., Wang, L., Li, W., You, J., Huang, B., 2017. Appl. Catal. B Environ. 209, 566–578. With permission.*

by various methods degrade dye contents into biodegradable product. The ternary nano-hybrid $ZnS-Ag_2S-RGO$ composite (Reddy et al., 2015a), $ZnO-RuO_2/RGO$ (Reddy et al., 2015b), and reduced graphene-oxide/titanium dioxide/zinc oxide ($rGO/TiO_2/ZnO$) have great efficiency towards photocatalytic oxidation of rhodamine B and methylene blue. The metal oxides of Zn−In (ZnIn-MMO) in combination with $g-C_3N_4$ show a higher performance for degradation of rhodamine B to that of pure $g-C_3N_4$ and ZnIn-MMO under visible light (Lan et al., 2015).

The core shell Ag_2S/Ag_3PO_4 composites were prepared by varying the content of Ag_2S. The composite with 5% and 50% Ag_2S shows highest activity in visible light and NIR-light-driven, respectively (Tian et al., 2017) (Fig. 6.8).

The one pot precipitation, hydrolysis, and UV photoreduction was used for preparation of plasmonic Bi and BiOCl sheet decorated $ZnSn(OH)_6$ that consist of cubic structure, and 14-facets polyhedral and octahedral system. The modification improves the abstraction of photon, charge segregation, and migration performance, hence increases the photooxidation of rhodamin B 81 times to that of pure $ZnSn(OH)_6$ (Wang et al., 2017).

Hussein and Abass (2010) tried decolorization of dye without light and without catalyst, and showed that decolorization occurs only in presence of both the catalyst and light. There are various binary, ternary, and quaternary metal oxides like TiO_2, ZnO, SnO_2 (Gubbala et al., 2009), WO_3, ZnS, ZrO_2, CdO, HgO, PbO, PbO_2, Fe_2O_3, $BiVO_4$, $SrTiO_3$ (Modak and Ghosh, 2015), $PbCrO_4$, $PbMoO_4$, $PbWO_4$, etc. exhibiting photocatalytic activity because of their low band gap, durability, stability, cost effectiveness, reusability, nontoxic nature, etc. Different methods have been used for synthesis of various photocatalysts such as hydrothermal-assisted sol-gel technique (Khaghanpour and Naghibi, 2017), sol-gel dip coating method (Hamid and Rahman, 2003), electrochemical deposition process (Samad et al., 2016), Stober method (Ameta et al., 2013), precipitation method (Solanki et al., 2015a), combustion method (Ma et al., 2014), etc. (Table 6.1).

TABLE 6.1   Band Gaps of Different Semiconductors

Semiconductor	Band Gap (eV at 300K)
ZnS (Wurtzite)	3.91
ZnS (Zinc blende)	3.54
SnO_2	3.60
TiO_2	3.20
ZnO	3.03
WO_3	2.60
CdS	2.42
Fe_2O_3	2.20
CdO	2.10
Cu_2O	2.10
CdSe	1.70
AlSb	1.58
CdTe	1.56
GaAs	1.42

As most of the photocatalysts have wide band gap (E_g) and these are active in UV region only. Solar spectrum consists of only about 4% of UV light (wavelength less than 387 nm). In this context, lots of investigations have been made to extend the photo-response of semiconductors into the visible region (Ali et al., 2012) and making use of 55% of the total solar radiation (Nyamukamba et al., 2017). Semiconductors provide promising solutions for environmental pollution problems. Attempts have been made to enhance the photocatalytic activity of these materials using various techniques. Semiconductors with wide band gap and white color do not absorb in the visible range of spectrum. So a need arises to modify these photocatalysts. Different methods have been used to modify the photocatalysts, like surface modification, composite or coupling materials, metal and nonmetal doping, surface sensitization by organic dyes and metal complexes, etc. These modifications can be done by various techniques like doping with metal or nonmetal (Rajkumar and Singh, 2015; Medina-Ramírez et al., 2014), codoping, combining different metal oxides (composites), sensitization, etc.

6.3 MODIFICATIONS

Enhancement in performance of a semiconductor can be obtained by four basic ways (Marschall and Wang, 2014) (Fig. 6.9). These are:

- by formation of localized state just above the valence band,
- using semiconductor with low band gap,

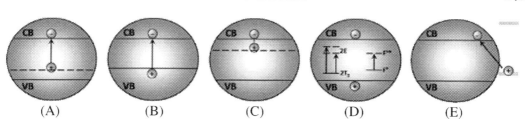

(A)   (B)   (C)   (D)   (E)

FIGURE 6.9   (A) Localized states above valence band. (B) Semiconductor with narrowed Eg. (C) Localized states below conduction band. (D) Color centers formed in the Eg and (E) Surface modification. *Adapted from Marshall R., Wang, L., 2014. Catal. Today, 225, 111–135. With permission.*

- by formation of localized state just below the conduction band,
- by color centre formation in band gap,
- and by surface modification.

Therefore, techniques involved in modifications are:

1. doping with metal and/or nonmetal
2. codoping with various combination of donor and acceptor materials
3. coupling photocatalysts (composite)
4. sensitization
5. substitution
6. miscellaneous

Wide band gap photocatalysts are colorless with high electron-hole recombination. The light harvesting span of substance, electron-hole pair life time, and number of reactive sites present on the semiconductor play an important role in photocatalytic performance (Johar et al., 2015) (Fig. 6.9).

6.3.1 Doping

Doping is one such modification of photocatalyst that reduces the band gap between valence band and conduction band by adding impurities in an otherwise pure semiconductor. Metals as well as nonmetals can be doped on a semiconductor. Each type of dopant has a unique impact on the crystal lattice of the semiconductor. Metal and nonmetal doping enhances the photo-responsiveness of the photocatalyst to the visible region by creating new energy levels (also called the impurity state), between the valence band and the conduction band (Figs. 6.10 and 6.11). These new levels reduce the band gap, and the excited electrons are shifted from the impurity state to the conduction band. Doping is considered as a part of band gap engineering, which involves introduction of an electron or hole in the photocatalytic semiconductor. Metal dopants like Co, Cr (Bae et al., 2007), W (Li et al., 2012), Cu, Zn (Khairy and Zakaria, 2014), Mo (Li et al., 2015), Mn, Fe, Ni (Feng et al., 2015), etc. give rise to a new band below the conduction band whereas, nonmetals like N (Zhang et al., 2013), P (Kuo et al., 2014), F, Si, S, Cl, Se, Br, I (Guo et al., 2015a), etc. create a new band above the valence band. It helps to avoids recombination by enhancing trapping of electrons (Xiao et al., 2008). As a consequence, photocatalytic activity also increases. Metal dopants improve the morphology, surface area, photocatalytic

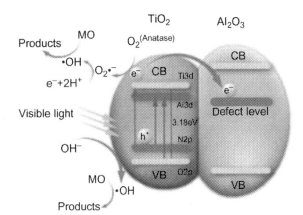

FIGURE 6.10   Effect of metal doping on conduction band.

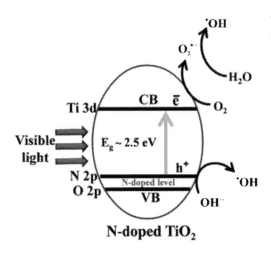

FIGURE 6.11   Effect of nonmetal (nitrogen) doping on valence band.

performance, electronic, and magnetic properties of photocatalytic semiconductors (Patil et al., 2013).

There are three possibilities in which a nonmetal reacts with oxides of photocatalyst. First, dopant could hybrid with oxide of photocatalyst. Second, dopant could replace the oxygen site and third, the addition of dopant in oxygen deficient site and oxygen sub-stoichiometry, which acts as a blocker for reoxidation (Rehman et al., 2009). Li et al. (2005) noticed the formation of oxygen vacancy state due to the presence of nonmetal carbon dopant due to the formation of Ti^{3+} species between the valence and the conduction bands in the TiO_2 band structure. Size of dopants is an important factor, which decides mode of substitution of oxygen from photocatalyst. Guo et al. (2015b) showed that the smaller radius species like C, N, or F substitute the O atom in the TiO_2 terminate surface whereas, larger radius species such as P, S, Cl, S, and Br replace the O in the SrO-terminated surface.

Generally, oxides of photocatalyst and nonmetal dopants have approximately close energy and, therefore, band gap of catalyst decreases and it becomes active in absorbing visible light (Wang et al., 2012a). Band gap engineering method also alters the morphology, structure, electrical and optical properties of a semiconductor (Mai et al., 2009), which affects the light absorbance, charge-carrier mobility, redox potential, and surface area (Hu et al., 2011). Le et al. (2011) reported that incorporation of donor and acceptor species directly affects electronic structure and in turn, reactivity of photocatalyst. The doping reduces the band gap of photocatalyst, i.e. dispersion of the CB and VB of photocatalysts shows a positive effect after doping. Such modification results in the red-shift of absorption spectra edge (Yun et al., 2013). This type of bathochromic shift means a decrease in band gap or an addition of intra band gap state due to doping resulting into the harnessing of more photons from visible light. The addition of dopants in photocatalyst exhibits enhanced performance because of the following reasons (Ran et al., 2015):

- it avoids electron-hole recombination,
- provides enhanced surface area,
- may lead to an increase in pore size of a sample,
- high crystalline state,
- and extends sensitivity towards a broad range of spectrum.

6.3.2 Codoping

Codoping involves different combinations of donor and acceptor species, which reduces the band gap by raising the valence band edge and also by lowering the conduction band edge. It helps in achieving activity towards a broad range of solar spectrum. Codoping of both metal and nonmetal has also been reported in some cases (Fig. 6.12).

Yan et al. (2013) found that like simple doping with a metal or nonmetal, this technique also helps in solving issues like carrier recombination solubility limit, low carrier mobility, and nonresponse to visible light in a host compound. Codoping with nonmetal/metal such as C-Mo, N-W (Folli et al., 2015), N-Nb (Folli et al., 2016), N-Ta, and Ce-F enhances

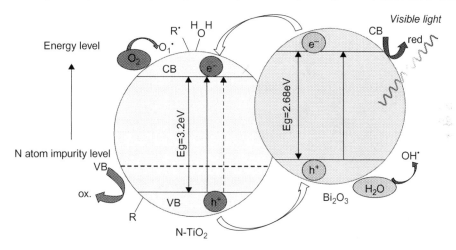

FIGURE 6.12   Effect of codoping.

the efficiency but puts very little effect on the position of conduction band edges (Liu, 2012). Huang et al. (2014) also reported that the combination of Ce and F affects the microstructure and optical band gap in Bi_2WO_6. Codoping with metal (Sm) and nonmetal (N) reduces the crystal growth, recombination of charge carrier, and the transformation from anatase to rutile phase (Ma et al., 2010). Ga and Cu codoped ZnS synthesized by hydrothermal method possesses stability under a long duration of irradiation and exhibited 58 times higher efficiency than that of pure ZnS (Kimi et al., 2015).

The N/F/anatase titania with {001} facets developed by hydrogel method, exhibited improved performance as it increases a driving force of the electron. It further increases the electron injection efficiency from LUMO of dye and conduction band of a semiconductor. Due to positive-shift flat band (Yu et al., 2014) (Fig. 6.13).

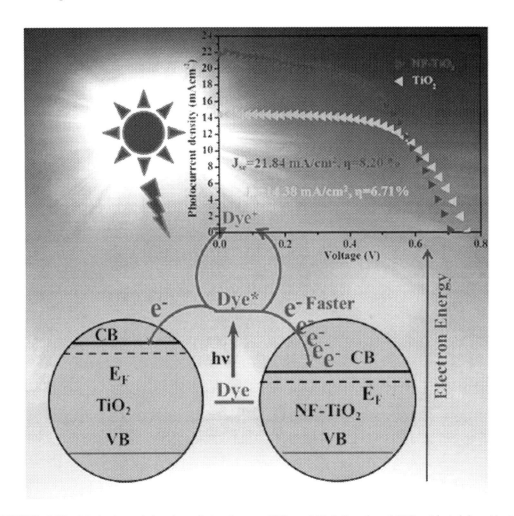

FIGURE 6.13 Mechanism of dye degradation in pure TiO_2 and N & F codoped TiO_2. *Adapted from Yu, J., Yang, Y., Fan, R., Li, L., Li, X., 2014. J. Phys. Chem. C., 118(17), 8795—8802. With permission.*

Nonmetal and plasmonic metal decorated titania showed photoresponse towards UV, visible light, and direct sunlight, which makes it useful in elimination of various organic compounds (Zhang et al., 2011). Wang et al. (2009) used biomolecule l-cysteine as a common source of carbon, nitrogen, and sulfur to prepare C-N-S-tridoped titanium dioxide nanocrystals by hydrothermal method. Here, sulfur substituted oxygen, nitrogen existed in bond with O and Ti as N-Ti-O and Ti-O-N, and carbon deposited on surface of titania by forming a mixed layer of carbonate species. It was observed that biomolecule l-cysteine also controlled the morphology and final crystal phases. C, N, S-Tridoped TiO_2 exhibited six times higher efficiency than pure and commercial photocatalyst (Zhou and Yu, 2008).

6.3.3 Coupled Semiconductors or Composites

Composite formation, i.e. coupling of semiconductors, is another efficient method to make semiconductors photoresponsive in the visible region of the spectrum. Semiconductors that are chosen to prepare a composite must have variable band gaps. A large band gap semiconductor is usually coupled with a small band gap semiconductor having a more negative conduction band level. Consequently, the electrons in the conduction bands are inserted from the small band gap semiconductor to the large band gap semiconductor.

Various combinations of photocatalyst have been coupled such as $FeTiO_3/TiO_2$, Ag_3PO_4/TiO_2, $W_{18}O_{49}/TiO_2$, CdS/TiO_2, $CdSe/TiO_2$, $NiTiO_3/TiO_2$, $CoTiO_3/TiO_2$, Fe_2O_3/TiO_2 (Rawal et al., 2013), ZnO/CdS, TiO_2/SnO_2 (Liu et al. 2007), ZnO/TiO_2, and ZnO/Ag_2S (Johar et al., 2015). Such composites give higher photocatalytic activity than single ones due to the synergistic effect. It was reported that Cr_2O_3 coupled with SnO_2 has been found to exhibit superior photocatalytic activity under visible-light irradiation because of good crystalline nanoparticles, smaller crystal size, and a stronger response to visible light. The sample degrades 98% of rhodamine B in 60 min under irradiation of visible light (Bhosale et al., 2013).

The photocalaytic activity can be further increased by doping coupled photocatalysts by various combinations of metals and nonmetals. Samadi et al. (2015) codoped TiO_2 and SiO_2 nanocomposite with Nd^{3+} and Zr^{4+} and observed the effect of dopant on distribution and monotonous coating of TiO_2/SiO_2 thin film. Doping creates uniform distribution of particles with low agglomeration, and spongy microstructure. There is active decomposition of methyl orange due to reduction in charge recombination. Coupling may also involve the modification of the surface of mesoporous SnO_2 by other metal oxides such as by metal oxides like TiO_2 or Al_2O_3. (Ramasamy and Lee, 2010).

6.3.4 Substitution

Another method to change activity of photocatalyst is by substitution of one metal by another. Borse et al. (2012) reported the substitution of Ti^{4+} at the Fe^{4+} site in a $CaFe_2O_4$-$MgFe_2O_4$ bulk hetero-junction (BH) lattice photocatalyst and reported enhanced photocurrent. The Ti ion doped sample also showed increased quantum yield for photodecomposition of H_2O-CH_3OH mixture due to efficient charge separation.

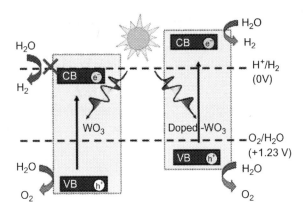

FIGURE 6.14 Mechanism of electron transfer in WO_3 and doped WO_3. *Adapted from Wang, F., Valentin, C.D., Pacchioni, G., 2012. J. Phys. Chem. C. 116 (16): 8901–8909. With permission.*

Wang et al. (2012c) studied change in the band gap on substituting W metal in WO_3 by various metals such as Mo, Ti, Zr, and Hf. They reported that this substitution of W in WO_3 by another metal having the same valency such as Cr and Mo reduces the band gap because of shifting of CB edge to the lower side. On the other hand, incorporation of metal of lower valency such as Ti, Zr, and Hf lead to an increase in band gap due to shifting of CB edge higher. The narrowing of band gap on substitution of O by S occurs due to formation of localized occupied states above the VB and shifts the CB minimum upward—two effects that go in the right direction (Fig. 6.14).

Effect on band edge or electronic properties was also studied by Long et al. (2009) by using Si, Ge, Sn, and Pb dopants. They observed that same dopant with different form i.e. anatase or rutile gives different electronic property. Thus, different dopants may provide the same effect on band gap. e.g. Si-, Ge doped, and Sn-; and Pb-doped rutile TiO_2 shows a decrease in band gap of about 0.1 with approximately 0.55 eV whereas, anatase form reduces band gap by 0.20 and 0.15 eV and broadening by 0.06 and 0.02 eV, respectively.

6.3.5 Sensitization

Another promising method for surface modification of photocatalyst is sensitization. Dyes or complexes can be used in photocatalytic systems, including solar cells due to their redox properties and sensitivity towards visible light. On exposure to visible light, dyes or complexes can inject electrons to the conduction band of the semiconductor so as to initiate a catalytic reaction (Fig. 6.15).

Sensitizers are organic and inorganic compounds that get adsorbed on the surface of semiconductor by chemisorption or physisorption. The phenomenon of coating the surface of a semiconductor by sensitizers is sensitization. Sensitizers act as antenna that absorb light and transfer it to a semiconductor, which helps in improving the excitation process.

Sensitizers can be categorized on basis of the occurrence:

- synthetic sensitizer
- natural sensitizers
- self sensitizer

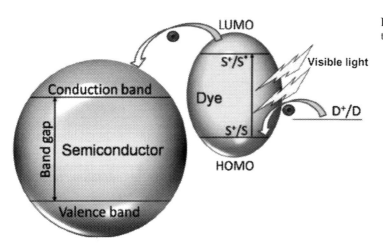

FIGURE 6.15 Role of a sensitizer in photocatalysis.

Synthetic sensitizer includes commercial dyes like porphyrin dye (Mathew et al., 2014), Ru complexes (Lee et al., 2013), Ru-polypyridyl-complex sensitizers, e.g., cis-dithiocyanato bis(4,4'-dicarboxy-2,2'-bipyridine)ruthenium(II) (Hara and Koumura, 2009), quinoxaline-based organic sensitizer (Chang et al., 2011), etc.

The rate of photocatalysis depends on the sensitivity of the photocatalyst to light radiations. In order to enhance the photosensitivity of the semiconductor some chemical substances having chromophores like dyes or natural pigments etc. could be used. These chemical substances are called photosensitizers. A photosensitizer absorbs light and transfers energy or electrons to the semiconductor. Dyes can be used in photocatalytic systems and in solar cells due to their redox properties and sensitivity towards visible light. On exposure to visible light, dyes can inject electrons to the conduction band of the semiconductor so as to initiate a catalytic reaction.

Natural sensitizers such as Monascus red (Lee et al., 2014), *Pastinaca sativa*, *Beta vulgaris* (Hemmatzadeh and Mohammadi, 2013), carotenoids, polyphenols (Hug et al., 2014), chlorophyll (Solanki et al., 2015a) and its derivative (Wang et al., 2010), anthocyanin (Lee et al., 2013), *Festucaovina*, *Hibiscus sabdariffa*, *Tageteserecta*, *Bougainvillea spectabilis*, and *Punicagranatum peel* (Hernández-Martínez et al., 2012), strawberry and turmeric powder dyes (Mohammed et al., 2015) are extracted from seeds, leaves, flower petals, fruits, barks, etc., insects, animals, and ores (Venil et al., 2013); and therefore, natural sensitizers are also said natural dyes or pigment. Performance depends on the type of natural pigment. Synthetic dyes are expensive so, natural dyes are preferred over synthetic dyes as they are (Narayan, 2012):

- abundant raw material,
- easy to extract,
- low cost,
- and eco-friendly.

These natural organic dyes are potential candidates to replace some of the man-made dyes used as sensitizers (Jasim, 2012). Natural pigments are photosensitizer because of their structural and optical properties (Alhamed et al., 2012). Kushwaha et al. (2013)

isolated extract from the leaves of teak (*Tectona grandis*), tamarind (*Tamarindus indica*), eucalyptus (*Eucalyptus globulus*), and the flower of crimson bottle brush (*Callistemon citrinus*) and used it for enhancing visible light photosensitivity of dye based solar cells.

The natural dyes show less efficiency as compared to synthetic dyes. Improved efficiency depends on various factors like (Hemmatzadeh and Mohammadi, 2013):

- method of extraction of dye,
- the acidity of solvent used for extraction, and,
- different compounds of solvents on the optical absorption spectra.

It was observed that a mixture of ethanol and water as an extracting solvent and also the acidity of dye solution influence the performance and optical activity.

The natural dye not only accelerates the light absorption efficiency but it also increases sustainability (Hug et al., 2014), and electron transfer rate (Rochkind et al., 2015; Solanki et al., 2015b). Benjamin et al. (2011) used extracts of natural pigments like chlorophyll and anthocyanin to sensitize ZnO and used it for the photocatalytic degradation of dyes. It was found that sensitization of photocatalyst by natural pigment reduces its band gap, which leads to utilization of visible and infrared radiations more efficiently.

Flowering plants of order *Caryophyllales* contain nitrogen-containing betalain pigments having desired light absorbing and antioxidant properties. These are used as coloring agents or sensitizers. These are associated with various copigments, which are responsible for changes in their light absorption properties. The carboxylic functional groups present in pigments easily bind to the surface of photocatalyst like TiO$_2$ (Hernández-Martínez et al., 2013). Saati et al. (2011) showed that the anthocyanin pigment extracted from red rose flower petals is not only used as sensitizer in many dye sensitized solar cells, but it also acts as an antioxidant to protect the fat content of fermented milk such as yogurt. Sorghum red pigment (natural dye extracted from sorghum shell) showed maximum absorption wavelength, and photostability of this pigment depends on the pH value (Zhang et al., 2013). The extract from *Saraca indica* leaf serves as reducing agent in CuO nanoparticles synthesis. This extract also acts as capping agent or stabilizing agent, which stabilizes the catalyst. CuO nanoparticle synthesis by green pathway, i.e., by using extracts of *S. indica* help in determining size and morphology (Prasad et al., 2017).

The natural dye molecules have a tendency to degrade in the presence of light, just like other hazardous sensitizers. After degradation, it leaves no traces of organic species in solution. Sensitizer degradation leads to deactivation of the supported catalyst, when it is recovered. In such cases, the problem can be solved by retreatment of the recovered catalysts with fresh dye (Zyouds et al., 2011).

However, synthetic dyes show more efficiency than that of natural dyes. Synthetic dyes are difficult to synthesize. They have a high cost, are toxic, and have less yield (Okoli et al., 2012). On the other hand, natural dye extraction is easy, low cost, temperature compatible, completely biodegradable, and nontoxic. But most of the sensitizers are susceptible to photocorrosion or degradation in aqueous solutions and therefore, are not suitable for photocatalytic water treatment applications (Pelaez et al., 2011). The quantum dots, plasmonic metal nanostructures, and carbon nanostructures are also included in synthetic photosensitizers, which are coupled with wide-band gap transition metal oxides to design better visible-light active photocatalysts (Chen and Wang, 2014) (Fig. 6.16).

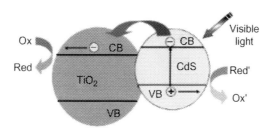

FIGURE 6.16 Charge transfer between CdS and TiO₂. *Adapted from Chen, H., Wang, J., 2014. J. Nanotechnol., 5, 696–710. With permission.*

It has been noticed that semiconductors such as CdSe, CdTe, and CdS also act as self sensitizers to sensitize other semiconductor TiO_2 (Shen et al., 2015), $SrTiO_3$, $BaTiO_3$, and $CaTiO_3$ (Pan et al., 2015).

6.3.6 Miscellaneous

Titania is a perfect crystalline metal oxide with band gap greater than 3.1 eV. Such wide band gap makes it active in the ultraviolet (UV) range, and it has a potential for clean energy generation and environmental remediation. Yaghoubi et al. (2015) noticed the formation of midgap-states-induced energy gap in titania when synthesizing it with solution-phase method. The undoped mixed-phase TiO_2 nanoparticles achieve the band gap of about 2.2 eV and also possess oxygen vacancy and high surface area, accelerating its photoreactivity.

6.3.6.1 Mechanism

The electron in valence band (VB) gets excited to conduction band (CB) on absorption of a suitable wavelength of light. The gap between VB and CB is known as band gap. Therefore, minimum energy required for excitation of an electron is equal to energy of band gap. Thus, photocatalytic oxidation or reduction may be initiated by absorption of a photon having energy equal to or more than the band gap of photocatalyst. Excitation of an electron leads to formation of a pair of charge carriers ($e^- - h^+$ pair) on the surface of photocatalyst, i.e. hole and electron in valence band and conduction band, respectively. At this stage, there are two possibilities:

1. Charge carrier produced again recombine and release energy in the form of heat.
2. Charge carrier may react further with an electron donor or electron acceptor on the photocatalyst surface.

In the first case, no reaction takes place. In the second case, electron or hole reacts with dissolved oxygen or water to form $O_2^{\bullet-}$, $HOO^{\bullet}$ or $OH^{\bullet}$ free radicals. These reactive oxidizing species along with other species like H_2O_2 and O_2 play a major role in photooxidation reaction (Fig. 6.17).

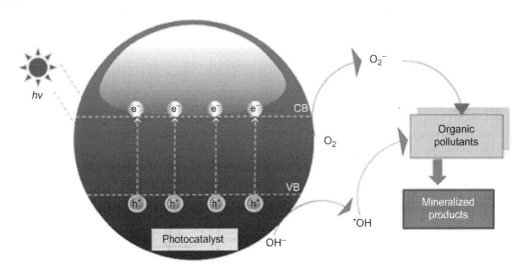

FIGURE 6.17   Photocatalytic oxidation by photocatalyst under solar irradiation.

6.4 WASTEWATER TREATMENT

Effluents from various sources like domestic, municipal, industrial, sludge, etc. consist of many nonbiodegradable organic compounds, xenobiotic chemicals, oil, petroleum waste, dyes, etc. An increase in industrialization has become a major source of pollution. Particularly, chemical, textile, paper, pesticides, fertilizer industries, etc. discharge their wastes in nearby water bodies without any primary treatment, which is a major reason for making water polluted. Water pollution is emerging as a serious threat to the environment.

Many conventional chemical, physical and biological methods like adsorption, sedimentation, coagulation, filtration, chlorination, ozonation, reverse osmosis, UVC disinfection activated sludge process, trickling filter, oxidation ditch, aerated lagoon, stabilization ponds, etc. have been used since decades (Pokharna and Shrivastava, 2013); but these are not adequate and/or appropriate to treat wastewater streams on a large scale as these require high capital investment, operation, maintenance cost, and a large area (Amin et al., 2014). Secondly, these methods do not degrade the effluents up to the desired extent that it can be recycled, and also require analysis time in a number of days. Therefore, there is a vital requirement for developing new hygienically friendly purification technologies.

Refinery wastewater consists of various concentrations of aliphatic and aromatic petroleum hydrocarbons, which are toxic and not possible to degrade by traditional treatment methods. Petroleum refineries' wastewater was treated using titania nanoparticles as the photocatalyst in UV/TiO_2 process by Saien and Shahrezaei (2012).

Researchers have supported the technique of photocatalytic decomposition process to degrade, decompose, eliminate the hazardous or nonhazardous hard-to-biodegrade organic, inorganic, pollutants or contaminants in industrial wastewater before they are released into mainstream water bodies. The photocatalytic decomposition process is also referred as photocatalytic oxidation or photocatalysis. This method generates highly reactive hydroxyl radicals, which oxidize the matter in solution and completely mineralize it into water, CO_2 and inorganic compounds (Reza et al., 2015).

It is classified as an advanced oxidation process (AOP) and it is carried out with the help of nanostructured metal oxides or sulfide semiconductors such as TiO_2, ZnO, SnO_2 (Song et al., 2014)., SiO_2, CeO_2 (Montini et al., 2016), ZnS (Fang et al., 2011), MnS, CdS, CuS, $SrTiO_3$, $BaTiO_3$, $PbTiO_3$ (Piskunov et al., 2004), hematite ($-Fe_2O_3$), Bi_2WO_6 (Rahimi et al., 2016), etc. AOP is an effective technique for the oxidation of resistant pollutants, such as phenolic compounds (Elango et al., 2015).

The combination of photocatalyst and UV light has the potential to oxidize organic pollutants or hazardous organic compounds into nontoxic or less harmful products, such as CO_2 and water (Chatterjee and Dasgupta, 2005). It can also destroy certain microorganisms i.e., bacteria and some viruses in the secondary wastewater treatment. Malato et al. (2009) has excellently given other reviews including ours, and reviewed the decomposition and degradation of wastewater contaminant, removal of toxic organic compound, and disinfection of water. Sharma et al. (2010) utilized TiO_2 photocatalyst in combination, with ferrate [Fe(VI)] for many industrial applications, including the degradation of recalcitrant contaminants in water and wastewater treatment. AOPs have proved to be a cost efficient method for conversion of hazardous substances or contaminents, even if present in minute amount (ppm or ppb level) into benign components. It also avoids harmful consequences of pollutants that are highly toxic, persistent, and difficult to treat (Baruah et al., 2012).

6.4.1 Dye Degradation

Recently, photocatalysis has attracted the attention of the scientific community all over the world for organic waste contaminant degradation, and also to solve or overcome the issues related to energy and the environment (Ibhadon and Fitzpatrick, 2013) (Fig. 6.18). Dye molecules have different structures and are very less or not biodegradable at all. The elimination of dye molecules from wastewater requires the complete breakdown of the conjugated unsaturated bonds in molecules (Pinotti et al., 1997). The decontamination of wastewater or degradation of dye contents by photocatalytic degradation of organic dyes using photocatalyst has been proved to be an excellent method (Gaya and Abdullah, 2008). Nanoparticles play a crucial role in heterogeneous catalysis as they complete mineralize hazardous pollutants to harmless materials (Yan et al., 2014) (Fig. 6.18).

Major investigations have been carried out on photocatalytic degradation of organic pollutant by using TiO_2 (Ohko et al., 1997). The nanocomposite of titania and carbon nanotube (CNT) shows an increase in degradation efficiency for methylene blue because CNT avoids the agglomeration of titania and thus, increases the surface area. Further, it leads to electron transfer in the composites (Wongaree et al., 2015). The nanocomposite absorbs light and creates a hole in CNT and transfers electrons to CB of titania. The positively charged hole reacts with a water molecule to produce OH ions and $^\bullet OH$ radicals. Simultaneously, photocatalyst reacts with dissolved oxygen to generate $O_2^{\bullet-}$ radical ions that are able to degrade methylene blue (Fig. 6.19).

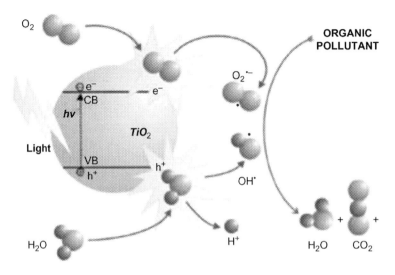

FIGURE 6.18 Removal of organic pollutants. *Adapted from Ibhadon, A.O., Fitzpatrick, P., 2013. Catalysts, 3, 189–218. With permission.*

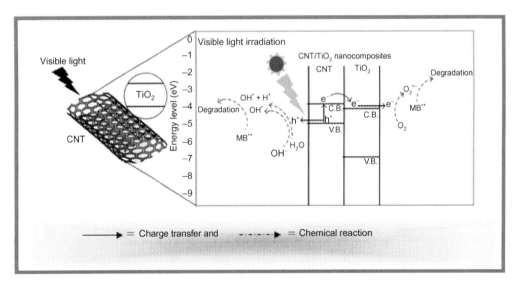

FIGURE 6.19 Photocatalytic mechanism of CNT/TiO$_2$ nanocomposites. *Adapted from Wongaree M, Chiarakorn, S., Chuangchote, S., 2015. J. Nanomater. doi.org/10.1155/2015/689306. With permission.*

The rate of photodegradation depends on the kind of photocatalytic surface. Smirnova et al. (2015) investigated two pathways for photodegradation of acridine yellow and these are:

1. *Fast rate of photodegradation*: When decomposition occurs on the surface of TiO_2 and TiO_2/SiO_2, then the half time of degradation has been reported to be 4 and 15 min, respectively.
2. *Slow rate of photodegradation*: It involves half time of more than 90 min, when reaction oxidation of protonated via amino group acridine yellow on silica surface is in the air.

They observed that degradation of acridine yellow on the surfaces of mesoporous TiO_2, SiO_2, and TiO_2/SiO_2 films take place in the following trend under UV irradiation:

$$SiO_2 < TiO_2/SiO_2 < TiO_2$$

Luna-Flores et al. (2017) incorporated sensitizer zinc phthalocyanine in commercial TiO_2 material and observed changes in optical and structural property. The synergetic effect of sensitizer and photocatalyst accelerates the electron injection into the semiconductor and simultaneously, it prevents the back electron transfer, i.e., recombination of the charge carrier. Sensitized semiconductor exhibited better degradation of rhodamine B under visible light irradiation than that of bare or unsensitized semiconductor. The toxicity of wastewater was also examined by the inhibitory effect on the sprouting of chia seeds (*Salvia hispanica*). They showed that byproducs formed by degradation of dye products do not lead to an acute toxicity (Fig. 6.20).

The industrial solvent trichloroethylene is a polar and toxic chemical that adversely affects the soil, surface of water, and also ground water. Bak et al. (2015) used nano-ZnO-laponite porous balls with 20% porosity for the elimination of trichloroethylene. Removal occurs due to generation of hydroxide by photons.

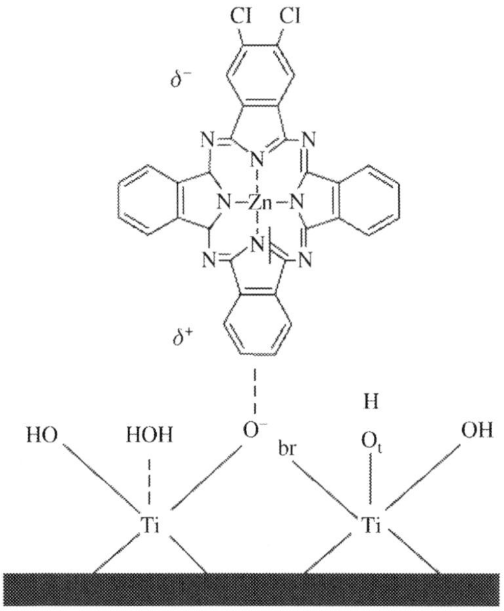

FIGURE 6.20 Preferential binding of zinc phthalocyanine onto the TiO_2 surface. *Adapted from Luna-Flores A., Valenzuela, M.A., Luna-López, J.A., Hernández de la Luz, A. D., Muñoz-Arenas, L.C., Méndez-Hernández, M., Sosa-Sánchez, J.L., 2017. Int. J. Photoenergy. doi.org/10.1155/2017/1604753. With permission.*

It is not always that the addition or substitution of dopant results in a decrease in band gap. Band gap may also increase in some cases, because of increase in particle size, which can be explained by the quantum confinement effect. The ZnS has band gap of 3.21 eV, which is increased by 0.21 eV i.e., 3.42 eV after being doped with Ag nanoparticles. ZnS-Ag nanoballs (doped) degraded organic dye methylene blue more effectively in comparison to the bare ZnS (Sivakumar et al., 2014). Cationic dye crystal violet was degraded by ZnO nanoparticles under visible irradiation (Habib et al., 2013).

According to them, dye contents are very sensitive to visible light. Therefore, on exposing it to light in the presence of a photocatalyst, electron transfer process occurs between dye and photocatalyst (semiconductor or SC).

- Firstly, dye molecule adsorbed on SC get excited by absorption of light.
- Then, excited dye transfers its electron in CB of SC.
- This excited electron reacts with O_2 (dissolved oxygen) to form active oxygen anion radicals.
- On excitation of photocatalyst, electron-hole pair formation takes place, which is responsible for production of $^\bullet OH$, $HO_2^\bullet$, H_2O_2, and also O^- reactive oxygen species.
- These active species react with dye to degrade or mineralize it.

Ali et al. (2013) reported the reaction of a hole with a hydroxide ion and water molecule to generate a stronger oxidizing agent, $^\bullet OH$ radical, which will assist in removal of organic pollutants from wastewater.

$$h^+ + OH^- \rightarrow \quad OH + H^+ \tag{6.1}$$

$$h^+ + H_2O \rightarrow \quad OH \tag{6.2}$$

Vattikuti and Byon (2016) used hydrothermal method for coating molybdenum disulfide nanospheres on monodispersed Al_2O_3. The core shell structure influences removal of water pollutant rhodamine B and enhances the chemical stability.

SnO_2 has excellent properties like mechanical hardness, uniformity, low resistivity, stability to heat treatment, optical transmittance, and piezoelectric behavior (Ali and Tajammul, 2013). Lin et al. (2017) reported that this photocatalyst can be synthesized in various shapes like nanorods, nanotubes, nanosheets, nanowires and so on. They form microflower three dimensional SnO_2 hierarchical nanostructures by hydrothermal method and exhibited good performance in wastewater purification. Panchal et al. (2014) carried out the photocatalytic degradation of erythrosine-B using semiconductor SnO_2 under visible irradiation. The effect of different parameters like pH, concentration of dye, amount of photocatalyst and light intensity on the rate of photodegradation of dye was observed and optimal conditions were obtained.

Bhattacharjee and Ahmaruzzaman (2015) fabricated quantum dot SnO_2 by using natural biomolecule amino acid, serine via green chemical pathway i.e., microwave heating method. Quantum dot SnO_2 lead to a blue shift in the band gap energy, i.e., 4.5 eV and was employed as photocatalyst in the degradation of an organic dye, eosin Y, by solar irradiation. Coupled ZnO/SnO_2 nanocomposites were prepared by using coprecipitation method by Tanasa et al. (2013). They used different molar ratios of 1:10, 1:2, 2:1, and 10:1 and observed that the presence of ZnO in SnO_2 improves its performance under UV irradiation. The photocatalytic activity of the samples was tested by examining the eosin Y degradation from wastewaters. Different parameters such as crystallite size, surface area, absorption edge, TOC values, time of reaction, and catalyst concentration were also taken into consideration. Similarly, improved efficiency was observed for $SnO_2/CNTs$ nanocomposites (Wang et al., 2012b), because the electron generated by photon absorption showed separately transferring to the CNx, has a high degree of defects. Hence, the radiative recombination of the charge carrier was hampered and the photocatalytic activity presented significant enhancement for the SnO_2/CNx photocatalysts.

pH is one of the crucial factors as the acidic or basic condition of solution decides the rate of degradation. It is so because pH affects the charge on the surface of metal oxide (Ghaly et al., 2011). Surface area is a core deciding factor for performance of a photocatalyst. As number of active site increases on increasing surface area; resulting into decrease in electron-hole recombination rate. As a consequence, photocatalytic activity is enhanced. The dye degradation is observed at absorption maximum of dye by observing change in the rate of decolorization with respect to the intensity of absorption peak in visible region. At this wavelength, dye decolorizes gradually and finally color disappears, which shows the complete degradation of dye (Kansal et al., 2009, Kalsoom et al., 2013). The degradation of textile dyes forms formate, acetate, and oxalate as dominant aliphatic intermediate whereas nitrate, sulfate, and chloride anions form after complete mineralization (Mahmoodi, 2014). The effluents also have various ions like SO_4^{2-}, Cl^-, NO_3^-, CH_3COO^-, HCO_3^-, and $H_2PO_4^-$ and they impart major influence on photocatalytic degradation. The anions like SO_4^{2-}, and Cl^- were found to accelerate the rate, but the presence of CH_3COO^-, HCO_3^-, and $H_2PO_4^-$ adversely affects the rate of degradation (Alahiane et al., 2014).

6.4.2 Antimicrobial Activity

Wastewater may also contain some harmful microorganisms along with organic pollutants. Photocatalysts support the destruction of such microrganisms (Benabbou et al., 2007) like bacteria, virus, and fungi, and it can also decompose the cell itself. Toxin is released after killing of the cell. This released toxin also gets decomposed by photocatalytic action (Liou and Chang, 2012). Sangchay (2017) synthesized WO_3-doped TiO_2 coating on charcoal, activated through microwave-assisted sol-gel method. They observed enhanced photocatalyic activity in degradation of methylene blue solution and also against the bacteria *E. coli*. 1% WO_3-doped TiO_2 coated activated charcoal exhibited excellent antimicrobial activity as compared to pure titania.

The aqueous solution, having *Giardia lamblia,* has been sterilized by titania under UV-irradiation (Lee et al., 2004). Adhesion of photocatalyst on surface of substrate assists in photocatalytic decomposition of acid orange 7, along with microorganisms like bacteria *E. coli* (E. Coli) and viruses like Herpes simplex virus (HSV-1) (Hajkova et al., 2007). The pure titanium dioxide and Ag modified TiO_2 nanocomposite were applied for disinfection of *Bacillus atrophaeus* (Lee et al., 2011). BiOBr photocatalyst also degrades the organic pollutant, and bacterial infection leading to detoxification and disinfection of water (Ahmad et al., 2016).

6.4.3 Organic Pollutants Elimination

Photocatalyst alone is not the only factor, which is responsible for removal or degradation of aqueous pollutants. With the presence of an electron acceptor, i.e. molecular oxygen, phenol red was 60% bleached only whereas in presence of hydrogen peroxide, it undergoes more than 90% degradation under visible light. Temperature is also an important factor, which influences the photodegradation of organic pollutants (Wahab and Hussain, 2016). Oxidants have an important role in degradation of an organic pollutant. Sohrabi and Akhlaghian (2016) observed 18.26% degradation of phenol whereas, 27.87% decomposition was there with the proper amount of H_2O_2 using iron-modified titanium dioxide, i.e., Fe_2O_3/TiO_2. They also reported an adverse effect of temperature, if it is more than room temperature.

Removal for substituted phenols has been studied using metal oxides like TiO_2 or ZnO, but the major problem is to separate the catalyst from the solution (Ahmed et al., 2010). Cobalt ferrite ($CoFe_2O_4$) is a semiconductor with magnetic property and narrow band gap and therefore, it can be easily isolated from a solution by applying an external magnetic field (Golsefidi et al., 2016). Zeynolabedin and Mahanpoor (2017) observed 96% degradation of water pollutant 2-nitrophenol by using Cu slag supported $CoFe_2O_4$. The dechlorination of 2,4-dichlorophenol by $N-TiO_2$ releases chlorine as chloride ions under irradiation of visible light (Gionco et al., 2016). The complete dechlorination takes 2 h with doped catalyst, whereas bare ZnO takes about 3 h.

Furfural is an organic compound that is used as a solvent in petrochemical installations, oil refineries, and other chemical industrial plants to separate hydrocarbon. It is nonbiodegradable, highly resistant, and has a slow degradation rate. Therefore, it is an important source of water pollution (Borghei and Hosseini, 2008). Furfural in wastewater has a major adverse impact on living organisms. It gets adsorbed into the skin, and exerts a negative effect on the nervous system, destroys cell membranes, and inhibits the metabolism (Ghosh et al., 2010; Mousavi-Mortazavi and Nezamzadeh-Ejhieh, 2016; Cuevas et al., 2014; Manz et al., 2016). It is toxic in nature, and hence, makes its degradation, mineralization, and removal from effluent essential.

Granular activated carbon is porous and posseses a low density. Titania supported on granular activated carbon was synthesized by simple sol-gel method, which has total pore volume and a specific surface area of 0.13 cm^3/g and 35.91 m^2/g, respectively. It shows degradation of furfural of about 95% within 80 min and it can be reused for a minimum of 4 times with only 2% reduction in furfural removal efficiency (Ghasemi et al., 2016). The total pore volume of a composite is lesser than the supporter GAC and therefore, titania gets completely embedded in pores of the supporter (Table 6.2).

TABLE 6.2 Different Parameters Values for GAC and TiO_2/GAC

Parameter	BET (m^2/g)	Total pore volume (cm^3/g)	Average pore diameter (nm)	BET constant
GAC value	226.78	0.182	3.22	146
TiO_2/GAC value	35.91	0.130	14	95

Adapted from Ghasemi, B., Anvaripour, B., Jorfi, S., Jaafarzadeh, N., 2016. Int. J. Photoenergy. With permission.

Lee et al. (2015) studied photocatalytic degradation of phenoxyacetic acid (PAA) by redox reaction. The photooxidation of ethanol by vanadium doped TiO_2 photocatalyst gives carbon dioxide as a major product and acetaldehyde, formic acid, and carbon monoxide as side products under visible irradiation (Klosek and Raftery, 2001).

6.4.4 Removal of Heavy Metal

Samadi et al. (2014) reported immobilization of Cu-titania/chitosan on polycarbonate substrate. The immobilized thin film was reliable, reproducible, and inexpensive to synthesize by simple sol-gel method. It showed excellent performance for the removal of heavy metal such as Pb^{2+} and Cr^{6+} from aqueous media. Thus, immobilization of a catalyst by various means provides a solution to problems of separation, recycling, and deactivation.

The maghemite (γ-Fe_2O_3) and titanium oxide were prepared by hydrothermal and coprecipitation process. These were then mixed in 1:10, 1:20, 1:60, 1:80, and 1:100 ratios and covered by polyvinyl alcohol (PVA)-alginate beads. The enclosed composite removes toxic metal, Pb(II) to the extent of 100% from the aqueous solution in 100 min at neutral condition under sunlight (Majidnia and Idris, 2016).

6.4.5 Degradation of Oil in Wastewater

Oil spills are the biggest untold nonbiodegradable hazardous factor that pollute the water and disturb whole ecosystems of water bodies. Photocatalytic processes may rapidly degrade crude oil or other hydrocarbon containments released from petroleum and oil refineries. It is a low cost and fast technique to convert the contaminants in water, carbon dioxide, and biodegradable organics. Thus, it has wide applications in remediating oil pollution near the seashore, in the middle of the ocean, at a refinery, or at a water cleaning facility.

Ziolli, Jardim (2002) reported that crude oil compounds in contaminated waters do not mineralize by photolysis but by heterogeneous photocatalytic processes; their degradation is around 90%. Titania in anatase form is proved to be best for oil spill remediation (Narayan, 2010). Total organic carbon (TOC) and polycyclic aromatic hydrocarbons (PAH) in oil sludge were eliminated by heterogeneous photocatalysis using titanium dioxide in the presence of hydrogen peroxide under solar radiation (Topare et al., 2015). Polycyclic aromatic hydrocarbons get completely eliminated in 96 h, and 91% reduction of total organic carbon in 144 h using 5 mg of TiO_2 (Rocha et al., 2014). Photo-Fenton homogeneous and heterogeneous systems in the presence of H_2O_2, treat oily wastewater more effectively as compared to a system without Fenton (Mustafa et al., 2014).

Photocatalysis break down the hydrocarbon spill into oil, leaving only biodegradable compounds behind. Yeber et al. (2012) used bacteria *Pseudomonas aeruginosa* based titania film for removal of oil (99%) and total organic carbon (TOC mg/L) (78.6%). The biodegradability of oily wastewater from the restaurant was enhanced on treatment TiO_2 oxidation system in the presence of vacuum ultraviolet (VUV, 185 nm) (Kang et al., 2010).

Toluene represents the aromatics in the oil spill, which are decomposed by N doped TiO$_2$ nanotube thin film coated onto a levee at the water surface level (Hsu et al., 2008). Maghemite nanoparticles (-Fe$_2$O$_3$) prepared by precipitation method showed 90% toluene elimination from sea water in 2 h under sunlight (Roushenas et al., 2014).

6.5 IMMOBILIZATION

The use of titania has limitations such as that it cannot be recollected or separated from water, but immobilization of TiO$_2$ can solve this problem. The introduction of silane coupling agents that silylate the titanium dioxide nanotube improves its surface property. Grafting it on the surface of polyurethane (PU) membrane helps in immobilization of semiconductors. It's not easy to separate photocatalysts after treatment of wastewater, but it also provides high activation and catalytic performance for degradation of methyl orange. It is found stable for a longer time, if the catalyst is washed with ethanol (Lin et al., 2017). It is interesting to note that immobilized PU/TiO$_2$ sustains stable photocatalytic properties, even if reused many times with no significant loss of catalyst (Fig. 6.21)

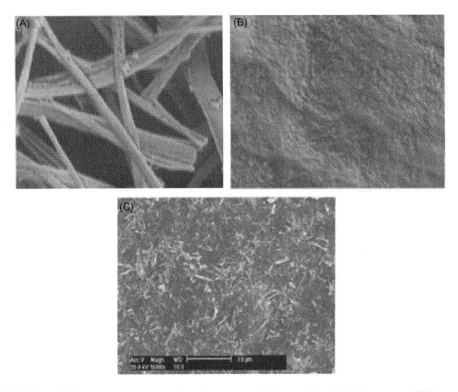

FIGURE 6.21 (A) Titania nanotube treated by silanization, (B) Immobilized PU/titania, and (C) Immobilized PU/titania after use. *Adapted from Lin, L., Wu, Q., Gong, X., Zhang, Y., 2017. J. Anal. Methods Chem. doi.org/10.1155/ 2017/9629532. With permission.*

Titanium dioxide nanoparticles were synthesized though the hydrolysis and condensation of titanium tetrachloride by Nyamukamba et al. (2016). These were further localized on polyacrylonitrile (PAN) and were heated to convert:

- polymer nanofibers to carbon nanofibers, and
- amorphous TiO_2 to crystalline TiO_2.

The immobilized photocatalyst was able to degrade organic contaminants in water. The degradation was approximately 60% in 210 min because of direct contact among the TiO_2 photocatalyst and methyl orange.

The reusability of photocatalyst is an important factor for its industrial use as it reduces its cost and increases the availibilty and reliability for photoctalysis and environmental safety by limiting the disposal of secondary pollutants into the environment. In order to ascertain the reusability of photocatalyst materials, it could be immobilized or can be provided magnetic property (Fig. 6.22).

Composite photocatalysts $NiFe_2O_4$-SiO_2, $NiFe_2O_4$-TiO_2, $NiFe_2O_4$-SiO_2-TiO_2, and TiO_2 with various magnetic properties have a great application in elimination of contaminants. The

(A)

(B)

(C)

(D)

FIGURE 6.22 Different nanofibers. (A) Neat PAN nanofibers. (B) Neat carbon nanofibers. (C) CNF/TiO2-SR nanofibers. (D) PAN/TiO2-EM. *Adapted from Nyamukamba, P., Okoh, O., Tichagwa, L., Greyling, C., 2017. Int. J. Photoenergy. doi:10.1155/2017/3079276. With permission.*

reduction efficiency for Cr(VI) in aqueous solution was observed to be higher for $NiFe_2O_4$-SiO_2-TiO_2 than in magnetic titanium dioxide without silica interlayer $NiFe_2O_4$-TiO_2. The order of reduction of contaminants was observed as follows (Ojemaye et al., 2017):

TiO_2	>	$NiFe_2O_4$-SiO_2-TiO_2	>	$NiFe_2O_4$-TiO_2
(96.7% in 240 min)		(96.5% 300 min)		(60% in 300 min)

Although it was observed that titania shows slightly more efficiency but it is associated with recovery problem. But $NiFe_2O_4$-SiO_2-TiO_2 has less activity (0.2% less) due to the presence of interlayer of silica between the magnetic core and photocatalyst such as titanium dioxide, but it also makes the recovery of photocatalyst easier because of its magnetic property.

6.6 EFFECT OF MORPHOLOGY

Chen et al. (2015) synthesized TiO_2/stellerite composite by scattering TiO_2 on the surface of stellerite treated by HCl, NaOH, or NaCl via sol-gel method. It was observed that the HCl and NaCl increased the degradation efficiency of methyl orange due to pore formation at the satellite surface, and also induces the microscopic changes whereas, NaOH decreases the activity of photocatalyst by distorting the morphology of stellerite due to formation of silicates (Fig. 6.23).

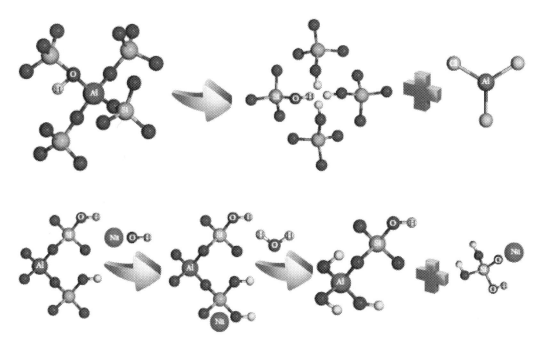

FIGURE 6.23 Dealuminization and desilication of stellerites. *Adapted from Chen, H., Wang, J., Wang, H., Yang, F., Zhou, J.N., Fu, J., Yang, J., Yuan, Z., Zheng, B., 2015. J. Nanomter. doi:10.1155/2015/7015-89. With permission.*

The performance of the semiconductor gets influenced by the structure of co-catalyst added. The preferential Pt(111) nanoparticles supported on TiO_2 shows 1.6 times higher efficiency than polyoriented Pt supported TiO_2. It is due to variation in the count of surface atoms at corners and edges between the Pt(poly) and Pt(111) nanoparticles. The preferential Pt(111) nanoparticles enhance the transfer efficiency of photo-induced electrons from the conduction band of TiO_2 to Pt nanoparticles. Thus, the surface structure of Pt decides the performance of the semiconductor (Cui et al., 2016).

6.7 OTHER APPLICATIONS

Photocatalysts are very versalite materials because of their attractive properties such as optical transmittance, uniformity, low resistivity, mechanical hardness, stability to heat treatment, hydrophobic and piezoelectric behavior. Therefore, they become a suitable candidate for devices such as for gas sensor application, Li-ion batteries, photovoltaic, super capacitor, light emitting diode, display devices and solar cell (Birkel et al., 2012; Ali and Tajammul, 2013), (Kumara et al., 2013), DSSCs electrochemical water splitting (Swierk et al., 2015), antibacterial (Joost et al., 2015), antifogging (Takagi et al., 2001), self cleaning (Banerjee et al., 2015), etc.

Photocatalysis is the most widely used field of green chemistry as it uses light, which is freely available and does not pollute the atmosphere. Many chemical and physical processes are used in wastewater treatment, but photocatalytic treatment of organic pollutants present in wastewater has proved to one of the most efficient and ecofriendly techniques. Other methods either produce secondary pollutants or release harmful compounds in the environment, while photocatalytic remediation of wastewater is a step towards clean chemistry and it is one of the most important advanced oxidation processes. Time is not far off, when photocatalysis will prove its worth and will find a dominant position in the treatment of wastewaters in years to come.

References

Ahmad, A., Meng, X., Yun, N., Zhang, Z., 2016. Preparation of hierarchical BiOBr microspheres for visible light-induced photocatalytic detoxification and disinfection. J. Nanomater. Available from: http://dx.doi.org/10.1155/2016/1373725.

Ahmed, S., Rasul, M.G., Martens, W.N., Brown, R., Hashib, M.A., 2010. Heterogeneous photocatalytic degradation of phenols in wastewater: a review on current status and developments. Desalination 261, 3–18.

Akurati, K.K., Vital, A., Dellemann, J.P., Michalow, K., Graule, T., Ferri, D., et al., 2008. Flame-made WO_3/TiO_2 nanoparticles: relation between surface acidity, structure and photocatalytic activity. App.Catal. B Environ. 79, 53–62.

Alahiane, S., Qourzal, S., Ouardi, M.E., Abaamrane, A., Assabbane, A., 2014. Factors influencing the photocatalytic degradation of reactive yellow 145 by TiO_2-coated non-woven fibers. Am. J. Anal. Chem. 5, 445–454.

Alhamed, M., Issa, A.S., Doubal, A.W., 2012. Studying of natural dyes properties as photo-sensitizer for dye sensitized solar cells (DSSC). J. Electron Devi. 16, 1370–1383.

Ali, A.M., Muhammad, A., Shafeeq, A., Asghar, H.M.A., Hussain, S.N., Sattar, H., 2012. Doped metal oxide (ZnO) and photocatalysis: a review. J. Pak. Inst. Chem. Eng. 40 (1), 11–19.

Ali, M.A., Idris, M.R., Quayum, M.E., 2013. Fabrication of ZnO nanoparticles by solution combustion method for the photocatalytic degradation of organic dye. J. Nanostruct. Chem. Available from: http://dx.doi.org/10.1186/2193-8865-3-36.

Ali, S., Tajammul, M.S., 2013. Effect of doping on the structural and optical properties of SnO_2 thin films fabricated by aerosol assisted chemical vapor deposition, 6th vacuum and surface sciences conference of Asia and Australia (VASSCAA-6) 2013. J. Phys. Conf. Ser. Available from: http://dx.doi.org/10.1088/1742-6596/439/1/012013.

Ameta, R., Sharma, D., Ordia, M., 2013. Use of advanced oxidation technology for removal of azure B. Acta Chim. Pharm. Indica 3, 94–100.

Amin, M.T., Alazba, A.A., Manzoor, U., 2014. A review of removal of pollutants from water/wastewater using different types of nanomaterials. Adv. Mater. Sci. Eng. Available from: http://dx.doi.org/10.1155/2014/825910.

Aramendía, M.A., Marinas, A., Marinas, J.M., Moreno, J.M., Urbano, F.J., 2005. Photocatalytic degradation of herbicide fluroxypyr in aqueous suspension of TiO_2. Catal. Today 101, 187–193.

Bae, S.W., Borse, P.H., Hong, S.J., Jang, J.S., Lee, J.S., 2007. Photophysical properties of nanosized metal-doped TiO_2 photocatalyst working under visible light. J. Korean Phys. Soc. 51, S22–S26.

Bak, S.A., Song, M.S., Nam, I.T., Lee, W.G., 2015. Photocatalytic oxidation of trichloroethylene in water using a porous ball of nano-ZnO and nanoclay composite. J. Nanomater. Available from: http://dx.doi.org/10.1155/2015/160212.

Banerjee, S., Dionysiou, D.D., Pillai, S.C., 2015. Self-cleaning applications of TiO_2 by photo-induced hydrophilicity and photocatalysis. App. Catal. B Environ. 176–177, 396–428.

Baruah, S., Pal, S.K., Dutta, J., 2012. Nanostructured zinc oxide for water treatment. Nanosc. Nanotechnol.-Asia 2, 90–102.

Benabbou, A.K., Derriche, Z., Felix, C., Lejeune, P., Guillard, C., 2007. Photocatalytic inactivation of Escherischia coli. Effect of concentration of TiO_2 and microorganism, nature, and intensity of UV irradiation. Appl. Catal. B Environ 76 (3-4), 257–263.

Benjamin, S., Vaya, D., Punjabi, P.B., Ameta, S.C., 2011. Enhancing photocatalytic activity of zinc oxide by coating with some natural pigments. Arab. J. Chem. 4, 205–209.

Bessekhouad, Y., Chaoui, N., Trzpit, M., Ghazzal, N., Robert, D., Weber, J.V., 2006. UV–vis versus visible degradation of Acid Orange II in a coupled CdS/TiO_2 semiconductors suspension. J. Photochem. Photobiol. A Chem. 183, 218–224.

Bhattacharjee, A., Ahmaruzzaman, M., 2015. Facile synthesis of SnO_2 quantum dots and its photocatalytic activity in the degradation of eosin Y dye: a green approach. Mater. Lett. 139, 418–421.

Bhosale, R., Pujari, S., Muley, G., Pagare, B., Gambhire, A., 2013. Visible-light-activated nanocomposite photocatalyst of Cr_2O_3/SnO_2. J. Nanostruct. Chem. Available from: http://dx.doi.org/10.1186/2193-8865-3-46.

Birkel, A., Lee, Y.-G., Koll, D., Meerbeek, X.V., Frank, S., Choi, M.J., et al., 2012. Highly efficient and stable dye-sensitized solar cells based on SnO_2 nanocrystals prepared by microwave-assisted synthesis. Energy Environ. Sci. 5, 5392–5400.

Borghei, S.M., Hosseini, S.N., 2008. Comparison of furfural degradation by different photooxidation methods. Chem. Eng. J. 139 (3), 482–488.

Borse, P.H., Kim, J.Y., Lee, J.S., Lim, K.T., Jeong, E.D., Bae, J.S., et al., 2012. Ti-dopant-enhanced photocatalytic activity of a $CaFe_2O_4/MgFe_2O_4$ bulk heterojunction under visible-light irradiation. J. Korean Phys. Soc. 61 (1), 73–79.

Chan, S.H.S., Wu, T.Y., Juan, J.C., Teh, C.Y., 2011. Recent developments of metal oxide semiconductors as photocatalysts in advanced oxidation processes (AOPs) for treatment of dye waste-water. J. Chem. Technol. Biotechnol. 86, 1130–1158.

Chang, D.W., Lee, H.J., Kim, J.H., Park, S.Y., Park, S.M., Dai, L., et al., 2011. Novel quinoxaline-based organic sensitizers for dye-sensitized solar cells. Org. Lett. 13 (15), 3880–3883.

Chatterjee, D., Dasgupta, S., 2005. Visible light induced photocatalytic degradation of organic pollutants. J. Photochem. Photobiol. C Rev. 6 (2-3), 186–205.

Chen, H., Wang, J., Wang, H., Yang, F., Zhou, J.N., Fu, J., et al., 2015. Preparation of stellerite loading titanium dioxide photocatalyst and its catalytic performance on methyl orange, 2015. J. Nanomater. Available from: http://dx.doi.org/10.1155/2015/7015-89.

Chen, H., Wang, L., 2014. Nanostructure sensitization of transition metal oxides for visible-light photocatalysis. Beil. J. Nanotechnol. 5, 696–710.

Cuevas, M., Quero, S.M., Hodaifa, G., L'opez, A.J.M., Sanchez, S., 2014. Furfural removal from liquid effluents by adsorption onto commercial activated carbon in a batch heterogeneous reactor. Ecol. Eng 68, 241–250.

Cui, E., Hou, G., Shao, R., Guan, R., 2016. Facet-dependent activity of Pt nanoparticles as cocatalyst on TiO_2 photocatalyst for dye-sensitized visible-light hydrogen generation. J. Nanomater. 2016. Available from: http://dx.doi.org/10.1155/2016/3469393.

Curri, M.L., Comparelli, R., Cozzoli, P.D., Mascolo, G., Agostiano, A., 2003. Colloidal oxide nanoparticles for the photocatalytic degradation of organic dye. Mater. Sci. Eng. C. 23, 285–289.

Das, R., 2014. Application photocatalysis for treatment of industrial waste water—a short review. Open Acc. Lib. J. 1, 1–17.

Dlamini, L.N., Krause, R.W., Kulkarni, G.U., Durbach, S.H., 2011. Synthesis and characterization of titania based binary metal oxide nanocomposite as potential environmental photocatalysts. Mater. Chem. Phys. 129 (1-2), 406–410.

Doma, R., Borse, P.H., Chob, C.R., Leec, J.S., Yud, S.M., Yoond, J.H., et al., 2012. Synthesis of $SrFe_{12}O_{19}$ and $Sr_7Fe_{10}O_{22}$ systems for visible light photocatalytic studies. J. Ceram. Proc. Res. 13, 451–456.

Elamin, N., Elsanousi, A., 2013. Synthesis of ZnO nanostructures and their photocatalytic activity. J. Appl. Ind. Sci. 1, 32–35.

Elango, G., Kumaran, S.M., Kumar, S.S., Muthuraja, S., Roopan, S.M., 2015. Green synthesis of SnO_2 nanoparticles and its photocatalytic activity of phenolsulfonphthalein dye. Spectrochim. Acta A 145, 176–180.

Fabiyi, M.E., Skelton, R.L., 2000. Photocatalytic mineralization of methylene blue using buoyant TiO_2-coated polystyrene beads. J. Photochem. Photobiol. A Chem. 132, 121–128.

Fang, X., Zhai, T., Gautam, U., Li, L., Wu, L., Bando, Y., et al., 2011. ZnS nanostructures: from synthesis to applications. Prog. Mater. Sci. 56 (2), 175–287.

Feng, Y., Ji, W.X., Huang, B.J., Chen, X.L., Li, F., Li, P., et al., 2015. The magnetic and optical properties of 3d transition metal doped SnO_2 nanosheets. RSC Adv. 5, 24306–24312.

Folli, A., Bloh, J.Z., Lecaplain, A., Walker, R., Macphee, D.E., 2015. Properties and photochemistry of valence-induced-Ti^{3+} enriched (Nb,N)-codopedanatase TiO_2 semiconductors. Phys. Chem. Chem. Phys. 17, 4849–4853.

Folli, A., Bloh, J.Z., Macphee, D.E., 2016. Band structure and charge carrier dynamics in (W, N)-codoped TiO_2 resolved by electrochemical impedance spectroscopy combined with UV–vis and EPR spectroscopies. J. Electroanal. Chem. 780, 367–372.

Fu, P., Zhang, P., 2010. Uniform dispersion of Au nanoparticles on TiO_2 film via electrostatic self-assembly for photocatalytic degradation of bisphenol A. Appl. Catal. B Environ. 96, 176–184.

Gaya, U.I., Abdullah, A.H., 2008. Heterogeneous photocatalytic degradation of organic contaminants over titanium dioxide: a review of fundamentals, progress and problems. J. Photochem. Photobiol. C 9 (1), 1–12. 2008.

Ghaly, M.Y., Jamil, T.S., El-Seesy, I.E., Souaya, E.R., Nasr, R.A., 2011. Treatment of highly polluted paper mill wastewater by solar photocatalytic oxidation with synthesized nano TiO_2. Chemi. Eng. J. 168, 446–454.

Ghasemi, B., Anvaripour, B., Jorfi, S., Jaafarzadeh, N., 2016. Enhanced photocatalytic degradation and mineralization of furfural using $UVC/TiO_2/GAC$ composite in aqueous solution. Int. J. Photoenergy. Available from: http://dx.doi.org/10.1155/2016/2782607.

Ghosh, U.K., Pradhan, N.C., Adhikari, B., 2010. Pervaporative separation of furfural from aqueous solution using modified polyurethaneureamembrane. Desalination 252 (1-3), 1–7.

Gionco, C., Fabbri, D., Calza, P., Paganini, M.C., 2016. Synthesis, characterization, and photocatalytic tests of N-doped zinc oxide: a new interesting photocatalyst. J. Nanomater. Available from: http://dx.doi.org/10.1155/2016/4129864.

Golsefidi, M.A., Yazarlou, F., Nezamabad, M.N., Nezamabad, B.N., Karimi, M., 2016. Effects of capping agent and surfactant on the morphology and size of $CoFe_2O_4$ nanostructures and photocatalyst properties. J. Nanostruct. 6, 121–126.

Gondal, M.A., Dastageer, M.A., Khalil, A., 2009. Synthesis of nano-WO_3 and its catalytic activity for enhanced antimicrobial process for water purification using laser induced photo-catalysis. Catal. Commun. 11, 214–219.

Gubbala, S., Russell, H.B., Shah, H., Deb, B., Jasinski, J., Rypkemac, H., et al., 2009. Surface properties of SnO_2 nanowires for enhanced performance with dye-sensitized solar cells. Energy Environ. Sci. 2, 1302–1309.

Guo, W., Guo, Y., Dong, H., Zhou, X., 2015a. Tailoring the electronic structure of -Ga$_2$O$_3$ by non-metal doping from hybrid density functional theory calculations. Phys. Chem. Chem. Phys. 17 (8), 5817−5825.

Guo, Y., Qiu, X., Dong, H., Zhou, X., 2015b. Trends in non-metal doping of the SrTiO$_3$ surface: a hybrid density functional study. Phys. ChemChem Phys. 17 (33), 21611−21621.

Habib, M.A., Muslim, M., Shahadat, M.T., Islam, M.N., Ismail, I.M.I., Islam, T.S.A., et al., 2013. Photocatalytic decolorization of crystal violet in aqueous nano-ZnO suspension under visible light irradiation. J. Nanostruct. Chem . Available from: http://dx.doi.org/10.1186/2193-8865-3-70.

Hajkova, P., Spatenka, P., Horsky, J., Horskaand, I., Kolouch, A., 2007. Photocatalytic effect of TiO$_2$ films on viruses and bacteria. Plasma Processes Polym. 4, S397−S401.

Hamid, M.A., Rahman, I.A., 2003. Preparaton of titanium dioxide (TiO$_2$) thin films by sol gel dip coating method. Malay. J. Chem. 5, 086−091.

Hara, K., Koumura, N., 2009. Organic dyes for efficient and stable dye-sensitized solar cells. Mater. Matters 4 (4), 92.

Hemmatzadeh, R., Mohammadi, A., 2013. Improving optical absorptivity of natural dyes for fabrication of efficient dye-sensitized solar cells. J. Theor. Appl. Phys. 7, 57.

Hernández-Martínez, A.R., Estevez, M., Vargas, S., Quintanilla, F., Rodríguez, R., 2012. Natural pigment-based dye-sensitized solar cells. J. Appl. Res. Technol. 10 (1), 38−47.

Hernández-Martínez, A.R., Estevez, M., Vargas, S., Rodriguez, R., 2013. Stabilized conversion efficiency and dye-sensitized solar cellsfrom Beta vulgaris pigment. Int. J. Mol. Sci. 14, 4081−4093.

Houas, A., Lachheb, H., Ksibi, M., Elaloui, E., Guillard, C., Herrmann, J.M., 2001. Photocatalytic degradation pathway of methylene blue in water. Appl. Catal. B Environ. 31, 145−157.

Hsu, Y.Y., Hsiung, T.L., Wang, H.P., Fukushima, Y., Wei, Y.L., Chang, J.E., 2008. Photocatalytic degradation of spill oils on TiO$_2$ nanotube thin films. Marine Pollut. Bull. 57, 873−876.

Hu, X.Y., Fan, J., Zhang, K.L., Wang, J.J., 2012. Photocatalytic removal of organic pollutants in aqueous solution by Bi$_4$Nb$_x$Ta$_{1-x}$O$_8$I. Chemosphere. 87, 1155−1160.

Hu, Y., Liu, H., Rao, Q., Kong, X., Sun, W., Guo, X., 2011. Effects of N precursor on the agglomeration and visible light photocatalytic activity of N-doped TiO$_2$ nanocrystalline powder. J Nanosci. Nanotechnol. 11 (4), 3434−3444.

Hua, M., Zhang, S., Pan, B., Zhang, W., Lv, L., Zhang, Q., 2012. Heavy metal removal from water/wastewater by nanosized metal oxides: a review. J. Hazard. Mater. 211−212, 317−331.

Huang, H., Liu, K., Chen, K., Zhang, Y., Zhang, Y., Wang, S., 2014. Ce and F Comodification on the crystal structure and enhanced photocatalytic activity of Bi$_2$WO$_6$ Photocatalyst under visible light irradiation. J. Phys. Chem. C 118, 14379−14387.

Huang, Q.Z., Xiong, Y., Zhang, Q., Yao, H.C., Li, Z.J., 2017. Noble metal-free MoS$_2$ modified Mn$_{0.25}$Cd$_{0.75}$S for highly efficient visible-light driven photocatalytic H$_2$ evolution. Appl. Catal. B Environ. 209, 514−522.

Hug, H., Bader, M., Mair, P., Glatzel, T., 2014. Biophotovoltaics: Natural pigments in dye-sensitized solar cells. Appl. Energy 115, 216−225.

Hur, S.G., Kim, T.W., Hwang, S.J., Park, H., Choi, W., Kim, S.J., et al., 2005. Synthesis of new visible light active photocatalysts of Ba (In$_{1/3}$Pb$_{1/3}$M'$_{1/3}$)O$_3$ (M' = Nb, Ta): A band gap engineering strategy based on electronegativity of a metal component. J. Phys. Chem. B. 109 (36), 17346.

Hussein, F.H., Abass, T.A., 2010. Solar photolysis and photocatalytic treatment of textile industrial wastewater. Int. J. Chem. Sci. 8 (3), 1409−1420.

Ibhadon, A.O., Fitzpatrick, P., 2013. Heterogeneous photocatalysis: Recent advances and applications. Catalysts 3, 189−218.

Jang, J.S., Borse, P.H., Lee, J.S., Jung, O.S., Cho, C.R., Jeong, E.D., et al., 2009. Synthesis of nanocrystalline ZnFe$_2$O$_4$ by polymerized complex method for its visible light photocatalytic application: an efficient photo-oxidant. Bull. Korean Chem. Soc. 30 (8), 1738−1742.

Jasim, K.E., 2012. Natural dye-sensitized solar cell based on nanocrystalline TiO$_2$. Sains Malay. 41, 1011−1016.

Johar, M.A., Afzal, R.A., Alazba, A.A., Manzoor, U., 2015. Photocatalysis and bandgap engineering using ZnO nanocomposites. Adv. Mater. Sci. Eng. 2015. Available from: http://dx.doi.org/10.1155/2015/934587.

Joost, U., Juganson, K., Visnapuu, M., Mortimer, M., Kahru, A., Nõmmiste, E., et al., 2015. Photocatalytic antibacterial activity of nano-TiO$_2$ (anatase)-based thin films: effects on Escherichia coli cells and fatty acids. J. Photochem. Photobiol. B. 142, 178−185.

Kalsoom, U., Ashrf, S.S., Meetani, M.A., Rauf, M.A., Bhatti, H.N., 2013. Mechanistic study of a diazo dye degradation by soybean peroxidase. Chem. Central J. 2013 (7), 93.

Kambur, A., Pozan, G.S., Boz, I., 2012. Preparation, characterization and photocatalytic activity of TiO_2-ZrO_2 binary oxide nanoparticles. Appl. Catal. B Environ. 115–116, 149–158.

Kang, J.X., Lu, L., Zhan, W., Liu, D.Q., 2010. Photocatalytic pretreatment of oily wastewater from the restaurant by a vacuum ultraviolet/TiO_2 system. J. Hazard. Mater. 186 (1), 849–854.

Kansal, S.K., Kaur, N., Singh, S., 2009. Photocatalytic degradation of two commercial reactive dyes in aqueous phase using nanophotocatalysts. Nanoscale Res. Lett. 4 (7), 709–716.

Karunakaran, C., Dhanalakshmi, R., 2008. Semiconductor-catalyzed degradation of phenols with sunlight. Sol. Energy Mater. Solar Cells 92, 1315–1321.

Karunakaran, C., Senthilvelan, S., 2005. Photocatalysis with ZrO_2: Oxidation of aniline. J. Mol. Catal. A Chem. 233, 1–8.

Karunakaran, C., Senthilvelan, S., 2006. Fe_2O_3-photocatalysis with sunlight and UV light: oxidation of aniline. Electrochem. Commun. 8, 95–101.

Khaghanpour, Z., Naghibi, S., 2017. Perforated ZnO nanoflakes as a new feature of ZnO achieved by the hydrothermal-assisted sol–gel technique. J. Nanostruct. Chem. 7, 55–59.

Khairy, M., Zakaria, W., 2014. Effect of metal-doping of TiO_2 nanoparticles on their photocatalytic activities toward removal of organic dyes. Egypt. J. Pet. 23, 419–426.

Kimi, M., Yuliati, L., Shamsuddin, M., 2015. Preparation of high activity Ga and Cu doped ZnS by hydrothermal method for hydrogen production under visible light irradiation. J. Nanomater. 2015. Available from: http://dx.doi.org/10.1155/2015/195024.

Kiriakidis, G., Binas, V., 2014. Metal oxide semiconductors as visible light photocatalysts. J. Korean Phys. Soc. 65, 297–302.

Klosek, S., Raftery, D., 2001. Visible light driven V-Doped TiO_2 photocatalyst and its photooxidation of ethanol. J. Phys. Chem. B. 105 (14), 2815–2819.

Kumara, N.T.R.N., Kooh, M.R.R., Lim, A., Petra, M.I., Voo, N.Y., Lim, C.M., et al., 2013. DFT/TDDFT and experimental studies of natural pigments extracted from black tea waste for DSSC application. Int. J. Photoenergy 2013. Available from: http://dx.doi.org/10.1155/2013/109843.

Kuo, C.Y., Wu, C.H., Wu, J.T., Chen, Y.R., 2014. Synthesis and characterization of a phosphorus-doped TiO_2 immobilized bed for the photodegradation of bisphenol A under UV and sunlight irradiation. React. Kinet. Mech. Catal. 114 (2), 753–766.

Kushwaha, R., Srivastava, P., Bahadur, L., 2013. Natural pigments from plants used as sensitizers for TiO_2 based dye-sensitized solar cells. J. Energy. Available from: http://dx.doi.org/10.1155/2013/654953.

Lachheb, H., Puzenat, E., Houas, A., Ksibi, M., Elaloui, E., Guillard, C., et al., 2002. Photocatalytic degradation of various types of dyes (Alizarin S, Crocein Orange G, Methyl Red, Congo Red, Methylene Blue) in water by UV-irradiated titania. Appl. Catal. B Environ. 39, 75–90.

Lan, M., Fan, G., Yang, L., Li, F., 2015. Enhanced visible-light-induced photocatalytic performance of a novel ternary semiconductor coupling system based on hybrid Zn–In mixed metal oxide/g-C_3N_4 composites. RSC Adv. 5, 5725–5734.

Le, N.T., Nagata, H., Aihara, M., Takahashi, A., Okamoto, T., Shimohata, T., et al., 2011. Additional effects of silver nanoparticles on bactericidal efficiency depend on calcination temperature and dip-coating speed. Appl. Environ. Microbiol. 77, 5629–5634.

Lee, C.H., Shie, J.L., Tsai, C.Y., Yang, Y.T., Chang, C.Y., 2013. Photocatalytic decomposition of indoor air pollution using dye-sensitized TiO_2 induced by anthocyanin and Ru complexes. J. Clean Energy Technol. 1 (2), 115–119.

Lee, J.H., Kang, M., Choung, S.J., Ogino, K., Miyata, S., Kim, M.S., et al., 2004. The preparation of TiO_2 nanometer photocatalyst film by a hydrothermal method and its sterilization performance for Giardia lamblia. Water Res. 38, 713–719.

Lee, J.H., Park, N.G., Shin, Y.J., 2011. Nano-grain SnO_2 electrodes for high conversion efficiency SnO_2–DSSC. Solar Energy Mater. Solar Cells 95, 179–183.

Lee, J.W., Kim, T.Y., Han, S., Lee, S.H., Park, K.H., 2014. Influence of polar solvents on photovoltaic performance of Monascus red dye-sensitized solar cell. Spectrochim Acta A. 126, 76–80.

Lee, K.M., Hamid, S.B.A., Lai, C.W., 2015. Mechanism and kinetics study for photocatalytic oxidation degradation: a case study for phenoxyacetic acid organic pollutant. J. Nanomater. Volume 2015. Available from: http://dx.doi.org/10.1155/2015/940857.

Li, N., Teng, H., Zhang, Li., Zhou, J., Liu, M., 2015. Synthesis of Mo-doped WO3 nanosheets with enhanced visible-light-driven photocatalytic properties. doi:10.1039/C5RA17098B.

Li, W.X., 2013. Photocatalysis of oxide semiconductors. J. Aust. Ceram. Soc. 49, 41−46.

Li, Y., Hwang, D.S., Lee, N.H., Kim, S.J., 2005. Synthesis and characterization of carbon-doped titania as an artificial solar light sensitive photocatalyst. Chem. Phy. Lett. 404 (1−3), 25−29.

Li, Y., Zhou, X., Chen, W., Li, L., Zen, M., Qin, S., et al., 2012. Photodecolorization of Rhodamine B on tungsten-doped TiO_2/activated carbon under visible-light irradiation. J. Hazard. Mater. 227−228, 25−33.

Lin, C.F., Wu, C.H., Onn, Z.N., 2008. Degradation of 4-chlorophenol in TiO_2, WO_3, SnO_2, TiO_2/WO_3 and TiO_2/SnO_2 systems. J. Hazard. Mater. 154, 1033−1039.

Lin, L., Wu, Q., Gong, X., Zhang, Y., 2017. Preparation of TiO_2 Nanotubes loaded on polyurethane membrane and research on their photocatalytic properties. J. Anal. Methods Chem. Available from: http://dx.doi.org/10.1155/2017/9629532.

Liou, J.W., Chang, H.H., 2012. Bactericidal effects and mechanisms of visible light-responsive titanium dioxide photocatalysts on pathogenic bacteria. Arch. Immunol. Ther. Exp. (Warsz) 60, 267−275.

Liu, G., Li, X., Zhao, J., Horikoshi, S., Hidaka, H., 2000. Photooxidation mechanism of dye alizarin red in TiO_2 dispersions under visible illumination: an experimental and theoretical examination. J. Mol. Catal. A Chem. 153, 221−229.

Liu, J., 2012. Band gap narrowing of TiO_2 by compensated codoping for enhanced photocatalytic activity. J. Nat. Gas Chem. 21, 302−307.

Liu, X., Dang, R., Dong, W., Huang, X., Tang, J., Wang, G., 2017. A sandwich-like heterostructure of TiO_2 nanosheets with MIL-100 (Fe): a platform for efficient visible-light-driven photocatalysis. Appl. Catal. B Environ. 209, 506−513.

Liu, Z., Sun, D.D., Guo, P., Leckie, J.O., 2007. An efficient bicomponent TiO_2/SnO_2 nanofiber photocatalyst fabricated by electrospinning with a side-by-side dual spinneret method. Nano Lett. 7, 1081−1085.

Long, R., Dai, Y., Meng, G., Huang, B., 2009. Energetic and electronic properties of X- (Si, Ge, Sn, Pb) doped TiO_2 from first-principles. Phys. Chem. Chem. Phys. 11 (37), 8165−8172.

Luna-Flores, A., Valenzuela, M.A., Luna-López, J.A., Hernández de la Luz, A.D., Muñoz-Arenas, L.C., Méndez-Hernández, M., et al., 2017. Synergetic enhancement of the photocatalytic activity of TiO_2 with visible light by sensitization using a novel push-pull zinc phthalocyanine. Int. J. Photoenergy. Available from: http://dx.doi.org/10.1155/2017/1604753.

Ma, X., Xue, L., Yin, S., Yang, M., Yan, Y., 2014. Preparation of V-doped TiO_2 photocatalysts by the solution combustion method and their visible light photocatalysis activities. J. Wuhan Univ. Technol. Mater. Sci. Ed. 29 (5), 863−868.

Ma, Y., Zhang, J., Tian, B., Chen, F., Wang, L., 2010. Synthesis and characterization of thermally stable Sm, N co-doped TiO_2 with highly visible light activity. J. Hazard. Mater. 182 (1-3), 386−393.

Magdalane, C.M., Kaviyarasu, K., Vijaya, J.J., Siddhardha, B., Jeyaraj, B., 2016. Photocatalytic activity of binary metal oxide nanocomposites of CeO_2/CdO nanospheres: Investigation of optical and antimicrobial activity. J. Photochem. Photobiol. B: Biol. 163, 77−86.

Mahmoodi, N.M., 2014. Binary catalyst system dye degradation using photocatalysis. Fibers Polym. 15 (2), 273−280.

Mai, L., Huang, C., Wang, D., Zhang, Z., Wang, Y., 2009. Effect of C doping on the structural and optical properties of sol-gel TiO_2 thin films. Appl. Surf. Sci. 255, 9285−9289.

Majidnia, Z., Idris, A., 2016. Synergistic effect of maghemite and titania nanoparticles in PVA-alginate encapsulated beads for photocatalytic reduction of Pb(II). Chem. Eng. 203 (4), 425−434.

Malato, S., Fernández-Ibáñez, P., Maldonado, M.I., Blanco, J., Gernjak, W., 2009. Decontamination and disinfection of water by solar photocatalysis: recent overview and trends. Catal. Today. 147, 11−59.

Manz, K.E., Haerr, G., Lucchesi, J., Carter, K.E., 2016. Adsorption of hydraulic fracturing fluid components 2-butoxyethanol and furfural onto granular activated carbon and shale rock. Chemosphere 164, 585−592.

Marschall, R., Wang, L., 2014. Non-metal doping of transition metal oxides for visible light photocatalysis. Catal. Today 225, 111−135.

Mathew, S., Yella, A., Gao, P., Humphry-Baker, R., Curchod, B.F.E., Astani, N.A., et al., 2014. Dye-sensitized solar cells with 13% efficiency achieved through the molecular engineering of porphyrin sensitizers. Nature Chem. 6 (3), 242−247.

Medina-Ramírez, I., Hernández-Ramírez, A., Maya-Treviño, M.L., 2014. Synthesis methods for photocatalytic materials. Photocatal. Semiconductors 69−102.

Modak, B., Ghosh, S.K., 2015. Origin of enhanced visible light driven water splitting by (Rh, Sb)-$SrTiO_3$. Phys. Chem. Chem. Phys. 17 (23), 15274−15283.

Mohamed, R.M., Aazama, E.S., 2012. Enhancement of photocatalytic activity of $ZnO-SiO_2$ by nano-sized Ag for visible photocatalytic reduction of Hg(II). Desalin. Water Treat. doi:10.1080/19443994.2012.708559.

Mohammed, A.A., Ahmad, A.S.S., Azeez, W.A., 2015. Fabrication of dye sensitized solar cell based on titanium dioxide (TiO_2). Adv. Mater. Phys. Chem. 5, 316−367.

Montini, T., Melchionna, M., Monai, M., Fornasiero, P., 2016. Fundamentals and catalytic applications of CeO_2-based materials. Chem. Rev. 116 (10), 5987−6041.

Mousavi-Mortazavi, S., Nezamzadeh-Ejhieh, A., 2016. Supported iron oxide onto an Iranian clinoptilolite as a heterogeneous catalyst for photodegradation of furfural in a wastewater sample. Desalin. Water Treat. 57 (23), 10802−10814.

Mustafa, Y.A., Alwared, A.I., Ebrahim, M., 2014. Heterogeneous photocatalytic degradation for treatment of oil from wastewater. Al-Khwarizmi Eng. J. 10 (3), 53−61.

Mutamim, N.S.A., Noor, Z.Z., Hassan, M.A.A., Olsson, G., 2012. Application of membrane bioreactor technology in treating high strength industrial wastewater: a performance review. Desalination 305, 1−11.

Nagaveni, K., Sivalingam, G., Hegde, M.S., Madras, G., 2004. Solar photocatalytic degradation of dyes: high activity of combustion synthesized nano TiO_2. Appl. Catal. B Environ. 48, 83−93.

Narayan, M.R., 2012. Review: dye sensitized solar cells based on natural photosensitizers. Renew. Sustain. Energy Rev. 16, 208−215.

Narayan, R., 2010. Titania: a material-based approach to oil spill remediation? Mater. Today 13 (9), 58−59.

Neppolian, B., Wang, Q., Yamashita, H., Choi, H., 2007. Synthesis and characterization of ZrO_2-TiO_2 binary oxide semiconductor nanoparticles: application and interparticle electron transfer process. Appl. Catal. A Gen. 333, 264−271.

Nyamukamba, P., Mamphweli, S., Petrik, L., 2017. Silver/carbon codoped titanium dioxide photocatalyst for improved dye degradation under visible light. Int. J. Photoenergy. Available from: http://dx.doi.org/10.1155/2017/3079276.

Nyamukamba, P., Okoh, O., Tichagwa, L., Greyling, C., 2016. Preparation of titanium dioxide nanoparticles immobilized on polyacrylonitrile nanofibres for the photodegradation of methyl orange. Int. J. Photoenergy. Available from: http://dx.doi.org/10.1155/2016/3162976.

Ohko, Y., Hashimoto, K., Fujishima, A., 1997. Kinetics of photocatalytic reactions under extremely low-intensity uv illumination on titanium dioxide thin films. J. Phys. Chem. A 101 (43), 8057−8062.

Ojemaye, M.O., Okoh, O.O., Okoh, A.I., 2017. Performance of $NiFe_2O_4$-SiO_2-TiO_2 magnetic photocatalyst for the effective photocatalytic reduction of Cr (VI) in aqueous solutions. J. Nanomater. Available from: http://dx.doi.org/10.1155/2017/5264910.

Okoli, L.U., Ozuomba, J.O., Ekpunobi, A.J., Ekwo, P.I., 2012. Anthocyanin-dyed TiO_2 electrode and its performance on dye-sensitized solar cell. Res. J. Recent Sci. 1 (5), 22−27.

Pan, J.H., Shen, C., Ivanova, I., Zhou, N., Wang, X., Tan, W.C., et al., 2015. Self-template synthesis of porous perovskite titanate solid and hollow submicrospheres for photocatalytic oxygen evolution and mesoscopic solar cells. ACS Appl. Mater. Interfaces 27 (27), 14859−14869.

Panchal, S., Jhala, Y., Soni, A., Vyas, R., 2014. Photocatalytic degradation of erythrosin-B in the presence of tin dioxide. Acta. Chim. Pharm. Indica 4, 68−77.

Patil, G.E., Kajale, D.D., Shinde, S.D., Wagh, V.G., Gaikwad, V.B., Jain, G.H., 2013. Synthesis of Cu-Doped SnO_2 thin films by spray pyrolysis for gas sensor application. Adv. Sensing Technol. Smart Sensors Measur. Instrum. 1, 299−311.

Pelaez, M., de la Cruz, A.A., O'Shea, K., Falaras, P., Dionysiou, D.D., 2011. Effects of water parameters on the degradation of microcystin-LR under visible light-activated TiO_2 photocatalyst. Water Res. 45, 3787−3796.

Pinotti, A., Bevilacqua, A., Zaritzky, N., 1997. Optimization of the flocculation stage in a model system of a food emulsion waste using chitosan as polyelectrolyte. J. Food Eng. 32, 69−81.

Piskunov, S., Heifets, E., Eglitis, R.I., Borstel, G., 2004. Bulk properties and electronic structure of $SrTiO_3$, $BaTiO_3$, $PbTiO_3$ perovskites: an ab initio HF/DFT study. Comput. Mater. Sci. 29, 165–178.

Pokharna, S., Shrivastava, R., 2013. Photocatalytic treatment of textile industry effluent using titanium oxide. Int. J. Rec. Res. Rev. VI (2), 9–17.

Polisetti, S., Deshpande, P.A., Madras, G., 2011. Photocatalytic activity of combustion synthesized ZrO_2 and ZrO_2–TiO_2 mixed oxides. Ind. Eng. Chem. Res. 50, 12915–12924.

Prasad, K.S., Patra, A., Shruthi, G., Chandan, S., 2017. Aqueous extract of saraca indica leaves in the synthesis of copper oxide nanoparticles: finding a way towards going green. J. Nanotechnol. Available from: http://dx.doi.org/10.1155/2017/7502610.

Priya, M.H., Madras, G., 2006. Photocatalytic degradation of nitrobenzenes with combustion synthesized nano-TiO_2. J. Photochem. Photobiol. A Chem. 178, 1–7.

Qamar, M., Gondal, M.A., Yamani, Z.H., 2009. Synthesis of highly active nanocrystalline WO_3 and its application in laser-induced photocatalytic removal of a dye from water. Catal. Commun. 10, 1980–1984.

Rahimi, R., Pordel, S., Rabbani, M., 2016. Synthesis of Bi_2WO_6 nanoplates using oleic acid as a green capping agent and its application for thiols oxidation. Nanostruct. Chem. 6, 191–196.

Rajkumar, R., Singh, N., 2015. To study the effect of the concentration of carbon on ultraviolet and visible light photo catalytic activity and characterization of carbon doped TiO_2. J. Nanomed. Nanotechnol. 6, 260. Available from: http://dx.doi.org/10.4172/2157-7439.1000260.

Ramasamy, E., Lee, J., 2010. Ordered mesoporous SnO_2 – based photoanodes for high-performance dye-sensitized solar cells. J. Phys. Chem. C 114, 22032–22037.

Ran, L., Zhao, D., Gao, X., Yin, L., 2015. Highly crystalline Ti-doped SnO_2 hollow structured photocatalyst with enhanced photocatalytic activity for degradation of organic dyes. CrystEngComm 17, 4225–4237.

Rawal, S.B., Bera, S., Lee, D., Jang, D.J., Lee, W.I., 2013. Design of visible-light photocatalysts by coupling of narrow bandgap semiconductors and TiO_2: Effect of their relative energy band positions on the photocatalytic efficiency. Catal. Sci. Technol. 3, 1822–1830.

Reddy, K.R., Nakata, K., Ochiai, T., Murakami, T., Tryk, D.A., Fujishima, A., 2011. Facile fabrication and photocatalytic application of Ag nanoparticles-TiO_2 nanofiber composites. J. Nanosci. Nanotechnol. 11 (4), 3692–3695.

Reddy, A.D., Ma, R., Yong, C.M., Kim, T.K., 2015a. Reduced graphene oxide wrapped ZnS–Ag_2S ternary composites synthesized via hydrothermal method: applications in photocatalyst degradation of organic pollutants. Appl. Surf. Sci. 324, 725–735.

Reddy, A., Ma, R., Kim, T.K., 2015b. Efficient photocatalytic degradation of methylene blue by heterostructured ZnO–RGO/RuO_2 nanocomposite under the simulated sunlight irradiation. Ceram. Int. 41, 6999–7009.

Rehman, S., Ullah, R., Butt, A.M., Gohar, N.D., 2009. Strategies of making TiO_2 and ZnO visible light active. J. Hazard. Mater. 170 (2-3), 560–569.

Reza, K.M., Kurny, A., Gulshan, F., 2015. Parameters affecting the photocatalytic degradation of dyes using TiO_2: a review. Appl. Water Sci. 1–10. Available from: http://dx.doi.org/10.1007/s13201-015-0367-y.

Rocha, O.R.S., Duarte, M.M.M., Dantas, R.F., Duarte, M.M.L., Silva, V.L., 2014. Oil sludge treatment by solar TiO_2-photocatalysis to remove polycyclic aromatic hydrocarbons (PAH). Braz. J. Petrol. Gas. 8 (3), 089–096.

Rochkind, M., Pasternak, S., Paz, Y., 2015. Using dyes for evaluating photocatalytic properties: a critical review. Molecules 20, 88–110.

Roushenas, R., Yusop, Z., Majidnia, Z., Nasrollahpour, R., 2014. Photocatalytic degradation of spilled oil in sea water using maghemite nanoparticles. Desalin. Water Treat. 57 (13), 5837–5841.

Saati, E.A., Simon, B.W., Yunianta, Aulanni'am, 2011. Isolation of red rose anthocyanin pigment and its application to inhibit lipid oxidation in yoghurt. J. Agric. Sci. Technol. A 1, 1192–1195.

Saien, J., Shahrezaei, F., 2012. Organic pollutants removal from petroleum refinery wastewater with nanotitania photocatalyst and UV light emission. Int. J. Photoenergy. Available from: http://dx.doi.org/10.1155/2012/703074.

Samad, N.A., Lai, C.W., Hamid, S.B.A., 2016. Influence applied potential on the formation of self-organized ZnO nanorod film and its photoelectrochemical response. Int. J. Photoenergy. Available from: http://dx.doi.org/10.1155/2016/1413072.

Samadi, S., Khalilian, F., Tabatabaee, A., 2014. Synthesis, characterization and application of Cu–TiO_2/chitosan nanocomposite thin film for the removal of some heavy metals from aquatic media. J. Nanostruct. Chem. 4, 84.

Samadi, S., Yousefi, M., Khanlilian, F., Tabatabaee, A., 2015. Synthesis, characterization, and application of Nd, Zr–TiO$_2$/SiO$_2$ nanocomposite thin films as visible light active photocatalyst. J. Nanostruct. Chem. 5, 7–15.

Sang, Y., Liu, H., Umar, A., 2015. Photocatalysis from UV/Vis to near-infrared light: Towards full solar-light spectrum activity. ChemCatChem. 7, 559–573.

Sangchay, W., 2017. WO$_3$-doped TiO$_2$ coating on charcoal activated with increase photocatalytic and antibacterial properties synthesized by microwave-assisted sol-gel method. doi:10.1155/2017/7902930.

Sharma, V.K., Graham, N.J., Li, X.Z., Yuan, B.L., 2010. Ferrate (VI) enhanced photocatalytic oxidation of pollutants in aqueous TiO$_2$ suspensions. Environ. Sci. Pollut. Res. Int. 17, 453–461.

Shen, C., Wang, X., Jiang, X.F., Zhu, H., Li, F., Yang, J., et al., 2015. Fast charge separation at semiconductor sensitizer–molecular relay interface leads to significantly enhanced solar cell performance. J. Phys. Chem. C 119 (18), 9774–9781.

Sivakumar, P., Kumar, G.K.G., Sivakumar, P., Renganathan, S., 2014. Synthesis and characterization of ZnS-Ag nanoballs and its application in photocatalytic dye degradation under visible light. J. Nanostruct. Chem. Available from: http://dx.doi.org/10.1007/s40097-014-0107-0.

Sivalingam, G., Nagaveni, K., Hegde, M.S., Madras, G., 2003. Photocatalytic degradation of various dyes by combustion synthesized nano anatase TiO$_2$. Appl. Catal. B Environ. 45, 23–38.

Smirnova, N.P., Surovtseva, N.I., Fesenko, T.V., Demianenko, E.M., Grebenyuk, A.G., Eremenko, A.M., 2015. Photodegradation of dye acridine yellow on the surface of mesoporous TiO$_2$, SiO$_2$/TiO$_2$ and SiO$_2$ films: spectroscopic and theoretical studies. J Nanostruct. Chem. 5, 333–346.

Sohrabi, S., Akhlaghian, F., 2016. Surface investigation and catalytic activity of iron-modified TiO$_2$. J. Nanostruct. Chem. 6, 93–102.

Solanki, M.S., Ameta, R., Benjamin, S., 2015a. Sensitization of carbon doped tin (IV) oxide nanoparticles by chlorophyll and its application in photocatalytic degradation of toluidine blue. Int. J. Adv. Chem. Sci. Appl. 3 (3), 24–30.

Solanki, M.S., Trivedi, M., Ameta, R., Benjamin, S., 2015b. Preparation and use of chlorophyll sensitized carbon doped tin (IV) oxide nanoparticles for photocatalytic degradation of azure A. Int. J. Chem. Res. 5 (4), 1–11.

Song, J., Hua, L., Shen, Q., Wang, F., Zhang, L., 2014. Synthesis and characterization of SnO$_2$ nano-cystalline for dye sensitized solar cells. Key Eng. Mater. 602–603, 876–879.

Sotto, A., Kim, J., Arsuaga, J.M., del Rosario, G., Martínez, A., Nam, D., et al., 2014. Binary metal oxides for composite ultrafiltration membranes. J. Mater. Chem. A 2, 7054–7064.

Swierka, J.R., Méndez-Hernándezb, D.D., McCoola, N.S., Liddellb, P., Terazonob, Y., Pahk, I., 2015. Metal-free organic sensitizers for use in water-splitting dye-sensitized photoelectrochemical cells. Proc. Natl. Acad. Sci. U.S.A. 112 (6), 1681–1686.

Takagi, K., Makimoto, T., Hiraiwa, H., Negishi, T., 2001. Photocatalytic, antifogging mirror. J. Vac. Sci. Technol. A 19, 2931–2935.

Tanasa, D.E., Piuleac, C.G., Curteanu, S., Popovici, E., 2013. Photodegradation process of eosin Y using ZnO/SnO$_2$ nanocomposites as photocatalysts: Experimental study and neural network modeling. J. Mater. Sci. 48, 8029–8040.

Tian, J., Yan, T., Qiao, Z., Wang, L., Li, W., You, J., et al., 2017. Anion-exchange synthesis of Ag$_2$S/Ag$_3$PO$_4$ core/shell composites with enhanced visible and NIR light photocatalytic performance and the photocatalytic mechanisms. Appl. Catal. B Environ. 209, 566–578.

Topare, N., Joy, M., Joshi, R.R., Kshirsagar, L., 2015. Treatment of petroleum industry wastewater using TiO$_2$/UV photocatalytic process. J. Indian Chem. Soc. 92 (2), 219–222.

Vattikuti, S.V., Byon, C., 2016. Synthesis and structural characterization of Al$_2$O$_3$-coated MoS$_2$ spheres for photocatalysis applications. J. Nanomater. Available from: http://dx.doi.org/10.1155/2015/978409.

Venil, C.K., Zakaria, Z.A., Ahmad, W.A., 2013. Bacterial pigments and their applications. Proc. Biochem. 48, 1065–1079.

Vijayasankara, K., Hebalkara, N.Y., Kimb, H.G., Borsea, P.H., 2013. Controlled band energetics in Pb-Fe-Nb-O metal oxide composite system to fabricate efficient visible light photocatalyst. J. Ceram. Proc. Res. 14, 557–562.

Vinodgopal, K., Bedja, I., Kamat, P.V., 1996a. Nanostructured semiconductor films for photocatalysis: Photoelectrochemical behavior of SnO$_2$/TiO$_2$ composite systems and its role in photocatalytic degradation of a textile azo dye. Chem. Mater. 8, 2180–2187.

Vinodgopal, K., Wynkoop, D.E., Kamat, P.V., 1996b. Environmental photochemistry on semiconductor surfaces: Photosensitized degradation of a textile azo dye, Acid Orange 7, on TiO_2 particles using visible light. Environ. Sci. Technol. 30, 1660−1666.

Wahab, H.S., Hussain, A.A., 2016. Photocatalytic oxidation of phenol red onto nanocrystalline TiO_2 particles. J. Nanostruct. Chem. 6, 261−274.

Wang, D.H., Jia, L., Wu, X.L., Lu, L.Q., Xu, A.W., 2012a. One-step hydrothermal synthesis of N-doped TiO_2/C nanocomposites with high visible light photocatalytic activity. Nanoscale 4 (2), 576−584.

Wang, H., Zhang, L., Chen, Z., Hu, J., Li, S., Wang, Z., et al., 2014. Semiconductor heterojunction photocatalysts: Design, construction, and photocatalytic performances. Chem. Soc. Rev. 43 (15), 5234−5244.

Wang, H., Yuan, X., Wu, Y., Tu, G., Sheng, C., Deng, Y., et al., 2017. Plasmonic Bi nanoparticles and BiOCl sheets as cocatalyst deposited on perovskite-type $ZnSn(OH)_6$ microparticle with facet-oriented polyhedron for improved visible-light-driven photocatalysis. Appl. Catal. B Environ. 209, 543−553.

Wang, L., Shen, L., Zhu, L., Jin, H., Bing, N., Wang, L., 2012b. Preparation and photocatalytic properties of SnO_2 coated on nitrogen-doped carbon nanotubes. J. Nanomater. Available from: http://dx.doi.org/10.1155/2012/794625.

Wang, F., Valentin, C.D., Pacchioni, G., 2012c. Doping of WO_3 for photocatalytic water splitting: hints from density functional theory. J. Phys. Chem. C 116 (16), 8901−8909.

Wang, X.F., Tamiaki, H., Wang, L., Tamai, N., Kitao, O., Zhou, H., et al., 2010. Chlorophyll-a derivatives with various hydrocarbon ester groups for efficient dye-sensitized solar cells: Static and ultrafast evaluations on electron injection and charge collection processes. Langmuir 26 (9), 6320−6327.

Wang, Y., Huang, Y., Ho, W., Zhang, L., Zou, Z., Lee, S., 2009. Biomolecule-controlled hydrothermal synthesis of C-N-S-tridoped TiO_2 nanocrystalline photocatalysts for NO removal under simulated solar light irradiation. J. Hazard. Mater. 169 (1-3), 77−87.

Wongaree, M., Chiarakorn, S., Chuangchote, S., 2015. Photocatalytic improvement under visible light in TiO_2 nanoparticles by carbon nanotube incorporation. J. Nanomater. . Available from: http://dx.doi.org/10.1155/2015/689306.

Wu, L., Yu, J.C., Fu, X., 2006. Characterization and photocatalytic mechanism of nanosized CdS coupled TiO_2 nanocrystals under visible light irradiation. J. Mol. Catal. A Chem. 244, 25−32.

Xia, H., Zhuang, H., Zhang, T., Xiao, D., 2008. Visible-light-activated nanocomposite photocatalyst of Fe_2O_3/SnO_2. Mater. Lett. 62, 1126−1128.

Xia, H.L., Zhuang, H.S., Zhang, T., Xia, D.C., 2007. Photocatalytic degradation of Acid Blue 62 over CuO-SnO_2 nanocomposite photocatalyst under simulated sunlight. J. Environ. Sci. 19, 1141−1145.

Xiao, Q., Si, Z., Zhang, J., Xiao, C., Tan, X., 2008. Photoinduced hydroxyl radical and photocatalytic activity of samarium-doped TiO_2 nanocrystalline. J. Hazard Mater. 150 (1), 62−67.

Xu, J., Xiao, X., Stepanov, A.L., Ren, F., Wu, W., Cai, G., et al., 2013. Efficiency enhancements in Ag nanoparticles-SiO_2-TiO_2 sandwiched structure via plasmonic effect-enhanced light capturing. Nanoscale. Res. Lett. Available from: http://dx.doi.org/10.1186/1556-276X-8-73.

Yaghoubi, H., Li, Z., Chen, Y., Ngo, H.T., Bhethanabotla, V.R., Joseph, B., et al., 2015. Toward a visible light-driven photocatalyst: The effect of midgap-states-induced energy gap of undoped TiO_2 nanoparticles. ACS Catal. 5 (1), 327−335.

Yan, H., Wang, X., Yao, M., Yao, X., 2013. Band structure design of semiconductors for enhanced photocatalytic activity: the case of TiO_2. Prog. Nat. Sci. Mater. Int. 23, 402−440.

Yan, Z., Gong, W., Chen, Y., Duan, D., Li, J., Wang, W., et al., 2014. Visible-light degradation of dyes and phenols over mesoporous titania prepared by using anthocyanin from red radish as template. Int. J. Photoenergy. Available from: http://dx.doi.org/10.1155/2014/968298.

Yeber, M., Paul, E., Soto, C., 2012. Chemical and biological treatments to clean oily wastewater: Optimization of the photocatalytic process using experimental design. Desalin. Water Treat. 47 (1-3), 295−299.

Yu, J., Yang, Y., Fan, R., Li, L., Li, X., 2014. Rapid Electron injection in nitrogen- and fluorine-doped flower-like anatase TiO_2 with {001} dominated facets and dye-sensitized solar cells with a 52% increase in photocurrent. J. Phys. Chem. C 118 (17), 8795−8802.

Yun, J.N., Zhang, Z.Y., Zhao, W., 2013. First-principles study of Sc-doping effect on the stability, electronic structure and photocatalytic properties of Sr_2TiO_4. Thin Solid Films 542, 276−280.

Zerjav, G., Arshad, M.S., Djinovic, P., Zavasnik, J., Pintar, A., 2017. Electron trapping energy states of TiO_2-WO_3 composites and their influence on photocatalytic degradation of bisphenol A. Appl. Catal. B Environ. 209, 273–284.

Zeynolabedin, R., Mahanpoor, K., 2017. Preparation and characterization of nano-spherical $CoFe_2O_4$ supported on copper slag as a catalyst for photocatalytic degradation of 2-nitrophenol in water. J. Nanostruct. Chem. 7, 67–74.

Zhang, D., Gong, J., Ma, J.J., Han, G., Tong, Z., 2013. A facile method for synthesis of N-doped ZnO mesoporous nanospheres and enhanced photocatalytic activity. Dalton Trans. 42, 16556–16561.

Zhang, J., Ma, Y., Du, Y., Jiang, H., Zhou, D., Dong, S., 2017. Carbon nanodots/WO_3 nanorods Z-scheme composites: Remarkably enhanced photocatalytic performance under broad spectrum. Appl. Catal. B Environ. 209, 253–264.

Zhang, Q., Lima, D.Q., Lee, I., Zaera, F., Chi, M., Yin, Y., 2011. A highly active titanium dioxide based visible-light photocatalyst with nonmetal doping and plasmonic metal decoration. Angew. Chem. Int. Ed. 50 (31), 7088–7092.

Zhang, T., Wang, X., Zhang, X., 2014. Recent progress in TiO_2-mediated solar photocatalysis for industrial wastewater treatment. Int. J. Photoenergy. Available from: http://dx.doi.org/10.1155/2014/607954.

Zhou, M., Yu, J., Liu, S., Zhai, P., Jiang, L., 2008. Effects of calcination temperatures on photocatalytic activity of SnO_2/TiO_2 composite films prepared by an EPD method. J. Hazard Mater. 154, 1141–1148.

Zhou, M., Yu, J., 2008. Preparation and enhanced daylight-induced photocatalytic activity of C, N, S-tridoped titanium dioxide powders. J. Hazard. Mater. 152 (3), 1229–1236.

Ziolli, R.L., Jardim, W.F., 2002. Photocatalytic decomposition of seawater-soluble crude oil fractions using high surface area colloid nanoparticles of TiO_2. J. Photochem. Photobiol. A Chem. 147 (3), 205–212.

Zou, L., Luo, Y., Hooper, M., Hu, E., 2006. Removal of VOCs by photocatalysis process using adsorption enhanced TiO_2-SiO_2 catalyst. Chem. Eng. Proc. 45, 959–964.

Zyoud, A., Zaatar, N., Saadeddin, I., Helal, M.H., Campet, G., Hakim, M., et al., 2011. Alternative natural dyes in water purification: anthocyanin as TiO_2-sensitizerin methyl orange photo-degradation. Solid State Sci. 13 (6), 1268–1275.

C H A P T E R

7

Sonolysis

Ricardo A. Torres-Palma and Efraim A. Serna-Galvis

University of Antioquia, Medellín, Colombia

7.1 INTRODUCTION

Since the early 20th century, the ultrasound has gained considerable attention due to the enhancement of chemical and physical effects in many processes. In the 1930s, it was discovered that sonication induces polymers degradation (Thompson and Doraiswamy, 1999). The first experimental evidence for the formation of hydroxyl radicals in aqueous solution was reported by Parke and Taylor (1956). Spin trapping techniques allowed for verification of the formation of hydrogen and hydroxyl radicals from water sonication (Makino et al., 1982). Then the "hot spot" theory was proposed to explain the thermal and radical phenomena promoted by ultrasound. This theory was widely accepted and consequently, the proliferation of ultrasound applications to the removal of organic pollutants from water considerably increased (Thompson and Doraiswamy, 1999).

Due to the ability to produce radical species (mainly hydroxyl radical, $^{\bullet}OH$) that promote the degradation of refractory substances in water, ultrasound was considered an advanced oxidation process (AOP). Currently, this process is an alternative AOP with potential to eliminate chemical and microbiological pollutants in wastewaters as a tertiary treatment technology. Ultrasound has unique advantages compared to other AOPs, such as no addition of reagents, ease in handling, and differential or selective degradation according to the pollutant nature (Xiao et al., 2014b). Owing the interesting perspectives of sonochemistry applied to water treatment, this chapter presents fundamental and applicative aspects of water remediation using ultrasound.

7.2 PRINCIPLES OF THE PROCESS

Ultrasound refers to sound waves with frequencies above the detected by human ear. It is ranged from 20–10,000 kHz. Typically, ultrasound is divided into three regions

Advanced Oxidation Processes for Wastewater Treatment
DOI: https://doi.org/10.1016/B978-0-12-810499-6.00007-3

according to frequency: (1) low, (2) high, and (3) very high (Table 7.1). Low and high ultrasound frequencies are used in chemical processes, whereas very high frequency is applied in medical diagnostics.

When high-intensity ultrasound waves interact with dissolved gases in liquid medium, it is promoted acoustic cavitation (i.e., the formation, growth, and implosive collapse of bubbles) (Suslick and Fang, 1999). Ultrasound waves consist of compression and expansion cycles. During the expansion, waves having the sufficient intensity to exceed the molecular forces of liquid generate bubbles. These bubbles continually absorb energy from alternating compression and expansion ultrasound cycles. Thus, bubbles grow (by diffusion of vapor or gas from liquid medium) until they reach a critical size and then collapse (Fig. 7.1) (Suslick and Fang, 1999).

The bubble collapse acts as a localized "hot spot" with singular conditions of temperature ($\sim$5000K), pressure ($\sim$1000 atm), and short-life. Hence, chemical processes (bond

TABLE 7.1   Frequency Ranges of Ultrasound

Ultrasound Range (kHz)	Name
20–100	Low frequency
200–1000	High frequency
5000–10,000	Very high frequency

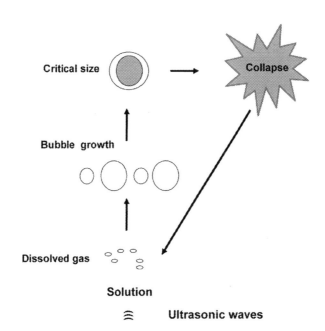

FIGURE 7.1   Cavitation phenomenon induced by ultrasound. *Adapted from Ashokkumar, M., Lee, J., Kentish, S., Grieser, F., 2007. Ultrasonics Sonochem. 14(4), 470–475. With permission.*

cleavage) occur; e.g., molecules of water and gases are broken (sonochemistry). Water is decomposed into a hydrogen atom and hydroxyl radical (Eq. 7.1). Oxygen and nitrogen produces atomic species (Eqs. 7.2 and 7.3), and volatile organic substances in the system are pyrolyzed. Additionally, some reactions of radicals occur (Eqs. 7.4–7.11) (Adewuyi, 2001). Sonochemistry is often accompanied by emission of light, and this phenomenon is called sonoluminescence (Suslick 1989; Suslick and Fang, 1999).

$$H_2O + \quad \rightarrow H + \quad OH \tag{7.1}$$

$$O_2 + \quad \rightarrow 2O \tag{7.2}$$

$$N_2 + \quad \rightarrow 2N \tag{7.3}$$

$$H + O_2 \rightarrow HOO \tag{7.4}$$

$$O + H_2O \rightarrow 2 \quad OH \tag{7.5}$$

$$H + H \rightarrow H_2 \tag{7.6}$$

$$OH + \quad OH \rightarrow H_2O_2 \tag{7.7}$$

$$N + \quad OH \rightarrow NO + H \tag{7.8}$$

$$NO + \quad OH \rightarrow HNO_2 \tag{7.9}$$

$$2 HOO \rightarrow H_2O_2 + O_2 \tag{7.10}$$

$$H + HOO \rightarrow H_2O_2 \tag{7.11}$$

The sonochemical process is rationalized as a three zones system: (1) bulk of solution, (2) interface of cavitation bubble, and (3) inner of bubble (Fig. 7.2). The hydrophilic substances are placed in the bulk of the solution, hydrophobic nonvolatile compounds accumulate in the interfacial zone, while volatile substances enter the cavitation bubble (Gogate, 2008). Thus, the nature of a pollutant determines its degradation route in the

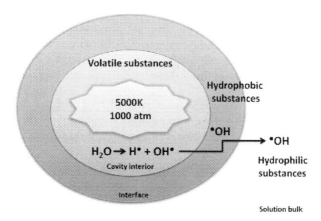

FIGURE 7.2 Zones of sonochemical system, where pollutants are degraded (inner of bubble, interface, and solution bulk). *Adapted from Adewuyi, Y.G., 2001. Ind. Eng. Chem. Res., 40, 4681–4715. With permission.*

sonochemical process. The removal of those hydrophilic compounds is promoted by hydroxyl radicals that reach the solution bulk after bubble collapse. Hydrophobic nonvolatile compounds are eliminated in the interfacial zone by radical attacks and/or thermal reactions. In its turn, volatile pollutants are pyrolyzed inside of bubbles (Fig. 7.2) (Pétrier and Francony, 1997).

On the other hand, the degradation of several pollutants under ultrasound action follows a pseudo-first order of reaction. Therefore, the rate of removal (r) is expressed as:

$$r = -\,dC/dt = kC \qquad\qquad (7.12)$$

where C is the concentration of the pollutant and k is the pseudo-first order constant. The integration of Eq. (7.12) leads to Eq. (7.13).

$$\ln C/C_i \,= kt \qquad\qquad (7.13)$$

where C_i is the initial pollutant concentration and t is the treatment time. Thus, the k value can be determined as the slope from the plot of $\ln(C/C_i)$ versus t, using the experimental data. Furthermore, the kinetics of sonochemical degradation can be modeled taking into account variations on the initial concentration and nature of the pollutant (see Section 7.4.5).

7.3 TYPES OF MAIN REACTORS (REACTION SYSTEMS)

Bath and probe (also called horn) are the two most common reaction systems for the generation of acoustic cavitation in water (Fig. 7.3). Ultrasound is introduced into a system by direct contact of the ultrasonic source with the liquid medium (direct sonication) or by dipping a vessel containing the solution to treat (indirect sonication). In either case, the ultrasound waves are generated through a piece that incorporates a transducer (the part of system that transforms electrical power into waves) coupled to a vibrating plate for the bath and a tip for the probe. Additionally, all systems should be equipped with a cooler/heater device to stabilize the bulk solution temperature (Berlan and Mason 1992; Sathishkumar et al., 2016).

From the two main systems (i.e., bath and probe), diverse configurations (by modification of the amount and positions of ultrasonic wave sources) and geometries of reactors have been developed. Also, different modes (batch or continuous flow) of the ultrasonic reaction systems have been considered (Mason, 2001; Sutkar and Gogate, 2009). Supplementary information about design aspects of sonochemical reactors, advantages, and limitations of the diverse systems can be found in the references (Berlan and Mason, 1992; Gogate, 2008; Sutkar and Gogate, 2009; Sathishkumar et al., 2016).

On the other hand, two ultrasound irradiation ways (i.e., continuous or pulsed) are possible. Continuous way is the most frequently used. However, pulsed wave ultrasound has been gaining attention because it can enhance elimination of pollutants (mainly hydrophobic substances) by increasing their accumulation on bubble interface during the intervals between pulses, causing more molecule degradation during the collapse (Xiao et al. 2013, 2014b).

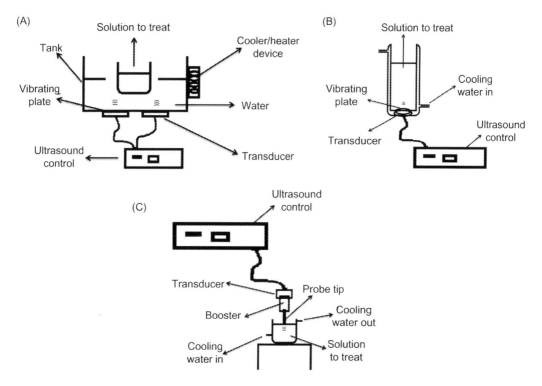

FIGURE 7.3 Basic schemes of the main types of reactors for the sonochemical process. (A) Bath system (indirect sonication), (B) Bath system (direct sonication), and (C) Probe system (direct sonication). *Adapted from Berlan, J., Mason, T.J., 1992. Ultrasonics, 30, 203–212; Sathishkumar, P., Mangalaraja, R.V., Anandan, S., 2016. Renew. Sustain. Energy Rev., 55, 426–454. With permission.*

7.4 THE EFFECT OF SONOCHEMICAL OPERATIONAL PARAMETERS

In addition to the reaction system (type, design, irradiation way, etc.), sonochemistry and its application to water remediation also depends on operational parameters such as the ultrasonic frequency, dissolved gases, acoustic power, temperature of the liquid, and the initial concentration of the pollutant.

7.4.1 Ultrasound Frequency

The ultrasonic frequency affects the cavitation process by modifying the bubble size and collapse time of the cavity. The increase of frequency makes the collapse time shorter and the bubble size smaller (Fig. 7.4) (Lim et al., 2011). A higher frequency induces more cavitation events per time unit, and the flux of gases and volatile substances toward the bubble is increased (Beckett and Hua, 2001; Yang et al., 2008). However, the highest radicals formation occurs at frequencies around 200–350 kHz (Kang et al., 1999). Consequently, a larger accumulation of hydrogen peroxide coming from the combination of radicals (Eq. 7.7, 7.10–7.11) is observed at these frequencies (Table 7.2).

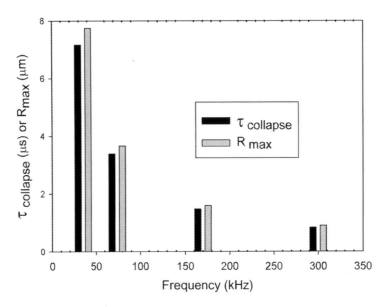

FIGURE 7.4   Effect of frequency on both the time of cavity collapse ($_{collapse}$) and the maximum radius of cavitation bubble (R_{max}). *Adapted from Lim, M., Son, Y., Karim, J., 2011. Ultrasonics Sonochem. 18, 460–465. With permission.*

However, the effect of frequency in the degradation depends on the contaminant nature. For hydrophilic and moderate hydrophobic substances (e.g., coomassie brilliant blue and phenol), their major removals are coincident with the range of high hydroxyl radical formation (i.e., 200–350 kHz), whereas the degradation of very hydrophobic or volatile substances (e.g., octyl-benzenesulfonate and carbon tetrachloride) is more favored at higher values of frequency (Table 7.2).

The ultrasonic treatment of phenol and CCl_4 in the range 20–800 kHz showed that the former is faster degraded at 200 kHz, while the carbon tetrachloride disappearance is enhanced on increasing frequency. As the CCl_4 degradation takes place inside of the cavitation bubble, its removal is favored with the increasing frequency because the flux of this volatile substance toward the bubble and the number of cavitation collapses are greater at higher frequencies. In contrast, phenol is degraded by attacks of hydroxyl radical (out of cavitation bubbles), thus, its elimination is faster at the frequency of higher production of •OH (Pétrier and Francony, 1997).

Additional information about physical principles of ultrasound frequency effects can be found in the literature (Beckett and Hua, 2001; Brotchie et al., 2009; Merouani et al., 2013).

7.4.2 Dissolved Gas

The properties of dissolved gases, such as their ratio of specific heats (γ, also named polytropic index), thermal conductivity (Q), and solubility in water, influence the

TABLE 7.2 Effect of Ultrasonic Frequency During Sonochemical Treatment of Organic Pollutants

EFFECT OF ULTRASOUND FREQUENCIES ON THE ACCUMULATION OF HYDROGEN PEROXIDE IN WATER

Frequency (kHz)	20	200	500	800
H_2O_2 Accumulated (mmol/L)	0.080	0.600	0.220	0.158

FREQUENCY EFFECT ON POLLUTANTS DEGRADATION

Tested Frequencies (kHz)	Frequency of Maximum Degradation (kHz)	Pollutant	References
20, 200, 500, and 800	200	Phenol	Pétrier and Francony (1997)
20, 200, 500, and 800	200	4-Chlorophenol	Jiang et al. (2006)
300, 500, 600, and 800	300	Bisphenol A	Torres et al. (2008a,b,c)
20, 300, and 446	300	Alachlor	Wayment and Casadonte (2002)
200, 350, 620, and 1000	350	Coomassie brilliant blue	Rayaroth et al. (2015)
205, 358, 618, and 1078	358	MTBE	Kang et al. (1999)
205, 358, 618, and 1071	358	1,4-Dioxane	Beckett and Hua (2000)
205, 358, 618, and 1078	618	1,1,1-Trichloroethane	Colussi et al. (1999)
20, 205, 358, 500, 618, and 1078	618	Hexachloroethane	Hung and Hoffmann (1999)
205, 358, 618, and 1078	618	Tetrachloroethylene	Colussi et al. (1999)
206, 354, 620, 803, and 1062	620	Octyl-benzenesulfonate	Yang et al. (2008)
20, 200, 500, and 800	800	Carbon tetrachloride	Pétrier and Francony (1997)

Source: Adapted from Pétrier, C., Francony, A., 1997. Ultrasonics Sonochem. 4, 295–300. With permission.

sonochemical activity (Hua and Hoffmann, 1997). Table 7.3 presents the γ and Q values for the commonly used gases in sonochemical processes. The first property is related to the maximum implosion temperature, while the second property determines the rate of heat transfer to the surrounding liquid (Adewuyi, 2001; Rooze et al., 2013). Hence, the gases with high γ and low Q values favor a high temperature during the implosion of cavitation bubbles. Additionally, a higher gas solubility provides more nucleation sites for cavitation, leading to a higher number of bubbles in the medium (Hua and Hoffmann, 1997; Merouani et al., 2015).

Typically air, oxygen, and noble gases (mainly helium, argon, and krypton) have been used in sonochemistry. Air or oxygen, as dissolved gases, produce additional radicals from interaction of O_2 with $H^{\bullet}$(Eq. 7.4) or $^{\bullet}O$ with H_2O (Eq. 7.5). Furthermore, the formation of hydroxyl radicals using oxygen is higher than when air is the saturating gas. For nonvolatile pollutants, the degradation and the production of H_2O_2 in the presence of

TABLE 7.3   Properties of Commonly Used Gases in the Sonochemical Processes

Gas[a]	(Dimensionless)	Q (mW/K per m)
Air	1.40	26.5
Oxygen	1.41	25.9
Helium	1.63	252.4
Argon	1.66	30.6
Krypton	1.66	17.1

[a]*Solubility of these gases in water follows the order: Kr > Ar > He > O_2 > Air.*
Source: Adapted from Hua and Hoffmann, 1997. Environ. Sci. Technol., 31, 2237–2243. With permission.

oxygen are faster than when air is used. This can be also associated to the higher solubility of O_2 (Table 7.3).

For noble gases such as helium, argon, and krypton, the production of hydroxyl radical in water follows the order: Kr > Ar > He (Hua and Hoffmann, 1997), which is reflected on the degradation of pollutants. This is because krypton and argon gases have larger specific heat ratios and solubility, and lower thermal conductivities than helium (Lifka et al., 2003). For example, the comparative treatment of the pharmaceutical fluoxetine at 600 kHz using He and Ar as dissolved gases showed that both the pollutant degradation and H_2O_2 accumulation in argon are significantly higher than in helium presence (Fig. 7.5). Similar results have been reported in the case of p-nitrophenyl acetate, where its sonochemical degradation was faster in Ar than in He (Lifka et al., 2003; Hua et al., 1995).

On the other hand, several works have compared the degradation of pollutants in the presence of air and argon. The ratio between rate constants or initial rates (denoted as k_{air}/k_{Ar}) for different compounds is presented in Table 7.4. A value of k_{air}/k_{Ar} higher than one indicates that pollutant removal is better in air, whereas a ratio lower than one shows that degradation in argon is favored. Merouani et al. (2015) studied the effect of frequency on the rate of •OH formation for argon and air in the range 200–1200 kHz. They found that at frequencies above 500 kHz, hydroxyl radical is produced faster in argon; whereas at frequencies below 500 kHz, the •OH formation rate is higher in air.

From Table 7.4, it can be observed that hydrophilic (e.g., tannic acid, gallic acid, crystal violet) or moderate hydrophobic substances (e.g., bisphenol-A, ibuprofen, fluoxetine), which are susceptible to be degraded by •OH in the bulk of solution or bubble interface, show a faster degradation at 200–400 kHz range using air as dissolved gas, or under the argon at higher frequencies (500–800 kHz). This is associated to the larger production of hydroxyl radical in such conditions. Otherwise, high hydrophobic or volatile substances (e.g., pentachlorophenol, PFOS, PFOA, alachlor, trichloroethylene, etc.) are faster degraded in argon, which is related with the fact that Ar has stronger thermal effects than air at 200–1100 kHz frequency range (Merouani et al., 2015).

In addition to air, oxygen, helium, argon, and krypton as saturating gases, nitrogen and carbon dioxide or mixtures of oxygen and argon have been considered. N_2 gas promotes

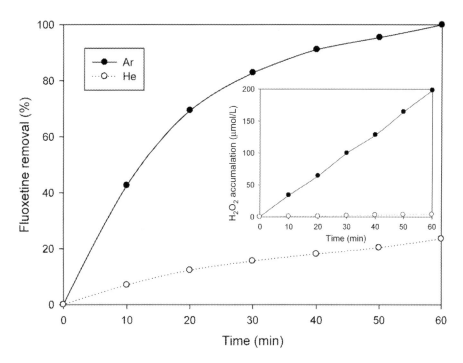

FIGURE 7.5 Fluoxetine degradation using He and Ar as saturating gases. *Adapted from Serna-Galvis, E.A., Silva-Agredo, J., Girldo-Aguirre, A.L., Torres-Palma, R.A., 2015. Sci. Total Environ, 524–525, 354–360. With permission.*

TABLE 7.4 Ratio between Sonochemical Rate Constants (or Initial Rates) Under Air and Argon for Different Compounds

Compound	Frequency (kHz)	k_{air}/k_{Ar} Ratio	References
Tannic acid	200	2.56	Nagata et al. (1996)
Gallic acid	200	2.12	Nagata et al. (1996)
Pentachlorophenol	200	0.92	Rooze et al. (2013)
PFOS	200	0.43	Moriwaki et al. (2005)
PFOA	200	0.48	Moriwaki et al. (2005)
Bisphenol-A	300	1.03	Torres et al. (2008a,b,c)
4-Chlorophenol	300	>1	Rooze et al. (2013)
Ibuprofen	300	1.38	Méndez-Arriaga et al. (2008)
Alachlor	300	0.48	Rooze et al. (2013)
1,4-Dioxane	358	1.85	Rooze et al. (2013)
2,4-Dichlorophenol	489	0.57	Rooze et al. (2013)
Trichloroethylene	520	0.62	Rooze et al. (2013)
Fluoxetine	600	<1	Serna-Galvis et al. (2015)
2-Methyl-isoborneol	640	0.44	Song and O'Shea (2007)
Crystal violet	800	0.88	Guzman-Duque et al. (2011)

cavitation events; however, the generated atomic nitrogen scavenges the hydroxyl radical (Eqs. 7.8 and 7.9) inducing the formation of HNO_2 (Mead et al., 1976) and this can decrease the efficiency of removal of contaminants. Meanwhile, the absence of cavitation under CO_2 saturated conditions, due to its low solubility, hinders the degradation of pollutants (Henglein and Gutierrez, 1988; Guzman-Duque et al., 2011). In contrast, the use of argon/oxygen mixtures (such as Ar/O_2: 90/10, 80/20, 70/30, 75/25, and 60/40) improves the degradation of organic compounds (Hart and Henglein, 1985; Henglein and Gutierrez, 1988; Schramm and Hua 2001; Beckett and Hua, 2000). Further information about these topics can be found in literature (Beckett and Hua, 2000; Schramm and Hua 2001; Rooze et al., 2013; Guzman-Duque et al., 2011; Mead et al., 1976; Hart and Henglein, 1985; Henglein and Gutierrez, 1988).

7.4.3 Power Input

The power is the energy emitted by the source (vibrating plate or probe) in the form of ultrasound waves (Adewuyi, 2001). Sometimes, this parameter is divided by the solution volume to indicate the power density (Lifka et al., 2003). The increase in the ultrasonic power induces more sonochemical activity in the liquid medium. For all kinds of substances (volatiles, hydrophobics, and hydrophilics), the response to the effect of the ultrasonic power is similar (Table 7.5).

A higher input power leads to a faster degradation of the pollutants due to more cavitation events and hydroxyl radical production, which is evidenced by a greater accumulation of H_2O_2, when the power is augmented (Fig. 7.6). However, it is important to remark that after a certain value, an improvement in the pollutant degradation is not possible with the increasing of the ultrasonic power (Lifka et al., 2003; Tezcanli-Guyer and Ince, 2003). This is due to the fact that cavitation bubble size increases with ultrasound power; however, after a limit value the bubble radius is approximately constant (Brotchie et al., 2009).

7.4.4 Effect of Bulk Temperature

The maximum temperature generated during the bubble collapse depends on the liquid bulk temperature; therefore, this parameter modifies the sonochemical activity (Thompson and Doraiswamy, 1999). The effects of temperature have been studied at both low (<100 kHz) and high ultrasound frequencies (200–1000 kHz). Generally, at low frequencies, the increase in temperature has a detrimental effect on the pollutants degradation (Table 7.6) (Psillakis et al., 2004a; Jiang et al., 2006; Goskonda et al., 2002; Zhang et al., 2011). This is attributed to the fact that when the liquid temperature is increased, its viscosity and/or surface tension decreases and the vapor pressure increases; thus, bubbles contain more water vapor than gas, which cushion bubble implosion. Additionally, the increasing temperature favors the degassing of the liquid phase, thus diminishing the number of gas nuclei available for cavitation bubble formation (Zhang et al., 2011; Adewuyi, 2001). Therefore, for applications in which thermal effects are the main cause of

TABLE 7.5 Effect of Ultrasound Power (or Power Density) on Degradation of Pollutants of Diverse Nature

Pollutant Nature	Pollutant	Potency or Potency Density	k or r or Removal[a]	References
Hydrophilic	Coomassie brilliant blue	3.5 W/mL	0.0510 min^{-1}	Rayaroth et al. (2015)
		9.8 W/mL	0.0928 min^{-1}	
		19.6 W/mL	0.1335 min^{-1}	
Hydrophobic Nonvolatile	Fluoxetine	20 W	0.2651 mol/L per min	Serna-Galvis et al. (2015)
		40 W	0.9078 mol/L per min	
		60 W	1.2853 mol/L per min	
Volatile	Naphthalene	45 W	39.4%	Psillakis, Goulaet al. (2004)
		75 W	60.6%	
		150 W	94.2%	

[a]k: rate constant and r: degradation rate.

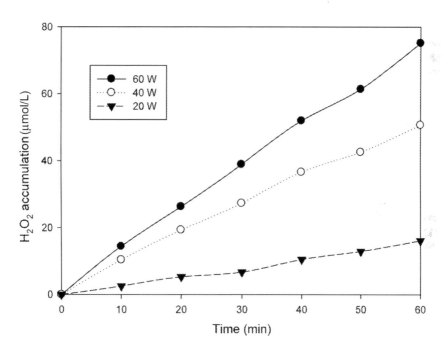

FIGURE 7.6 Effect of ultrasound power/power density on the formation of radicals (expressed as the accumulation of H_2O_2).

TABLE 7.6   Effect of Temperature (T) on Sonochemical Degradation of Pollutants

Compound	Ultrasound Frequency (kHz)	T Values (C)	Effect of T Increasing on Degradation	References
2,4-Dichlorophenol	20	12.5, 18, and 30	Decreasing	Goskonda et al. (2002)
4-Chlorophenol	20	10–45	Decreasing	Jiang et al. (2006)
Dextran	20	30, 50, and 70	Decreasing	Lorimer et al. (1995)
Diazinon	20	25 and 35	Decreasing	Zhang et al. (2011)
Dimethyl phthalate	80	21 and 50	Decreasing	Psillakis et al. (2004b)
Rhodamine B	300	25–55	Increasing	Merouani et al. (2010)
4-Chlorophenol	500	10–60	Degradation has a maximum at 40 C	Jiang et al. (2006)

the degradation (such as frequencies < 100 kHz), a low operating temperature is recommended (Sutkar and Gogate 2009).

In contrast to the observations for low frequencies, at high frequencies when the liquid temperature increases, the cavitation bubbles formed in the process have a predominant gaseous nature (i.e., bubbles contain more gas than water vapor) (Jiang et al., 2006). Then, the number of cavitation bubbles and the rate of radicals production are higher; which can favor the elimination of pollutants (rhodamine B). However, after a certain temperature of bulk, there is less dissolved gas in the cavitation bubbles and they have a more vaporous nature. Under such conditions, similar to the observed at low frequencies, the increasing temperature declines the pollutants degradation (Fig. 7.7), (Jiang, et al., 2006; Lifka et al., 2003).

7.4.5 Pollutant Concentration

The concentration of pollutants varies depending on the water source. Then, this parameter is usually evaluated during the removal of organic pollutants upon sonochemical action. The chemical nature of compounds also influences such parameter (Fig. 7.8). For volatile substances, the degradation rate (r) decreases with the increase of the initial concentration of organic pollutants (Jiang et al., 2002a,b; De Visscher et al. 1996; Goel et al., 2004). Meanwhile, the removal of hydrophilic and hydrophobic nonvolatile compounds is increased at higher initial concentrations in the aqueous media (Merouani et al., 2010; Okitsu et al., 2005; Serna-Galvis et al., 2015; Torres et al., 2008a,b,c; Villaroel et al., 2014; Villegas-Guzman et al., 2015).

The decreasing in the degradation rate constants at higher initial concentrations of the volatile compounds is explained by their influence on the cavitation bubble. As concentration of the volatile substances is increased, more molecules have opportunity of diffusing into the cavitation bubble. Then, the temperature and pressure within the cavitation bubbles decrease and this weakens the cavitational collapse; thus, resulting in a lower efficiency of sonochemical degradation (Jiang et al., 2002a,b). On the contrary, when the

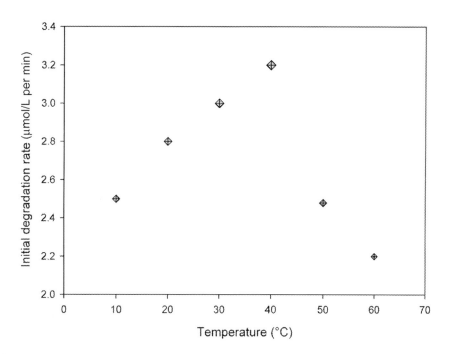

FIGURE 7.7   Variation in the removal rate of 4-chlorophenol with the temperature of liquid bulk at 500 kHz of ultrasonic frequency. *Adapted from Jiang, Y., Petrier, C., Waite, T.D., 2006. Ultrasonics Sonochem., 13, 415–422. With permission.*

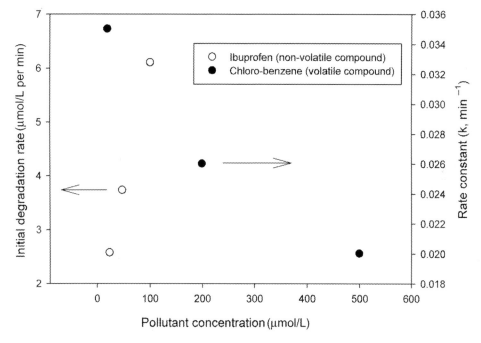

FIGURE 7.8   Effect of initial concentration of pollutants during the sonochemical degradation of chloroben-zene and ibuprofen. *Adapted from Méndez-Arriaga, F., Torres-Palma, R., Petrier, C., Esplugas, S., Gimanez, J., Pulgarin, P., 2008. Water Res. 42, 4243–4248; Jiang, Y., Petrier, C., Waite, T.D., 2002. Ultrasonics Sonochem., 9, 317–323.*

concentration of nonvolatile pollutants is higher, more molecules of them are close to the cavitation bubble interface, and the reaction with radicals and thermal processes are more favored, which resulted in an increase in the degradation rate of these compounds (Serna-Galvis et al., 2015).

The degradation kinetic of volatile pollutants can be modeled taking into account the dependence on the initial concentration (C_i). De Visscher et al. (1996) proposed that the rate of degradation (r) can be represented as:

$$r = kC_i$$
$$\text{With} \quad k = k_o e^{-aC_i} \tag{7.14}$$

where k_o is the rate constant at infinite dilution (very low concentration of the pollutant) and reflects the degradability of the volatile compounds on exposure to ultrasound, whereas the a parameter represents the kinetic factor associated to the inhibitory effect of the volatile substance due to its own presence in the cavitation bubble (De Visscher et al., 1996; Goel et al., 2004). This kinetic model assumes that the concentration of the volatile compound in the bulk liquid phase is proportional to the concentration in the bubble, and the reaction rate is proportional to the reaction rate in the cavitations during collapse (De Visscher et al., 1996).

The equation for k can be linearized as follows:

$$\ln k = -aC_i + \ln k_o \tag{7.15}$$

The k values are the pseudo-first order rate constant at the different initial pollutant concentrations (C_i). Therefore, from the plot of ln k versus C_i, the kinetic parameters (k_o and a) are obtained. The comparison between the experimental data (k_{exp}) and the predicted rate constant using the model (k_{calc}) for different initial concentrations of ethylbenzene is given in Table 7.7, where very low error percentages are observed (<1%). Additionally, the linearly shows an excellent correlation coefficient ($R^2 = 0.997$). This indicates an excellent fit between the experimental results and the model. However, it should be mentioned that at high concentrations of volatile compounds (e.g., benzene at 100 mg/L), some deviations of the model are observed (De Visscher et al., 1996; Goel et al., 2004).

The application of this model to the sonochemical degradation of styrene and the comparison of its kinetic parameters (Table 7.7) with those obtained for ethylbenzene shows that k_o and a of styrene are higher. This indicates that styrene is more susceptible than ethylbenzene to sonodegradation and the increasing of styrene concentration affects more significantly its own degradation than in the case of ethylbenzene, which is associated to the more volatile nature of styrene.

On the other hand, for nonvolatile substances, kinetics of degradation (as a function of initial concentration) can be adjusted to a model based on a Langmuir-Hinshelwood-type mechanism, which considers the movement of pollutant molecules from the bulk solution to the interfacial region of bubbles. This model proposed by Okitsu et al. (2005) assumes that: (1) pollutant molecules in the interface are proportional to the initial concentration in the bulk of solution, (2) the pollutant is in an adsorption/desorption equilibrium between the interfacial zone and the bulk, and (3) a high concentration of radicals exists at the interfacial region and pollutants rapidly react with them (Okitsu et al., 2005; Chiha et al., 2010).

The degradation rate (r) is represented by the following equation:

TABLE 7.7 Comparison between the Experimental and the Predicted Values of Sonochemical Rate Constants

Compound	C_i (mmol/L)	k_{exp} (min^{-1})	k_{calc} (min^{-1})	Error (%)	k_o (min^{-1})	A (mmol^{-1} per L)
					Parameters	
Ethylbenzene	0.47	0.0137	0.01369	0.22	0.1619	0.3607
	0.94	0.0115	0.01154	0.33		
	1.89	0.0082	0.00819	0.12		
Styrene	0.97	0.01267	0.01268	0.13	0.4600	1.328
	0.49	0.02409	0.02399	0.39		
	0.25	0.03292	0.03300	0.26		

Source: Adapted from Goel, M., Hongqiang, H., Muzumdar, A.S., Ray, M. B., 2004. Water Res., 38, 4247–4261. With permission.

TABLE 7.8 Comparison between the Experimental and the Predicted Values of Fluoxetine Upon Ultrasonic Action

C_i (mol/L)	r_{exp} (mol/L per min)	r_{calc} (mol/L per min)	Error (%)
2.9	0.110	0.1097	0.19
8.7	0.270	0.2775	2.80
17.4	0.470	0.4491	4.43
40.5	0.690	0.6939	0.57
162.0	0.995	1.0024	0.75

Source: Adapted from Serna-Galvis, E.A., Silva-Agredo, J., Giraldo-Aguirre, A.L., Torres-Palma, R.A., 2015. Sci. Total Environ., 525–525, 354–360. With permission.

$$r = kKC_i / 1 + KC_i \qquad (7.16)$$

where, C_i is the initial concentration of the pollutant, k and K are the rate and the equilibrium constant, respectively. The Okitsu et al. Eq. (7.16) can be rewritten in a linear form as Eq. (7.17) (other forms of linearization are also possible (Chiha et al., 2010)), which shows that the kinetic parameters (k and K) can be obtained from the experimental. A comparison of the degradation rate between the experimental values (r_{exp}) and the predicted (r_{calc}) using the model has been presented (Table 7.8). Low percentages of error are observed (<5%). In addition, the linearization shows a high correlation coefficient ($R^2 = 0.999$). This indicates the excellent fit between the experimental results and the results obtained by the model (Okitsu et al., 2005). This kinetic analysis has also been applied to acetaminophen, dicloxacillin, and parathion (Villaroel et al., 2014; Villegas-Guzman et al., 2015; Yao et al., 2010); thus demonstrating its versatility to model for the degradation of different kinds of nonvolatile organic pollutants.

Further information about theoretical aspects of De Visscher et al. and Okitsu et al. models can be found in the literature (De Visscher et al., 1996; Goel et al., 2004; Okitsu et al., 2005; Chiha et al., 2010).

7.5 EFFECT OF THE CHEMICAL POLLUTANT NATURE AND ITS TRANSFORMATIONS UPON SONOCHEMICAL PROCESS

In the sonochemical systems, there are three zones for the pollutants degradation: (1) the bulk of solution, where attack of radicals take place, (2) the interfacial zone, where radical attack and/or thermal processes occur, and (3) inside of the bubble through pyrolysis. The predominant elimination route depends on the nature (structure and physico-chemical properties) of the pollutants. Furthermore, the analysis of degradation products provides insights on the main transformations induced by the ultrasound (Sivasankar and Moholkar, 2009). Therefore here, the effect of the chemical nature and the transformations of the pollutants upon sonochemical action are presented.

7.5.1 Structural Effects and Physico-Chemical Properties

7.5.1.1 Small Chlorinated Hydrocarbons (SCHs)

Small chlorinated hydrocarbons are organic solvents frequently used as industrial degreasing agents for the metals, paint stripping, and dry cleaning. They are recognized as water pollutants, and high frequency ultrasound has been applied to eliminate them. The treatment of three representative SCHs (dichloromethane, trichloromethane, and tetrachloromethane) by sonochemistry have their degradation constants (k) follow the order (Table 7.9):

$$CCl_4 > CHCl_3 > CH_2Cl_2 \tag{7.17}$$

These SCHs are volatile substances and as the number of chlorine atoms increases, their ability to escape from aqueous medium to the gas phase (which is related to the Henry constant, H) is also higher. Due to its largest H value, CCl_4 faster diffuse into the bubbles and undergo pyrolytic route (Hung and Hoffmann, 1999). Additionally, the bond C−Cl is weaker than C−H; thus, the increasing number of chlorine atoms in the SCHs' structures eases the pyrolysis (Moldoveanu, 2010).

TABLE 7.9   Characteristics of Small Chlorinated Hydrocarbons (SCHs)

Compound	k (min^{-1})	H (atm m^3/mol)
Dichloromethane	0.016	1.97×10^{-3}
Trichloromethane (Chloroform)	0.028	5.30×10^{-3}
Tetrachloromethane (Carbon tetrachloride)	0.044	2.42×10^{-2}

Source: Adapted from Hung, H.M., Hoffmann, M.R., 1999. J. Phys. Chem A, 103(10), 2734–2739. With permission.

Other SCHs commonly detected in groundwater, because their poor degradability under natural conditions, are the 1,1,1-trichloroethane (1-TCE) and trichloroethylene (TCE). The application of the ultrasound process to eliminate 1-TCE and TCE shows rate constants of 0.0821 and 0.0668 min^{-1}, respectively. In addition, their degradation was not affected by the presence of tert-butanol, indicating that these SCHs were removed by direct thermal process inside of cavitation bubbles (Yim et al., 2001). Due to the structural configuration of 1-TCE, it has lower intermolecular forces than TCE and therefore, 1,1,1-trichloroethane is more volatile and faster removed sonochemically.

7.5.1.2 Monocyclic Aromatic Compounds (MACs)

Many MACs (e.g., phenol, aniline, toluene, and styrene) are widely used as industrial precursors (raw materials) of dyes, pesticides, and polymers. As a consequence, this class of compounds is frequently found in industrial wastewater. The ultrasound process has shown to be efficient for the treatment of diverse MACs. A relationship between the initial rates of sonochemical degradation of these substances and their physicochemical parameters has been reported (Nanzai et al., 2008). For MACs with skeleton mainly composed by C and H or C,H, and Cl (such as chlorobenzene, ethylbenzene, and propylbenzene), their degradation rates (r) correlate well with the Henry constant (H). Thus, the MAC having the greatest H value presents the highest degradation rate. Moreover, similar to the observed for SCH, those volatile MACs containing more chlorine atoms are more susceptible to sonodegradation. For example, 1,4-dichlorobenzene is eliminated faster than chlorobenzene (Jiang et al., 2002a,b).

Meanwhile, MACs such as phenol, aniline, and benzoic acid, which have polar groups in their structures, are hydrophobic nonvolatile compounds. Thus, they tend to accumulate at the bubble interface and their degradation rates can be related to the Log P (also named Log K_{ow}), a parameter of hydrophobicity. MACs with higher Log P are more hydrophobic and present larger r values (Fig. 7.9).

On the other hand, the treatment of a mixture containing both a volatile and a hydrophobic nonvolatile MAC shows that in the first stage of the process the degradation of the volatile compound predominates and the transformation of the nonvolatile only starts, when the concentration of the volatile compound is very low. This clearly shows that the presence of a volatile compound hinders the degradation of nonvolatile substances (Petrier et al., 1998).

7.5.1.3 Polycyclic Aromatic Hydrocarbons (PAHs)

PAHs are derived from incomplete combustion of fossil fuels and enter natural waters through leaching. PAHs are persistent pollutants due to their high chemical stability and biodegradation resistance (Rubio-Clemente et al., 2014). Representative PAHs (i.e., naphthalene, acenaphthylene, and phenanthrene) have been treated by ultrasound. It is reported that the degradation follows the order: naphthalene > acenaphthylene > phenanthrene (Psillakis et al., 2004a).

These PAHs are semi-volatile pollutants (Stogiannidis and Laane, 2015). The H values for naphthalene, acenaphthylene, and phenanthrene are 4.4×10^{-4}, 1.1×10^{-4}, and 4.23×10^{-5}, respectively (National Library of Medicine, 2017). As the aromatic rings number on PAHs increases, the $-$ interactions (dispersion forces) are higher making

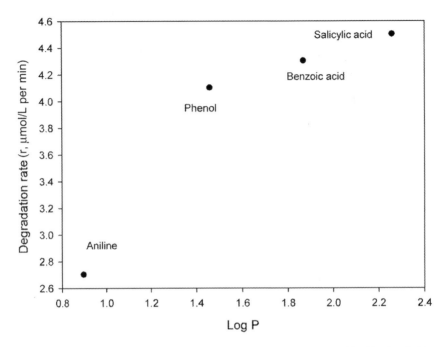

FIGURE 7.9 Relationship between Log P and sonochemical degradation rates (r) of hydrophobic nonvolatile MACs. *Adapted from Nanzai B., Okitsu, K., Takenaka, N., Bandow, H., Maeda, Y., 2008. Ultrasonics Sonochem. 15, 478–483. With permission.*

such substances less volatile. Therefore, phenanthrene and acenaphthylene (which have three rings) are less susceptible toward degradation by sonochemical process than naphthalene (having two rings). In fact, the degradation in the presence of 1-butanol (a known radical scavenger) shows that acenaphthylene and phenanthrene are more inhibited than naphthalene. This indicates that in naphthalene, removal via thermal is predominant; while for the other two PAHs both the radical and pyrolytic routes are involved (Psillakis et al., 2004a).

7.5.1.4 Perfluoroalkyl Sulfonates (PFAS) and Perfluoroalkyl Acids (PFAA)

Perfluoroalkyl compounds (PFAS and PFAA) have a wide variety of industrial applications, e.g., polymers, coatings for paper textiles and metals, film-forming foams, and thermally stable lubricant production. In environmental water, they are extremely persistent, bioaccumulative in humans, and toxic to animals and wildlife. Perfluorinated compounds are highly hydrophobic pollutants with very low vapor pressure. Consequently, the sonolysis of these substances occurs primarily at the bubble interface through thermal reactions (Moriwaki et al., 2005; Fernandez et al. 2016). Indeed, representative perfluoroalkyl compounds such as perfluorooctanoic acid and perfluorooctanoic sulfonate are not degraded by the Fenton process (system based on the $^{\bullet}$OH action, which is generated from a reaction between Fe (II) and H_2O_2) (Moriwaki et al., 2005), indicating their refractory character toward even strong oxidizing hydroxyl radical.

Analysis of structural effects in the sonochemical treatment of PFAS and PFAA shows that the larger the perfluoroalkyl chain is, the higher the degradation rate is. This is because an enhancement in the hydrophobic character occurs with the chain length increasing of the perfluoroalkyl substances. Moreover, the comparison of PFAS and PFAA with similar chain length indicates that the compounds having carboxylates are degraded slightly faster than those with sulfonates, and it is attributed to the lower thermal stability of PFAA (Fernandez et al., 2016).

7.5.1.5 Phthalate Acid Esters (Phthalates)

Phthalates have important uses as plasticizers in the polymeric industries. In recent years, presence of phthalates in water have gained environmental attention, even at low concentration levels they can act as disruptors of the endocrine system in humans and wildlife (Psillakis et al., 2004a; Xu et al., 2015). The phthalates differ in the alkyl chains and an increasing of such structural groups makes them more hydrophobic (Psillakis et al., 2004a). Consequently, they are more susceptible to sonochemical degradation, which is clearly illustrated by the ultrasonic treatment of mono-methyl phthalate (MMP), dimethyl phthalate (DMP), diethyl phthalate (DEP), and di-n-butyl phthalate (DBP) (Table 7.10) (Xu et al., 2015). The sonolytic degradation of a mixture of these phthalates shows that the rate constants are lower than in the individual treatment, due to the competitive effect among them. However, the degradation of the most hydrophobic phthalate (DBP) is less affected than the hydrophilic one (MMP).

7.5.1.6 Textile Dyes

There are many classes of textile dyes, and the most used classes are azo, indigo, xanthene, phthalocyanine, and tri-phenylmethane. The textile industry is one of the main contributors in terms of the color and organic load in water bodies (Rayaroth et al., 2015; Guzman-Duque et al., 2011). Textile dyes are hydrophilic substances. They are mainly degraded by hydroxyl radicals in the bulk of solution (at low pollutant concentrations) and close to the cavitation bubbles interface (at high concentrations).

Presence of functional groups that increase the hydrophilic character of the dye (e.g., sulfonate, amine, or carbonyl) decreases the efficiency of the sonochemical degradation

TABLE 7.10 Characteristics of Phthalates Upon Sonochemical Degradation

Compound Structure	Log P (Hydrophobicity)	Removal[a] (%) for Individual Degradation	Removal[a] (%) for Degradation in the Mixture	Inhibition Measured as Removal (%)
MMP	1.13	41.82	13.96	27.86
DMP	1.60	66.04	40.24	25.8
DEP	2.47	75.88	66.60	9.28
DBP	4.9	85.94	81.56	4.38

[a]Percentage of removal after 90 min of treatment.
Source: Adapted from Xu, L.J., Chu, W., Graham, N., 2015. J. Hazard. Mater., 288, 43–50, With permission.

(Ince and Tezcanli-Güyer, 2004; Rehorek et al., 2004). A comparative degradation of acid orange 7 and reactive orange 16 shows that the former has a larger degradation rate constant. Due to the higher sulfonation of the reactive orange 16 molecule, it is more hydrophilic and its degradation is slower than the acid orange 7.

7.5.1.7 Organophosphorus Pesticides (OPPs)

Organophosphorus pesticide (OPPs) is a class of substances frequently encountered in surface water and groundwater coming from agro-industrial processes. OPPs are toxic and can be bioaccumulated (Zhang et al., 2010). Due to their hydrophobic and nonvolatile nature, they are mainly degraded in the interfacial region of cavitation bubbles. All the structural factors that increase the hydrophobic character of OPPs make them more viable for degradation by ultrasound. The treatment of structurally related OPPs: chlorpyrifos (Log P: 4.96), and diazinon (Log P: 3.81) shows a higher removal of the former. This is associated with the longer hydrophobicity of chlorpyrifos caused by the presence of chlorine atoms in its aromatic substituent (Zhang et al., 2011).

7.5.1.8 Pharmaceuticals

Pharmaceuticals are biologically active substances used for the treatment of diseases in humans and animals. Many of them have been designed to resist the biodegradation. Currently, pharmaceuticals are associated with the development of adverse effects in aquatic organisms (Khetan and Collins, 2007). Pharmaceuticals comprise a big set of compounds (analgesics, antimicrobials, hormones, antihypertensives, etc.) with a high structural diversity.

The sonochemical degradation of pharmaceuticals with different sizes and functional groups shows a positive relationship between the degradation rates and Log P (Fig. 7.10) Xiao et al. (2014b). The most hydrophobic compounds are faster degraded. A similar connection between degradation rate and hydrophobicity is also reported for degradation of other pharmaceuticals (Isariebel et al., 2009; Xiao et al., 2014a).

7.5.2 Sonochemical Transformations of Pollutants and Their Implications

The degradation of volatile compounds by pyrolysis upon sonochemical action mainly leads to formation of smaller molecules such as short chain hydrocarbons and carbon monoxide or carbon dioxide. For example, the sonolysis of chlorobenzene initially generates acetylene and 1,3-butadiyne, which are transformed in further stages to CO, CH_4, and CO_2 (Fig. 7.11) (Jiang et al., 2002a,b).

In the case of very hydrophobic and nonvolatile compounds, both thermal and radical routes can be involved in the degradation. This is illustrated through the sonochemical destruction of 1,4-dioxane (Fig. 7.12) (Beckett and Hua, 2000). The thermal route induces the dioxane ring cleavage to produce small molecules such as ethane, formaldehyde, and carbon monoxide. The ring opening is also promoted via radical. However, it generates short chain oxidized species (e.g., methoxyacetic acid, glycolic acid, formaldehyde, and formic acid), which could be biocompatible.

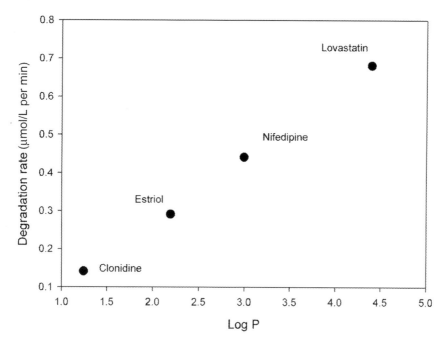

FIGURE 7.10 Relationship between sonochemical degradation rate and Log P of diverse pharmaceuticals. *Adapted from Xiao, R., Wei, Z., Chen, D., Weavers, L.K., 2014b. Environ. Sci. Technol, 48, 9675–9683.With permission.*

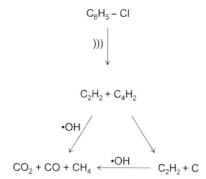

FIGURE 7.11 Scheme of chlorobenzene degradation by ultrasound. *Adapted from Jiang, Y., Petrier, C., Waite, T.D., 2002. Ultrasonics Sonochem., 9, 317–323. With permission.*

Meanwhile, highly hydrophobic and nonvolatile compounds such as PFOS and PFOA, which are refractory to hydroxyl, are degraded by the shortening of perfluoroalkyl chain by thermal reactions. This sonochemical action leads to the lowering of the toxicity associated to PFOS and PFOA, which contributes to the remediation of environmental pollution (Moriwaki et al., 2005).

On the other hand, moderate hydrophobic or hydrophilic compounds are mainly degraded by $^{\bullet}$OH action. Due to its electrophilic character, the hydroxyl radical has strong

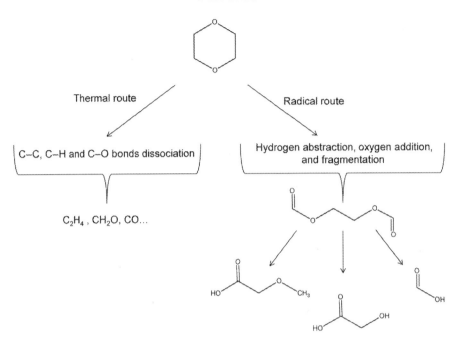

FIGURE 7.12 Scheme of 1,4-Dioxane degradation upon ultrasonic action. *Adapted from Beckett, M.A., Hua, I., 2000. Environ. Sci. Technol., 34, 3944–3953. With permission.*

reactivity to electron rich moieties such as benzene rings, double bonds, amines, and thioethers (Pignatello et al., 2006). Therefore, oxidized forms and/or broken species of these organic moieties are produced. In fact, the successive attack of •OH at aromatic structures leads to cleavage of rings.

The hydroxyl radical experiments additions to the aromatic rings; and then mono-hydroxylated compounds, 4-isopropenyl phenol, and hydroquinone are initially generated during the sonochemical treatment of compounds such as bisphenol-A. The further action of the process transforms the primary products into aliphatic acids such as formic, acetic, and oxalic acids. These latter substances are recalcitrant toward ultrasonic process due to their very high hydrophilic nature; however, many of them are biodegradable.

Similar to moderate hydrophobic compounds, in the case of hydrophilic substances, the hydroxyl radical attacks their electron-rich regions such as aromatic rings and atoms in reduced form (e.g., nitrogen of azo moiety), promoting the oxidation of the pollutant and/ or the breakdown to smaller molecules (Fig. 7.13).

In general, the ultrasound process converts toxic substances (e.g., pesticides, endocrine disruptors, dyes, and pharmaceuticals) into highly oxidized products such as carboxylic acids of short chain (Torres et al., 2008a,b,c; Torres et al., 2009; Villaroel et al., 2014). In most cases, such compounds are not toxic and biodegradable, so they can be easily eliminated by a biological treatment or mineralized by combination of sonochemistry with other systems (e.g., UV irradiation, addition of iron (II)) (Torres et al., 2008a,b,c; Guzman-Duque et al., 2011). Also, it is recognized that ultrasonic process induces transformations

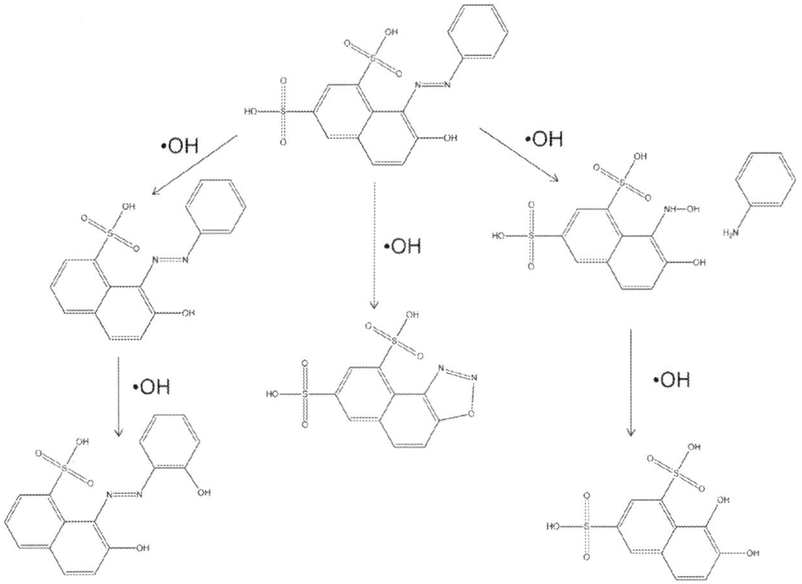

FIGURE 7.13 Degradation schema of orange g dye upon ultrasonic action. *Adapted from Cai, M., Jin, M., Weavers, L.K., 2015. Ultrasonics Sonochem. 18, 1068–1076. With permission.*

on active nucleus of -lactam antibiotics, decreasing the antimicrobial potency and consequently diminishing the risk of bacterial resistance development (Villegas-Guzman et al., 2015; Serna-Galvis et al., 2016).

It should be mentioned that the transformation of pollutants by pyrolytic and radical routes involves intermediary mechanistic steps, the understanding of such could be deepened by the reader in the literature (Moldoveanu, 2010; Pignatello et al., 2006).

7.6 INFLUENCE OF WATER MATRIX IN THE POLLUTANTS DEGRADATION

7.6.1 Effect of pH

The pH value is one of the most important characteristics of water, which must be taken into account in sonochemical treatment, especially for pollutants with acid-basic functions. In such case, the pH of water has significant implications on the chemical structure of the substrate and therefore, on the efficiency of the degradation by the ultrasonic process (Guzman-Duque et al., 2011).

The neutral (noncharged) species of pollutants accumulate more easily at the interface of cavitation bubbles in comparison with their ionic forms (anion or cation) (Jiang et al., 2002a,b), as a consequence the sonochemical degradation of neutral structures is favored. Due to the equilibrium between neutral and ionic forms, molecules with carboxylic acids, amine and phenolic groups have strong modifications in the degradation kinetics with the pH variation of water. pH values below their pKa favor the degradation of compounds with carboxylic or phenolic groups, while compounds with amine groups are faster degraded at pH values above of their pKa (Table 7.11). In contrast, for substances without acid—base character (e.g., alachlor pesticide) or with a predominant form at the tested pH range (e.g., crystal violet dye), the variation of pH has a poor effect on the degradation by ultrasonic action (Torres et al. 2009; Guzman-Duque et al. 2011).

The pH effect is clearly illustrated by the case of 4-nitrophenol and aniline. The degradation rate of 4-nitrophenol decreases with the increase in pH, whereas the removal rate of aniline exhibits a maximum under alkaline conditions (Fig. 7.14) (Jiang et al., 2002a,b). 4-

TABLE 7.11   Effect of pH Changes on Sonochemical Degradation of Pollutants

Compound	pKa (Functional Group)	Considered pH Range	Degradation More Favored at pH	References
Ibuprofen	4.90 (R-COOH)	3.0—11.0	3.0	Méndez-Arriaga et al. (2008)
Diclofenac	4.35 (R-COOH)	3.0—11.0	3.0	Naddeo et al. (2010)
Fluoxetine	10.05 ($R_1R_2NH_2^+$)	3.0—11.0	11.0	Serna-Galvis et al. (2015)
Hydrazine	7.93 ($HN=NH_2^+$)	2.0—8.7	>8.0	Nakui et al. (2007)

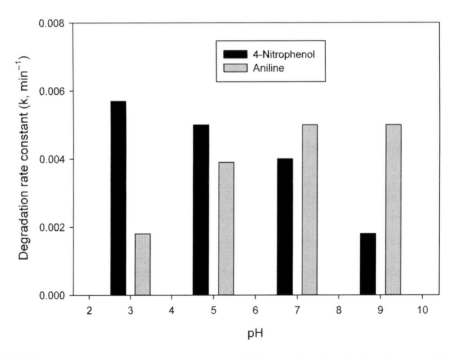

FIGURE 7.14 Variation of reaction rate constants at different pH values during the degradation of 4-nitrophenol and aniline by ultrasound. *Adapted from Jiang Y., Petrier, C., Waite, T.D., 2002. Ultrsonics Sonochem. 9 (6), 317—323. With permission.*

Nitrophenol has a pKa: 7.08 and hence, this compound is deprotonated at pH > pKa (anionic form, Table 7.12), which is more hydrophilic than its neutral form. Thus, the anionic species of 4-nitrophenol is less susceptible to degradation. Contrary, aniline has a pKa: 4.6, and the increase in pH above pKa generates its neutral form, which is more hydrophobic than the cationic one. Additionally, the hydroxyl radical reacts faster with aniline in neutral form (k: 1.4×10^{10}/M per s) than with anilinium cation (k: 4.8×10^9/M per s). This explains the improvement of aniline degradation with the increasing pH.

On the other hand, in the case of levofloxacin, which has both acid and amine groups, higher degradation occurs at pH between 6 and 7. At such a pH range, the zwitterionic form is in equilibrium with the neutral, which favors its elimination (Guo et al. 2010). For azo dyes containing sulfonate groups, a strong acidification of solution (i.e., pH < 3.0) can protonate this moiety, making such substances more hydrophobic, and enhancing their sonodegradation (Ince and Tezcanli-Güyer, 2004). Therefore, in general, the pH values that promote a more hydrophobic form of organic pollutants favor their sonochemical degradations.

7.6.2 Sonochemical Degradation in Presence of Inorganic Components

In the application of advanced oxidation processes, such as ultrasound, the effect of inorganic components of water matrix, particularly the anion has to be considered, because of their scavenger ability toward radicals. Other processes such as Fenton, photo-Fenton, TiO_2-photocatalysis, and UV/H_2O_2 are negatively affected by inorganic anions (Torres et al., 2007a,b). However, the effect of anions on the sonochemical treatment is different.

Chloride (Cl^-), sulfate (SO_4^{2-}), and bicarbonate (HCO_3^-) ions are among the anions most frequently found in wastewater, industrial water, and natural water. Due to their more hydrophilic nature, chloride, and sulfate anions do not interfere on the degradation of some target molecules (Guzman-Duque et al., 2011; Villaroel et al., 2014). In other cases, they can accelerate the sonolysis of organic substances by a "salting-out" effect (the presence of anions pushes the organics toward the bubble interface leading to an enhancement of the pollutants degradation). The sonochemical degradation of catechol and resorcinol was improved by the presence of NaCl and Na_2SO_4 (Fig. 7.15) (Uddin et al., 2016). The effect of anions followed the order of the Hofmeister series (i.e., sulfate > chloride), as SO_4^{2-} has a stronger interaction with water than Cl^-. On the other hand, a similar sono-degradation enhancement by salting-out effect was reported for the treatment of fluoxetine in the presence of nitrate anion (Serna-Galvis et al., 2015).

Bicarbonate anion has a dual behavior that is determined by the relative concentrations of both the pollutant and anion. At low concentrations of the target organic compound and high concentrations of bicarbonate, the removal of the organic pollutant is increased. Instead, at high concentrations of the pollutant, bicarbonate anion has a competitive effect. The sonochemical treatment of bisphenol-A under different concentrations of the pollutant and bicarbonate illustrates the aforementioned situations (Fig. 7.16) (Pétrier et al., 2010). These effects are related with the formation of the carbonate radical and its interaction with the pollutants.

TABLE 7.12 Structural Changes with pH Variation

Compound	Structural Changes with pH Variation	References
p-Nitrophenol	 Neutral form → (pH > pKa) → Anionic form	Jiang et al. (2002a,b)
Aniline	 Cationic form → (pH > pKa) → Neutral form	Jiang et al. (2002a,b)
Levofloxacin	 Zwitterion ⇌ (pH: 6 – 7) → Neutral form	Guo et al. (2010)
Azo dyes	 Anionic form → (H^+) → Neutral form	Ince and Tezcanli-Güyer (2004)

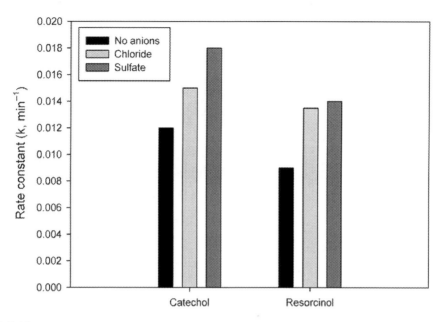

FIGURE 7.15 Effect of chlorine and sulfate anions on the sonochemical degradation of catechol and resorcinol. *Adapted from Uddin, M.H., Nanzai, B., Okitsu, K., 2016. Ultrasonics Sonochem., 28, 144–149. With permission.*

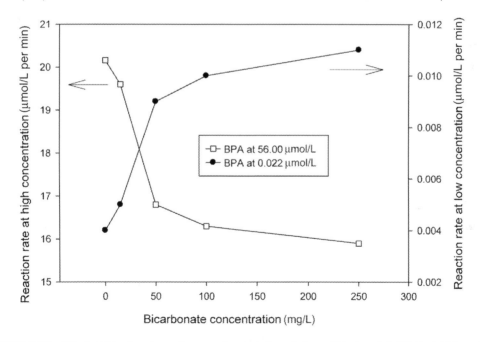

FIGURE 7.16 Effect of bicarbonate on the sonochemical degradation of bisphenol-A (BPA) at high and low concentrations. *Adapted from Pétrier, C., Torres-Palma, R., Combet, E., Sarantakos, G., Baup, S., Pulgarin, C., 2010. Ultrasonics Sonochem. 17, 111–115. With permission.*

The carbonate radical $(CO_3^{\bullet -})$ is a strong oxidant $(E = 1.78\ V)$ that results from the reaction of hydroxyl radical with bicarbonate or carbonate anions (Eqs. 7.18–7.20). Carbonate radical is able to degrade organic compounds through electron transfer or hydrogen abstraction.

$$HCO_3^- +\ OH \rightarrow HCO_3 +\ OH^- \tag{7.18}$$

$$HCO_3 \rightarrow CO_3^- + H^+ \tag{7.19}$$

$$CO_3^{2-} +\ OH \rightarrow CO_3^- +\ OH^- \tag{7.20}$$

At relative high concentrations of pollutant, bicarbonate competes with the hydroxyl radicals and a detrimental effect is observed. When the concentration of the target substance is much lower than bicarbonate, the produced carbonate radicals can react with the pollutant improving its degradation. In fact, carbonate radical can migrate to the bulk solution, to a greater extent than hydroxyl radicals, and enhances the degradation of hydrophilic pollutants (Guzman-Duque et al., 2011).

The presence of bromide anion (Br^-) in the water also increases the sonochemical degradation of pollutants (Moumeni and Hamdaoui, 2012; Chiha et al., 2011). The sonolytic destruction is augmented with increasing bromide concentration. At high concentrations of these anions, they reach the interface of bubble and react with hydroxyl radicals, generating other radical species such as $Br^{\bullet}$ and $Br_2^{\bullet -}$ (Eqs. 7.21 and 7.22), which can act as degrading agents of pollutants. Although, hydroxyl radical is scavenged by bromide, the presence of radical species in the bulk increases, favoring the removal of pollutants. The elimination of malachite green (at 5 mg/L) is around 2.7 folds faster in the presence of 5000 mg/L of bromide ion (Moumeni and Hamdaoui, 2012).

$$Br^- +\ OH \rightarrow Br +\ OH^- \tag{7.21}$$

$$Br + Br^- \rightarrow Br_2^- \tag{7.22}$$

Recent works (Lee et al., 2016; Hamdaoui and Merouani, 2017) show that the periodate in water matrix (IO_4^-) is sono-activated generating periodyl $(IO_4^{\bullet})$ and iodyl $(IO_3^{\bullet})$ radicals (Eqs. 7.23–7.25). These radicals are able to enhance the ultrasonic degradation of organic pollutants such as the brilliant blue R and perfluorooctanoic acid.

$$IO_4^- +\ OH \rightarrow IO_4 +\ OH^- \tag{7.23}$$

$$IO_4^- +\ H \rightarrow IO_3^- +\ OH \tag{7.24}$$

$$IO_3^- +\ OH \rightarrow IO_3 +\ OH^- \tag{7.25}$$

Other inorganic substances that improve the pollutants degradation are: Fe (II), TiO_2, $S_2O_8^{2-}$, HSO_5^-, and O_3. Their presence leads to a combined action of ultrasound with other advanced oxidation processes.

Real waters containing organic pollutants and a high presence of inorganic substances have been successfully treated by the sonochemical process. A comparative removal of the fluoxetine pollutant in deionized water and natural mineral water has been developed (Fig. 7.17). The pollutant degradation in natural mineral water was slightly lower than in deionized water. The small difference is associated with the high concentration of anions (particularly

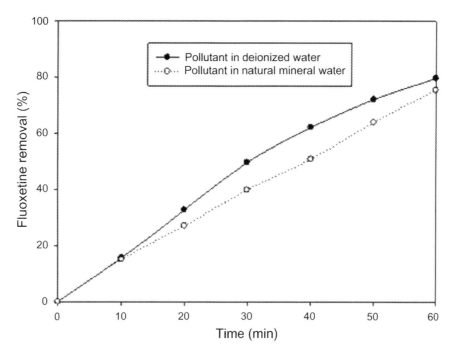

FIGURE 7.17 Comparative removal of pollutants in deionized water and natural mineral water by ultrasound. *Adapted from Serna-Galvis, E.A., Silva-Agredo, J., Giraldo-Aguirre, A.L., Torress-Palma, Sci. Total Environ., 524–525, 354–360, 2015. With permission.*

bicarbonate) (Serna-Galvis et al., 2015). Interestingly, the sonochemical treatments of bisphenol-A or 4-cumylphenol are accelerated in mineral natural waters (Torres et al., 2007a,b; Chiha et al., 2011), which is attributed to the role of bicarbonate. Moreover, the addition of bromide anion to mineral water and seawater significantly enhanced the sonolytic removal of malachite green, due to the action of bromine radicals (Moumeni and Hamdaoui, 2012).

7.6.3 Sonochemical Degradation of Pollutants in Presence of Other Organic Components

Organic substances in water can modify the sonodegradation rate of the target contaminants. The effect is dependent on the nature of both the pollutant and matrix components (Table 7.13) (Serna-Galvis et al., 2016; Xiao et al., 2014a). The degradation of the pharmaceuticals acetaminophen and dicloxacillin is not affected by the presence of glucose (a very hydrophilic organic component) even at concentrations higher than the target compound (Villegas-Guzman et al., 2015; Villaroel et al., 2014). Due to their nature, these pharmaceuticals are closer to cavitation bubble and therefore, hydroxyl radical first reacts with the pollutants, and degradation is not affected by the glucose presence.

Similarly, mannitol (a typical pharmaceutical excipient) has no effect during the sonochemical treatment of oxacillin. This antibiotic is a hydrophobic and nonvolatile

TABLE 7.13　Effect of Organic Matrix Components on Pollutants Degradation Upon High Frequency Ultrasound According to Their Nature

Pollutant Nature	Nature of Other Organic Components	Effect
Hydrophilic	Hydrophilic	Competitive
	Hydrophobic	Competitive
Hydrophobic	Hydrophilic	Noninterfering
	Hydrophobic	Competitive

compound, whereas the excipient has a hydrophilic character. Thus, oxacillin is selectively degraded in the interfacial zone. On contrary, when more hydrophobic substances, such as 1-hexanol, are in water matrix, the elimination of lesser hydrophobic compounds (e.g., acetaminophen or fluoxetine) is strong because of the competence by $^{\bullet}OH$.

Presence of CCl_4 in water matrix can improve the sonochemical removal of pollutants (Zheng et al., 2005; Eren and Ince, 2010; Merouani et al., 2010); such is the case of rhodamine B (5 mg/L), which is 21-fold faster degraded, when 200 mg/L of carbon tetrachloride is added (Merouani et al., 2010). The effect is based on the generation of hypochlorous acid (HOCl), chlorine (Cl_2), and chlorine-type radicals ($Cl^{\bullet}$, $^{\bullet}CCl_3$) coming from sonolysis of CCl_4 (Eqs. 7.26–7.28), which react with the target organic contaminants, in addition to the hydroxyl radical action.

$$CCl_4 + \quad \rightarrow \quad CCl_3 + Cl \tag{7.26}$$

$$2Cl \rightarrow Cl_2 \tag{7.27}$$

$$Cl_2 + H_2O \rightarrow HOCl + HCl \tag{7.28}$$

Finally, it is worthy to indicate that the treatment by ultrasound of complex real waters (e.g., groundwater or wastewater effluent) containing other organic components shows that the process removes target pollutants. However, the degradation kinetics, in general terms, are slower than in pure water (Xiao et al., 2014b; Cheng et al., 2008).

7.7 COMBINATION OF SONOCHEMISTRY WITH OTHER PROCESSES

Although sonochemistry successfully degrades a wide number of organic compounds, in several cases, it has low levels of mineralization (<10%), as generated products mainly have hydrophilic and nonvolatile nature, which limits their complete elimination. Besides, because of the elevated electrical energy consumption, the operation of ultrasound demands relative high economical costs. Therefore, short treatment times are a need. However, in most cases, these difficulties can be overcome by combining ultrasound with other advanced oxidation processes or biological systems.

Ultrasound can be combined with UV-C light (sono-photolysis), ferrous ions (sono-Fenton), or ferrous ions and light (sono-photo-Fenton), titanium dioxide and light (sono-photocatalysis), ozone (sono-ozonolysis), and electrochemistry (sono-electrochemistry).

Such combinations involve the additional production of nonselective radicals, which accelerate the contaminants degradation and lead to synergistic effects in the mineralization (Table 7.14). The combination of ultrasound with more than one process (e.g., ultrasound/photo-Fenton/TiO_2-photocatalysis; ultrasound/UV-light/ozone) is also possible. These cases are discussed in details by the authors (Torres et al., 2008a,b,c; Méndez-Arriaga et al., 2009; Tezcanli-Güyer and Ince, 2004).

TABLE 7.14 Examples of Ultrasound Combination with Other Advanced Oxidation Processes

Combination	Additional Ways of Oxidizing Agents Production	Pollutant	Synergy in the Mineralization Extent[a] (Treatment Time)	References
Ultrasound/UV-C light	$UV\text{-}C + H_2O_2 \rightarrow 2\,^{\bullet}OH$	Acid orange 7	7.0, (60 min)	Tezcanli-Güyer and Ince (2004)
		Bisphenol-A	8.25, (180 min)	Torres et al. (2007a,b)
Ultrasound/Fenton	$Fe^{2+} + H_2O_2 \rightarrow Fe^{3+} + {}^{\bullet}OH + OH^-$ $Fe^{3+} + H_2O_2 \rightarrow Fe^{2+} + {}^{\bullet}OOH + H^+$	Bisphenol-A	2.64, (240 min)	Torres et al. (2008c)
Ultrasound/Photo-Fenton	$Fe^{2+} + H_2O_2 \rightarrow Fe^{3+} + {}^{\bullet}OH + OH^-$ $Fe^{3+} + H_2O + light \rightarrow Fe^{2+} + {}^{\bullet}OH + H^+$	Ibuprofen	1.12, (240 min)	Méndez-Arriaga et al. (2009)
	$Fe^{3+} + H_2O_2 \rightarrow Fe^{2+} + {}^{\bullet}OOH + H^+$	Bisphenol-A	4.67, (240 min)	Torres et al. (2008c)
Ultrasound/TiO_2 + UV light	$TiO_2 + UV\ light \rightarrow TiO_2\ (h^+ \ldots e^-)$ $h^+ + H_2O \rightarrow {}^{\bullet}OH + H^+$	Ibuprofen	3.44, (240 min)	Méndez-Arriaga et al. (2009)
	$e^- + O_2 \rightarrow {}^{\bullet}O_2^-$ $e^- + H_2O_2 \rightarrow {}^{\bullet}OH + OH^-$	Bisphenol-A	3.5, (120 min)	Torres et al. (2008b)
	TiO_2 particles provide extra nuclei for cavitation bubble formation			
Ultrasound/Ozone	$O_3 +))) \rightarrow O_2 + {}^{\bullet}O$ ${}^{\bullet}O + H_2O \rightarrow O_2 + 2\,^{\bullet}OH$	Acid orange 7	1.23, (60 min)	Tezcanli-Güyer and Ince (2004)
	$2O_3 + H_2O_2 \rightarrow 3O_2 + 2\,^{\bullet}OH$	Phenol	1.65, (300 min)	Sathishkumar et al. (2016)
Ultrasound/Electrochemistry	$BDD_{(anode)} + H_2O \rightarrow {}^{\bullet}OH + H^+ + e^-$	Crystal violet	1.32, (120 min, using BBD anode)	Guzman-Duque et al. (2016)
	$IrO_{2(anode)} + 2Cl^- \rightarrow Cl_2 + 2e^-$ $Cl_2 + H_2O \rightarrow HOCl + HCl$		1.12, (120 min, using IrO_2 anode)	

[a]Synergy: TOC removed by combination processes/ (TOC removed by individual process); a value higher than one indicates synergistic effects, a value lower than one means an antagonistic effect, and a value equal to one indicates additive effects of combination processes.

TABLE 7.15   Degradation Enhancement by Combining Ultrasound with Persulfate and Peroxymonosulfate

System	Production of Radical Sulfate	Effect of Combination	References
Ultrasound/ Persulfate	$S_2O_8^{2-} +))) \rightarrow 2SO_4^{\bullet-}$ $SO_4^{\bullet-} + OH^- \rightarrow SO_4^{2-} + {}^{\bullet}OH$	Ammonium perfluorooctanoate removal is increased from 35.5% to 51.2%	Hao et al. (2014)
Ultrasound/ Peroxymonosulfate	$HSO_5^- +))) \rightarrow SO_4^- + {}^{\bullet}OH$ $SO_4^{\bullet-} + OH^- \rightarrow SO_4^{2-} + {}^{\bullet}OH$	Degradation of cresol red is drastically improved	Soumia and Petrier (2016)

Recently, the joint action of sonochemistry and persulfate or peroxymonosulfate has also been considered. These systems promote the formation of sulfate radical ($SO_4^{\bullet-}$, E = 2.6 V) (Hao et al., 2014; Soumia and Petrier, 2016), which is able to degrade refractory substances through hydroxyl radical (e.g., perfluoroalkyl compounds and cyanuric acid, (Moriwaki et al., 2005; Yin et al., 2016; Manoj et al., 2002; Liu, 2014)). Thus, the presence of sulfate radical, in addition to hydroxyl radical, significantly increases the pollutants degradation. Some examples of the combinations, ultrasound/persulfate and ultrasound/peroxymonosulfate, are given (Table 7.15).

In addition to the combinations with advanced oxidation processes, the ultrasound can be coupled to biological systems (which are low cost treatments). Although sonochemical process has low mineralizing ability, it is capable to transform many pollutants into biodegradable substances. Therefore, after application of ultrasound, microorganisms can eliminate the sono-generated products. For example, the sonication of a solution of oxacillin during 2 h leads to elimination of the antibiotic and its antimicrobial activity; the subsequent application of a biological process (using nonadapted microorganisms from a municipal wastewater treatment plant) completed the mineralization (Serna-Galvis et al., 2016). Similarly, the ultrasonic degradation of fluoxetine generates biotreatable products, which are mineralized by action of aerobic microorganisms (Serna-Galvis et al. 2015).

Additional information about the combination of sonochemistry with other processes can be found in the following references (Sathishkumar et al., 2016; Mahamuni and Adewuyi, 2010; Pang et al., 2011; Papoutsakis et al., 2015).

Finally, it can be concluded that sonochemical effects are strongly influenced by operational parameters (frequency, dissolved gas, and power input, especially), which must be optimized to obtain the best performance of the process for degradation of pollutants. The alone action of ultrasound preferentially favors the elimination of compounds with high hydrophobic or volatile nature. This fact allowed the selective degradation of some pollutants in complex matrices by ultrasound action. Additionally, kinetic models for the sonodegradation based on the nature of substances can be proposed.

The application of ultrasound for treatment of water polluted with organic substances lead to environmental benefits (e.g., transformation of hazardous materials into biocompatible or innocuous compounds). However, the process has slow degradation rates and elevated economical costs of operation (caused by high electrical energy consumption). Such limitations of sonochemical treatment are significantly overcome by combining it with other processes (AOPs or biological systems).

The future outlooks in sonochemistry and its combination with other processes should be focused on applications to real waters containing mixtures of organic and inorganic substances, determining both kinetics and mechanisms of pollutants degradation, in addition to toxicity and biodegradability of byproducts. Contaminated waters, where the pollutant has volatile or hydrophobic nature, are of special interest because of the selectivity of the technology for this type of organic compound. Also, from an engineering point of view, the developments in reactors design, scale-up, and ultrasonic transduction efficiency, to maximize the effects in the area of contaminants degradation by ultrasound are still needed.

References

Adewuyi, Y.G., 2001. Sonochemistry: Environmental science and engineering applications. Ind. Eng. Chem. Res. 40, 4681–4715.

Beckett, M.A., Hua, I., 2000. Elucidation of the 1,4-dioxane decomposition pathway at discrete ultrasonic frequencies. Environ. Sci. Technol. 34, 3944–3953.

Beckett, M.A., Hua, I., 2001. Impact of ultrasonic frequency on aqueous sonoluminescence and sonochemistry. J. Phys. Chem. A 105, 3796–3802.

Berlan, J., Mason, T.J., 1992. Sonochemistry: from research laboratories to industrial plants. Ultrasonics 30 (4), 203–212.

Brotchie, A., Grieser, F., Ashokkumar, M., 2009. Effect of power and frequency on bubble-size distributions in acoustic cavitation. Phys. Rev. Lett. 102, 1–4.

Cheng, J., Vecitis, C.D., Park, H., Mader, B.T., Hoffmann, M.R., 2008. Sonochemical degradation of perfluorooctane sulfonate (PFOS) and perfluorooctanoate (PFOA) in landfill groundwater: environmental matrix effects. Environ. Sci. Technol. 42, 8057–8063.

Chiha, M., Merouani, S., Hamdaoui, O., Baup, S., Gondrexon, N., Pétrier, C., 2010. Modeling of ultrasonic degradation of non-volatile organic compounds by Langmuir-type kinetics. Ultrasonics Sonochem. 17, 773–782.

Chiha, M., Vecitis, C.D., Park, H., Mader, B.T., Hoffmann, M.R., 2011. Sonolytic degradation of endocrine disrupting chemical 4-cumylphenol in water. Ultrasonics Sonochem. 18 (5), 943–950.

Colussi, A.J., Hung, H.-M., Hoffmann, M.R., 1999. Sonochemical degradation rates of volatile solutes. J. Phys. Chem. A 103, 2696–2699.

De Visscher, A., Van Eenoo, P., Drijvers, D., Van Langenhove, H., 1996. Kinetic model for the sonochemical degradation of monocyclic aromatic compounds in aqueous solution. J. Phys. Chem. 100, 11636–11642.

Eren, Z., Ince, N.H., 2010. Sonolytic and sonocatalytic degradation of azo dyes by low and high frequency ultrasound. J. Hazard. Mater. 177 (1–3), 1019–1024.

Fernandez, N.A., Rodriguez-Freire, L., Keswani, M., Sierra-Alvarez, R., 2016. Effect of chemical structure on the sonochemical degradation of perfluoroalkyl and polyfluoroalkyl substances (PFASs). Environ. Sci. Water Res. Technol. 2, 975–983.

Goel, M., Hongqiang, H., Mujumdar, A.S., Ray, M.B., 2004. Sonochemical decomposition of volatile and non-volatile organic compounds - a comparative study. Water Res. 38, 4247–4261.

Gogate, P.R., 2008. Cavitational reactors for process intensification of chemical processing applications: a critical review. Chem. Eng. Process. Process Intens. 47 (4), 515–527.

Goskonda, S., Catallo, J.W., Junk, T., 2002. Sonochemical degradation of aromatic organic pollutants. Waste Manage. 22, 351–356.

Guo, W., Shi, Y., Wang, H., Yang, H., Zhang, G., 2010. Sonochemical decomposition of levofloxacin in aqueous solution. Water Environ. Res. 82 (8), 696–700.

Guzman-Duque, F., Pétrier, C., Pulgarin, C., Peñuela, G., Torres-Palma, R., 2011. Effects of sonochemical parameters and inorganic ions during the sonochemical degradation of crystal violet in water. Ultrasonics Sonochem. 18, 440–446.

Guzman-Duque, L.F., Petrier, C., Pulgarin, C., Penuela, G., Herrera-Calderon, E., Torres-Palma, R.A., 2016. Synergistic coupling between electrochemical and ultrasound treatments for organic pollutant degradation as a function of the electrode material (IrO_2 and BDD) and the ultrasonic frequency (20 and 800 kHz). Int. J. Electrochem. Sci. 11, 7380–7394.

Hamdaoui, O., Merouani, S., 2017. Improvement of sonochemical degradation of brilliant blue R in water using periodate ions: implication of iodine radicals in the oxidation process. Ultrasonics Sonochem. 37, 344–350.

Hao, F., Guo, W., Wang, A., Leng, Y., Li, H., 2014. Intensification of sonochemical degradation of ammonium perfluorooctanoate by persulfate oxidant. Ultrasonics Sonochem. 21 (2), 554–558.

Hart, E.J., Henglein, A., 1985. Free radical and free atom reactions in the sonolysis of aqueous iodide and formate solutions. J. Phys. Chem. 89, 4342–4347.

Henglein, A., Gutierrez, M., 1988. Sonolysis of polymers in aqueous solution. New observations on pyrolysis and mechanical degradation. J. Phys. Chem. 92 (13), 3705–3707.

Hua, I., Hoffmann, M.R., 1997. Optimization of ultrasonic irradiation as an advanced oxidation technology. Environ. Sci. Technol. 31, 2237–2243.

Hua, I., Hoechemer, R.H., Hoffmann, M.R., 1995. Sonolytic hydrolysis of p-nitrophenyl acetate: the role of supercritical water. J. Phys. Chem. 99, 2335–2342.

Hung, H.M., Hoffmann, M.R., 1999. Kinetics and mechanism of the sonolytic degradation of chlorinated hydrocarbons: frequency effects. J. Phys. Chem. A 103, 2734–2739.

Ince, N.H., Tezcanli-Güyer, G., 2004. Impacts of pH and molecular structure on ultrasonic degradation of azo dyes. Ultrasonics 42, 591–596.

Isariebel, Q.P., Carine, J.L., Ulises-Javier, J.H., Anne-Marie, W., Henri, D., 2009. Sonolysis of levodopa and paracetamol in aqueous solutions. Ultrasonics Sonochem. 16, 610–616.

Jiang, Y., Pétrier, C., Waite, T.D., 2002a. Effect of pH on the ultrasonic degradation of ionic aromatic compounds in aqueous solution. Ultrasonics Sonochem. 9 (3), 163–168.

Jiang, Y., Pétrier, C., Waite, T.D., 2002b. Kinetics and mechanisms of ultrasonic degradation of volatile chlorinated aromatics in aqueous solutions. Ultrasonics Sonochem. 9, 317–323.

Jiang, Y., Petrier, C., Waite, T.D., 2006. Sonolysis of 4-chlorophenol in aqueous solution: Effects of substrate concentration, aqueous temperature and ultrasonic frequency. Ultrasonics Sonochem. 13, 415–422.

Kang, J.W., Hung, H.M., Lin, A., Hoffmann, M.R., 1999. Sonolytic destruction of methyl tert-butyl ether by ultrasonic irradiation: the role of O_3, H_2O_2, frequency, and power density. Environ. Sci. Technol. 33, 3199–3205.

Khetan, S.K., Collins, T.J., 2007. Human pharmaceuticals in the aquatic environment: a challenge to green chemistry. Chem. Rev. 107, 2319–2364.

Lee, Y.-C., Chen, M.-J., Huang, C.-P., Kuo, J., Lo, S.-L., 2016. Efficient sonochemical degradation of perfluorooctanoic acid using periodate. Ultrasonics Sonochem. 31, 499–505.

Lifka, J., Ondruschka, B., Hofmann, J., 2003. Review - the use of ultrasound for the degradation of pollutants. Eng. Life Sci. 3, 253–262.

Lim, M., Son, Y., Khim, J., 2011. Frequency effects on the sonochemical degradation of chlorinated compounds. Ultrasonics Sonochem. 18, 460–465.

Liu, G., 2014. Recalcitrance of cyanuric acid to oxidative degradation by OH radical: theoretical investigation. RSC Adv. 4, 37359–37364.

Lorimer, J.P., Mason, T.J., Cuthbert, T.C., Brookfield, E.A., 1995. Effect of ultrasound on the degradation of aqueous native dextran. Ultrasonics Sonochem. 2, S55–S57.

Mahamuni, N.N., Adewuyi, Y.G., 2010. Advanced oxidation processes (AOPs) involving ultrasound for waste water treatment: a review with emphasis on cost estimation. Ultrasonics Sonochem. 17 (6), 990–1003.

Makino, K., Mossoba, M.M., Riesz, P., 1982. Chemical effects of ultrasound on aqueous solutions. Evidence for •OH and •H by spin trapping. J. Am. Chem. Soc. 104, 3537–3539.

Manoj, P., Varghese, R., Manoj, V.M., Aravindakumar, C.T., 2002. Reaction of sulphate radical anion ($SO4^{•−}$) with cyanuric acid: a potential reaction for its degradation? Chem. Lett. 2002 (1), 74–75.

Mason, T.J., 2001. The design of ultrasonic reactors for environmental remediation. Adv. Sonochem. 6, 247–268.

Mead, E., Sutherland, R., Verrall, R., 1976. The effect of ultrasound on water in the presence of dissolved gases. Can. J. Chem 54, 1114–1120.

Méndez-Arriaga, F., Torres-Palma, R., Pétrier, C., Esplugas, S., Gimenez, J., Pulgarin, C., 2008. Ultrasonic treatment of water contaminated with ibuprofen. Water Res. 42, 4243–4248.

Méndez-Arriaga, F., Torres-Palma, R.A., Pétrier, C., Esplugas, S., Gimenez, J., Pulgarin, C., 2009. Mineralization enhancement of a recalcitrant pharmaceutical pollutant in water by advanced oxidation hybrid processes. Water Res. 43 (16), 3984–3991.

Merouani, S., Hamdaoui, O., Saoudi, F., Chiha, M., 2010. Sonochemical degradation of rhodamine B in aqueous phase: effects of additives. Chem. Eng. J. 158, 550–557.

Merouani, S., Hamdaoui, O., Rezgui, Y., Guemini, M., 2013. Effects of ultrasound frequency and acoustic amplitude on the size of sonochemically active bubbles-theoretical study. Ultrasonics Sonochem. 20, 815–819.

Merouani, S., Ferkous, H., Hamdaoui, O., Rezgui, Y., Guemini, M., 2015. New interpretation of the effects of argon-saturating gas toward sonochemical reactions. Ultrasonics Sonochem. 23, 37–45.

Moldoveanu, S.C., 2010. Pyrolysis of Organic Molecules: Applications to Health and Environmental Issues, First ed. Elsevier B.V, Great Britain.

Moriwaki, H., Takagi, Y., Tanaka, M., Tsuruho, K., Okitsu, K., Maeda, Y., 2005. Sonochemical decomposition of perfluorooctane sulfonate and perfluorooctanoic acid. Environ. Sci. Technol. 39, 3388–3392.

Moumeni, O., Hamdaoui, O., 2012. Intensification of sonochemical degradation of malachite green by bromide ions. Ultrasonics Sonochem. 19 (3), 404–409.

Naddeo, V., Belgiorno, V., Kassinos, D., Mantzavinos, D., Meric, S., 2010. Ultrasonic degradation, mineralization and detoxification of diclofenac in water: Optimization of operating parameters. Ultrasonics Sonochem. 17 (1), 179–185.

Nagata, Y., Hirai, K., Bandow, H., Maeda, Y., 1996. Decomposition of hydroxybenzoic and humic acids in water by ultrasonic irradiation. Environ. Sci. Technol. 30, 1133–1138.

Nakui, H., Okitsu, K., Maeda, Y., Nishimura, R., 2007. The effect of pH on sonochemical degradation of hydrazine. Ultrasonics Sonochem. 14 (5), 627–632.

Nanzai, B., Okitsu, K., Takenaka, N., Bandow, H., Maeda, Y., 2008. Sonochemical degradation of various monocyclic aromatic compounds: Relation between hydrophobicities of organic compounds and the decomposition rates. Ultrasonics Sonochem. 15, 478–483.

National Library of Medicine, 2017. ChemIDplus Advanced [WWW Document]. ChemIDplus Adv. <http://chem.sis.nlm.nih.gov/chemidplus/>.

Okitsu, K., Iwasaki, K., Yobiko, Y., Bandow, H., Nishimura, R., Maeda, Y., 2005. Sonochemical degradation of azo dyes in aqueous solution: a new heterogeneous kinetics model taking into account the local concentration of OH radicals and azo dyes. Ultrasonics Sonochem. 12, 255–262.

Pang, Y.L., Abdullah, A.Z., Bhatia, S., 2011. Review on sonochemical methods in the presence of catalysts and chemical additives for treatment of organic pollutants in wastewater. Desalination 277 (1–3), 1–14.

Papoutsakis, S., Miralles-Cuevas, S., Gondrexon, N., Baup, S., Malato, S., Pulgarin, C., 2015. Coupling between high-frequency ultrasound and solar photo-Fenton at pilot scale for the treatment of organic contaminants: an initial approach. Ultrasonics Sonochem. 22, 527–534.

Parke, A.V.M., Taylor, D., 1956. The chemical action of ultrasonic waves. J. Chem. Soc. 4442–4450.

Pétrier, C., Francony, A., 1997. Ultrasonic waste-water treatment: incidence of ultrasonic frequency on the rate of phenol and carbon tetrachloride degradation. Ultrasonics Sonochem. 4, 295–300.

Petrier, C., Jiang, Y., Lamy, M.F., 1998. Ultrasound and environment: Sonochemical destruction of chloroaromatic derivatives. Environ. Sci. Technol. 32, 1316–1318.

Pétrier, C., Torres-Palma, R., Combet, E., Sarantakos, G., Baup, S., Pulgarin, C., 2010. Enhanced sonochemical degradation of bisphenol-A by bicarbonate ions. Ultrasonics Sonochem. 17 (1), 111–115.

Pignatello, J.J., Oliveros, E., Mackay, A., 2006. Advanced oxidation processes for organic contaminant destruction based on the Fenton reaction and related chemistry. Crit. Rev. Environ. Sci. Technol. 36, 1–84.

Psillakis, E., Goula, G., Kalogerakis, N., Mantzavinos, D., 2004a. Degradation of polycyclic aromatic hydrocarbons in aqueous solutions by ultrasonic irradiation. J. Hazard. Mater. 108, 95–102.

Psillakis, E., Mantzavinos, D., Kalogerakis, N., 2004b. Monitoring the sonochemical degradation of phthalate esters in water using solid-phase microextraction. Chemosphere 54, 849–857.

Rayaroth, M.P., Aravind, U.K., Aravindakumar, C.T., 2015. Sonochemical degradation of coomassie brilliant blue: effect of frequency, power density, pH and various additives. Chemosphere 119, 848–855.

Rehorek, A., Tauber, M., Gübitz, G., 2004. Application of power ultrasound for azo dye degradation. Ultrasonics Sonochem. 11, 177–182.

Rooze, J., Rebrov, E.V., Schouten, J.C., Keurentjes, J.T.F., 2013. Dissolved gas and ultrasonic cavitation — a review. Ultrasonics Sonochem. 20, 1—11.

Rubio-Clemente, A., Torres-Palma, R.A., Peñuela, G.A., 2014. Removal of polycyclic aromatic hydrocarbons in aqueous environment by chemical treatments: a review. Sci. Total Environ. 478, 201—225.

Sathishkumar, P., Mangalaraja, R.V., Anandan, S., 2016. Review on the recent improvements in sonochemical and combined sonochemical oxidation processes — a powerful tool for destruction of environmental contaminants. Renew. Sustain. Energy Rev. 55, 426—454.

Schramm, J.D., Hua, I., 2001. Ultrasonic irradiation of dichlorvos: decomposition mechanism. Water Res. 35, 665—674.

Serna-Galvis, E.A., Silva-Agredo, J., Giraldo-Aguirre, A.L., Torres-Palma, R.A., 2015. Sonochemical degradation of the pharmaceutical fluoxetine: Effect of parameters, organic and inorganic additives and combination with a biological system. Sci. Total Environ. 524—525, 354—360.

Serna-Galvis, E.A., Silva-Agredo, J., Giraldo-Aguirre, A.L., Flórez-Acosta, O.A., Torres-Palma, R.A., 2016. High frequency ultrasound as a selective advanced oxidation process to remove penicillinic antibiotics and eliminate its antimicrobial activity from water. Ultrasonics Sonochem. 31, 276—283.

Sivasankar, T., Moholkar, V.S., 2009. Physical insights into the sonochemical degradation of recalcitrant organic pollutants with cavitation bubble dynamics. Ultrasonics Sonochem. 16, 769—781.

Song, W., O'Shea, K.E., 2007. Ultrasonically induced degradation of 2-methylisoborneol and geosmin. Water Res 41, 2672—2678.

Soumia, F., Petrier, C., 2016. Effect of potassium monopersulfate (oxone) and operating parameters on sonochemical degradation of cationic dye in an aqueous solution. Ultrasonics Sonochem. 32, 343—347.

Stogiannidis, E., Laane, R., 2015. Source characterization of polycyclic aromatic hydrocarbons by using their molecular indices: an overview of possibilities. In: Whitacre, D.M. (Ed.), Reviews of Environmental Contamination and Toxicology. Springer International Publishing, Switzerland, pp. 49—133.

Suslick, K.S., 1989. The chemical effects of ultrasound. Sci. Amer. 260 (2), 80—86.

Suslick, S., Fang, M., 1999. Acoustic cavitation and its chemical consequences. Phil. Trans. R. Soc. Lond. A 357, 335—353.

Sutkar, V.S., Gogate, P.R., 2009. Design aspects of sonochemical reactors: Techniques for understanding cavitational activity distribution and effect of operating parameters. Chem. Eng. J. 155, 26—36.

Tezcanli-Guyer, G., Ince, N.H., 2003. Degradation and toxicity reduction of textile dyestuff by ultrasound. Ultrasonics Sonochem. 10, 235—240.

Tezcanli-Güyer, G., Ince, N.H., 2004. Individual and combined effects of ultrasound, ozone and UV irradiation: a case study with textile dyes. Ultrasonics Sonochem. 42 (1—9), 603—609.

Thompson, L.H., Doraiswamy, L.K., 1999. Sonochemistry: science and engineering. Ind. Eng. Chem. Res. 38, 1215—1249.

Torres, R.A., Abdelmalek, F., Combet, E., Pétrier, C., Pulgarin, C., 2007a. A comparative study of ultrasonic cavitation and Fenton's reagent for bisphenol A degradation in deionised and natural waters. J. Hazard. Mater. 146 (3), 546—551.

Torres, R.A., Pétrier, C., Combet, E., Moulet, F., Pulgarin, C., 2007b. Bisphenol A mineralization by integrated ultrasound-UV-iron (II) treatment. Environ. Sci. Technol. 41 (1), 297—302.

Torres, R.A., Pétrier, C., Combet, E., Carrier, M., Pulgarin, C., 2008a. Ultrasonic cavitation applied to the treatment of bisphenol A. Effect of sonochemical parameters and analysis of BPA by-products. Ultrasonics Sonochem. 15, 605—611.

Torres, R.A., Nieto, J.I., Combet, E., Pétrier, C., Pulgarin, C., 2008b. Influence of TiO_2 concentration on the synergistic effect between photocatalysis and high-frequency ultrasound for organic pollutant mineralization in water. Appl. Catal. B Environ. 80 (1—2), 168—175.

Torres, R.A., Sarantakos, G., Combet, E., Pétrier, C., Pulgarin, C., 2008c. Sequential helio-photo-Fenton and sonication processes for the treatment of bisphenol A. J. Photochem. Photobiol. A Chem. 199 (2—3), 197—203.

Torres, R.A., Mosteo, R., Pétrier, C., Pulgarin, C., 2009. Experimental design approach to the optimization of ultrasonic degradation of alachlor and enhancement of treated water biodegradability. Ultrasonics Sonochem. 16, 425—430.

Uddin, M.H., Nanzai, B., Okitsu, K., 2016. Effects of Na_2SO_4 or NaCl on sonochemical degradation of phenolic compounds in an aqueous solution under Ar: positive and negative effects induced by the presence of salts. Ultrasonics Sonochem. 28, 144—149.

Villaroel, E., Silva-Agredo, J., Petrier, C., Taborda, G., Torres-Palma, R., 2014. Ultrasonic degradation of acetaminophen in water: Effect of sonochemical parameters and water matrix. Ultrasonics Sonochem. 21, 1763–1769.

Villegas-Guzman, P., Silva-Agredo, J., Giraldo-Aguirre, A.L., Flórez-Acosta, O., Petrier, C., Torres-Palma, R., 2015. Enhancement and inhibition effects of water matrices during the sonochemical degradation of the antibiotic dicloxacillin. Ultrasonics Sonochem. 22, 211–219.

Wayment, D.G., Casadonte, D.J., 2002. Frequency effect on the sonochemical remediation of alachlor. Ultrasonics Sonochem. 9, 251–257.

Xiao, R., Diaz-Rivera, D., Weavers, L.K., 2013. Factors influencing pharmaceutical and personal care product degradation in aqueous solution using pulsed wave ultrasound. Ind. Eng. Chem. Res. 52 (8), 2824–2831.

Xiao, R., He, Z., Diaz-Rivera, D., Pee, G.Y., Weavers, L.K., 2014a. Sonochemical degradation of ciprofloxacin and ibuprofen in the presence of matrix organic compounds. Ultrasonics Sonochem. 21, 428–435.

Xiao, R., Wei, Z., Chen, D., Weavers, L.K., 2014b. Kinetics and mechanism of sonochemical degradation of pharmaceuticals in municipal wastewater. Environ. Sci. Technol. 48, 9675–9683.

Xu, L.J., Chu, W., Graham, N., 2015. Sonophotolytic degradation of phthalate acid esters in water and wastewater: influence of compound properties and degradation mechanisms. J. Hazard. Mater. 288, 43–50.

Yang, L., Sostaric, J.Z., Rathman, J.F., Weavers, L.K., 2008. Effect of ultrasound frequency on pulsed sonolytic degradation of octylbenzene sulfonic acid. J. Phys. Chem. B 112, 852–858.

Yao, J.J., Gao, N.Y., Li, C., Li, L., Xu, B., 2010. Mechanism and kinetics of parathion degradation under ultrasonic irradiation. J. Hazard. Mater. 175, 138–145.

Yim, B., Okuno, H., Nagata, Y., Maeda, Y., 2001. Sonochemical degradation of chlorinated hydrocarbons using a batch and continuous flow system. J. Hazard. Mater. 81, 253–263.

Yin, P., Hu, Z., Song, X., Liu, J., Lin, N., 2016. Activated persulfate oxidation of perfluorooctanoic acid (PFOA) in groundwater under acidic conditions. Int. J. Environ. Res. Public Health 13 (602). Available from: http://dx.doi.org/10.3390/ijerph13060602.

Zhang, Y., Xiao, Z., Chen, F., Ge, Y., Wu, J., Hu, X., 2010. Degradation behavior and products of malathion and chlorpyrifos spiked in apple juice by ultrasonic treatment. Ultrasonics Sonochem. 17, 72–77.

Zhang, Y., Hou, Y., Chen, F., Xiao, Z., Zhang, J., Hu, X., 2011. The degradation of chlorpyrifos and diazinon in aqueous solution by ultrasonic irradiation: Effect of parameters and degradation pathway. Chemosphere 82, 1109–1115.

Zheng, W., Maurin, M., Tarr, M.A., 2005. Enhancement of sonochemical degradation of phenol using hydrogen atom scavengers. Ultrasonics Sonochem. 12, 313–317.

Further Reading

Ashokkumar, M., Lee, J., Kentish, S., Grieser, F., 2007. Bubbles in an acoustic field: an overview. Ultrasonics Sonochem. 14 (4), 470–475.

Cai, M., Jin, M., Weavers, L.K., 2015. Analysis of sonolytic degradation products of azo dye Orange G using liquid chromatography- diode array detection-mass spectrometry. Ultrasonics Sonochem. 18, 1068–1076.

Mason, T.J., Pétrier, C., 2004. Ultrasound processes. In: Parsons, S. (Ed.), Advanced oxidation processes for water and wastewater treatment. IWA Publishing, Padstow, pp. 185–208.

Microwave/Hydrogen Peroxide Processes

Afsane Chavoshani[1], Mohammad M. Amin[1], Ghorban Asgari[2], Abdolmotaleb Seidmohammadi[2] and Majid Hashemi[3]

[1]Isfahan University of Medical Sciences, Isfahan, Iran [2]Hamadan University of Medical Sciences, Hamadan, Iran [3]Kerman University of Medical Sciences, Kerman, Iran

8.1 INTRODUCTION

Microwave radiations are the part of electromagnetic radiations with frequency ranging from 300 MHz to 30 GHz. The major applications of the microwave are in communications and heating (Haque, 1999, Banik et al., 2003, Yuan et al., 2006). Historically, the development of the microwave for the heating of food is more than 50 years old. Later on in the 1970s, the construction of the microwave generator, the magnetron, was improved and simplified. Consequently, the prices of domestic microwave ovens fell considerably. However, the design of the oven chamber or cavity, which is crucial for the heating characteristics, was not significantly improved until the 1980s.

Microwave technology has been used in inorganic chemistry since the late 1970s, while it has only been implemented in organic chemistry since the mid-1980s. The development of the technology for organic chemistry has been rather slow as compared to combinatorial chemistry and computational chemistry. This slow acceptance of the technology has been principally attributed to its lack of controllability and reproducibility, safer aspects, and generally, a low degree of understanding of the basics of microwave dielectric heating. Since the mid-1990s, the number of publications in this field has increased significantly. The reason for this increase may be attributed to the availability of commercial microwave equipment intended for organic chemistry and the development of the solvent-free technique, which has improved the safety aspects, but it is mostly due to an increased interest in shorter reaction times (Lidström et al., 2001). Studies have shown that microwave heating is able to favorably change kinetic and selectivity of reactions (Raner et al., 1993).

8.1.1 Microwave Chemistry

Among chemical reactions based on microwaves, it can be cited for the synthesis of different organic and inorganic materials, polymerization, and nanoparticles selective sorption, reduction–oxidation reactions such as advanced oxidation processes (AOPs). Microwave chemistry is based on the fact that all materials (compounds, solvents, or reactants) are able to absorb microwave radiation and produce heat (Lew et al., 2002). In the majority of commercial, scientific, and medical applications microwave ovens with a frequency of 2.45 GHz are used (Kesari et al., 2010).

Interaction of different compounds with a microwave is based on two special mechanisms:

- dipole interactions and,
- ionic conduction.

Both these reactions require effective coupling between target compounds and an electric field related to oscillating microwave. Dipole interactions happened for polar molecules. When passing a microwave through polar molecules, both polar ends of these molecules start to reorient and will constantly align themselves with a magnetic field, to which they are influenced (Lidström et al., 2001) (Fig. 8.1).

This molecular rotation led to increasing intermolecular interactions, resulting in increasing temperature. Whenever a molecule is more polar, coupling with microwave radiation will be relatively better. Ion conduction fundamental is similar to polar interactions and has little difference with it. Ions are charged particles dispersed in solutions and could couple with an electric field related to microwave, and increase their movement dimension to increase the temperature. Therefore, it is expected that ion concentration has a remarkable effect on microwave heating efficiency.

Polar molecules are molecules containing atoms, whose electronegativity (tendency to attract electrons) is different from each other. This difference is the cause for electric cloud asymmetric distribution around molecules, and therefore accumulation of negative or positive charges may be more on one side than the other. This phenomenon is an important factor for the formation of polar molecules (Lidström et al., 2001).

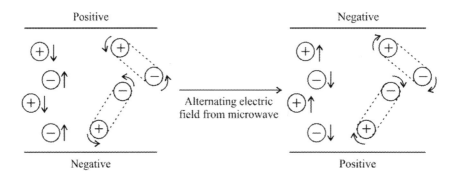

FIGURE 8.1   Effect of alternating electric field from microwave on polar molecules and ions.

The ability of a solvent for extraction of negative or positive ions from each other, under microwave radiation, is named dielectric constant (") (Zlotorzynski, 1995). Materials with higher dielectric constant have higher microwave absorption and heat production than others (Lidström et al., 2001).

8.1.2 Losses Factor or Tan δ

The effect of both processes of polarization and ion current, under microwave radiation, is defined as losses factor or tan δ. The amount of this factor is calculated according to the following equation:

$$\tan \delta = \frac{\mathrm{d}''}{\mathrm{c}''} \tag{8.1}$$

Where $_d''$ and $_c''$ are dielectric constants from dipole polarization and ion current, respectively.

Dielectric constant and loss tangent values of some solvents are summarized in Table 8.1.

TABLE 8.1 Dielectric Constant and Loss Tangent Values for Some Solvents Related to Organic Synthesis (Lidström et al., 2001)

Solvent	Dielectric Constant (ε_d'')	Loss Tangent (tan δ)
Hexane	1.9	0.02
Benzene	2.3	0.03
Carbon tetrachloride	2.2	0.04
Chloroform	4.8	0.091
Acetic acid	6.1	0.091
Ethyl acetate	6.2	0.174
Tetra hydrofuran	7.6	0.059
Methylene chloride	9.1	0.047
Acetone	20.6	0.042
Ethanol	24.6	0.054
Methanol	32.7	0.941
Acetonitlile		0.659
Dimethyl formamride	36.7	0.062
Dimethyl surfoxide (DMSO)		0.161
Formic acid		0.722
Water	80.4	0.123

8.1.3 Characteristic of Heating Microwave

Common heating sources in chemical reactions include oil bath, electric heating, gas flame, or water heating by convection, conduction, or radiation (Fig. 8.2). Only dielectric compounds are able to absorb microwave radiation and produce heat. In spite of fast material heating, by microwave radiation, the materials are quickly cooled on removing microwave radiations.

One of the problems about common heating is that a heating process happens from outside to inside, and therefore, the hottest part of the system is formed in reactor walls. But under microwave heating, first reactant and then reactor walls are heated, and reactants are quickly cooled by conductive method after stopping radiation. Other characteristics of the microwave and conventional heating have been shown in Table 8.2.

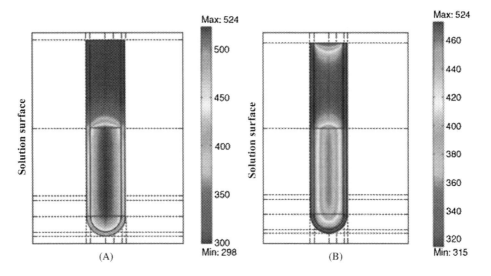

FIGURE 8.2 Thermal profile (in Kelvin) of a reaction process after 60 s: (A) the effect of microwave radiation, and (B) the effect of oil bath.

TABLE 8.2 Characteristics of Microwave and Conventional Heating (Kappe, 2004).

Microwave Heating	Conventional Heating
Depth	Shallow
Fast	Slow
Volumetric heating	Surface heating
Selective	Non-selective
Dependent on material type	Independent on material type

8.2 WASTEWATER TREATMENT

Wastewater treatment is an undeniable part of human society. The wastewaters are produced by numerous sectors including industries, agriculture, municipalities, etc. The accumulation of wastewater results in several environmental problems, health issues, safety hazards, and prevents sustainable development in terms of the reuse of wastewater.

Many of these wastewaters include some refractory compounds such as lignocelluloses, which are highly stable against enzymatic or bacterial attacks (Taherzadeh and Karimi, 2008).

Nowadays, there is a continuously increasing worldwide concern for the development of alternative water reuse technologies, which are mainly focused on agricultural and industrial wastewaters. Advanced oxidation processes (AOPs)-assisted with a microwave are considered a highly competitive wastewater treatment technology for the removal of organic pollutants that are not treatable by other conventional techniques due to their high chemical stability and/or low biodegradability. Although chemical oxidation for complete mineralization is usually expensive, its combination with a microwave or microwave/oxidants is widely reported to reduce operating costs.

Using microwave-assisted with other processes (MAP) is a simple technology that has been given special consideration because of high efficiency. It is known that photon energy is lonely 0.0016 eV. The amount of this energy is not sufficient for breaking of chemical bonds. This energy is even lower than Browning energy, and therefore, a microwave alone will not lead to chemical oxidations (Kappe, 2004; Stuerga et al., 1993; ÁMingos, 1992).

Chemical oxidations by microwave are based on non-thermal effects between a microwave and dielectric (microwave absorbing sheets) (Jacob et al., 1995). The effect of microwave irradiation in organic synthesis is based on thermal and nonthermal effects (super heating or hot spots, and selective absorption of radiation by polar substances) (de la Hoz et al., 2005).

During the last few years, a microwave has been used alone as well as assisted with other catalysts. A microwave usually increases the rate of reaction of other AOPs and for this reason, a microwave is applied along with oxidations such as $MW/UV/TiO_2$, MW/UV, and $MW/UV/H_2O_2$ (Remya and Lin, 2011a; Horikoshi et al., 2011; Asgari et al., 2013).

The removal of 4-chlorophenol under H_2O_2 and microwave alone was negligible, but the removal efficiency was increased significantly under microwave/H_2O_2 (Movahedyan et al., 2009).

The reason of this phenomenon is due to the existence of superheating and the production of hydroxyl radical in microwave/H_2O_2. Different radicals are produced under a microwave based on a catalyst and organic matter types (Zhao et al., 2011; Movahedyan et al., 2009).

H_2O_2 has been typically used in advanced oxidation processes and in situ chemical oxidation (ISCO) because it is a strong agent for oxidation and radical production. H_2O_2 is also economic and its decomposition led to safe materials like H_2O and O_2. H_2O_2 is able to participate in processes such as O_3/H_2O_2, UV/H_2O_2, or Fe^{2+}/H_2O_2. The required H_2O_2 concentration in reactions can be very different and it may change from micro mole to mole. The amount of its concentration is quite dependent on the type of advanced oxidation process, type of treatment, required efficiency, and chataracteristics of the pollutants. Hydroxyl radical with standard oxidation potential (E_o) of 2.7 V is produced by H_2O_2 dissociation (Asgari et al., 2013; Asgari et al., 2014; Chavoshani et al., 2016).

Hydrogen peroxide is a multiuse oxidant applied in many treatment systems. It is one of the inexpensive oxidizers usually used in AOPs as well as in situ chemical oxidation (ISCO) because it is a strong oxidizing agent and a potent source of $^\bullet$OH radicals at economical level of oxidation (Wu and Englehardt, 2012). H_2O_2 can also be used to remove cyanides, chromium (VI), oxidation of sulfur compounds, and for elimination of some inorganic nitrogen compounds. Hydrogen peroxide can be applied directly or with a catalyst. The use of hydrogen peroxide as an oxidant has a number of advantages and these are:

- It is cheaper, readily available, easy to store, relatively safe to handle and environmentally friendly since it slowly decomposes into oxygen and water.
- It is a very cost effective source of hydroxyl radicals as two hydroxyl radicals are formed for each molecule of H_2O_2 photolyzed.
- Peroxyl radicals are generated after $^\bullet$OH attack on most organic substrates; thus, facilitating oxidation reactions (Legrini et al., 1993; Sahoo et al., 2011).

A microwave led to decomposition of H_2O_2 in a microwave/H_2O_2 process producing hydroxyl radical ($^\bullet$OH). The following reactions are attributed to H_2O_2 decomposition under a microwave (Zhao et al., 2011; Yang et al., 2009).

$$H_2O_2 + OH \longrightarrow HO_2 + H_2O \tag{8.2}$$

$$H_2O_2 + OH \longrightarrow O_2^- + H_2O + H^+ \tag{8.3}$$

$$H_2O_2 + HO_2 \longrightarrow OH + H_2O + O \tag{8.4}$$

$$H_2O_2 + O_2^- \longrightarrow OH + OH^- + O_2 \tag{8.5}$$

$$HO_2^- + OH \longrightarrow HO_2 + OH^- \tag{8.6}$$

According to previous studies, efficiency of microwave/H_2O_2 is affected by many factors such as intensity, pH, pollutants concentration, H_2O_2 concentration, radical scavengers, and others.

8.2.1 Energy Intensity

Materials are able to absorb microwave energy by two mechanisms, i.e. polarization and ion current. The amount of energy absorbed and selected optimal energy depend on reactants and solvents under a microwave system (Clark and Sutton, 1996). Therefore, basically, selected optimal energy is different for each system. It is notable that increasing intensity will not be equal to increasing efficiency of a process, but in most of the studies with increasing energy intensity, removal efficiency has been increased (Zhang et al., 2007, Eskicioglu et al., 2008; Robinson et al., 2010). Efficiency removal of pentachlorophenol was increased from 31.82 to 90.82% under 180 and 600 Watt energy (intensity), respectively (Asgari et al., 2013). Movahedyan et al. (2009) also showed that the degradation rate of

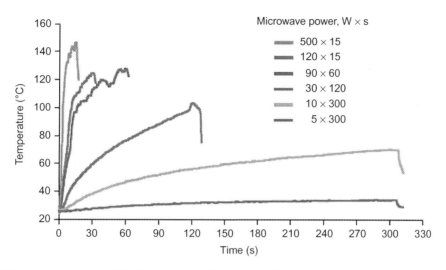

FIGURE 8.3 Temperature profiles for microwave flash-heated allylic substitutions in acetonitrile using different microwave powers. *Adapted from Larhed, M., Hallberg, A., 2001. Drug Discovery Today, 6, 406–416. With permission.*

p-chlorophenol was increased along with the increase in the microwave power. It was observed that percent degradation of p-chlorophenol at 180 min irradiation time was 93% in 600 W and 85.4% in the same time when the microwave was adjusted to the power of 180 W. Yang et al. (2009) made similar observations during removal of pharmaceutical formulations by microwave/Fenton process.

Depending on the type of reactants, use of higher intensity of microwave energy may result in superheating and sometimes, it is referred to as boiling retardation, or boiling delay. Larhed and Hallberg (2001) conducted reactions in acetonitrile (boiling point = 81 C–82 C) using different microwave power or intensity. It has been observed that at the reaction temperature (145 C) under microwave power 500 W, the boiling point of acetonitrile has been increased more than its standard value (Fig. 8.3).

Temperature profiles were recorded using fluoroptic probes. The reaction between H_2O_2 and organic matter is an exothermic process and heat produced in the reactor quickly increases the temperature of the system (Veschetti et al., 2000). It should be noticed in dissociation of H_2O_2 by microwave that H_2O_2 is unstable and will break down into H_2O and O_2. Heating a H_2O_2 solution would cause this breakdown to occur more quickly, perhaps with much foaming. This phenomenon is a disturbing factor under higher microwave powers and it should be considered (Chavoshani, 2012).

8.2.2 pH

pH is one of the most important factors in organic compounds degradation (Movahedyan et al., 2009). Several studies on MW/H_2O_2 system showed that high substrate oxidation efficiencies usually appeared at acidic and neutral pH (Eskicioglu et al., 2008; Wang et al., 2009). They observed that the highest substrate removal rate was pH 6, while the lowest one was at pH 11 and it was inferred to be due to self-decomposition of hydrogen peroxide since this is very limited and its scavenging effect of hydroxyl radicals in alkaline media (Sanz et al., 2003). Microwave-assisted degradation of carbofuran in the presence of granular active carbon (GAC), zero valent iron (ZVI), and H_2O_2 was studied by Remya and Lin (2011b); the influence of reaction temperature and pH was observed in the microwave-assisted GAC system. The degradation efficiency was increased with the increasing pH, whereas for the increase in pH from 2 to 4 resulted in decrease in the carbofuran degradation efficiency in the microwave-assisted-ZVI and H_2O_2 systems. The reason for this decrease in the degradation efficiency was not very clear. Contrary to these results the degradation of rhodamine B and methylene blue in MW/H_2O_2 system was found to be very competitive at extreme alkaline pH (Hong et al., 2012). This is interesting to note. The tests of radical species and control experiments were conducted. Degradation of rhodamine B at strong alkaline pH in MW/H_2O_2 system was considered to mainly undergo oxygen oxidation, and the enhancement of MW thermal effect. It provided a possible, effective way to degrade pollutants at highly alkaline circumstances, and it might promote the development of MW/H_2O_2 technology in alkaline wastewater treatment. Asgari et al. (2013) observed that pentachlorophenol removal efficiency under microwave/H_2O_2 at pH 3, 7 and 11 was 53.82, 56.82, and 63.82%, respectively. (Movahedyan et al., 2009) also opined that the degradation reactions were enhanced in an alkaline medium.

Organic compounds like phenolic compounds, are in anionic state at alkaline pH and absorb microwave more than at other pHs. Activated oxygen produced from hydrogen peroxide plays a key role under alkaline pH removal of organic compounds, except radical scavengers that may lead to the limitation of this condition (Hong et al., 2012; Movahedyan et al., 2009; Asgari et al., 2013). Multivariate analysis of sludge disintegration by the microwave/H_2O_2 pretreatment process, make it clear that a high pH appears to be effective in reducing percentage of sludge removal (Wang et al., 2009);whereas low pH was found to be favorable for the release of heavy metals and resulted in a decrease in volatile suspended solid/total suspended solid (VSS/TSS) for the treated sludge particles (Ya-Wei et al., 2015).

8.2.3 Pollutants Concentration

It is clear that oxidation processes assisted with microwave have high efficiency in removal of larger concentration of organic compounds. Phenols, aromatic hydrocarbons, and PCBs with concentrations 1000, 5000, and 10,000 mg/L, can be removed from raw oil in 98%, 96%, and 92%, respectively (Roshani and vel Leitner, 2011). Asgari et al. (2013) also observed that microwave/H_2O_2 process is able to remove 100, 200, 300, 400, 500, 750, and 1000 mg/L concentrations of pentachlorophenol (PCP) with efficiency 93%, 82%, 92%, 91%, 90%, 89%, 88%, and 87%, respectively. Increasing PCP concentration had a negligible role in decreasing process efficiency (approximately 7%). Consequently, selection of

1000 mg/L as the optimal dose was not economic and PCP concentration of 100 mg/L used for further experiments. These amounts are much higher than commonly adopted in attractive AOPs (Shiying et al., 2009) due to the high efficiency of processes integrated with microwave and non-thermal effects in these processes. HPLC data indicated that major organic pollutants are mineralized and there is a lack of toxic intermediates and byproducts (Han et al., 2004; Asgari et al., 2013).

8.2.4 H_2O_2 Concentration

The efficiency of this microwave/H_2O_2 system is influenced by an ionic and polarization amount of the system. It is expected that with increasing H_2O_2, a polar molecule with the potential of oxidation reduction 1.87, the efficiency of microwave/H_2O_2 will be increased. H_2O_2 is a polar molecule and it has not only the ability to absorb microwave radiation, but also the ability to produce hydroxyl radical ($^\bullet$OH with 2.8 eV) and other radicals. This characteristic led to more efficiency of H_2O_2 (Zhao et al., 2011). Results of Asgari et al. (2013) have shown that PCP removal efficiency under microwave/H_2O_2 for 0.01, 0.02, 0.1, 0.2, and 0.3 mol/L of hydrogen peroxide was 23.82%, 63.82%, 81.82%, 90.82%, and 90.82% at 60 min, respectively. PCP removal was almost stable at doses of 0.2 and 0.3 mol/L. Therefore, an H_2O_2 dose of 0.2 mol/L was selected and used as the optimal dose.

Recently, the microwave/H_2O_2 system was used as a tool for pretreatment of activated sludge. It has been emphasized that optimal H_2O_2 dosing is a function of multivariables. Wang et al. (2009) found that this optimal dosing is a function of temperature, pH, COD, and TSS of pollutants. It is favorable to model the relationship of H_2O_2 concentration with temperature, COD, pH, TSS, and the removal amount of pollutants (Ya-Wei et al., 2015). High H_2O_2 concentration led to interference in chemical oxygen demand (COD) removal, and therefore, it is suitable to select optimal H_2O_2 dosage based on ratios of H_2O_2/total chemical oxygen demand (TCOD) in the AOP of microwave and H_2O_2 (Chavoshani et al., 2016; Wang et al., 2009).

It has been reported that H_2O_2 residual in AOPs led to interference in the (COD) test; and it is able to hinder biological treatment of wastewater. In laboratory scale, a large amount of H_2O_2 residual can be estimated to be about 70%−80% of its initial concentration, and allocates high mole concentration to itself, leading to overestimation of the COD measurements. The amount of hydrogen peroxide interference depends on the type of pollutants. An average of COD after the reaction, COD before the reaction, removed COD, and false COD was 56206, 40650, 26707.57, and 13943 mg/L respectively (Chavoshani et al., 2016). This removal of COD has been very low. Reports on mixed waste chemicals existing in solid waste leachate, indicate that apart from residual H_2O_2, other factors such as chloride, bromide, iodide, ferrous and ferric ions, sulfide, and manganese led to interference and errors in the COD test. In fact, these factors resulted in false COD (Rezaeei et al., 2008). More efforts are required for optimization of H_2O_2 dosing in order to make it cost effective technology through reducing both H_2O_2 costs and residual H_2O_2 in sludge (Wang et al., 2009).

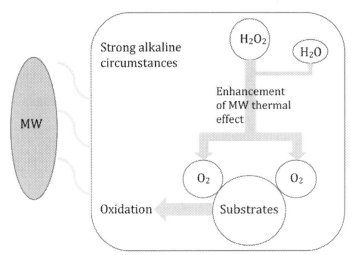

FIGURE 8.4 Degradation mechanism in alkaline MW/H$_2$O$_2$ system. *Adapted from Hong, J., Yuan, N., Wang, Y., Qi, S., 2012. Chem. Eng. J. 191, 364–368. With permission.*

8.2.5 Radical Scavengers

Formation of hydroxyl radical (OH$^\bullet$) under microwave/H$_2$O$_2$ is an important factor for removal of organic materials for deciding the role of OH$^\bullet$ radicals. Tert-butyl ether (TBA) was used as a radical scavenger in oxidations processes. Asgari et al. (2013) showed that tert-butyl ether led to the removal of PCP from 93.82% to 90.82%, which means that PCP removal was decreased as much as 3%.

A number of attempts were made to confirm the hydroxyl radical production during the microwave process. Contrary to this belief, rhodamine B removal under microwave/H$_2$O$_2$ has shown that the participation of $^\bullet$OH is negligible in this process, because when methanol was added as a radical scavenger, removal efficiency showed no significant change. It seems that, in addition to $^\bullet$OH radicals, other intermediate radicals play the main role in microwave/H$_2$O$_2$ process; O$_2$ was generated as the main species from the decomposition of H$_2$O$_2$ in alkaline MW/H$_2$O$_2$ system, which then oxidizes the substrate (Fig. 8.4). Meanwhile, MW thermal effect could enhance the decomposition of H$_2$O$_2$, generating more O$_2$ to accelerate the degradation of the substrate (Hong et al., 2012).

8.3 ENHANCEMENT OF SLUDGE ANAEROBIC BIODEGRADABILITY

Huge amounts of sewage sludge are produced yearly with improving wastewater treatment facilities. Over 6.25 million tons of dry sludge was produced in 2013 in China and it is increasing with the annual growth rate of 13% (Yang et al., 2015). Management of sludge is a very sensitive and expensive responsibility. Today, sludge disposal has become a problem of sludge management because up to 50% of the current operating costs of a wastewater treatment plant are allocated to sludge treatment. Although different disposal routes (incineration, landfill, and agricultural applications) are possible for sludge management, due to environmental health hazards and economic costs, the selection of other methods needs further studies (Appels et al., 2008). Reduce, reuse, and recycle (commonly known as 3Rs) are the three basic options available for sludge treatment. Among the 3Rs' treatment options, reducing the amount of sludge produced is regarded as the ideal method for solving the problems associated with sludge treatment.

Sewage sludge can be considered as a renewable energy resource, The sewage sludge has calorific values ranging from 11 to 17 MJ/kg, in Poland that is equal to burning about 443,000 tons of raw coal (Cao and Pawłowski, 2013). About 60% of all sewage sludge was applied to land as a soil amendment and fertilizer for growing crops in the United States (Lu et al., 2012). This type of compost played a key role for sustainable nutrient management for a variety of vegetables and plants (Clarke and Smith, 2011). Anaerobic digestion (AD) of sludge as an efficient and sustainable technology for sludge treatment has been proposed. The benefits of AD are including mass reduction, odor removal, pathogen reduction, less energy use, and more significantly, energy recovery in the form of methane (Pilli et al., 2011). Degradable organic compounds were converted to methane (CH_4) and carbon dioxide (CO_2) in the absence of elemental oxygen and the presence of anaerobic bacteria process in AD of sludge.

The stabilization and disintegration of anaerobic sludge occur in four stages: hydrolysis, acidogenesis, acetogenesis, and methanogenesis by three different groups of bacteria. The first group involves hydrolytic and acidogenic bacteria, which hydrolyze the complex substrates (carbohydrates, lipids, proteins, etc.) to dissolved monomers (sugars, fatty acids, amino acids, etc.) and further to CO_2, H_2, organic acids, and alcohols (Pilli et al., 2011). Due to competition between sulfate-reduction bacteria and methane-forming bacteria, depends on various factors like pH, temperature, HRT, C/N ratio, VS/COD reduction, lack of process stability, low loading rates, slow recovery after failure, and specific requirements for waste composition, digestion of anaerobic sludge and biogas production is a relatively slow process (Ali Shah et al., 2014).

Much efforts have been devoted to develop sludge disintegration and solubilization techniques that feed back to the biological processing step for further biodegradation (Øegaard, 2004). This type of process is known as lysis—cryptic growth (Zhou et al., 2013). In most treatment technologies, breaking the cell walls of the sludge particles is a key problem, due to the protection offered by the extracellular polymeric substances (EPS). Therefore, pretreatment technologies can be used to disrupt the cell walls, which will also help to improve biogas production, recover materials such as proteins, volatile fatty acids (VFAs), nitrogen, and phosphorus, and improve sludge dewatering. There are many sludge pretreatment technologies available and these are physical, chemical, mechanical, or biological hydrolysis, or a combination of these methods (Wang et al., 2015). The microwave (MW) technology has gained widespread popularity as an effective thermal method for sludge treatment (Tyagi and Lo, 2013). Another promising application of MW technology is in pyrolysis, which involves heating dried sludge at temperatures above 300 C in the absence of water and oxygen, to produce valuable gases and oils (Manara and Zabaniotou, 2012).

The major driving force behind the rise in the use of MW technologies is the fact that heating with MW energy is superior to conventional heating, mainly in terms of its ability to heat rapidly and selectively; thereby accelerating reaction rates. MW heating minimizes hazardous product formation and emissions, which renders this technique environmentally friendly.

A combination of MW and chemical pretreatment techniques has been suggested as a sludge pretreatment method due to increasing solubilization of the volatile suspended solids (VSS) (Tyagi and Lo, 2013). Nowadays, application of microwave technology and its combined processes for sludge pretreatment has received increased attention. Some of these technologies include microwave with pressure seal, microwave under ambient pressure conditions, microwave-acid, microwave-alkali, and microwave-H_2O_2 (MW/H_2O_2) process. These technologies have been used to improve sludge disintegration and sludge disinfection. Among them, MW/H_2O_2 has been indicated to be the most efficient.

Oxidation power of H_2O_2 can be enhanced using a microwave. When H_2O_2 and microwave are combined, H_2O_2 absorbs MW radiation and breaks down into $^{\bullet}$OH radicals, which reduce the pollutants in the sample.

8.3.1 Volatile Fatty Acids Production By MW/H_2O_2

Volatile fatty acids (VFA) are short-chain fatty acids consisting of six or fewer carbon atoms that can be distilled at atmospheric pressure. They are including acetic acid, propionic acid, and butyric acid. These acids have a wide range of applications such as in the production of bioplastics, bioenergy, and the biological removal of nutrients from wastewater. At present, commercial production of VFA is mostly accomplished by chemical routes. However, the use of nonrenewable petrochemicals as raw materials and the increasing price of oil have renewed the interest in biological routes of VFA production. In biological VFA production, pure sugars such as glucose and sucrose have been commonly employed as the main carbon source, which raises the ethical concern on the use of food to produce chemicals. This issue can be resolved by utilizing organic-rich wastes such as sludge generated from wastewater treatment plants, food waste, organic fraction of municipal solid waste, and industrial wastewater for VFA production. Such transformation of waste into VFA also provides an alternative route to reduce the ever increasing amount of waste generated (Lee et al., 2014).

In general, the production of VFA from waste is an anaerobic process involving hydrolysis and acidogenesis (the latter is also known as acidogenic fermentation) Bengtsson et al. (2008) or dark fermentation (Su et al., 2009). In hydrolysis, complex organic polymers in waste are broken down into simpler organic monomers by the enzymes excreted from the hydrolytic microorganisms (Lee et al., 2014).

Subsequently, acidogens ferment these monomers mainly into VFA such as acetic, propionic, and butyric acids. Both processes involve a complex consortium of obligate and facultative anaerobes, such as bacteriocides, clostridia, bifidobacteria, streptococci, and enterobacteriaceae. However, it is common practice that the hydrolysis and acidogenesis are conducted simultaneously in a single anaerobic reactor.

In the past decades, numerous efforts had been devoted to maximize the production of VFA by exploring different types of wastes and regulating the operating conditions of the anaerobic reactor. It is now realized that proper process control can manipulate the type of VFA produced, which is critical to the performances of the downstream applications such as the production of polyhydroxyalkanoates (biodegradable plastics), electricity, biogas, and the biological removal of phosphorus and nitrogen from wastewater(Lee et al., 2014).

MW/H_2O_2 pretreatment is a suitable way for maximization of VFA from anaerobic sludge. The VFAs are a product of hydrolysis and acidification, and a precursor for methane production (Devi et al., 2014).

Liao et al. found that the combined microwave-hydrogen peroxide-sulfuric acid enhanced advanced oxidation process (MW/H_2O_2/H^+-AOP), and is able to increase VFA concentration. The highest VFA concentration was obtained for soluble substrates with acid treatment only. The VFA concentrations of the resulting solution could be controlled by amounts of acid used in the MW/H_2O_2-AOP process(Liao et al., 2006).

In the study of Eskicioglu et al. (2008), VFA removal under different reactions has been reported: acetic acid removal by H_2O_2 (0.9 g/L), MW-60 C (0.8 g/L), MW-60 C/H_2O_2(1.9 g/L), MW-80 C(1 g/L), MW-80 C/H_2O_2(1.3 g/L), MW-100 C(0.7 g/L), MW-100 C/H_2O_2(1 g/L), MW-120 C(0.7 g/L), and MW-120 C/H_2O_2(1.1 g/L); propionic acid removal by H_2O_2 (0.6 g/L), MW-60 C (0.6 g/L), MW-60 C/H_2O_2(1.1 g/L), MW-80 C (0.7 g/L), MW-80 C/H_2O_2(0.9 g/L), MW-100 C(0.6 g/L), MW-100 C/H_2O_2(0.8 g/L), MW-120 C(0.7 g/L), and MW-120 C/H_2O_2(0.8 g/L); and butyric acid removal by H_2O_2 (0.0 g/L), MW-60 C (0.4 g/L), MW-60 C/H_2O_2(0.0 g/L), MW-80 C(0.0 g/L), MW-80 C/H_2O_2(0.2 g/L), MW-100 C(0.0 g/L), MW-100 C/H_2O_2(0.4 g/L), MW-120 C(0.0 g/L). and MW-120 C/H_2O_2(0.2 g/L). These results show that the heights VFA removal was attributed to acetic acid under MW-60 C/H_2O_2.

8.3.2 Change of Biological Nutrient in Anaerobic Sludge

Biological nutrient removal (BNR) from wastewater has been widely used in the world. Several methods have been reported to remove nutrients (ortho-phosphorus, ammonium, protein, carbohydrate and oil) from anaerobic sludge (Tong and Chen, 2009).

It has been reported in several studies that MW/H_2O_2 process is able to remove nutrients from anaerobic sludge. Hydrogen peroxide, as a strong oxidizer, converts some of the sensitive organic compounds to CO_2 and water in addition to many others oxidized to other organics containing stable functional groups; therefore it will provide some level of permanent stabilization. Table 8.3 indicates the effects of MW and MW/H_2O_2 treatments on total organic matter composition in terms of TS, total COD, total proteins, total sugars, total humic acids, as well as supernatant NH_3-N concentrations.

Based on this table, hydrogen peroxide oxidized a portion of sludge in H_2O_2 and MW/H_2O_2 treated samples and caused reductions in the total organic solids and biopolymers. The TWAS sample, treated with H_2O_2 (1 g H_2O_2/g TS), lost $19 \pm 0\%$, $18 \pm 5\%$, $11 \pm 3\%$, $34 \pm 3\%$, and $16 \pm 2\%$ of TS, COD, proteins, sugars, and humic acid concentrations, respectively, through oxidation compared to controls. The concentration of organic compounds present in the TWAS samples decreased further when H_2O_2 was combined with MW irradiation especially at temperatures above 80 C. At MW temperature of 120 C, MW/H_2O_2 samples experienced additional 2%, 12%, 27%, 52%, and 45% reductions in their TS, COD, proteins, sugars, and humic acid concentrations, respectively, compared to the H_2O_2 treated sample. The level of permanent stabilizations in similar samples (MW/H_2O_2 at 120 C) were $20 \pm 0\%$, $28 \pm 2\%$, $35 \pm 0\%$, $68 \pm 0\%$, and $53 \pm 1\%$ for TS, COD, proteins, sugars and humic acids compared to the original raw TWAS (control) sample. It is possible that the elevated MW temperatures increased the decomposition of H_2O_2 into $^\bullet OH$ radicals and enhanced the oxidation process when both MW and H_2O_2 are applied. Among organic substances studied, the sugars and humic acids experienced the highest level of reductions after H_2O_2 (16%−34%) and MW/H_2O_2 (53%−68%) treatments at 120 C (Eskicioglu et al., 2008).

All pretreatment scenarios contained higher NH_3-N in their supernatants compared to that of control. Both H_2O_2 treatment and MW irradiation resulted in significant (2−4 fold) ammonia release into solution over the control (Table 8.3). These results are in agreement with previous studies that concluded both MW and MW/H_2O_2 treatments enhance the release of ammonia depending on the MW temperature and H_2O_2 concentration (Wong et al., 2006; Yin et al., 2007). Among the pretreatment temperatures tested, the TWAS

TABLE 8.3   General Characteristics of Raw and Pretreated TWAS[a]

Parameters	TWAS (Control)	H_2O_2 (Treated)	T = 60 C MW	T = 60 C MW/H_2O_2	T = 80 C MW	T = 80 C MW/H_2O_2	T = 100 C MW	T = 100 C MW/H_2O_2	T = 120 C MW	T = 120 C MW/H_2O_2
pH	6.4	6.1	6.2	6.5	6.3	6.5	6.3	6.6	6.2	6.7
TS (%, w/w)	6.4	5.2	6.8	5.2	6.4	5.3	6.4	5.3	5.8	5.1
VS	7.4	3.8	5.1	3.8	4.7	3.9	4.6	3.8	4.7	3.7
VS/TS (%)	73.4	72.8	74.7	73	73.7	72.9	72.9	72.8	72.8	72.8
NH_3-N (mg/L)	106	284	451	318	341	323	304	301	185	373
TOTAL FRACTION (G/L)										
COD	77.9	63.6	81.4	74.3	73.6	68.6	80.7	62.9	82.1	55.7
Proteins	12.3	10.9	14.8	7	15	9.5	16	10.2	17.7	8
Sugars	16.9	11.2	10.8	11.7	10.5	9.2	11.8	7.9	12.7	5.4
Humic acids	8.3	6.9	4.4	7.7	6.2	6.1	6	4.5	5.4	3.9
SOLUBLE FRACTION (<0.45 μ) (G/L)										
COD	2.7	10.6	10	10.8	10.9	12.5	13.2	13.2	12.1	13.6
Proteins	0.1	1	0.4	1.1	0.6	1.4	1	2.3	2	2.5
Sugars	0.2	1.6	0.7	1.6	1.7	1.8	2.3	2.6	2.3	1.9
Humic acids	0.3	1.5	1.4	1.6	1.9	1.9	1.9	1.7	1.1	1.4
VOLATILE FATTY ACID (VFA) (G/L)										
Acetic	1.5	0.9	0.8	1.9	1	1.3	0.7	1	0.7	1.1
Propionic	0.9	0.6	0.6	1.1	0.7	0.9	0.6	0.8	0.7	0.8
Butyric	0.5	0	0	0.4	0	0.2	0	± 0.4	0	0.2

[a] *Adapted from Eskicioglu, C., Prorot, A., Marin, J., Droste, R.L., Kennedy, K.J., 2008. Water Res., 42(18), 4674−4682. With permission.*

sample pretreated at 60 C contained the highest ammonia concentration which was 451 mg/L. The other TWAS samples pretreated at 80 C, 100 C, and 120 C had NH_3-N concentrations of 341, 304, and 185 mg/L, respectively.

These results are in agreement with previous studies that concluded both MW and MW/H_2O_2 treatments enhance the release of ammonia depending on the MW temperature and H_2O_2 concentration.

It was expected that by using a combination of MW and H_2O_2, polymeric network disintegration could be increased. As it can be observed from Figs. 8.5A–D, all MW, H_2O_2, and MW/H_2O_2 treatments increased the amount of soluble COD, soluble proteins, soluble sugars, and soluble humic acid fractions in the TWAS samples compared to those of control. The level of COD solubilizations without the presence of H_2O_2 were $3 \pm 0\%$, $12 \pm 0\%$, $15 \pm 0\%$, $16 \pm 0\%$, and $15 \pm 1\%$ for control, MW-60, MW-80, MW-100, and MW-120 samples, respectively.

Furthermore, H_2O_2, $MW-60/H_2O_2$, $MW-80/H_2O_2$, $MW-100/H_2O_2$, and $MW-120/$ H_2O_2 samples achieved $17 \pm 2\%$, $15 \pm 1\%$, $18 \pm 0\%$, $21 \pm 1\%$, and $24 \pm 1\%$ COD solubilizations, respectively, indicating that a synergistic disintegration effect was observed when both treatments were combined, and resulted in smaller organic compounds (Fig. 8.5A). Both proteins and humic acids can be found in polymeric network as extra- and intra-cellular compounds. The increase of concentration of these compounds in the soluble phase of WAS after both MW and MW/H_2O_2 treatments is evidence of disintegration and solubilization floc structure as MW temperature increased. In general, the COD solubilization trend observed in this table was consistent with the data obtained in other MW/H_2O_2 studies on sludge. The results in Fig. 8.5A–D revealed that H_2O_2 and MW/H_2O_2 treatments did not only reduce the amount of total organic residues in TWAS, but also reduced the particulate fractions by converting them into soluble (<0.45 mm) organics. Among the organic substances studied, higher levels of solubilizations were achieved in humic acids and sugars, which were $37 \pm 1\%$ and $35 \pm 2\%$ for the MW/H_2O_2 sample at 120 C (Fig. 8.5C, D) (Eskicioglu et al., 2008).

Also, the influence of combined pretreatment on the concentration of suspended solids, SCOD, protein, carbohydrate, Suspended solid (SS) percentage, and COD solubilization has been perforemd. It was observed that at optimal specific energy input (18,600 kJ/kg TS), the SCOD concentration, protein, and carbohydrate in the medium were found to increase with all types of MW pretreatment. The SCOD, protein, and carbohydrate release at this energy level were found to be 7.4, 12.3, and 13.8 g/L; 0.7, 1.5, and 1.7 g/L, and 0.05, 0.15, and 0.16 g/L, respectively.

The increase in the soluble organics was mainly due to synergistic effects brought about by the combined action of MW, H_2O_2, and acidic condition. Among different treatments, combined pretreatment was proved to be effective compared to MW alone (Eskicioglu et al., 2008). COD solubilizations for MW, MW/H_2O_2, and MW/H_2O_2/acid at the MW energy of 18,600 kJ/kg TS, were recorded to be 30.2%, 50.3%, and 56.1%, respectively. The highest solubilization (56.1%) was noted with MW/H_2O_2/acid combination. This could be due to the fact that the hydroxyl radical formation was higher at a lower pH, which increased the efficiency of H_2O_2. The currently achieved COD solubilization (56%) at an energy input of 18,600 kJ/kg TS was relatively greater than that achieved in other studies, in which greater energy input was used to obtain similar solubilization.

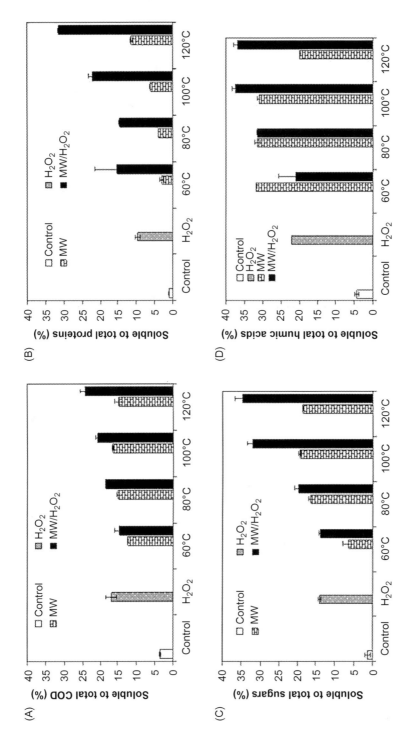

FIGURE 8.5 Particulate organics solubilization achieved on TWAS samples in terms of: (A) COD, (B) proteins, (C) sugars, and (D) humic acids after MW and MW/H_2O_2 treatments. *Adapted from Eskicioglu, C., Prorot, A., Marin, J., Droste, R.L., Kennedy, K.J., 2008. Water Res. 42(18), 4674–4682. With permission.*

Similarly, the SS concentration in the medium was found to decrease in all types of MW disintegration of sludge. The SS concentrations for MW, MW + H_2O_2, and MW/ H_2O_2/acid were recorded to be 15, 13.3, and 12.7 g/L, respectively at the MW specific energy of 18,600 kJ/kg TS. The corresponding SS reductions at this energy level were calculated to be 20.9%, 29.5%, and 33.1%, respectively. The addition of H_2O_2 dosage resulted in a significant increase in MW assisted SS reduction from 20.9% to 29.5%. This could be because of rapid formation of the MW-mediated hydroxyl radicals from H_2O_2. The acidification of sludge (pH 5) further increased the SS reduction from 29.5% to 33.1%.

8.3.3 Biochemical Methane Potential Assays

The biochemical methane potential (BMP) test is increasingly being recognized as a viable method to explore and determine the feasibility of a material to serve as a substrate in AD. BMP assays provide an array of information on the substrate including how fast and how much of the material can be degraded under optimal conditions as well as the potential methane yield from the material. A primary output of BMP assays is cumulative methane production curves, where the cumulative specific methane production is plotted against time. The patterns these curves follow are far from trivial and have meaningful implications on the degradation of the substrate. The kinetics of the different stages of the AD process (hydrolysis, acidogenesis, acetogenesis, and methanogenesis) and ultimately the shape of the methane production curves is primarily controlled by the biodegradability characteristics of the substrate, the production of inhibitory intermediate fermenters, and the performance of the methanogenic bacterial population (Ware and Power, 2017).

An anaerobic biodegradability assay has been performed with MW, MW/H_2O_2, and MW/H_2O_2/acid-pretreated and the control sludge during 30 days. Among them, the MW/H_2O_2/acid pretreated sample has produced the highest methane yield of 323 mL/g VS at the end of 30 days. The MW/H_2O_2-pretreated sample had a maximum methane production of 288 mL/g VS, while for MW-pretreated sample, the methane production was 175 mL/g VS. The methane production of the control sludge was low (33 mL/g VS). This could be because of the occurrence of a lesser or inconsequential amount of freely accessible substrate.

The methane production for the control sludge had a lag period of eight days, and this prolonged lag phase may be because of the deliberate hydrolysis of the substrate. Two separate exponential methane production phases were observed for the MW, MW/H_2O_2, and MW/ H_2O_2 /acid-pretreated sludges. The first exponential phase persisted for 6−10, 5−7, and 4−6 days of operation in the MW, MW/H_2O_2, and MW/H_2O_2/acid-pretreated sludge, respectively. The initial exponential increase in the biogas production is because of the presence of the readily available soluble organic compounds that are converted into biogas under anaerobic digestion. Among various samples, the MW/H_2O_2 /acid-pretreated sludge exhibited an early start of the first exponential phase; thereby, indicating a better hydrolysis of the substrate. The first exponential phase was followed by a transitional lag period. The second exponential increase was associated with the degradation of the nonsoluble particulate organic matter and it extended from 7−16, 8−17, and 11−20 days for the MW/H_2O_2/acid, MW/H_2O_2, and MW-pretreated sludges, respectively.

Two different models were used for MW, (the MW/H_2O_2, MW/H_2O_2/acid-pretreated samples), to simulate the two observed exponential phases. The first one was the exponential model and was applied to model the first exponential increase in methane production (ranging from 0 to 5 days of digestion). The experimental data were found to be in agreement with the goodness of fit prominent for the combined MW-pretreated samples. The higher values of R^2, reveal the goodness of fit for all the experimental data with the proposed models and thereby, indicating a positive correlation between the model and experiment. A second model is the logistic model, and it was used to model the second exponential increase in the biogas production (10–30 days). The model fit for the control sludge was made with a modified Gompertz equation, as no phase separation was evident in the control. The kinetic parameters for all the models were evaluated using nonlinear regression. It was observed that the rate constant K for the MW/H_2O_2/acid-pretreated sample was higher (0.77 day^{-1}) than the control (0.2054 day^{-1}) and MW (0.426 day^{-1}). The ultimate methane yield (P) for the control was found to be very low (150.3 mL/g VS) because of the lack of easy accessibility of the substrate to the methanogens. The high rate constant for MW/H_2O_2/acid was attributed to more SCOD release. Similarly, the maximum rate of methane production and methane production potential was found to be higher. In contrast, the lag time for the MW/H_2O_2/acid was less (2.3 days) than the other treatment methods and the control (8.3 days) because of the readily available substrate for degradation

Great amounts of residual H_2O_2 in pretreated sludge after MW/H_2O_2 pretreatment resulted in inhibitory effect on sludge anaerobic digestion in the initial days of BMP tests. Besides, more refractory compounds were generated after pretreatment with high H_2O_2 dosage (0.6 and 1.0 g H_2O_2/g TS). These negative effects gave rise to a long lag phase, low methane production rate, and no enhancement of cumulative methane production, even though high COD solubilization was achieved after MW/H_2O_2 pretreatment. The residual H_2O_2 had a serious inhibitory effect on metabolic activity of methanogens and inhibited hydrolysis-acidification stage of anaerobic digestion mildly. Low dosage of H_2O_2 such as 0.2 g H_2O_2/g TS in MW/H_2O_2 pretreatment process was appropriate for enhancing anaerobic digestion. The cumulative methane production was increased by 29.02% compared with the control. Moreover, decomposition of the residual H_2O_2 by catalyst can relieve the inhibitory effect on enhanced anaerobic digestion (Liu et al., 2015a,b).

The estimation of biological methane production at mesophilic temperature (33 C ± 2 C) is a suitable information source about both initial and accumulative methane production. Comparison of results showed that reduction of organic matters and methane productions under MW/H_2O_2 has been remarkable (Eskicioglu et al., 2008).

The first order anaerobic biodegradation constants (K, d^{-1}) as well as the coefficient of determination (R^2) is used to determine the relationship between measured and predicted accumulative biogas production (Eskicioglu et al., 2008).

Among three different sludge samples, that is, anaerobic digested sludge, primary sludge, and activated sludge, it was clear that the highest MW toxicity was observed in activated sludge, anaerobic digested sludge, and primary sludge respectively; while, conventional heating sludge samples did not indicate any toxicity. Based on previous studies, this toxicity has been produced due to leaching sludge substances to soluble phase (<0.45 m) after MW irradiation (Eskicioglu et al., 2006).

Results have shown the trend of the accumulative biogas curve under different processes of MW, MW/H_2O_2, and control systems has an increasing trend because in all of the pretreatment types, all biogas curves had three separate phases (from 1−12, 12−20, and 20−30 days) indicating three different substances and biogas accumulations (Fig. 8.6) (Eskicioglu et al., 2008).

8.3.4 Effect Of H_2O_2 on Anaerobic Sludge Pretreatment

H_2O_2 as an oxidant has a favorable effect on anaerobic sludge pretreatment. It was clear that among other oxidants, H_2O_2 had been a cost effective oxidant (Ya-Wei et al., 2015). The $^\bullet$OH (with 2.8 eV) produced from H_2O_2 could react rapidly and non-selectively with many refractory pollutants. This reduces the treatment time and enhances the solubilization plausibly (Kato et al., 2014; Steriti et al., 2014). The effects of the H_2O_2 concentration on COD solubilization, soluble protein, and carbohydrate have been investigated (Eswari et al., 2016) and it was clear that up to 0.3 mg/g SS of H_2O_2 dosage, the concentration of protein, carbohydrate, and SCOD release have been increased and the corresponding values were observed to be 1.52, 0.045, and 12.35 g/L, respectively.

A further increase in H_2O_2 concentration decreases the release of soluble organics. For example, an increase in H_2O_2 dosage from 0.3 to 0.4 mg/g SS causes a decrease in COD solubilization from 12.35 to 11.56 g/L. A similar kind of decrease was also observed for the release of soluble organics. This was because of the fact that higher concentrations of H_2O_2 lead to the scavenging of $^\bullet$OH radicals and, as a result, the release of $HO_2^\bullet$ will occur by the following equations:

$$H_2O_2 + \ ^\bullet OH \rightarrow HO_2 + H_2O \tag{8.7}$$

$$HO_2 + \ ^\bullet OH \rightarrow H_2O + O_2 \tag{8.8}$$

It is important to note that $HO_2^\bullet$ is less reactive than $^\bullet$OH and thus an increasing concentration of hydrogen peroxide resulted in a diminished reaction rate (Herney-Ramirez et al., 2010). Therefore, 0.3 mg/g SS of H_2O_2 was found to be the optimum dosage for sludge solubilization when it was combined with MW.

Although MW/H_2O_2 pretreatment of sludge for enhancing anaerobic digestion has been investigated in several studies, there are still many questions that need to be answered. These are:

- What is the appropriate H_2O_2 dosage in pretreatment for enhancing sludge anaerobic digestion?
- What is the mechanism of the inhibitory effect?
- Is there a method to control the inhibitory effect and further improve the performance of sludge anaerobic digestion?

8.3.5 Inhibitory Effects on Microbial Methabolism

The residual H_2O_2 in pretreated sludge would possibly damage the microbial cells or inhibit the microbial metabolism of anaerobic digestion. The initial methane production

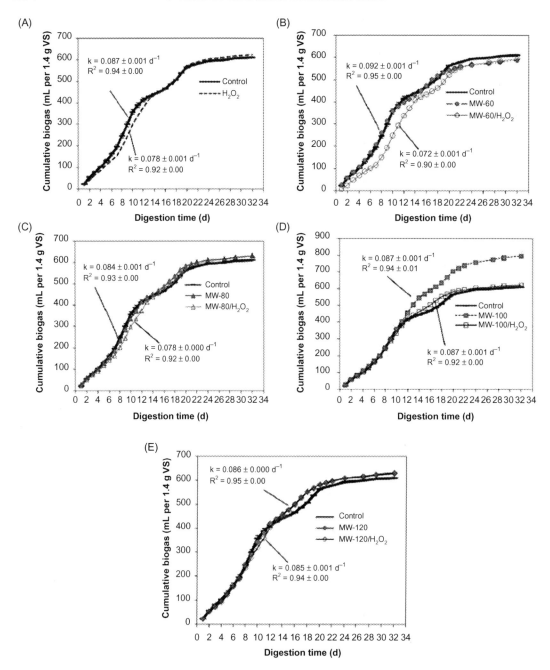

FIGURE 8.6 Accumulative biogas production (mL per 1.4 g volatile solids added) and first-order biodegradation constant (K) from: (A) Control and H_2O_2 samples, (B) MW-60 and MW-60/H_2O_2 samples, (C) MW-80 and MW-80/H_2O_2 samples, (D) MW-100 and MW-100/H_2O_2 samples, and (E) MW-120 and MW-120/H_2O_2 samples. *Adapted from Eskicioglu, C., Prorot, A., Marin, J., Droste, R.L., Kennedy, K.J., 2008. Water Res., 42(18), 4674–4682. With permission.*

could reveal the biodegradation rate and toxicity of tested samples relative to the control. The cumulative final methane production gave valuable information on ultimate biodegradability. H_2O_2, known as one of the toxic forms of oxygen, could damage protein, lipids, and DNA components in organisms. H_2O_2 was capable of inhibiting the growth of pathogenic bacteria (Zhao et al., 1998). As a highly active biocide at a concentration of 30 to 1000 mg/L, H_2O_2 can inhibit the bioactivity of microbial cells and has a lethal effect on many bacteria (Jiang and Yuan, 2013). Due to the inhibitory effect with excessive H_2O_2 addition in pretreatment, low methane yield rate was obtained at 0.6 and 1.0 g H_2O_2/g TS, respectively. The maximum methane yield rate was observed in the anaerobic digestion of pretreated sludge with the H_2O_2 dosage of 0.2 g H_2O_2/g TS, which was slightly higher than the control. After the 30 day BMP tests, the cumulative methane production (P) was significantly improved with MW/H_2O_2 pretreatment at low dosage of H_2O_2 (0.2 g H_2O_2/g TS), 29.02% higher than the control. Although the similar concentration of soluble organics was released, no enhancement of cumulative methane production was obtained with high H_2O_2 dosage (0.6 and 1.0 g H_2O_2/g TS) pretreatment. These results revealed that the refractory soluble organics were generated at high H_2O_2 dosage pretreatment. Increasing sludge solubilization does not always result in enhanced anaerobic digestion.

8.3.6 Regression Model Optimizing H_2O_2

A MW/H_2O_2 process for sludge pretreatment exhibited high efficiencies of releasing organics, nitrogen, and phosphorus, but large quantities of H_2O_2 residues were detected. A uniform design method was thus employed to further optimize H_2O_2 dosage by investigating effects of pH and H_2O_2 dosage on the amount of H_2O_2 residue and releases of organics, nitrogen, and phosphorus. A regression model was established with pH and H_2O_2 dosage as the independent variables, and H_2O_2 residue and releases of organics, nitrogen, and phosphorus as the dependent variables (Xiao et al., 2012).

Therefore, the released TOC, TP, TN, and the residual H_2O_2 after sludge pretreatment by the MW/H_2O_2 process were set as optimization indices. The optimization directions of the released TOC, TP, and TN were set as positive and that of the residual H_2O_2 was set as negative. Eq. (8.9) was established:

$$Z = Z_1 + Z_2 + Z_3 - Z_4 \tag{8.9}$$

where, Z_1, Z_2, Z_3, and Z_4 are the indices of TOC, TP, TN, and H_2O_2-res, respectively (Xiao et al., 2012). Based on regression model (Eq. 8.10), the optimal values of solution were obtained as $\times$ (pH value) = 11, y (H_2O_2: MLSS ratio) = 0.2. This result showed the optimized operational conditions of the MW/H_2O_2 advanced oxidation process for sludge pretreatment as follows: the pH value of the sludge was first adjusted to 11.0, then the sludge was heated to 80 C with microwave irradiation, and H_2O_2 was dosed at a H_2O_2: /MLSS ratio of 0.2, and finally the sludge was continuously heated to 100 C with microwave irradiation.

$$z\ H_2O_{2-res} = P_1 + P_2x + P_3x^2 + P_4x^3 + P_5x^4 + P_6x^5 + P_7 \ln y + P_8 \ln y^2 + P_9 \ln y^3 \tag{8.10}$$

To explain the regression model, an analysis of the H_2O_2 residual concentration (H_2O_2-res) was selected as an example. As shown in (Eq. 8.10), H_2O_{2-res} as the dependent variable (z) was fitted to the pH value and H_2O_2: MLSS ratio as independent variables x and y, respectively, and the values of $P_1 - P_9$ were $-8.4\,E + 0.5$, $3.7\,E + 0.5$, $-5.77\,E + 0.4$, $3.29\,E + 0.3$, and $1.99\,E - 0.2$, respectively.

Obviously, the z value (e.g. H_2O_2-res, TOC, and IC) can be predicted for non-test point combinations of the independent variables (x, y) within their defined ranges by Eq. (8.10).

In order to investigate the accuracy of the regression model, several combinations of independent variables (x, y) within the ranges were selected for validation tests, and the calculated values (z) were compared with the experimental values. The average of the ratios of the calculated values to the experimental values was 0.9946, indicating the high accuracy of the regression model. As shown in Fig. 8.7, the H_2O_{2-res} concentration decreased with the increased pH value and the decreased H_2O_2: MLSS ratio. However, the H_2O_2: MLSS ratio had more significant impact on the H_2O_{2-res} concentration. As shown in Fig. 8.7, the results clearly showed that all concentrations of TOC, NH_4^+-N, PO_4^{3-}-P, TN, and TP increased with increases in both the pH value and H_2O_2: MLSS ratio. At the same H_2O_2 dosage, a high pH value improved the release efficiencies of organic matter, nitrogen, and phosphorus. For the same pH value, concentrations of TOC, NH_4^+-N, PO_4^{3-}-P, TN, and TP were increased with increasing pH and H_2O_2: MLSS ratio.

Yu et al. (2010) found that, organic matter was increased with the increase of H_2O_2 dosage at the same temperature and then decreased as further increase of H_2O_2 dosage. Therefore, it was clear that optimization of H_2O_2 dosage is necessary for MW/H_2O_2 process. MW/H_2O_2 process was developed by using novel H_2O_2 dosing strategy.

The reasons of a more oxidative effect of H_2O_2 under MW/H_2O_2 and alkaline pH include prevention of catalase activity (H_2O_2: H_2O_2 oxidoreductase) over 40 C and alkaline conditions. Catalase presented in all aerobic living cells is able to breakdown H_2O_2 in water and oxygen; thus, it plays the protective role for cells against reactive oxygen species damages. Maximum catalase activity occurred at pH 7; the catalase active was suddenly decreased with increase in both pH and temperature, resulting in little decomposition of H_2O_2 by catalase (Fig. 8.8). H_2O_2 is unstable under alkaline pH and therefore, produces hydroxyl radicals ($^\bullet$OH) and superoxide ions ($O_2^{\bullet-}$) that are more effective on organic matter decomposition than H_2O_2 itself. Additionally, if the temperature is in the range of 60 C$-$80 C, the $^\bullet$OH production from H_2O_2 decomposition will also be increased.

Although some studies have focused on improving the lysis efficiency of sludge pretreated by microwave and combined processes, few studies are centered on decreasing the cost of microwave and combined processes for sludge pretreatment, such as saving dosage of the chemical by optimizing these processes. A new H_2O_2 dosing strategy for sludge pretreatment by MW/H_2O_2 was developed considering pretreatment sensitive catalase.

Xiao et al. (2012) observe that the effects of oxidation and disintegration caused by H_2O_2 and/or $^\bullet$OH have been increased at pH 11 by MW/H_2O_2. In this condition, cell wall and membrane reactions with radicals occurred in a variety of ways. Saponification between the alkali and fatty acids of cell membranes can change the permeability of cell membranes and damage the cell wall or structure of the membranes. In a summary, the

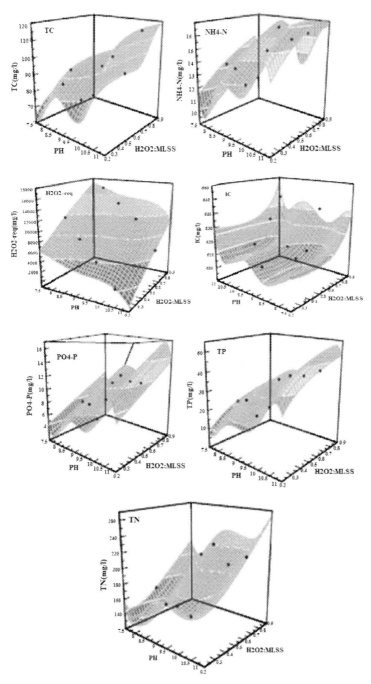

FIGURE 8.7 Changes in calculated H_2O_2, IC, TOC, NH_4^+-N, TN, PO_4^{3-}-P, and TP concentrations (z) with pH value (x) and H_2O_2:MLSS ratio (y) based on the regression model. *Adapted from Xiao, Q., Yan, H., Wei, Y. Wang, Y., Zheng, F., 2012. J. Environ. Sci., 24 (12), 2060−2067. With permission.*

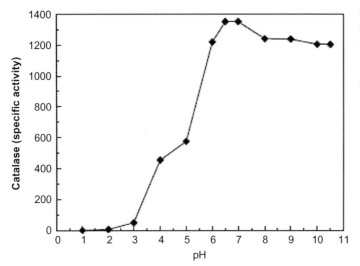

FIGURE 8.8 Catalase activity of activated sludge versus pH value. *Adapted from Xiao, Q., Yan, Q.H., Wei, Y., Wang, Y., Zeng, F., Zheng, J., 2012. Environ. Sci., 24(12), 2060–2067. With permission.*

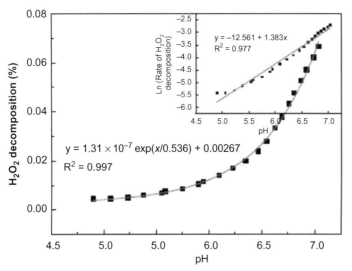

FIGURE 8.9 Rates of decomposition of H_2O_2 versus pH value. *Adapted from Xiao, Q., Yan, H., Wei, Y., Wang, Y., Zeng, F., Zheng, J., 2012. Environ. Sci., 24(12), 2060–2067, 2012. With permission.*

addition of alkali can not only promotes structure damage within the sludge and increases the release of particulate organic matter, but also promotes the decomposition of H_2O_2 into $^{\bullet}OH$ and O_2 (Fig. 8.9) (Erdincler and Vesilind, 2000).

Organic matters transform to inorganic matters, under MW/H_2O_2 process, due to the oxidative decomposition of organic substances by MW/H_2O_2 advanced oxidation process. Mineralization and dissolution are the evidences of CO_2 production and formation of soluble low molecular weight organics from particulate high molecular weight organics, respectively. Results of this study have shown that as H_2O_2: MLSS ratio increased, total organic carbon (TOC) concentration was higher than inorganic carbon (IC) concentration indicating that most of the organic compounds transformed to low molecular weight organic compounds except small parts of them, which are mineralized into CO_2.

With decreasing H_2O_2: MLSS ratio from 1 to 0.5, the percentage of H_2O_2-residues decreased from 63.5% to 41%, the SCOD decreased from 4158 to 3168 mg/L and SCOD release rate decreased from 38.2% to 29%, whereas the concentration of NH_4-N, PO_4^{3-}-P, and soluble total phosphorus (STP) all increased (e.g. NH_4^+-N, increased from 8.76 to 27.25 mg/L, and STP increased from 26.74 to 43.49 mg/L). When the H_2O_2: MLSS ratio was optimized at 0.2 and pH was kept at 11, the release SCOD concentration and rate from the sludge were 3223 mg/L and 29.5%, respectively.

TSS, H_2O_2 concentrations, and pH are known to be the critical factors controlling sludge disintegration at mild temperatures (100 C) under MW/H_2O_2; however, the effect of pH (from acidic to basic) and the interactions among these factors have not yet been fully recognized. Moreover, the majority of workers have focused on the treatment of target sludge, but neglecting the behavior of oxidant H_2O_2 during this process. In practical engineering application of this technology, it is always designed to achieve multiobjectives for the sludge disintegration unit.

It has been predicted that SCOD can be estimated with the Eq. (8.11):

$$SCOD = 2826.26 + 1040.65X_1 + 498.01X_2 + 439.24X_3 + 192.41X_1X_2 - 87.44X_1X_3 \\ + 158.54X_2X_3 + 191.75X_1^2 - 8.51X_2^2 - 78.17X_3^2 \tag{8.11}$$

Where, TSS, pH, and H_2O_2 to sludge ratio are represented by X_1, X_2, and X_3, respectively (Veschetti et al., 2000).

Multivariate analysis to investigate sludge disintegration is aimed:

- to investigate the effects of pH, H_2O_2 dosage and interactions among these factors during sludge disintegration and,
- to construct analyzing of the model to predict sludge solubilization and H_2O_2 usage; thus, providing a simple way to estimate optimal values for multiple sludge pretreatment objectives.

8.3.7 Effect of pH on Anaerobic Sludge Pretreatment

The pH of the MW/H_2O_2-pretreated sludge has been found to be in the range of 9.5 to 10. The increase in pH after pretreatment was due to the formation of ammonia (produced by the degradation of protein) (Luo et al., 2012). A similar observation was found by Rani et al. (2013). This highly alkaline nature of pretreated sludge makes it unattractive towards subsequent anaerobic digestion. Healthy anaerobic digestion happens in the pH range of 6.8 to 7.5 (Rani et al., 2013; Liu et al., 2008). In addition, the performance of H_2O_2 was observed to be effective at pH 5 and a final pH of 7 was attained after pretreatment. In order to make this alkaline sludge amenable for anaerobic digestion, it has to be neutralized with acid. This leads to an idea of combining acidic pH with MW/H_2O_2, with an interest focused towards the final pH of sludge.

Hong et al. (2012, 2015) have reported that $^\bullet OH$ radicals of H_2O_2 performed well under acidic pH, and as a result, the efficiency of H_2O_2 has been improved for a further disintegration process. Keeping this in mind, it was planned to maintain an acidic condition in the medium to enhance the disintegration efficiency of MW/H$_2$O$_2$ pretreatment. It is evident that the MW/H$_2$O$_2$ pretreatment efficiency was increased by decreasing the pH of the sludge. The soluble organics and soluble COD release were found to be higher at pH 2 and 3. However, the final pH of the pretreated sludge at this condition was found to be 3.12 and 4.18. This makes the process less attractive as it demands neutralization of the sludge for subsequent aerobic or anaerobic biodegradation. Rani et al. (2014) reported that the efficient degradation of organic matter was possible at the optimum pH of 6.3–7.8, whereas at lower and higher pH conditions, the methanogenic activity in anaerobic digestion will be slow and the biogas production is decreased.

It was found that the combined microwave/H$_2$O$_2$ treatment of sewage sludge followed by acid hydrolysis treatment (a two-stage process) resulted in significantly greater yields of soluble phosphates and ammonia; in which up to 61% of the total phosphorus was solubilized and 36% of the total Kjeldahl nitrogen (TKN) was solubilized as ammonia (Liao et al., 2006).

The dairy industry is one of the largest sources of the food processing industries that produces a huge amount of waste activated sludge with diverse characteristics. The main advantage of MW pretreatment is the effective disintegration of sludge biomass, which further enhances the biogas production efficiently. The main drawback of this pretreatment is high energy consumption. MW/H$_2$O$_2$ pretreatment was combined with the acid to maintain the sludge pH during disintegration in order to make it more amenable for subsequent anaerobic digestion. Thus, the effect of MW pretreatment using low specific energy to disintegrate the sludge particles has been investigated by applying microwave radiation and oxidation of organic molecules by H$_2$O$_2$ at different pH.

Kim et al. (2013) observed that the soluble fraction obtained under mild pretreatment conditions (pH = 9 and ultrasonication for 5 min) achieved 91 % methane yield of soluble fraction, while only 61% methane yield of soluble fraction was obtained at of pH = 13 and ultrasonication for 60 min.

Eswari et al. (2016) conducted a study on sludge anaerobic biodegradability by combined MW/H$_2$O$_2$ pretreatment in acidic conditions. They found that the application of MW/H$_2$O$_2$ in acidic condition was led to decreasing specific energy consumption.

8.3.8 Fate of Organic Matters

One of the major factors affecting anaerobic sludge pretreatment efficiency by MW/H$_2$O$_2$ is study of the fate of organic matters. The activated sludge can be classified into four parts as residual solids, particles in suspension (after 30 min settling), soluble substances, and CO$_2$ depending on the achievement of disintegration, solubilizaton, and mineralization under MW/H$_2$O$_2$. Study of Wang et al. (2009) shows the changes of these four parts using their distribution in TCOD (Fig. 8.11).

It has been observed that nearly all TCOD contents in untreated sludge are distributed on solids that are similar to the sludge pretreated by ozonation. Along with the increasing H$_2$O$_2$/TCOD ratio in the activated sludge pretreated by MW/H$_2$O$_2$, the contents of TCOD on particles, soluble substances, and mineralization also increased, but TCOD distribution on solids decreased. Contents of TCOD on particles, soluble organic matters, and

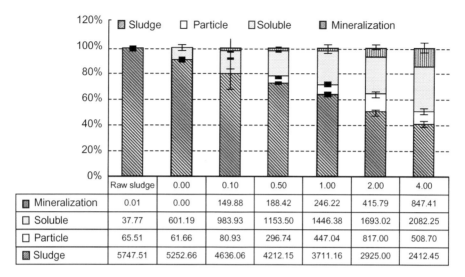

	Raw sludge	0.00	0.10	0.50	1.00	2.00	4.00
▦ Mineralization	0.01	0.00	149.88	188.42	246.22	415.79	847.41
☐ Soluble	37.77	601.19	983.93	1153.50	1446.38	1693.02	2082.25
☐ Particle	65.51	61.66	80.93	296.74	447.04	817.00	508.70
▨ Sludge	5747.51	5252.66	4636.06	4212.15	3711.16	2925.00	2412.45

FIGURE 8.10 The fate of organic matters in the WAS pretreated by the AOP of MW/H_2O_2 at different. *Adapted from Wang, Y., Wei, Y., Liu, J., 2009. J. Hazard. Mater., 169(1), 680–684. With permission.*

mineralization were found to increase from 1.04, 10.16, and 0 in waste activated sludge pretreated only by MW without H_2O_2, respectively, to 8.69%, 35.59%, and 14.48% of four $H_2O_2/TCOD$ ratios in the sludge pretreated by MW/H_2O_2. The contents of TCOD on residual solids were correspondingly decreased from 88.78% to 41.23%. With increasing $H_2O_2/TCOD$ ratio between 0.1 and 1, the amount of soluble organic matters and particles significantly increased, but the amount of mineralization was relatively small, less than 5%. However, the rate of mineralization increased faster than those of solubilization at H_2O_2 dosage (ratios of 2 $H_2O_2/TCOD$ and 4 $H_2O_2/TCOD$ ratio were 0.12 times to that at 0.1 of H_2O_2 /TCOD ratio was as high 5.65 times of that at 0.1 $H_2O_2/TCOD$ ratio) (Fig. 8.10) (Eskicioglu et al., 2008).

8.3.9 Morphological Changes of Sludge

Floc characteristics (morphology) is one of the tests monitored along the digestion test. Digested sludge shows a greater flow in the steady state and less elasticity in the linear viscoelastic region. This phenomenon is due to weak colloidal forces in the internal structure or a less rigid structure. Activated sludge and digested sludge are mainly composed of water, organic matters, microbial cells, and extracellular polymeric substances, which tend to aggregate, forming flocs. The characteristics of the adsorption ability, hydrophilicity/hydrophobicity, and content of the main components, such as proteins and polysaccharides of EPS critically affect the properties of microbial aggregates, e.g. mass transfer, surface characteristics, and stability.

By MW/H_2O_2 pretreatment, the color of sludge has been changed from original dark-brown to pale and the bulk solution has been changed to green after a 5 min treatment, gradually with the increased dosage of H_2O_2 (Fig. 8.11A). Compared to the agglomerated

FIGURE 8.11   Morphological changes of sludge: (A) Sludge suspension after pretreatment, (B) Raw sludge, (C) Sludge treated by microwave, and (D) Sludge treated by microwave/H_2O_2 at 4 g H_2O_2/g TCOD. *Adapted from Wang, Y., Wei, Y., Liu, J., 2009. J. Hazard. Mater., 169 (1), 680–684. With permission.*

floc in raw sludge shown in Fig. 8.11B, little morphological changes of sludge flocs occurred in the sludge pretreated by MW in absence of H_2O_2 (Fig. 8.11C), but the cells in the sludge were completely destroyed by MW/H_2O_2, with the breakage of the cell membrane and no whole cell was observed (Fig. 8.11D). Although the cell in the sludge pretreated by MW/H_2O_2 was totally broken, the percentage of organic matter released into the supernatant was not over 40%. It is regarded that MW heating process could break down particles with or without H_2O_2. However, it is also noted that sludge flocs in the sludge pretreated only by MW without H_2O_2 were not completely disintegrated because of short time heating below 100 C. Obviously, the MW/H_2O_2 is effective to disintegrate sludge and release cytoplasm into bulk solution through breaking the cell membrane (Wang et al., 2009).

8.3.10 Improvement Of EPS Extraction From Anaerobic Sludge

The efficacies of extracting extracellular polymeric substances (EPS) from aerobic, acidogenic, and methanogenic sludges using ethylendiaminetetraacetic acid (EDTA), cation exchange resin and formaldehyde under various conditions have been compared. Results show that formaldehye plus NaOH was the most effective in extracting EPS for all sludges. All EPS were mainly composed of carbohydrate, protein, and humic substance, plus small quantities of uronic acid and DNA. Carbohydrate is predominant in the acidgenic sludge. Nowadays, optimization of extraction procedures and/or development of a more effective extraction method is necessary (Fang and Liu, 2002).

Base on the study of Lio et al. (2007) MW/H_2O_2 pretreatment is a suitable method for increasing the amount of ESP extraction by the formaldehyde-NaOH extraction method from anaerobic sludge.

According to the analysis of EPS (Fig. 8.12A), large amounts of soluble EPS existed in the digested sludge of reactors that received the pretreatment compared with that of the control, whereas the concentrations of bound EPS were relatively the same for the different reactors. In the soluble EPS, the amounts of proteins and polysaccharides were obviously higher than those of the digested sludge of the control reactor (Fig. 8.12B). The concentrations of proteins and polysaccharides in the bound EPS did not present obvious differences (Fig. 8.12C). Moreover, it was found that pretreatment partly changed the proportion of proteins and polysaccharides in EPS (Fig. 8.12D). The ratio of proteins to

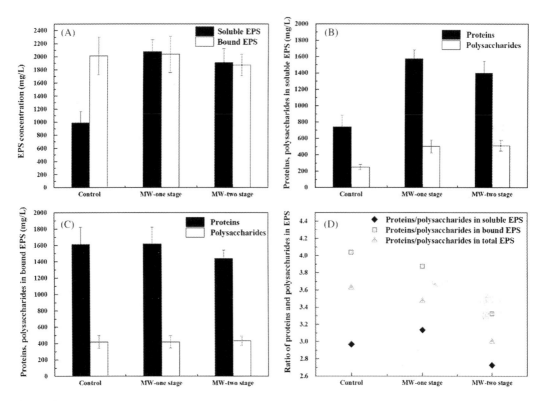

FIGURE 8.12 The concentration and components of EPS in digested sludge. ESP concentration (A), Proteins, polysaccharides in soluble ESP (B), Proteins, polysaccharides in bound ESP (C), Ratio proteins and polysaccharides in ESP (D). *Adapted from Liu, J., Yu, D., Zhang, J., Yang, M., Wang, Y., Wei, Y., Tong, J., 2016. Water Res., 98, 98—108. With permission.*

polysaccharides in bound EPS decreased when the feeding sludge underwent MW/H$_2$O$_2$ pretreatment. In addition, the ratio of proteins to polysaccharides in both soluble EPS and bound EPS presented the lowest value in the digested sludge from the MW-two stage reactor. These results imply that MW/H$_2$O$_2$ pretreatment strongly affected the concentration and components of EPS in digested sludge. Furthermore, it was observed that the molecular weight of the soluble organic matter in digested sludge was influenced by pretreatment. The pretreatment process resulted in a greater amount of soluble organic matter released at molecular weights lower than 20 kDa. In addition, the morphology of digested sludge was partly changed due to pretreatment, which was included in the supporting information. It seems that the microbial cells were less tightly embedded in sludge flocs when the pretreated sludge fed into the digesters.

8.3.11 Thermodynamic Analysis of WAS Hydrolysis

The thermal disintegration of sludge offers more advantages over other methods of sludge pretreatment. There are different thermal disintegrations, such as moist heat, dry heat, and radiation heat, which are available to disintegrate sludge (Wilson and Novak, 2009, Gianico et al., 2013). In thermal treatment, the possible solubilization cannot be achieved at a lower treatment time, whereas a prolonged treatment time leads to the formation of refractory organic compounds that limit the rate of degradation (Rani et al., 2013). In MW pretreatment, the treatment time is comparatively lower and the possible solubilization is achieved at a lower treatment time. Considering this fact, the MW pretreatment is considered to be more efficient than thermal treatment, the basis of the reaction activation energy. This can be done by using first-order kinetic modeling, which contributes to the traditional simplification of substrate biodegradation in terms of the SCOD release. The effect of MW irradiation on dairy sludge hydrolysis has been simplified to single first-order kinetics, based on the following equations (Eswari et al., 2016):

$$-dS/dT = kS \tag{8.12}$$

$$Ln\,S = -kt + b \tag{8.13}$$

where S denotes the SCOD concentration, k is the rate concentration, and b is the constant of integration.

By plotting ln S versus t, the slope and intercept could be obtained, which correspond to the values of k and b, respectively. The regression for ln S versus t can be obtained at different temperatures (Eswari et al., 2016) by the thermodynamic analysis. The activation energy was calculated from the following equation:

$$Ln = E_a/RT + \ln A \tag{8.14}$$

where A is the preexponential factor, E_a (kJ/mol) is the reaction activation energy, T (K) is the absolute temperature, and R is the gas constant (8.31 K/mol). The value of the preexponential factor has been found to be 0.498 and the reaction activation energy for SCOD release in the pretreated MW and control (thermally pretreated) has been evaluated to be 0.135 and 0.598 kJ/mol, respectively. The currently achieved activation, 0.135 kJ/mol, is lower than that achieved in other studies, where researchers achieved

20.19 kJ/mol (Luo et al., 2012) in the enzymatic treatment of sludge, which therefore suggests that MW irradiation increases the rate of chemical reaction largely by lowering the activation energy. Thus, the MW technique has been found efficient for effective sludge solubilization.

Contrary, thermal and chemical reactions from MW/H_2O_2 have some disadvantages. For example, the fraction of pretreated sludge, which is soluble but not convertible to methane, increases from 2.4% to 25.8% in sludge under severe pretreatment conditions. Due to MW or conventional heating pretreatment (at 96 C), slightly lower rate constants (k) of soluble organics degradation have been observed (Eskicioglu et al., 2006). Bougrier et al. (2007) also has reported that the soluble protein removal rate decreases from 75% for the control to only 68% due to thermal pretreatment at 190 C. The formation of refractory compounds such as melanoidines from Maillard reactions is the most possible explanation to account for these results (Pilli et al., 2015, Wilson and Novak, 2009).

8.3.12 Cost Analysis of Anaerobic Sludge Pretreatment

As low energy input is desirable, the calculation of specific energy (SE) becomes inevitable for evaluating the cost of pretreatment. SE was completed, based on Eq. (8.15) (Ebenezer et al., 2015a,b):

$$\text{Specific energy Kj/Kg TS} = \frac{Power\ of\ microwave\ KW\ \times Irradiation\ S}{Sample\ volume\ L\ \times Total\ solids\ Kg/L} \quad (8.15)$$

Theoretical energy balances and cost assessment have been performed for pilot scale reactors with the obtained experimental data to assess the potency of pretreatment methods. The estimation has been performed for 1 L of sludge based on the previous study. The theoretical calculation for methane was carried out based on the COD consumed with the following equation:

$$\begin{aligned} CH_4\text{production } m^3 &= \text{COD consumed kg} \times 0.35\ m^3/\text{kg COD} \\ &\quad \times \text{Biodegradability 0.28} \end{aligned} \quad (8.16)$$

Methane production assesses the level of biodegradability of a substrate (COD) consumed. Output energy (E_o) is calculated based on methane production and estimated by following the equation (Passos and Ferrer, 2014):

$$E_o = \quad CH_4 \quad V \quad (8.17)$$

where E_o is the output energy (KJ/day), CH_4 is the methane production rate ($m^3\ CH_4/m^3$ day),   is the lower heating energy value of methane (KJ/m^3 methane), V is the useful volume (m^3), and   is the energy conversion efficiency (90%).

To calculate the energy applied, the energy spent for sludge pretreatment by MW is arrived at by using COD solubilization as an index based on the following equation:

$$Q_{MW}\ kWh = \frac{P_{MW} \times T}{V \times SS} \quad (8.18)$$

where Q_{MW} is the energy spent for sludge pretreatment in kilowatt-hour, P_{MW} is the power of microwave (31 kWh), T is the time, and V is the volume of pilot scale plant (L). Energy spent for pumping and stirring in anaerobic digestion.

Q_p was calculated as per the calculation detailed by (Metcalf et al., 2003). The energy required in upsurging the temperature is calculated by following equation:

$$Q_T = P \times 100/Ps \times T_d - T_s \times 1/2 \times C_p \tag{8.19}$$

where p is the fresh dry sludge solids added per day (kg), Ps is the percentage dry solids in the fresh sludge (%), T_d and T_s are the temperatures of digester and sludge, respectively (C), and Cp is the specific heat constant (4200 J/ C). The heat required to make up for losses at the top, walls, and bottom (Q_L) is estimated by the following equation:

$$Q_L = CA \times \quad T \tag{8.20}$$

where C is the coefficient of heat flow ($J/m^2/h/$ C), A is the surface area (m^2), and   is the difference between tank temperature and temperature of outside material.

The total input energy is subtracted from the output energy to confirm the energy balance (E) and assay the net energy production, based on the following equation (Passos and Ferrer, 2014):

$$E = E_o - E_i, heat + E_i, electricity \tag{8.21}$$

Where   E is the energy balance, E_o is the output energy, E_i, heat is the input heat energy, and E_i, electricity is the input electricity.

One of the benefits associated with the anaerobic digestion of pretreated sludge is the production of fuel gas in the form of methane. Based on biodegradability (0.28), 38.3 m^3 of methane is obtained per liter of sludge. The COD solubilization of all pretreated sludge is the same; hence, the methane production also remains the same (381 kWh). The energy cost of methane is calculated as 0.187 € per kWh of energy, and it remains the same for all types of pretreatment. Similar to methane, the cost savings can be obtained by reducing the SS to be disposed. This could occur because anaerobic biodegradability remains the same for the substrate with a similar amount of COD solubilization (Saha et al., 2011).

The energy consumed for sludge pumping and mixing in the anaerobic digester is calculated to be 96 kWh for all treatment processes. Energy required to raise the temperature of the sludge to the digester temperature and energy required to make up losses through top, walls, and bottom of the digester is calculated to be 25 and 0.674 kWh, respectively. Gravity thickener is used for thickening of sludge. Energy consumed for this process is assessed to be 13.6 kWh. The energy spent for biogas purification and compression is 1.4 and 1.1 kWh, respectively. Therefore, the total input energy is estimated to be 749.8, 545.7, and 341.7 kWh for MW, MW/H_2O_2, and MW/H_2O_2 /acid pretreatment. The energy consumed for sludge pumping and mixing in the anaerobic digester is calculated to be 95.4 kWh for all treatment processes. The cost of H_2O_2 and H_2SO_4 used for the combined disintegration is also considered. In an economic analysis, the capital investment for sludge pretreatment is not considered. The results suggested that the MW/ H_2O_2/acid is the most economically feasible method with a net profit of 59.90 €/t of sludge. Pilot scale experiments with these optimized conditions and more realistic energy consideration are strongly recommended in the future.

8.3.13 Impact of MW Specific Energy on Anaerobic Sludge Pretreatment

The specific energy is considered as an essential factor for determining the economic feasibility and energy consumption of MW pretreatment.

It has been observed that the changes of soluble organic release (SCOD, protein, and carbohydrate) could be distinguished into two phases, i.e., a rapid release phase and a slower degradation phase. Phase 1 starts from a specific energy input of 0 to 18,600 kJ/kg TS while phase 2 continues from a specific energy input of 18,600 to 30,000 kJ/kg TS.

The specific energy increase from 0 to 18,600 kJ/kg TS led to a rapid release of soluble organics (SCOD, protein, and carbohydrate in phase 1 (Eswari et al., 2016).

As a result of cell wall disintegration, the extracellular and intracellular cytoplasmic materials both start to be released into the soluble phase (Rani et al., 2013). This could be the reason for an increment in the soluble organics in phase 1. The amount of changes of the concentrations of SCOD, protein, and carbohydrate are 7.4, 0.7, and 0.05 g/L, respectively. Further increasing the specific energy input beyond 18,600 kJ/kg TS not only decreases the solubilization percentage but also increases the treatment cost. In phase 2, degradation of soluble organics is decreased due to the loss of organics by evaporation (Ebenezer et al., 2015a,b). It is confirmed that high specific energy does not lead to increasing solubilization (Park et al., 2010).

In contrast to the pattern of soluble organic release, the SS removal exhibits a stabilizing pattern at phase 2. During phase 1, SS removal is gradual and from the initial value of 19 g/L, it is reduced to 15 g/L. Beyond the specific energy of 18,600 kJ/kg TS, the SS reduction is relatively slower and nearly stable. This could be because of the condensation of sludge. At elevated MW energy, the evaporation of water occurs, which results in the condensation of sludge. The increase in energy input also increases the temperature of the medium or sludge, leading to a loss of organics because of evaporation (Eswari et al., 2016). Many researchers have employed a cover to avoid evaporation loss (Eskicioglu et al., 2007, Rani et al., 2013).

On increasing the specific energy input beyond 18,600 kJ/Kg TS, the evaporation loss has been observed to be 20%–25%, which is relatively higher than that observed in the optimal specific energy input while only a slight increment in SS reduction was observed. The currently achieved SS reduction at the optimal specific energy of 18,600 kJ/kg TS is found to be 21%. By almost doubling the specific energy input to 30,000 kJ/kg TS, only a relatively slight increment in SS reduction (22.7%) is obtained. Therefore, it can be concluded that an increment in specific energy beyond 18,600 kJ/kg TS is not economically feasible. On the basis of these outcomes, it can be concluded that the specific energy input of 18,600 kJ/kg TS is considered optimum, and this optimal specific energy is employed for further studies (Eswari et al., 2016).

8.3.14 Release of Heavy Metals

The reduction of heavy metals in sewage sludge can be achieved by source control of discharge to sewer systems by removing metals from sludge. In source control, the major difficulty is identification of sources. Moreover, even with complete elimination of toxic metals from all industrial discharges to sewers, the problem remains because of metal content of domestic wastewaters. Several chemical methods for solubilization of heavy metals for sewage sludge have been suggested (Tyagi et al., 1988).

It is observed that MW/H_2O_2 is able to release metals from anaerobic sludge (Wang et al., 2014).

Wang et al. (2014) studied the release behavior of heavy metals during treatment of dredged sediment by MW/H_2O_2 (using a series of 150 mL round bottom flasks containing 5 ± 0.0005 g sediment and 50 mL ultrapure water) (Fig. 8.13).

Estimated recovery percent of heavy metals is given by Eq. (8.22) (Wang et al., 2014):

$$Recovery \% = \frac{F1 + F2 + F3 + F4}{TCHM} \qquad (8.22)$$

where F1, F2, F3, and F4 refer to the concentration of heavy metals in acid soluble/ exchangeable, reducible, oxidizable, and residual form, respectively. TCHM is the total concentration of heavy metals of raw/extracted sediment.

Under the four treatment methods (control, MW, H_2O_2, and MW/H_2O_2), amounts of recovery percent of four metals of Cu, Cd, Zn, and Pb have been calculated. A comparison between these groups indicates that the presence of H_2O_2 could remarkably enhance the release efficiency for the heavy metals.

H_2O_2 concentration higher than 0.2 M does not affect recovery of heavy metals. This is due to consumption of $^\bullet OH$ radical by H_2O_2 and generation of $HO_2^\bullet$ radical, which is less reactive. This reaction occurred follows: (Wang et al., 2014)

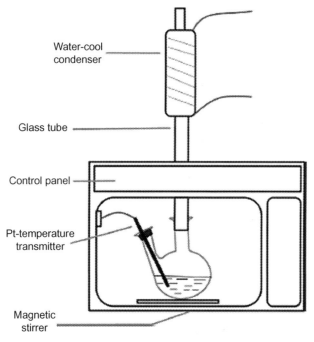

Water-cool condenser

Glass tube

Control panel

Pt-temperature transmitter

Magnetic stirrer

FIGURE 8.13 Schematic diagram of a reactor in microwave reactor. *Adapted from Wang, L., Yuan, X., Zhong, H., Wang, H., Wu, Z., Chen, X., Zeng, G., 2014. Chem. Eng. J. 258, 334–340. With permission.*

$$H_2O_2 + \ OH \rightarrow HO_2 + HO_2 \tag{8.23}$$

The mobility of heavy metals generally reduces in the following order:

$$F1 > F2 > F3 > F4 \tag{8.24}$$

Heavy metals in the form of F1 and F2 show high bioavailability and direct eco-toxicity. By increasing the oxidation—reduction potential (ORP) and decreasing the final pH of sediment slurry, MW/H_2O_2 enhances the release of heavy metals from sediment.

8.3.15 Effects of MW/H_2O_2 Pretreatment on Anaerobic Sludge Rheology

Rheology is the study of flow and deformation of materials under applied forces that are usually measured using a rheometer. The measurement of rheological properties is applicable to all materials (fluids, solids, and semi-solids such as sludge).

Liu et al. (2016a) have investigated the changes in the rheological characteristics of sludge under MW/H_2O_2 pretreatment during enhanced anaerobic digestion with microwave/H_2O_2 pretreatment. For AD, the rheological properties of sludge play key roles in processes such as mixing, pumping, and recirculating (Liu et al., 2016b).

The rheograms of raw sludge and pretreated sludge are presented in Fig. 8.14A. The shear stress increased nonlinearly with the increase of the shear rate, which suggested that both the raw sludge and pretreated sludge presented non-Newtonian flow. At a low shear rate, the apparent viscosity decreased with the increasing shear rate, demonstrating a shear-thinning property. The rheological properties of the sludge within the nonlinear viscoelastic region were influenced by the MW/H_2O_2 pretreatment. In particular, the apparent viscosity decreased obviously, indicating the improvement of sludge flowability. Moreover, both raw sludge and pretreated sludge presented thixotropic behavior for the formation of the hysteresis loop. Shear stresses of up-curves were higher than those of

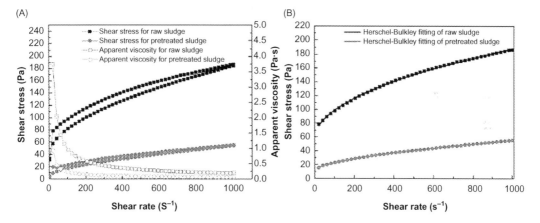

FIGURE 8.14 Flow cure of raw sludge and pretreated sludge. Rheograms of raw sludge and pretreated sludge (A), Herschel-Bulkely model of raw sludge and pretreated sludge (B). *Adapted from Liu, J., Yu, D., Zhang, J., Yang, M., Wang, Y., Wei, Y., Tong, J., 2016. Water Res. 98, 98—108. With permission.*

down-curves at the same shear rate. MW/H_2O_2 pretreatment resulted in the decrease of the hysteresis area from 11,121.57 Pa/s to 3567.94 Pa/s. The hysteresis area is useful to evaluate sludge thixotropy (Yen et al., 2002) and decrease in the area indicated that the thixotropy of sludge weakened. The thixotropic behavior of the sludge suggested that the kinetic processes of both breakdown and build-up exist when the flow is applied. The hysteresis area became small due to pretreatment, indicating that the kinetic processes could rapidly reach a steady state. In general, activated sludge shows high levels of thixotropy due to its internal structures (Liu et al., 2016a). The colloidal forces among particles tend to rebuild the structures (Eshtiaghi et al., 2013). Therefore, the decrease of sludge thixotropy with MW/H_2O_2 pretreatment is possibly the result of the weakness of colloidal forces among particles. Furthermore, the Herschel-Bulkley model was used to fit the flow curve, and a strong agreement ($R^2 > 0.999$) was obtained (Fig. 8.14B).

According to the Herschel-Bulkley model, the yield stress ($_y$) was taken into account in the equation, which is regarded to be an indication of the specific value of stress exerted on the sludge when it begins to flow. With MW/H_2O_2 pretreatment, the yield stress of sludge obviously decreased (Table 8.4). The consistency index (k) and flow behavior index (n) in the Herschel-Bulkley model are two other important parameters. The value of k decreased from 4.90 to 0.81, and the value of n increased from 0.48 to 0.58. The consistency index (k) is a measure of average firmness. The decrease of k is consistent with the decrease of apparent viscosity. The flow behavior index (n) is related to the type of fluids. Pseudo-plastic fluids present values of n lower than one. The increase of n revealed that the rheological properties of the pseudo-plastic fluid became weaker, suggesting that the internal structures of the sludge were broken. This is consistent with the release of soluble organic matter, such as proteins and polysaccharides, with treatment, which can be an indication of the degree of sludge flocs rupture (Liu et al., 2016).

One of the major advantages of MW/H_2O_2 is in increasing sludge flowability and weakness of sludge solid-like properties, which were beneficial for sludge pumping or mixing processes. The apparent viscosity of digested sludge in reactors combined with pretreatment was lower than the control reactor (Liu et al., 2016b). Anaerobic digestion is able to convert large sludge flocs into smaller ones, and cause the disintegration of macromolecular organic compounds, such as proteins and polysaccharides into micro-molecular organics, such as VFAs, which can be further used by methanogens for biogas production.

TABLE 8.4   Herschel-Bulkley Parameters of Different Sludge

Parameters	Raw Sludge	Pretreated Sludge	Control[a]	MW-one stage[b]	MW-Two Stage[c]
TS (w/w%)	7.63 ± 0.01	7.88 ± 0.01	5.09 ± 0.01	5.10 ± 0.01	5.15 ± 0.02
$_y$ (Pa)	54.51	11.08	10.48	5.10 ± 0.01	5.56
k (Pa s^n)	4.90	0.81	0.61	0.25	0.20
n	0.48	0.58	0.65	0.72	0.75

[a]Digested sludge with dilution from the control reactor.
[b]Digested sludge with dilution from the MW-one stage reactor.
[c]Digested sludge with dilution from the MW-two stage reactor.
Adapted from Liu, J., Yu, D., Zhang, J., Yang, M., Wang, Y., Wei, Y., Tong, J., 2016. Water Res. 98, 98−108. With permission.

It is well known that the rheology of sludge, i.e., apparent viscosity, is associated with the sludge solid concentration. The low apparent viscosity was perhaps due to the low solid concentration of digested sludge or may be the result of changes of digested sludge microstructures.

Microwave/H_2O_2 system with changes of digested sludge microstructures and morphology has been found as an environmental friendly method for decreasing and stabilization of environmental pollutions. High efficiency of this process is due to thermal effects and hydroxyl radical production. Optimizing H_2O_2:pollutant factors such as energy intensity, pH, pollutant concentration (NH_4^+-N, PO_4^{-3}-P, TP, TC, SCOD...), H_2O_2 concentration, H_2O_2 residual, radical scavengers' effects, and others is an engineering technique for increasing MW/H_2O_2 efficiency. This technique is not only for deceasing sludge viscosity, energy, and oxidant consumption, but also for breaking the cell walls, increasing biodegradability of wastewater and anaerobic sludge, methane and VFA production.

References

Alishah, F., Mahmood, Q., Maroofshah, M., Pervez, A., Ahmad Asad, S., 2014. Microbial ecology of anaerobic digesters: the key players of anaerobiosis. Sci. World J. 2014.

Ámingos, M.P.D., 1992. Superheating effects associated with microwave dielectric heating. J. Chem. Soc., Chem. Commun. 674–677.

Appels, L., Baeyens, J., Degreve, J., Dewil, R., 2008. Principles and potential of the anaerobic digestion of waste-activated sludge. Prog. Energy Combustion Sci. 34, 755–781.

Asgari, G., Seidmohammadi, A., Chavoshani, A., Rahmani, A.R., 2013. Microwave/H_2O_2 efficiency in pentachlorophenol removal from aqueous solutions. J. Res. Health Sci. 14, 36–39.

Asgari, G., Seidmohammadi, A., Chavoshani, A., 2014. Pentachlorophenol removal from aqueous solutions by microwave/persulfate and microwave/H_2O_2: A comparative kinetic study. J. Environ. Health Sci. Engg. 12, 1–7.

Banik, S., Bandyopadhyay, S., Ganguly, S., 2003. Bioeffects of microwave—A brief review. Bioresour. Technol. 87, 155–159.

Bengtsson, S., Hallqist, J., Werker, A., Welander, T., 2008. Acidogenic fermentation of industrial wastewaters: effects of chemostat retention time and pH on volatile fatty acids production. Biochem. Eng. J 40, 492–499.

Cao, Y., Pawłowski, A., 2013. Life cycle assessment of two emerging sewage sludge-to-energy systems: evaluating energy and greenhouse gas emissions implications. Bioresour. Technol. 127, 81–91.

Chavoshani, A., 2012. Study of advanced oxidation process effectivence: Microwave energy coupled with persulfate for pentachloro phenole removal in aqueous M. Sc. Dissertation. Hamedan University of Medical Sciences, Hamedan, Iran.

Chavoshani, A., Rostami, A., Golzari, F., Gholinia, A., 2016. The effect of H_2O_2 interference in chemical oxygen demand removal during advanced oxidation processes. Iran. J. Health Safety Environ. 3, 565–569.

Clark, D.E., Sutton, W.H., 1996. Microwave processing of materials. Ann. Rev. Mater. Sci. 26, 299–331.

Clarke, B.O., Smith, S.R., 2011. Review of 'emerging' organic contaminants in biosolids and assessment of international research priorities for the agricultural use of biosolids. Environ. Int. 37, 226–247.

De la Hoz, A., Diaz-Ortiz, Moreno, A., 2005. Microwaves in organic synthesis. Thermal and non-thermal microwave effects. Chem. Soc. Rev. 34, 164–178.

Devi, T.P., Ebenzer, A.V., Kumar, S.A., Kaliappan, S., Banu, J.R., 2014. Effect of deflocculation on the efficiency of disperser induced dairy waste activated sludge disintegration and treatment cost. Bioresour. Technol. 167, 151–158.

Ebenezer, A.V., Arulazhagan, P., Kumar, S.A., Yeom, I.-T., Banu, J.R., 2015a. Effect of deflocculation on the efficiency of low-energy microwave pretreatment and anaerobic biodegradation of waste activated sludge. Appl. Energy 145, 104–110.

Ebenezer, A.V., Kaliappan, S., Kumar, S.A., Yeom, I.-T., Banu, J.R., 2015b. Influence of deflocculation on microwave disintegration and anaerobic biodegradability of waste activated sludge. Bioresour. Technol. 185, 194–201.

Erdincler, A., Vesilind, P., 2000. Effect of sludge cell disruption on compactibility of biological sludges. Water Sci. Technol. 42 (9), 119–126.

Eshtiaghi, N., Markis, F., Yap, S.D., Baudez, J.-C., Slatter, P., 2013. Rheological characterisation of municipal sludge: a review. Water Res. 47, 5493–5510.

Eskicioglu, C., Kennedy, K.J., Droste, R.L., 2006. Characterization of soluble organic matter of waste activated sludge before and after thermal pretreatment. Water Res. 40, 3725–3736.

Eskicioglu, C., Kennedy, K.J., Droste, R.L., 2007. Enhancement of batch waste activated sludge digestion by microwave pretreatment. Water Environ. Res. 79, 2304–2317.

Eskicioglu, C., Prorot, A., Marin, J., Droste, R.L., Kennedy, K.J., 2008. Synergetic pretreatment of sewage sludge by microwave irradiation in presence of H_2O_2 for enhanced anaerobic digestion. Water Res. 42, 4674–4682.

Eswari, P., Kavitha, S., Kaliappan, S., Yeom, I.-T., Banu, J.R., 2016. Enhancement of sludge anaerobic biodegradability by combined microwave-H2. Environ. Sci. Pollut. Res. 23, 13467–13479.

Fang, H.H., Liu, H., 2002. Effect of pH on hydrogen production from glucose by a mixed culture. Bioresour. Technol. 82 (1), 87–93.

Gianico, A., Braguglia, C., Cesarini, R., Mininni, G., 2013. Reduced temperature hydrolysis at 134 C before thermophilic anaerobic digestion of waste activated sludge at increasing organic load. Bioresour. Technol. 143, 96–103.

Han, D.-H., Cha, S.-Y., Yang, H.-Y., 2004. Improvement of oxidative decomposition of aqueous phenol by microwave irradiation in UV/H2O2 process and kinetic study. Water Res. 38, 2782–2790.

Haque, K.E., 1999. Microwave energy for mineral treatment processes-a brief review. Int. J. Mineral Process. 57, 1–24.

Herney-Ramirez, J., Vicente, M.A., Madeira, L.M., 2010. Heterogeneous photo-Fenton oxidation with pillared clay-based catalysts for wastewater treatment: a review. Appl. Catal. B: Environ. 98, 10–26.

Hong, J., Yuan, N., Wang, Y., Qi, S., 2012. Efficient degradation of Rhodamine B in microwave-H_2O_2 system at alkaline pH. Chem. Eng. J. 191, 364–368.

Hong, J., Han, B., Yuan, N., Gu, J., 2015. The roles of active species in photo-decomposition of organic compounds by microwave powered electrodeless discharge lamps. J. Environ. Sci. 33, 60–68.

Horikoshi, S., Osawa, A., Abe, M., Serpone, N., 2011. On the generation of hot-spots by microwave electric and magnetic fields and their impact on a microwave-assisted heterogeneous reaction in the presence of metallic Pd nanoparticles on an activated carbon support. J. Phys. Chem. C 115, 23030–23035.

Jacob, J., Chia, L., Boey, F., 1995. Thermal and non-thermal interaction of microwave radiation with materials. J. Mater. Sci. 30, 5321–5327.

Jiang, G., Yuan, Z., 2013. Synergistic inactivation of anaerobic wastewater biofilm by free nitrous acid and hydrogen peroxide. J. Hazard. mater. 250, 91–98.

Kappe, C.O., 2004. Controlled microwave heating in modern organic synthesis. Angew. Chem. Int. Ed. 43, 6250–6284.

Kato, D.M., Elia, N., Flythe, M., Lynn, B.C., 2014. Pretreatment of lignocellulosic biomass using Fenton chemistry. Bioresour. Technol. 162, 273–278.

Kesari, K.K., Behari, J., Kumar, S., 2010. Mutagenic response of 2.45 GHz radiation exposure on rat brain. Int. J. Radiat. Biol. 86, 334–343.

Kim, D.-H., Cho, S.-K., Lee, M.-K., Kim, M.-S., 2013. Increased solubilization of excess sludge does not always result in enhanced anaerobic digestion efficiency. Biores. Technol. 143, 660–664.

Larhed, M., Hallberg, A., 2001. Microwave-assisted high-speed chemistry: a new technique in drug discovery. Drug Discov. Today 6, 406–416.

Lee, W.S., Chua, A.S.M., Yeoh, H.K., Ngoh, G.C., 2014. A review of the production and applications of waste-derived volatile fatty acids. Chem. Eng. J. 235, 83–99.

Legrini, O., Oliveros, E., Braun, A., 1993. Photochemical processes for water treatment. Chem. Rev. 93, 671–698.

Lew, A., Krutzik, P.O., Hart, M.E., Chamberlin, A.R., 2002. Increasing rates of reaction: microwave-assisted organic synthesis for combinatorial chemistry. J. Combin. Chem. 4, 95–105.

Liao, B.-Q., Kraemer, J.T., Bagley, D.M., 2006. Anaerobic membrane bioreactors: applications and research directions. Critical Rev. Environ. Sci. Technol. 36 (6), 489–530.

Lidström, P., Tierney, J., Wathey, B., Westman, J., 2001. Microwave assisted organic synthesis—a review. Tetrahedron 57, 9225–9283.

Lio, P.H., LO, K.V., Chan, W.I., Wong, W.T., 2007. Sludge reduction and volatile fatty acid recovery using microwave advanced oxidation process. J. Environ. Sci. Health Part A 42, 633–639.

Liu, J., Jia, R., Wang, Y., Wei, Y., Zhang, J., Wang, R., et al., 2015a. Does residual H_2O_2 result in inhibitory effect on enhanced anaerobic digestion of sludge pretreated by microwave-H_2O_2 pretreatment process? Environ. Sci. Pollut. Res. 1–10.

Liu, J., Jia, R., Wang, Y., Wei, Y., Zhang, J., Wang, R., et al., 2015b. Does residual H_2O_2 result in inhibitory effect on enhanced anaerobic digestion of sludge pretreated by microwave-H_2O_2 pretreatment process? Environ. Sci. Pollut. Res. 1–10.

Liu, J., Wei, Y., Li, K., Tong, J., Wang, Y., Jia, R., 2016. Microwave-acid pretreatment: a potential process for enhancing sludge dewaterability. Water Res. 90, 225–234.

Liu, J., Yu, D., Zhang, J., Yang, M., Wang, Y., Wei, Y., et al., 2016a. Rheological properties of sewage sludge during enhanced anaerobic digestion with microwave-H_2O_2 pretreatment. Water Res. 98, 98–108.

Liu, J., Wei, Y., Li, K., Tong, J., Wang, Y., Jia, R., 2016b. Microwave-acid pretreatment: a potential process for enhancing sludge dewaterability. Water Res. 90, 225–234.

Liu, X., Liu, H., Chen, J., Du, G., Chen, J., 2008. Enhancement of solubilization and acidification of waste activated sludge by pretreatment. Waste Manage. 28 (12), 2614–2622.

Lu, Q., He, Z.L., Stoffella, P.J., 2012. Land application of biosolids in the USA: a review. Appl. Environ. Soil Sci. 2012. 10.1155/2012/201462.

Luo, K., Yang, Q., Li, X.-M., Yang, G.-J., Liu, Y., Wang, D.-B., et al., 2012. Hydrolysis kinetics in anaerobic digestion of waste activated sludge enhanced by -amylase. Biochem. Eng. J. 62, 17–21.

Manara, P., Zabaniotou, A., 2012. Towards sewage sludge based biofuels via thermochemical conversion—a review. Renew. Sustain. Energy Rev. 16, 2566–2582.

Metcalf, E., Burton, F.L., Stensel, H.D., Tchobanoglous, G., 2003. Wastewater Engineering: Treatment and Reuse. McGraw Hill, New York.

Movahedyan, H., Mohammadi, A.S., Assadi, A., 2009. Comparison of different advanced oxidation processes degrading p-chlorophenol in aqueous solution. J. Environ. Health Sci., Engg. 6, 153–160.

Roshani, B., Vel Leitner, N.K., 2011. The influence of persulfate addition for the degradation of micropollutants by ionizing radiation. Chem. Eng. J. 168, 784–789.

Park, W.-J., Ahn, J.-H., Hwang, S., Lee, C.-K., 2010. Effect of output power, target temperature, and solid concentration on the solubilization of waste activated sludge using microwave irradiation. Bioresour. Technol. 101, S13–S16.

Passos, F., Ferrer, I., 2014. Microalgae conversion to biogas: thermal pretreatment contribution on net energy production. Environ. Sci. Technol. 48, 7171–7178.

Pilli, S., Bhunia, P., Yan, S., Leblanc, R., Tyagi, R., Surampalli, R., 2011. Ultrasonic pretreatment of sludge: a review. Ultrasonics Sonochem. 18, 1–18.

Pilli, S., Yan, S., Tyagi, R., Surampalli, R., 2015. Thermal pretreatment of sewage sludge to enhance anaerobic digestion: a review. Crit. Rev. Environ. Sci. Technol. 45, 669–702.

Raner, K.D., Strauss, C.R., Vyskoc, F., Mokbel, L., 1993. A comparison of reaction kinetics observed under microwave irradiation and conventional heating. J. Org. Chem. 58, 950–953.

Rani, R.U., Kumar, S.A., Kaliappan, S., Yeom, I., Banu, J.R., 2013. Impacts of microwave pretreatments on the semi-continuous anaerobic digestion of dairy waste activated sludge. Waste Manage. 33, 1119–1127.

Rani, R.U., Kumar, S.A., Kaliappan, S., Yeom, I.-T., Banu, J.R., 2014. Enhancing the anaerobic digestion potential of dairy waste activated sludge by two step sono-alkalization pretreatment. Ultrasonics Sonochem. 21 (3), 1065–1074.

Remya, N., Lin, J.-G., 2011a. Current status of microwave application in wastewater treatment—a review. Chem. Eng. J. 166, 797–813.

Remya, N., Lin, J.-G., 2011b. Microwave-assisted carbofuran degradation in the presence of GAC, ZVI and H_2O_2: Influence of reaction temperature and pH. Sep. Purif. Technol. 76, 244–252.

Rezaeei, A., Ghaneian, M.T., Hashemian, S.J., Mousavi, G., Ghanizadeh, G., 2008. Interference of potassium persulphate and hydrogen peroxide in the COD test. Wast Water 19, 77–81.

Robinson, J., Kingman, S., Irvine, D., Licence, P., Smith, A., Dimitrakis, G., et al., 2010. Understanding microwave heating effects in single mode type cavities—theory and experiment. Phys. Chem. Chem. Phys. 12, 4750–4758.

Saha, M., Eskicioglu, C., Marin, J., 2011. Microwave, ultrasonic and chemo-mechanical pretreatments for enhancing methane potential of pulp mill wastewater treatment sludge. Bioresour. Technol. 102, 7815–7826.

Sahoo, M., Sinha, B., Marbaniang, M., Naik, D., 2011. Degradation and mineralization of calcon using UV 365/H_2O_2 technique: Influence of pH. Desalination 280, 266–272.

Sanz, J., Lombrana, J., De Luis, A., Ortueta, M., Varona, F., 2003. Microwave and Fenton's reagent oxidation of wastewater. Environ. Chem. Lett. 1, 45–50.

Shiying, Y., Ping, W., Xin, Y., Guang, W., Zhang, W., Liang, S., 2009. A novel advanced oxidation process to degrade organic pollutants in wastewater: Microwave-activated persulfate oxidation. J. Environ. Sci. 21, 1175–1180.

Steriti, A., Rossi, R., Concas, A., Cao, G., 2014. A novel cell disruption technique to enhance lipid extraction from microalgae. Bioresour. Technol. 164, 70–77.

Stuerga, D., Gonon, K., Lallemant, M., 1993. Microwave heating as a new way to induce selectivity between competitive reactions. Application to isomeric ratio control in sulfonation of naphthalene. Tetrahedron 49, 6229–6234.

Su, H., Cheng, J., ZhoUu, J., Song, W., Cen, K., 2009. Improving hydrogen production from cassava starch by combination of dark and photo fermentation. Int. J. Hydrog. Energy 34, 1780–1786.

Taherzadeh, M.J., Karimi, K., 2008. Pretreatment of lignocellulosic wastes to improve ethanol and biogas production: a review. Int. J. Mol. Sci. 9, 1621–1651.

Tong, J., Chen, Y., 2009. Recovery of nitrogen and phosphorus from alkaline fermentation liquid of waste activated sludge and application of the fermentation liquid to promote biological municipal wastewater treatment. Water Res. 43, 2969–2976.

Tyagi, R., Couillard, D., Tran, F., 1988. Heavy metals removal from anaerobically digested sludge by chemical and microbiological methods. Environ. Pollut. 50, 295–316.

Tyagi, V.K., Lo, S.-L., 2013. Sludge: a waste or renewable source for energy and resources recovery? Renew. Sustain. Energy Rev. 25, 708–728.

Veschetti, E., Maresca, D., Cutilli, D., Santarsiero, A., Ottaviani, M., 2000. Optimization of H_2O_2 action in sewage-sludge microwave digestion using pressure vs. temperature and pressure vs. time graphs. Microchem. J. 67, 171–179.

Wang, K., Yin, J., Shen, D., Li, N., 2014. Anaerobic digestion of food waste for volatile fatty acids (VFAs) production with different types of inoculum: effect of pH. Bioresour. Technol. 161, 395–401.

Wang, L., Yuan, X., Zhong, H., Wang, H., Wu, Z., Chen, X., et al., 2014. Release behavior of heavy metals during treatment of dredged sediment by microwave-assisted hydrogen peroxide oxidation. Chem. Eng. J. 258, 334–340.

Wang, Y., Wei, Y., Liu, J., 2009. Effect of H_2O_2 dosing strategy on sludge pretreatment by microwave-H_2O_2 advanced oxidation process. J. Hazard. Mater. 169, 680–684.

Wang, Y., Xiao, Q., Liu, J., Yan, H., Wei, Y., 2015. Pilot-scale study of sludge pretreatment by microwave and sludge reduction based on lysis–cryptic growth. Bioresour. Technol. 190, 140–147.

Ware, A., Power, N., 2017. Modelling methane production kinetics of complex poultry slaughterhouse wastes using sigmoidal growth functions. Renew. Energy 104, 50–59.

Wilson, C.A., Novak, J.T., 2009. Hydrolysis of macromolecular components of primary and secondary wastewater sludge by thermal hydrolytic pretreatment. Water Res. 43, 4489–4498.

Wong, W., Chan, W., Liao, P., LO, K., Mavinic, D.S., 2006. Exploring the role of hydrogen peroxide in the microwave advanced oxidation process: solubilization of ammonia and phosphates. J. Environ. Eng. Sci. 5, 459–465.

Wu, T., Englehardt, J.D., 2012. A new method for removal of hydrogen peroxide interference in the analysis of chemical oxygen demand. Environ. Sci. Technol. 46, 2291–2298.

Xiao, Q., Yan, H., Wei, Y., Wang, Y., Zeng, F., Zheng, X., 2012. Optimization of H_2O_2 dosage in microwave-H_2O_2 process for sludge pretreatment with uniform design method. J. Environ. Sci. 24 (12), 2060–2067.

Ya-Wei, W., Cheng-Min, G., Xiao-Tang, N., Mei-Xue, C., Yuan-Song, W., 2015. Multivariate analysis of sludge disintegration by microwave–hydrogen peroxide pretreatment process. J. Hazard. Mater. 283, 856–864.

Yang, G., Zhang, G., Wang, H., 2015. Current state of sludge production, management, treatment and disposal in China. Water Res. 78, 60–73.

Yang, Y., Wang, P., Shi, S., Liu, Y., 2009. Microwave enhanced Fenton-like process for the treatment of high concentration pharmaceutical wastewater. J. Hazard. Mater. 168, 238–245.

Yen, P.-S., Chen, L., Chien, C., Wu, R.-M., Lee, D., 2002. Network strength and dewaterability of flocculated activated sludge. Water Res. 36, 539–550.

Yin, G., Liao, P.H., Lo, K.V., 2007. An ozone/hydrogen peroxide/microwave-enhanced advanced oxidation process for sewage sludge treatment. J. Environ. Sci. Health Part A 42, 1177–1181.

Yu, Y., Chan, W., Liao, P., Lo, K., 2010. Disinfection and solubilization of sewage sludge using the microwave enhanced advanced oxidation process. J. Hazard. Mater. 181, 1143–1147.

Yuan, S., Tian, M., Lu, X., 2006. Microwave remediation of soil contaminated with hexachlorobenzene. J. Hazard. Mater. 137, 878–885.

Zhang, Z., Shan, Y., Wang, J., Ling, H., Zang, S., Gao, W., et al., 2007. Investigation on the rapid degradation of congo red catalyzed by activated carbon powder under microwave irradiation. J. Hazard. Mater. 147, 325–333.

Zhao, D., Cheng, J., Hoffmann, M.R., 2011. Kinetics of microwave-enhanced oxidation of phenol by hydrogen peroxide. Front. Environ. Sci. Eng. China 5, 57–64.

Zhao, Z.-H., Sakagami, Y., Osaka, T., 1998. Toxicity of hydrogen peroxide produced by electroplated coatings to pathogenic bacteria. Can. J. Microbiol. 44, 441–447.

Zhou, A., Yang, C., Guo, Z., Hou, Y., Liu, W., Wang, A., 2013. Volatile fatty acids accumulation and rhamnolipid generation in situ from waste activated sludge fermentation stimulated by external rhamnolipid addition. Biochem. Eng. J. 77, 240–245.

Zlotorzynski, A., 1995. The application of microwave radiation to analytical and environ. chemistry. Crit. Rev. Anal. Chem. 25, 43–76.

Øegaard, H., 2004. Sludge minimization technologies-an overview. Water Sci. Technol. 49, 31–40.

Further Reading

American Public Health Association (APHA). 2005. Standard methods for the examination of water and wastewater. Washington, DC, USA.

Burton, G.F.L., Stensel, H.D., 2003. Wastewater Engineering: Treatment and Reuse. McGraw Hill, New York.

Kavitha, S., Kumar, S.A., Kaliappan, S., Yeom, I.-T., Banu, J.R., 2015. Achieving profitable biological sludge disintegration through phase separation and predicting its anaerobic biodegradability by non linear regression model. Chem. Eng. J. 279, 478–487.

Li, X., Yang, S., 2007. Influence of loosely bound extracellular polymeric substances (EPS) on the flocculation, sedimentation and dewaterability of activated sludge. Water Res. 41 (5), 1022–1030.

Liu, C., Shi, W., Li, H., Lei, Z., He, L., Zhang, Z., 2014. Improvement of methane production from waste activated sludge by on-site photocatalytic pretreatment in a photocatalytic anaerobic fermenter. Bioresour. Technol. 155, 198–203.

Liu, J., Jia, R., Wang, Y., Wei, Y., Zhang, J., Wang, et al., 2017. Environ. Sci. Pollut. Res. 24 (10), 9016–9025.

Markis, F., Baudez, J.-C., Parthasarathy, R., Slatter, P., Eshtiaghi, N., 2014. Rheological characterisation of primary and secondary sludge: impact of solids concentration. Chem. Eng. J. 253, 526–537.

Gamma-ray, X-ray and Electron Beam Based Processes

Marek Trojanowicz[1,2], *Krzysztof Bobrowski*[1,3],
Tomasz Szreder[1] *and Anna Bojanowska-Czajka*[1]

[1]Institute of Nuclear Chemistry and Technology, Warsaw, Poland [2]University of Warsaw,
Warsaw, Poland [3]Radiation Laboratory, University of Notre Dame,
Notre Dame, IN, United States

9.1 INTRODUCTION

The beginnings of scientific research on using chemical radiation methods for the treatment of waters and wastewaters dates back to the 1950s. Pioneering applications also began to appear in this period. During the first decade of post WWII rebuilding, there was a renaissance in the activity of all industrial fields. As this activity grew, it became increasingly clear that there was an important problem associated with the need to protect the environment (i.e., water resources). The effort was part of the general push for the peaceful use of atomic energy following its use in wartime. It was also consistent with the movement to improve the standard of living at the time.

Pioneering investigations were initiated at MIT in Boston, Massachusetts, using sewage sterilization by electron beams (EBs) and gamma irradiation (Dunn, 1953). Similar activity was followed soon after, in several other research institutions in the United States (Touhill et al., 1969; Ballantine et al., 1969; Compton et al., 1970) and in Canada (Grant et al., 1969). Some years later similar studies were undertaken in the Soviet Union (Brusentseva et al., 1978) and Japan (Nagai, 1976). Some activity in this field was inaugurated by certain European research groups (Gehringer and Krenmayr, 1973; Lesse and Suess, 1984; Perkowski et al., 1984). Within two decades it resulted in a rapidly growing number of original research reports in scientific and technical journals, with the highest activity in the years 1990–2010. During the same period several pilot and large-scale installations were

Advanced Oxidation Processes for Wastewater Treatment
DOI: https://doi.org/10.1016/B978-0-12-810499-6.00009-7

built in various countries for the treatment of waters and wastewaters, mostly for munici-pal purposes (Woodbridge et al., 1972; Cleland et al., 1984). It has to be emphasized at the very beginning of this article, that contrary to some special fields of nuclear technology, none of the radiation techniques discussed herein that are employed for waters and waste-waters purification induce energy levels, which could affect the natural one.

The basis of the application of ionizing radiation for the decomposition of pollutants occurring in waters and wastewaters is mostly associated with the reactions of these pollu-tants with the products of water radiolysis. Direct radiolysis of dissolved compounds (i.e., the pollutants) in diluted solutions does not take place. The applications of the radiation methods described in this chapter can be considered practical successes in the decades of developments in radiation chemistry. The results of these developments are documented in thousands of original research works and a wide selection of books or chapters in books (Cleland et al., 1984; Thomas, 1969; Draganic and Draganic, 1971; Buxton, 1982, 1987, 2008; von Sonntag, 1987; Bensasson et al., 1993; Richter, 1998; Bobrowski, 2012, 2017; Spinks and Woods, 1990) published in the last half-century.

The radiolysis of water is a relatively easy and convenient way to produce an enormous variety of highly reactive radical species that otherwise cannot readily be generated by ther-mal, chemical, electrochemical, and photochemical methods. Formation of the radical cation ($H_2O^{+\bullet}$) and electrons (e^-) represents the primary process (ionization) in the radiolysis of water and can be summarized by the Reaction 9.1 (Buxton, 2008):

$$H_2O \text{---}\sim\wedge\wedge\wedge\sim\text{---} \rightarrow H_2O^{+\bullet} + e^- \qquad\qquad 9.1$$

There is no unequivocal evidence that electronic excitation of water molecules (Reaction 9.2) plays any significant role in radiolysis of water.

$$H_2O \text{---}\sim\wedge\wedge\wedge\sim\text{---} \rightarrow H_2O* \qquad\qquad 9.2$$

Both processes occur on the time scale of electronic transition, i.e., $\sim 10^{-16}$ s. The water rad-ical cation ($H_2O^{+\bullet}$) immediately loses a proton (in $\sim 10^{-14}$ s) to neighboring water molecules (Reaction 9.3), and the electrons ejected in an ionization process (Reaction 9.1) lose gradually their energy and undergo hydration by $\sim 10^{-12}$ s, after thermalization (Reaction 9.4).

$$H_2O^+ + H_2O \rightarrow OH + H_3O^+ \qquad\qquad (9.3)$$

$$e^- \rightarrow e^-_{therm}/e^-_{prehyd} + nH_2O \rightarrow e^-_{aq} \qquad\qquad (9.4)$$

On the other hand, the electronically excited water molecules (H_2O*) may undergo homolytic dissociation (Reaction 9.5) in $\sim 10^{-13}$ s.

$$H_2O \rightarrow H + OH \qquad\qquad (9.5)$$

At this time, for low linear energy transfer (LET) radiation (X/γ-rays and fast electrons from accelerators), the products of Reactions (9.3, 9.4, 9.5) are distributed inhomogen-eously and located together in small and widely separated volumes called *spurs*. Before diffusing apart into the bulk of water, a fraction of them interacts with one another form-ing molecular and secondary radical products. These spur reactions together with their respective rate constants are listed in Table 9.1 (Buxton, 1987, 2008).

TABLE 9.1 Spur Reactions in Water and Their Rate Constants

Spur Reaction (Number)	Reaction Rate-Constant k (M^{-1}s^{-1})
$e_{aq}^- + e_{aq}^- \rightarrow H_2 + 2OH^{-\,a}$	5.4×10^9
$e_{aq}^- + OH \rightarrow OH^-$	3.0×10^{10}
$e_{aq}^- + H_3O^+ \leftrightarrows H + H_2O$	2.3×10^{10}
$e_{aq}^- + H \rightarrow H_2 + OH^{-\,a}$	2.5×10^{10}
$H + H \rightarrow H_2$	1.3×10^{10}
$OH + OH \rightarrow H_2O_2$	5.3×10^9
$OH + H \rightarrow H_2O$	3.2×10^{10}
$H_3O^+ + OH^- \rightarrow 2H_2O$	1.4×10^{11}

[a]The mass balance is ensured by hydration water.
With permission, adapted from Buxton, G.V., 1987. Water and homogeneous aqueous solutions. In: Rodgers, M.A.J. (Ed.), Radiation Chemistry. Principles and Application. Farhataziz. VCH Publishers, Inc., New York, pp. 321–350; Buxton, G.V., 2008. An overview of the radiation chemistry of liquids. In: Spotheim-Maurizot, M., Mostafavi, M., Douki, T., Belloni, J. (Eds.), Radiation Chemistry: From Basics to Applications in Material and Life Sciences. EDP Sciences, pp. 3–6.

The spur reactions and diffusion of reactants leading to their homogeneity are completed within 10^{-7} s. The radiolysis of water (for low LET irradiation ~ 0.23 eV/nm) is summarized by the Eq. 9.14:

$$H_2O \text{---}\wedge\wedge\rightarrow {}^\bullet OH(0.28),\ e_{aq}^-(0.28),\ {}^\bullet H(0.062),\ H_2(0.047),\ H_2O_2(0.073),\ H_3O^+(0.28) \qquad 9.14$$

where the numbers in parentheses represent radiation chemical yields (G-values) in units mol/J (Buxton, 2008; Spinks and Woods, 1990).

The remaining primary and secondary products, which diffuse outside the spurs, become homogeneously distributed and may react with added solutes acting as radical scavengers. The effective radiation chemical yields (G-values) of the primary species available for the reaction with scavengers depend on their concentrations since at high concentration they can interfere with the spur reactions.

The primary transients from water radiolysis (Eq. 9.14) are involved in the following acid/base equilibria represented by Eqs. 9.15 and 9.16:

$$H + HO^- \leftrightarrows e_{aq}^- + H_2O \qquad (9.15)$$

$$OH + HO^- \leftrightarrows O^{-} + H_2O \qquad (9.16)$$

Hydrogen atoms (${}^\bullet H$) and hydrated electrons (e_{aq}^-) exist in the acid/base equilibrium (Eq. 9.15) with $pK_a = 9.1$ (${}^\bullet H$ atom is a conjugate acid of e_{aq}^-). In very basic solutions ${}^\bullet H$ atoms are converted into e_{aq}^- with $k_{forward}$ (in equilibrium represented by Eq. 9.15) = 2.2×10^7 M^{-1} s^{-1}. Since protonation by water is very slow, the life-time of e_{aq}^- in pure water is quite long. On the other hand, at pH < 4, the diffusion-controlled reaction of e_{aq}^- with bulk protons (Reaction 9.8, Table 9.1) causes an increase of the yield of $^{\bullet}$H atoms with decreasing pH.

Hydroxyl radicals ($^{\bullet}$OH) exist in the acid/base equilibrium (Eq. 9.16) with pKa = 11.8 ($^{\bullet}$OH is a conjugate acid of the oxide radical ion ($O^{\bullet-}$)), with $k_{forward}$ (in equilibrium represented by Eq. 9.16) = 1.3×10^{10} M^{-1} s^{-1}) and $k_{reverse}$ (in equilibrium represented by Eq. 9.16) = 7.9×10^{7} s^{-1} (Weeks and Rabani, 1966). Since $O^{\bullet-}$ radical anion is very rapidly protonated by water, its radical reactions occur in a significant extent only at a pH > 12. For a better illustration, the initial radiation chemical yields (G) of the primary products of water radiolysis as function of pH are presented in Fig. 9.1 (Getoff, 1996).

Redox properties of the primary radicals from water radiolysis can be summarized as follows: (1) the $^{\bullet}$OH radical is a powerful one-electron oxidant with a reduction potential varying with the pH (from +1.90 V for the redox couple $^{\bullet}$OH/$^-$OH to +2.72 V for the redox couple $^{\bullet}$OH, H^+/H_2O), (2) the hydrated electron (e_{aq}^-) is a powerful reductant in neutral and alkaline solutions with a reduction potential -2.87 V while the $^{\bullet}$H atom becomes the major reductant in acidic solutions with a reduction potential -2.31 V. All potentials were measured versus the normal hydrogen electrode (NHE) (Buxton et al., 1988; Wardman, 1989). Those statements also lead to a very important conclusion that the radiation treatment may involve, depending on particulate solute (pollutant), both oxidative and reductive processes. Therefore, in contrary to many other typical advanced oxidation processes, the radiation techniques should be considered as advanced oxidation/reduction processes.

Since the primary radicals formed in the radiolysis of water possess both oxidative ($^{\bullet}$OH) and reductive properties (e_{aq}^-, H), it is of special importance to study the reactions of individual radicals. These conditions can be achieved via an appropriate design of the reaction system.

In order to eliminate the contribution of hydrated electrons (e_{aq}^-), the solution is saturated with nitrous oxide (N_2O). Hydrated electrons, e_{aq}^-, are converted into $^{\bullet}$OH radicals

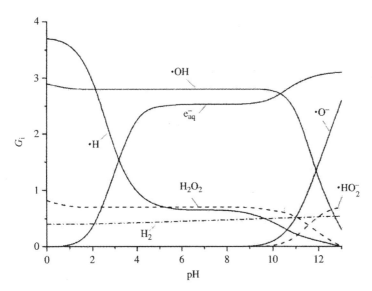

FIGURE 9.1 Effect of pH on G values of the primary products of water radiolysis. *With permission, adapted from Getoff, N., 1996. Radiation-induced degradation of water pollutants-state of the art. Radiat. Phys. Chem. 47:581–593.*

according to Reaction 9.17 with $k_{17} = 9.1 \times 10^9$ M^{-1} s^{-1} (Janata and Schuler, 1982), which nearly doubles the amount of $^\bullet$OH radicals available for reactions with scavengers.

$$e_{aq}^- + N_2O \rightarrow N_2 + O^{\bullet -} + H_2O \rightarrow N_2 + \ ^\bullet OH + \ ^-OH \qquad (9.17)$$

At pH ≤ 4, the diffusion-controlled reaction of e_{aq}^- with protons (Reaction 8) becomes important, resulting in a pH-dependent lowered yield of $^\bullet$OH radicals and a correspondingly increased yield of $^\bullet$H atoms.

A good approach to study reactions of e_{aq}^- (without contribution of $^\bullet$OH radicals) is irradiation of Ar/N$_2$-saturated solutions containing high concentration (> 0.1 mol/dm^3) of 2-methyl-2-propanol (*tert*-butanol). The highly reactive $^\bullet$OH radicals are scavenged by *tert*-butanol (Reaction 9.18) with $k_{18} = 6.0 \times 10^8$ M^{-1} s^{-1} (Buxton et al., 1988):

$$^\bullet OH + \ (CH_3)_3C\text{-}OH \rightarrow H_2O + \ ^\bullet CH_2(CH_3)_2C\text{-}OH \qquad (9.18)$$

The forming 2-hydroxy-2,2-dimethyl radical ($^\bullet$CH$_2$(CH$_3$)$_2$C-OH) is largely unreactive and its optical absorption spectrum does not interfere with other radical spectra located at $\lambda > 270$ nm. The irradiated system contains only e_{aq}^- and $^\bullet$H-atoms with the respective G-values 0.28 and 0.062 mol/J.

A convenient way to study reactions of $^\bullet$H atoms (without contribution of $^\bullet$OH radicals and e_{aq}^-) is irradiation of highly acidic Ar/N$_2$-saturated solutions containing a high concentration (> 0.1 M) of *tert*-butanol. The $^\bullet$OH radicals are removed in Reaction 16 and e_{aq}^- are converted to $^\bullet$H atoms via reaction with H$_3$O$^+$ (Reaction 9.8, Table 9.1), and thus their radiation chemical yield amounts to 0.34 mol/J.

In aerated and O$_2$-saturated solutions both $^\bullet$H atoms and e_{aq}^- react with molecular oxygen (O$_2$) according to Reactions 9.19 and 9.20 with $k_{19} = 2.1 \times 10^{10}$ and $k_{20} = 1.9 \times 10^{10}$ M^{-1} s^{-1} (Buxton et al., 1988):

$$^\bullet H + O_2 \rightarrow HO_2 \qquad (9.19)$$

$$e_{aq}^- + O_2 \rightarrow O_2^{\bullet -} \qquad (9.20)$$

Perhydroxyl radicals (HO$_2$) and superoxide radical anions (O$_2^{\bullet -}$) exist in the acid/base equilibrium (Eq. 9.21) with pKa $= 4.88$ (HO$_2$ radical is a conjugate acid of O$_2^{\bullet -}$ radical anion) (Bielski et al., 1985):

$$HO_2 \leftrightarrows H^+ + O_2^{\bullet -} \qquad (9.21)$$

Redox properties of these transients can be summarized as follows: (1) the HO$_2$ radical is a mild one-electron oxidant with a reduction potential varied with the pH (from $+0.79$ V for the redox couple HO$_2$/HO$_2^-$ to $+1.48$ V for the redox couple HO$_2$, H$^+$/ H$_2$O$_2$), (2) the O$_2^{\bullet -}$ radical anion is either a mild one-electron reductant with a reduction potential -0.33 V (for the redox couple O$_2$/O$_2^{\bullet -}$) or a mild one-electron oxidant with a reduction potential $+1.03$ V (for the redox couple O$_2^{\bullet -}$, H$^+$/HO$_2^-$. All potentials were measured versus the NHE (Wardman, 1989).

Interestingly, $^\bullet$OH radical is unreactive towards O$_2$, however, its conjugate base, the oxide radical anion (O$^{\bullet -}$) reacts with O$_2$ forming ozonide radical anion (O$_3^{\bullet -}$) (Reaction 9.22):

$$O^{\bullet -} + O_2 \leftrightarrows O_3^{\bullet -} \qquad (9.22)$$

These two radicals exist in an equilibrium with $k_{22} = 3.0 \times 10^9$ M^{-1} s^{-1} (Slegers and Tilquin, 2006), and $k_{-22} = 3.3 \times 10^3$ s^{-1} (Gall and Dorfman, 1969).

Radical reactions involving primary transients from water radiolysis can be summarized as follows: (1) the hydroxyl radical ($^\bullet$OH) undergoes the following types of reactions: electron transfer (ET), hydrogen atom abstraction, addition to C=C, C=N, and S=O bonds, addition to aromatic rings, and addition to electron-rich functional groups. These reactions reflect its electrophilic character and strong oxidizing properties (De Vleeschouver et al., 2007), (2) the hydrated electron (e_{aq}^-) undergoes the following types of reactions: reduction, dissociative electron attachment, addition to functional groups of high electron affinity, addition to aromatic rings containing electron-withdrawing substituents, and addition to conjugated double bonds (Bobrowski, 2012). The hydrated electron acts as a nucleophile in its reactions with organic molecules, (3) the hydrogen atom ($^\bullet$H) undergoes the following types of reactions: ET, hydrogen atom abstraction, addition to C=C bonds, addition to aromatic rings, addition to electron-rich functional groups, and homolytic substitution (Richter, 1998; Bobrowski, 2012). These reactions reflect its weak nucleophilic (De Vleeschouver et al., 2007), and strong reducing properties, however, lower than e_{aq}^-. Interestingly, radical reactions of $^\bullet$H atoms resemble those of $^\bullet$OH radicals.

Rate constants for thousands of reactions of these transients with molecules, ions, and radicals derived from inorganic and organic compounds have been compiled (Buxton et al., 1988; Hart et al., 1964; Neta et al., 1988; Haag and Yao, 1992). From the point of view of practical application of this technology for treatment of waters and wastewaters, the presence of numerous scavengers of radicals (which commonly occur in those media) is always a very essential element need to be taken into account during the irradiation process optimization. As it is showed by example rate-constants in Table 9.2, the radical reactions with many of them occur with a similar rate compared to reactions of target environmental pollutants.

TABLE 9.2 The Values of Rate-Constants (M^{-1}s^{-1}) for Reactions of Main Products of Water Radiolysis With Commonly Occurring Radical Scavengers in Natural Waters and Wastewaters

Compound	$^\bullet$OH	e_{aq}^-	$^\bullet$H
O$_2$	NR	1.9×10^{10}	$<1.0 \times 10^6$
HCO$_3^-$	8.5×10^6	$<1.0 \times 10^6$	$<1.0 \times 10^6$
CO$_3^{2-}$	3.9×10^8	3.9×10^5	NR
Cl$^-$	3.0×10^9	$<1.0 \times 10^6$	$<1.0 \times 10^5$
NO$_2^-$	1.1×10^{10}	3.5×10^6	7.1×10^8
NO$_3^-$	NR	9.7×10^9	1.4×10^6
DOC	2.0×10^8	NR	NR

DOC, dissolved organic carbon; NR, nonreacting.
With permission, adapted from Buxton, G.V., Greenstock, C.L., Helman, W.P., Ross, A.B., 1988. Critical review of rate constants for reactions of hydrated electrons, hydrogen atoms and hydroxyl radicals ($\bullet$OH/$\bullet$O-) in aqueous solution. J. Phys. Ref. Chem. Data 17:513–886.

9.2 SOURCES OF RADIATION—TECHNOLOGICAL INSTALLATIONS

The type of the ionization radiation is the key factor that determines the mechanism of the energy deposition in the irradiated matter. In general, beam of photons (X, γ), high energy electrons (electron accelerator beam, $^-$), heavy, charged particles and neutrons are considered in the radiation chemistry. Application of the first two types of the ionization irradiation in the water treatment industry is justified by economic, safety, and technological issues so the following description will be focused on them.

Interaction of photon radiation (X, γ), often called electromagnetic radiation with matter is described by photoelectric, Compton scattering, and electron-positron pair formation phenomena. In the photoelectric phenomenon an incident photon is absorbed by an electron bounded to atom (molecule). As a consequence, an electron is ejected leaving behind an ionized atom (molecule). The ejected electron carries out the excess energy equal to incident photon energy reduced by ionization potential. Photoelectric phenomenon prevails for relatively low photon energies (i.e., below 100 keV in water) (Spinks and Woods, 1990). Strongly bounded electrons (on the lowest atomic/molecular orbitals) are the most vulnerable to this effect. On the other hand, Compton scattering involves free or relatively weakly bounded electrons (valence electrons). As a result of the interaction, an incident photon vanishes and its energy is distributed between a recoiled electron and a scattered new photon that carries lower energy than the incoming one. Trajectory and energy distribution between electrons and photons in the system follows the momentum and energy conservation law. This effect prevails for moderate photon energies (i.e., from 100 keV to 10 MeV in water) (Spinks and Woods, 1990). This is the most common effect in the industrial radiation treatment (sterilization, degradation) of the condensed phase. Incident photons with energy above 1.02 MeV ($2m_0c^2$, twice of the electron invariant mass) can be converted to an electron–positron pair. Subsequently the positron annihilates with one of the electron on atomic (molecular) orbital leading to an ionized-electron deficient species. The pair formation occurs only in a close proximity of the nucleus (or charged particle). This is a major effect for high energy photons.

There is no difference in the mechanism of X and γ ionization radiation interaction with matter. Various symbols are used rather for distinguishing origin of both the radiation types. X-rays are produced in vacuum tubes as a result of X-ray fluorescence (discrete spectrum) or Bremsstrahlung radiation (continuous spectrum) generated by a collision of high energy electrons with a metal target. Industrial X-ray sources have relatively low photon energies from 100 eV to hundreds of keV and low dose rates up to about 10 Gy/h (Getoff, 2002). Thus, their potential water treatment application is very limited. On the other hand γ-radiation is produced by radioactive, unstable nuclei of various isotopes. The cobalt source containing ^{60}Co isotope (half-life 1925.20 $\pm$ 0.25 day (Unterweger, 2002), 1.17 and 1.33 MeV photons (Getoff, 1996)) is the most popular one. Very often mentioned in literature, the caesium source containing ^{137}Cs isotope is quite reactive, and hence difficult to handle. Its salts are soluble in water, which complicates their safe handling. Moreover, photons produced by ^{137}Cs have lower energies (0.66 MeV) (Getoff, 1996) than those produced by ^{60}Co (vide supra). It is characteristic for this kind of radiation that material attenuation coefficient is used instead of maximum range. The photon flux (I) passing the element with thickness (x) can be expressed as $I = I_0 \exp(-\mu x)$ (Spinks and Woods, 1990), where I_0 is the initial photon flux and μ the total linear attenuation coefficient very often replaced by mass attenuation coefficient ($\mu_m = \mu/\rho$; the ration of the linear attenuation coefficient and the mass density of material). Selected values of μ_m in water for 0.01, 1, and 10 MeV photons are equal to 5.329, 0.07072, and 0.02219 cm^2/g (Hubbell and Seltzer, 2004), respectively. Typically, industrial dose rates of γ-irradiation do not exceed several MGy/h (Getoff, 2002).

Energy transfer mechanism for all charged particles is similar. It involves electromagnetic interaction between a moving incident particle and a nuclei with associated atomic orbital electrons of the irradiated system. For the sake of simplicity these interactions are very often called elastic and inelastic collisions. In inelastic collision, the particle loses its energy for excitation and/or ionization of the atoms (molecules) of the medium. Loss of energy for the particle traveling along a track is described by the Bethe and Heitler equation (Spinks and Woods, 1990). There are some differences in final effects of the irradiation upon heavy charged particles and electrons. They are caused by a significant difference in their masses. In general, (1) an incident electron very often changes direction and its energy; (2) its resulting range is many times shorter than actually traveled path; (3) its relatively low kinetic energy causes distinct relativistic effect resulting in Bremsstrahlung and Cerenkov radiation.

It is worthy to mention that radiation chemistry upon ⁻, accelerated electron or γ-irradiation is essentially the same. This is due to the fact that the major chemical changes in γ-irradiated systems are caused by high energy electrons generated as a result of photon—matter interactions. On the other hand, some effects may be observed due to the difference in dose rates of radiation sources. Commercially available pulsed accelerators may work with several MGy/s dose rates. They are much higher than those that can be achieved by γ or ⁻ radiation from isotopes. High concentration of intermediates is expected for a system irradiated with a high dose rate. Thus, some reaction channels unavailable at low concentration of intermediates may open.

The ability of electrons to penetrate an irradiated medium is an important factor for industrial applications. The maximum range of electrons significantly decreases with their kinetic energy (Fig. 9.2A) as a consequence of increased interaction with matter. At very high, relativistic electron energies efficiency of interaction slightly increases again. In this case, it results in Bremsstrahlung and Cerenkov radiation. Stopping power $(1/\rho \times dE/dx)$ described by kinetic energy loss (dE) in the medium with mass density (ρ) per unit path length of particle (dx) clearly depicts this effect (Table 9.3).

There is no doubt that γ-photons are capable of deeper penetration in the medium, however, the penetration range of accelerated electrons is sufficient for many applications. Though, an increasing of electron energy improves their maximum range and thus the rate of water systems processing, there is a particular limitation that does not allow to increase their energy above the certain value. In general, industry uses up to 14 MeV EB since electrons with higher energies may activate irradiated material or accelerator construction elements. In most cases, electron accelerators are the best option in water treatment application. They provide desired dose rates, electron energies, and a high standard of safety (radiation can be shut down instantly).

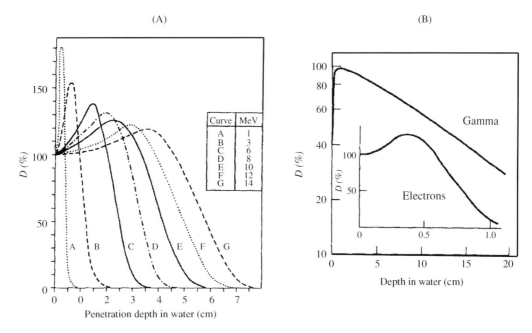

FIGURE 9.2 Depth-dose (%) distribution curves in water for electrons with different energies (A), and for gamma-rays from ^{60}Co source compared to 2 MeV electrons (B). *With permission, adapted from Getoff, N., 1996. Radiation-induced degradation of water pollutants-state of the art. Radiat. Phys. Chem. 47:581−593.*

TABLE 9.3 Selected Values of Stopping Power for Electrons in Liquid Water ($\rho = 0.998$ g/cm^3)

Electron Energy (MeV)	Stopping Power (MeV cm^2/g)
0.01	22.39
0.10	4.097
1.00	1.858
2	1.848
10	2.148
15	2.305

With permission, adapted from Report 90. J. Int. Commission Radiation Units Measurements, 14(1), 1−110, 2014. Report 90 (2014).

The properties of different kinds of ionizing radiation, which can be potentially employed for the purification of waters and wastewaters, first of all from organic pollutants, imply that the main sources applied for this purpose are ^{60}Co-sources as a source of gamma-rays, and accelerators as a source of high energy electrons. As it can be convincingly concluded from a vast number of original research papers, and also from numerous review papers on the subject listed in Table 9.4, the gamma-ray sources are used mostly

TABLE 9.4 Review Papers Published in Scientific Journals on Different Aspects of Irradiation Treatment of Waters and Wastewaters

Subject	Authors (Year of Publication)
REACTION RATE-CONSTANTS	
Rate constants of hydrated electron reactions with organic compounds	Hart et al. (1964)
Critical review of rate constants for reaction e_{aq}^-, H and $^\bullet$OH	Buxton et al. (1988)
Rate constants for reactions of inorganic radicals in aqueous solution	Neta et al. (1988)
Rate constants for reaction of hydroxyl radicals	Haag and Yao (1992)
RADIATION PROCESSING OF WATER AND WASTES	
Role of radiation in wastewater treatment	Ballantine (1971)
Sewage and wastewater processing with isotopic radiation	Gerrard (1971)
EB process designs and economic feasibility	Cleland et al. (1964)
Radiation treatment of liquid wastes	Pikaev and Shubin (1984)
Advancements of radiation induced degradation of pollutants	Getoff (1989)
High energy EB irradiation of water, wastewater	Kurucz et al. (1991a,b)
High energy EB irradiation	Cooper et al. (1992)
Radiation-induced degradation of water pollutants—state of the art	Getoff (1996)
Environmental application of ionizing radiation	Cooper et al. (1998)
Radiation processing of sewage and sludge	Borrely et al. (1998a,b)
Radiation induced pollutant decomposition in water	Gehringer and Matschiner (1998)
Radiation chemistry and the environment	Getoff (1999)
New developments in radiation processing in Russia	Pikaev (1999)
Current status of the application of ionizing radiation	Pikaev (2000)
Radiation-induced oxidation of pollutants	Cooper et al. (2001)
Radiation-induced degradation of water pollutants	Getoff (2002)
New data on EB purification of wastewater	Pikaev (2002)
Application of ionizing radiation for degradation of organic pollutants	Drzewicz et al. (2003)
Remediation of polluted waters and wastewater by radiation	Sampa et al. (2007)
Radiation induced degradation of organic pollutants	Wojarovits and Takacs (2016)
EB technology for environmental pollution control	Chmielewski and Han (2016)
Wastewater treatment with ionizing radiation	Wojnarovits and Takacs (2017)
TREATMENT OF COMPOUNDS IN PARTICULAR FIELDS OF APPLICATION	
Irradiation treatment of pharmaceutical and personal care products	Wang and Chu (2016)
Rate coefficients of $^\bullet$OH reactions with pesticides	Wojnarovits and Takacs (2014)
DIFFERENT CHEMICAL GROUPS OF DECOMPOSED POLLUTANTS	
The status of PCB radiation chemistry research	Curry and Mincher (1999)
Irradiation treatment of azo dye containing wastewater	Wojnarovits and Takacs (2008)
Radiation induced degradation of dyes	Rauf and Ashraf (2009)
Radiation-induced destruction of hydroxyl-containing amino acids	Sladkova et al. (2012)

for research purposes, whereas in technological applications, including pilot and large-scale treatment plants, the use of accelerators predominates for EB treatment.

With the use of typical, commercial ^{60}Co gamma sources the irradiation is carried out in a discrete mode (sample by sample), by placing the container with liquid samples inside the source, after appropriate chemical sample processing (changing pH, deaeration, additions of required reagents), if needed. In early works in the field, for this purpose also, flow systems have been assembled which are schematically shown in Fig. 9.3A and B (Hashimoto and Kawakami, 1979; Singh et al., 1985; Singh and Kremers, 2002). The cobalt sources were even installed in the field, directly at the groundwater intake in order to decompose harmful compounds, e.g., in case of elevated level of cyanide (Pastuszek et al., 1993). In case of much seldom used gamma radiation from a ^{137}Cs sources, commercial irradiators are used with a fixed central rod source, also in discrete mode (Song et al., 2008a,b). There were also some attempts to employ spent reactor fuel, which can be a source of gamma-rays with an average energy of 700 keV. A 100 kGy dose can be obtained in less than 13 h with use of a sample capsule (Fig. 9.3C), which was inserted into the spent fuel pool (Mincher et al., 1991).

Very rarely for discussed environmental purposes, X-ray irradiation was applied with the use of conventional X-ray sources, mostly due to a much smaller dose rate than provided by isotope gamma sources or accelerators. For instance, in attempt to decompose the pesticide diazinon, a 50% decomposition was reported at the dose about 160 Gy, with a dose-rate about 50 mGy/s (Trebse and Aron, 2003).

For technological applications, including the treatment of waters and wastewaters, with no doubts, beams of accelerated high energy electrons from different types of accelerators are most widely used. Their most important advantage is a much larger dose rate and the operation in on/off instant mode, however, with an essential drawback—limited penetration depth of irradiated solutions. In order to overcome most efficiently that drawback, different constructions of the irradiation flow installations were designed for EB irradiation of a stream of liquid. The cascade-type irradiator, where a thin layer of falling stream was formed, was employed in a widely described Electron Beam Research Facility in Wastewater Treatment Plant in Miami (Mak et al., 1997; Cooper et al., 1993; Lin et al., 1995; Kurucz et al., 1995a,b). It is schematically shown together with the floor plan of whole installation in Fig. 9.4.

With application of a 1.5 MeV accelerator operating at 50 mA (75 kW), the waste stream receives an average dose of about 6.5 kGy, and with employed flow-rate the wastewater spends about 0.1 s in front of the beam. This gives a dose-rate at least 65 kGy/s (Kurucz et al., 1995b). Examples of other efficient constructions are shown in Fig. 9.5, and they include the up-flow stream irradiator developed in Brazil (Duarte et al., 2002), and the nozzle type injector developed in South Korea for purification of textile dyeing (Pikaev et al., 2001). A very efficient design involves irradiation of wastewater in aerosol form, with additional merge with gaseous ozone, which in optimized conditions may enhance production of $^{\bullet}OH$ radicals (Pikaev et al., 1997a,b,c). In aerosol phase, the range of electron penetration increased by 20−50 times, compared with liquid phase (Podzorova et al., 1998). Therefore, even for highly polluted wastewater, the doses required for purification are relatively low (4−5 kGy). Degradation of pollutants takes place not only by reactions with products of water radiolysis, but also with the primary products of radiolysis of gas phase (oxygen atoms) and ozone (Han et al., 2016).

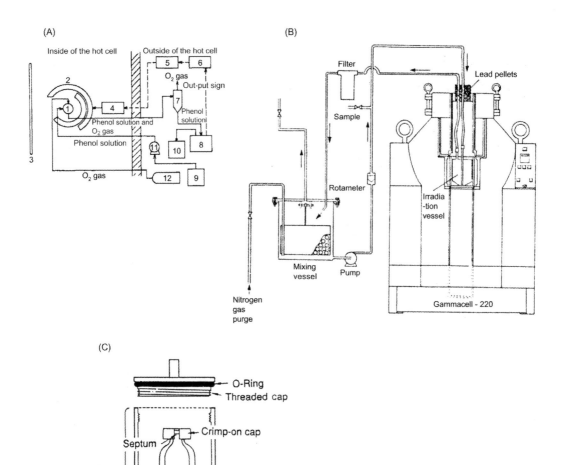

FIGURE 9.3 Schematic diagrams of flow systems developed for gamma-irradiation(A,B), and sample capsule for irradiation using spent reactor fuel (C) (*With permission, adapted from Mincher, B.J., D.H. Meikrantz, R.J. Murphy, G.L. Gresham, M.J. Connolly, 1991. Appl. Radiat. Isot. 42:1061–1066*). In (A) flow system for radiation treatment of wastewater with gamma-irradiation for ^{60}Co source, which was developed and optimized in Japan Atomic Energy Research Institute for phenol decomposition (*With permission, adapted from Hashimoto, S., W. Kawakami, 1979. Ind. Eng. Chem. Process Des. Dev. 18:169–274*): 1-Pyrex glass reactor 330 mL, 2-lead shields, 3-^{60}Co source which is the plate type of 700 mm with and 300 mm height, 4-servo-motor, 5-amplifier, 6-feedback control system for regulation of dose-rate, 7-gas separator, 8-spectrofluorimeter for phenol detection, 9-penol solution reservoir, 10-waste reservoir, 11-piston pump, 12-oxygen cylinder. In (B) a large-scale installation designed for gamma-irradiation with ^{60}Co source and employed for radiolytic decomposition of polychlorinated biphenyls (*With permission, adapted from Singh, A., W. Kremera, P. Smalley, G.S. Bennett, 1985. Radiat. Phys. Chem. 25:11–19; Singh, A., W. Kremers, 2002. Radiat. Phys. Chem. 65:467–472*). A 4.4 L stainless-steel vessel fits the Gammacell 220 source, and is connected to 80 L holding tank and purged with nitrogen.

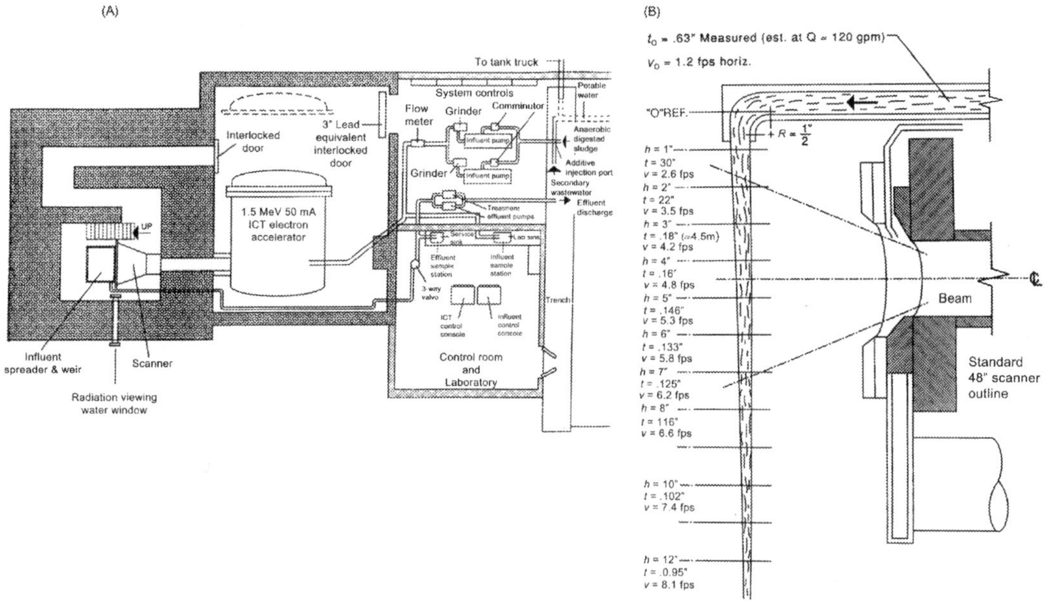

FIGURE 9.4 Floor plan (A) and cross sectional view of an aqueous waste stream falling over the weir delivery system (B) of the Miami Electron Beam Research Facility that became operational in 1984. *With permission, adapted from Kurucz, C.N., Waite, T.D., Cooper, W.J., 1995b. The Miami electron beam research facility: a large scale wastewater treatment application. Radiat. Phys. Chem. 45:299–308.*

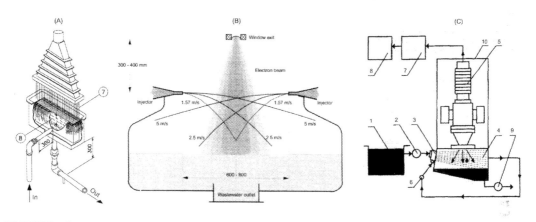

FIGURE 9.5 Schematic diagrams of different elements of pilot EB irradiation installations designed for radiolytic processing of waters and wastewaters: (A) Scheme of the overflow-type irradiation box, developed in Nuclear and Energy Research Institute (IPEN) in Sao Paulo, Brazil (*With permission, adapted from Duarte, C.L., M. H.O. Sampa, P.R. Rela, H. Oikawa, C.G. Silveira, A.L. Azevedo, 2002. Radiat. Phys. Chem. 63:647–651*): 7-electron beam, 8-liquid to be irradiated; (B) The irradiation chamber with two opposite nozzles for wastewater transport developed for industrial wastewater (*With permission, adapted from Pikaev, A.K., A.V. Ponomarev, A.V. Bludenko, V. N. Minin, L.M. Elizar'eva. Radiat, 2001. Phys. Chem. 61:81–87*); (C) The combined EB/ozone installation for treatment of wastewater in aerosol flow developed in Russian Academy of Sciences (*With permission, adapted from Pikaev, A.K., E.A. Podzorova, O.M. Bakhtin, 1997. Radiat. Phys. Chem. 49:155–157*): 1-reservoir of wastewater intake, 2-pump, 3-sprayer, 4-irradiation chamber, 5-electron accelerator, 6-turboblower, 7-power supply, 8-control desk, 9-pump, 10-biological shielding.

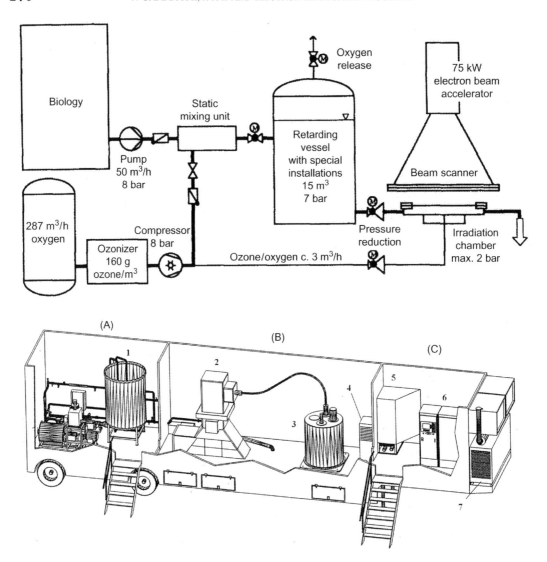

FIGURE 9.6   Electron-beam systems developed for the treatment of waters and wastewaters.

The schemes of other complete installations for the EB irradiation with a simultaneous ozonation, which was widely examined for environmental applications, is shown in Fig. 9.6. In order to ensure a constant level of the dissolved oxygen during the EB irradiation, a dual-tube bubbling column reactor was also developed (Kawakami et al., 1978).

Upper scheme: The prototype system for continuous ozone-electron beam treatment of wastewater developed in Austrian Research Centre Seibersdorf (*With permission, adapted from Gehringer, P., H. Fiedler, 1998. Radiat. Phys. Chem. 52, 345–349*). Lower scheme: Construction of a mobile EB treatment system for contaminated groundwater and industrial wastes for the field applications *(With permission, adapted from Nickelsen, M.G., D.C. Kajdi, W.J. Cooper, C.N. Kurucz, T.D. Waite, F. Gensel, H. Lorenzl, U. Sparka, 1998. Field application of a mobile 20 kW electron-beam treatment system on contaminated groundwater and industrial wastes. In: Cooper, W.J., R.D. Curry, K.E. O'Shea (Eds.), Environmental Application of Ionizing Radiation, Wiley, New York, pp. 451–466).*

Since 1990-ties there are designed and produced mobile EB systems in the United States (Nickelsen et al., 1998) (Fig. 9.6), and in South Korea (Kim et al., 2012), which can be transported to a particular heavy-polluted locations for the in situ remediation. The first one mentioned above contains an electron accelerator with 500 keV transformer, and at full power the system is rated at 20 kW. The maximum flow rate can be 150 L/min, and the doses up to 20 kGy can be obtained in a single pass. It was reported that this installation was employed in the field treatment of contaminated groundwater, and also in industrial and refining process of wastewaters over a 2-year period.

9.3 DISINFECTION OF WASTEWATERS

The possibility of efficient hygienization of waters, wastewaters, and sewage sludges is a significant attribute of radiation technologies, which was practically a priority objective in earliest environmental applications, connected, e.g., with construction of the first large-scale installations in early 1970s (Cleland et al., 1984; Trump et al., 1984). Also in this case, however, first works were conducted already in 1950s, and investigations on destruction of microorganisms in water, sewage, and sewage sludge by gamma-irradiation for ^{60}Co source in Oak Ridge National Laboratory in the United States. The 93% removal of bacteria was obtained for doses from 0.5 to 10 kGy for half-diluted wastewater from the primary setting tank of the local laboratories (Lowe Jr. et al., 1956). There were also some attempts reported on application of X-ray irradiation for the inactivation of bacteriophage (Cotton and Lockingen, 1963). Predominated routine methods used in wastewater treatment technology, (first of all chlorination) exhibit numerous side effects like producing trace amounts of toxic chlorinated products, odor of treated media, and also its efficiency was limited. Therefore, among the earliest applications of radiation technology one can find several reports on their efficiency in removal of bacterial pathogens, coliforms and salmonellae from waters and wastewaters of various origins. Some selected applications are listed in Table 9.5.

Practically, in all reported applications the use of gamma and EB treatment results in the decrease of microorganisms count about 3 orders of magnitude at their initial level 10^3-10^7 species/100 mL. In comparison study on the use of gamma and EB in large-scale installation, it was shown for coliform bacteria, that for the same absorbed dose 5 kGy the gamma irradiation reduces the microorganism population by 4 orders of magnitude, while with EB—3 orders of magnitude, only (Farooq et al., 1993). The application of those two kinds of treatment for disinfection of municipal raw sewage and secondary effluent from a pure oxygen extended aeration biological treatment system and chlorination, is shown also in Fig. 9.7A and B. In case of the EB irradiation, doses below 1 kGy (100 krad) do not affect the coliform population, and in both cases a pronounced dose-rate effect shows evidently a larger yield of gamma irradiation. The comparison of both types of irradiation was carried out also for inactivation of nine different bacteriophages and *Escherichia coli*

TABLE 9.5 Inactivation of Microorganisms in Different Samples by Ionizing Radiation

Examined Wastewater	Ionizing Radiation Used	Microorganisms	Initial Population	Inactivation (Orders of Magnitude)	Employed Dose	References
Effluents from STP: Primary Secondary	Gamma (^{60}Co)	E. coli	10^6/100 mL 5300/100 mL	3 orders 3 orders	5 kGy 5 kGy	Taghipour (2004)
Raw wastewater (influent to STP) Secondary wastewater effluent	EB and gamma (^{60}Co)	Coliform bacteria	4.6×10^7/ 100 mL 4.6×10^5/ 100 mL	~ 3 orders (EB) ~ 6 orders (γ) ~ 3 orders (EB) ~ 3.5 orders (γ)	5 kGy	Farooq et al. (1993)
Tap water Low protein solution	Gamma (^{60}Co)	Bacteriophage MS2 Feline calicivirus Enteric canine calicivirus	10^6/100 mL 10^3/100 mL 10^3/100 mL	3 orders 3 orders 3 orders	100 Gy 500 Gy 300 Gy	Husman et al. (2004)
Tap water	EB and gamma (^{60}Co)	E. coli Bacteriophage PHI X 174 Bacteriophage MS2 Bacteriophage B40-8	No data 9.97 ng/L 5.98 ng/L No data	3 orders	0.17 kGy (γ); 0.34 kGy (EB) 1 kGy (γ); 2.2 kGy (EB) 0.1 kGy (γ); ~50 Gy (EB) 0.7 kGy (γ); 0.5 kGy (EB)	Gehringer et al. (2003)
Raw domestic wastewater	EB	Total coliforms Fecal coliforms Total bacteria count	1.3×10^7/ 100 mL 7.3×10^6/ 100 mL	3 orders 3 orders 4 orders	~2.0 kGy ~3.0 kGy ~3.0 kGy	Sampa et al. (1995)
Buffer solution		Salmonella serotypes: S. derby, S. infants, S. typhimurium S. meleagridis	3.1×10^6/ 100 mL No data	3 orders	0.5–0.9 kGy	
Municipal wastewaters: Raw sewage Secondary effluent Chlorinated final effluent	EB	Coliforms	1.0×10^8/ 100 mL 2×10^6/ 100 mL 1.5×10^2/ 100 mL	~4.5 ~6 ~2	3.0 kGy	Borrely et al. (1998a,b)

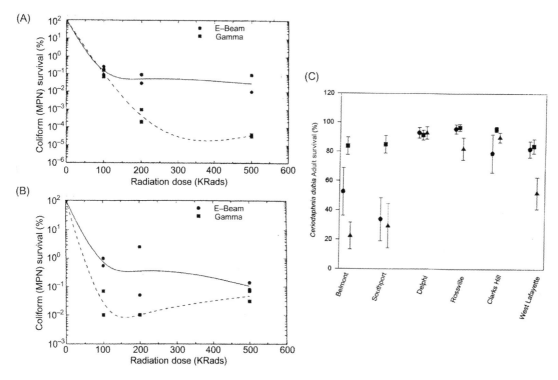

FIGURE 9.7 Effects of the dose-rate (A,B) and the obtained changes in toxicity (C) in disinfection of wastewaters by ionizing radiation. In A and B: comparison of the yield of inactivation of coliform in municipal raw sewage (A) and secondary effluent from sewage treatment plant (B) by the use of irradiation with electron beam and gamma-rays (*With permission, adapted from Farooq, S., C.N. Kurucz, T.D. Waite, W.J. Cooper, 1993. Wat. Res. 27:1177–1184*). In C: comparison of toxicity measured by survival of *Ceriodaphnia dubia* adults in wastewater effluent from six municipal treatments facilities after chlorination and gamma irradiation using doses from 500 to 700 Gy (*With permission, adapted from Thompson, J.E., E.R. Blatchley, 1999. Wat. Res. 33:2053–2058*): (●) undesinfected, (■) gamma-irradiated, (▲) chlorinated.

suspended in tap water (Gehringer et al., 2003). It was shown, that mostly •OH radicals are responsible for the virus inactivation under employed conditions. In the case of *E. coli*, the use of gamma-rays was found to be more efficient. However, the lack of a dose-rate effect was observed for two bacteriophage species MS2 and B40-8. Different radiation resistance was found for species PH1 X 174, and similarly to *E. coli*, inactivation by gamma-rays was more efficient.

The effects of gamma-inactivation were also examined in terms of induced toxicity of wastewater effluents, where postdisinfection samples were evaluated using *Ceriodaphnia dubia* chronic toxicity test (Thompson and Blatchley, 1999). For this purpose samples from six treatment facilities were exposed to gamma-rays and also to chlorination, and then survival of *C. dubia* was monitored for 7 days. Generally, it was found that gamma-irradiation either had no effect on toxicity, or in a half of examined samples actually improved a survival of monitored species (Fig. 9.7C). The gamma-irradiation of effluents from the wastewater treatment facility using a ^{60}Co-source (with a dose-rate 1.4 Gy/s) was compared also with UV disinfection in terms of *E. coli* inactivation (Taghipour, 2004). A similar efficiency for both processes was found, but in terms of the energy required and the energy cost, the cost of ionizing radiation inactivation was evaluated as a considerably higher than that for UV treatment. The estimated treatment costs for *E. coli* inactivation in primary wastewater using UV, EB, and gamma-radiation treatment were evaluated as 0.4, 1.25, and 25 US cents per m^3, respectively. It has to be admitted, however, that this estimation concerns disinfection, only, without chemical purification.

9.4 RADIOLYTIC DECOMPOSITION OF INDIVIDUAL COMPOUNDS

The basis for the application of radiation techniques for purification of waters and wastewaters, or degradation of pollutants to chemical forms (which are more susceptible to removal with other methods, e.g., biodegradation), is kinetics of their reactions with radicals produced by water radiolysis, or those formed in other radiation processes conducted for that purpose. The rate-constants for reactions with •OH radicals, hydrated electrons, and hydrogen atoms were known for thousands of organic and inorganic compounds and ions from fundamental radiation chemistry studies, already at the end of 1980s (Buxton et al., 1988) (Hart et al., 1964; Neta et al., 1988; Haag and Yao, 1992). Those vast collected data, only to a certain degree, can be utilized as supporting or indicating information for authentic technological applications. Decomposition in natural media requires knowledge of numerous physico-chemical factors' parameters on conducting those radiation processes which have to be taken substantially into account. These parameters comprise: a dose-rate of the employed source of ionizing radiation, pH of irradiated solution, content of oxygen and different matrix components in natural media, and identification of interradical reactions and transformations of numerous transient products.

Quite a number of attempts were already made towards simulation (kinetic modeling) of such processes based on the known kinetic data. They usually require several tens of reaction rate-constants for radical reactions, ionic equilibria, and redox reactions, which have to be taken into consideration (Cooper et al., 2001; Mak et al., 1997; Kawakami et al., 1978; Taghipour and Evans, 1997; Kurucz et al., 2002; Bojanowska-Czajka et al., 2007). Since the list of chemical compounds considered as anthropogenic pollutants increases with the passing of time, numerous research groups investigate the radiolytic decomposition of selected environmental pollutants. In numerous cases it is also associated with determination of rate-constants for radical reactions. The main objective is optimization of radiation conditions, and evaluation of the role of radical scavengers commonly occurring in natural waters. As especially valuable those studies should be considered, where effects of natural matrices of waters and wastes are involved. These studies are parallel to studies on efficiency of treatment of particular wastewaters or waters of a different origin, which is monitored with general methods of quality control such as TOC (total organic carbon), COD (chemical oxygen demand), BOD (biological oxygen demand), AOX (adsorbable organohalogenates), etc.

A quite long list of substances, for which the radiolytic decomposition was evaluated as for environmental pollutants, with the use of different radiation sources, is presented in Table 9.6. Although, this list does not pretend to be a complete presentation of the

TABLE 9.6 The Selected Publications on Application of Radiolytic Decomposition of Individual Environmental Pollutants With the Use of Ionizing Radiation

Substrate	Ionizing Radiation Used	Matrix	Initial Concentration	Dose Required (Yield, %)	References
DYES					
Acid Blue 62	Gamma (^{60}Co)	Std. solution	0.14 mM	1.3 kGy (90%)	Perkowski and Mayer (1989)
		Wastewater	6 mg/L	0.12 kGy (90%) deaerated solution	Perkowski et al. (2003)
Acid Orange 7	Gamma (^{60}Co)	Std. solutions	1.0 mM	6.7 kGy (90%)	Zhang et al. (2005)
Acid Yellow 99	Gamma (^{60}Co)	Std. solutions	10 M	1.2 kGy (90%)	Zimina et al. (2002)
Anthraquinone dyes	Gamma (^{60}Co)	Wastewater	Not given	0.8 kGy (82–99%)	Vysotskaya et al. (1986)
Apollofix-Red SF-28	Gamma (^{60}Co)	Std. solutions	85 M	0.26 kGy (90%)	Wojnarovits et al. (2005)
Azo dyes Reactive Black 5 Apollofix-Red	Gamma (^{60}Co)	Std. solutions	50 mg/L	Deaerated solutions 0.7 kGy (90%) 1.3 kGy (96%)	Solpan et al. (2003)
Azo dyes Acid Red 172 Reactive Red 21	EB	Std. solutions	50 mg/L	Aearated solutions 9 kGy (100%) 7.0 kGy (100%)	Pikaev et al. (1997a,b,c)
Anthraquinone dyes Brillant Blue BG Cerise N Violet BN	Gamma (^{137}Cs) together with chlorination	Std. solutions	25 mg/L	3.34 kGy (100%)	Craft and Eichholz (1971)
C.I. Direct Black 22	EB	Std. solutions	50 mg/L	6.0 kGy (100%)	Vahdat et al. (2010)
C.I.Reactive dyes Cibacron Brillant Red 3 B-A Cibacron Blue 3 G-A Cibacron Yellow 6 G	EB	Std. solutions	100 mg/L	3.0 kGy (96%) 3.0 kGy (73%) 3.0 kGy (97%)	Vahdat et al. (2012)
Methyl orange	Gamma (^{60}Co)	Std. solutions	0.5 mM	8.3 kGy (90%)	Chen et al. (2008)

(Continued)

TABLE 9.6 (Continued)

Substrate	Ionizing Radiation Used	Matrix	Initial Concentration	Dose Required (Yield, %)	References
Mixture of commercial dyes	Gamma (^{60}Co)	Wastewater	30–300 mg/L	5 kGy improves sorption on C-act	Perkowski (1989)
Mixture of commercial dyes (Acid Blue 62, Direct Yellow 44, Driect Brown 2	Gamma (^{60}Co), also with oxygenation and ozonation	Model wastewater	30–40 mg/L of each dye	5.0 kGy (67%) 5.0 kGy (93%) with ozonation	Perkowski and Kos (2003)
p-Phenylazoaniline	Gamma (^{60}Co)	Std. solutions	10 M	0.05 kGy (80%) deaerated solution	Krapfenbauer et al. (2000)
Reactive Black 5	Gamma (^{60}Co) with H$_2$O$_2$	Std. solutions	50 mg/L	1.0 kGy (90%)	Solpan and Guven (2002)
Reactive Blue XBR Reactive Red KE-3B	EB with H$_2$O$_2$	Std. solutions	800 mg/L	25 kGy (80%) 25 kGy (96%)	Wang et al. (2006)
HALOGENATED HYDROCARBONS					
Carbon tetrachloride	EB	Treated groundwater	51 M	5.0 kGy (98%)	Mak et al. (1997)
Chloroform	Gamma (^{60}Co)	Std. solutions	30 mg/L	50 kGy (90%)	Taghipour and Evans (1997)
	Gamma (^{60}Co), also with H$_2$O$_2$	Std. solutions	8.0 g/L	60 kGy (60%) 30 kGy with H$_2$O$_2$ (100%)	Wu et al. (2002)
Methylene chloride	Gamma (^{60}Co)	Std. solutions	0.37 mM	3.0 kGy (90%)	Mucka et al. (2003)
	EB	Treated groundwater	102 M	5.0 kGy (91.6%)	Mak et al. (1997)
Perchloroethylene (PCE)	Gamma (^{60}Co)	Std. solutions, Drinking water	10 mg/L 50 g/L	0.17 kGy (90%) 0.8 kGy (90%)	Proksch et al. (1987)
	EB with ozone	Tap water	0.5 mg/L	0.14 kGy (90%)	Gehringer and Eschweiler (2002)
	EB with ozone	Std. solutions	10 mg/L	0.6 kGy (90%) 0.12 kGy (90%) with ozonation	Gehringer and Matschiner (1998)

Pollutant	Treatment	Matrix	Dose (removal %)	Reference
	EB	Alkaline 2-propanol (1 vol.%)	64 kGy (63.6%)	Mucka et al. (1999)
	Gamma (^{60}Co)	Std. solutions	0.3 kGy (100%)	Yoon et al. (2002)
	Gamma (^{60}Co)	Std. solutions	100 Gy (71%)	Chung et al. (2002)
	Gamma (^{60}Co) with ozone and TiO$_2$	Std. solutions	100 mg/L	0.3 kGy (100%)	Jung et al. (2002a,b)
Tetrachloroethylene	Gamma (^{60}Co)	Std. solutions	15 mg/L	60 kGy (~95%)	Wu et al. (2002)
Trichloroethylene (TCE)	Gamma (^{60}Co) and EB also with ozone	Tap water	~100 g/L	0.2 kGy (90%) EB 0.11 kGy (90%) γ ~20 Gy (90%) ozone with γ or EB	Gehringer and Eschweiler (2002)
	EB	Tap water with 3.3 mM methanol, Secondary wastewater (with MeOH)	100 g/L	5 kGy (98%) 8.3 kGy (95%)	Cooper et al. (1993)
	Gamma (^{60}Co)	Std. solutions	10 mg/L	0.3 kGy (100%)	Yoon et al. (2002)
	Gamma (^{60}Co) with ozone and TiO$_2$	Std. solutions	100 mg/L	0.3 kGy (100%)	Jung et al. (2002a,b)
1.1.2-Trichloro-trifluoro-ethane	Gamma (^{60}Co)	Alkaline 2-propanol	80 mM	4.0 kGy (100%)	Nakagawa and Shimokawa (2002)

INDUSTRIAL POLLUTANTS

Pollutant	Treatment	Matrix	Dose (removal %)	Reference
Benzotriazole	EB	Std. solutions with persulfate	0.45 mg/L	15 kGy (64%)	Roshani and Leitner (2011)
Bisphenol A	Gamma (^{60}Co)	Std. solutions and wastewaters	50 M in water with N$_2$O	0.05 kGy (86%)	Peller et al. (2011)
	Gamma (^{60}Co)	Std. solutions	10 mg/L	8 kGy (98%)	Guo et al. (2012a,b)

(Continued)

TABLE 9.6 (Continued)

Substrate	Ionizing Radiation Used	Matrix	Initial Concentration	Dose Required (Yield, %)	References
2-Chloroaniline	Gamma (^{60}Co) with ozonation	Std. solution	0.1 mM	0.92 kGy (100%)	Winarno and Getoff (2002)
p-Chloronitrobenzene	Gamma (^{60}Co) also with biodegradation	Std solution	1.27 mM	17.6 kGy (98.7%)	Bao et al. (2009)
2,6-Dichloroaniline	Gamma (^{60}Co)	Std. solutions	mM level	<1 kGy (100%)	Homlok et al. (2012)
Polyoxyethylene n-nonyl phenyl ether	Gamma (^{60}Co)	Std. solutions	100 mg/L (as TOC)	10 kGy (N_2-sat.solution, coagulation)	Sakumoto and Miyata (1984)
Polyvinyl alcohol	Gamma (^{60}Co) with coagulation	Std. solutions	100 mg/L (as TOC)	10 kGy (N_2- sat.solution, coagulation)	Sakumoto and Miyata (1984)
Polyvinyl alcohol	Gamma (^{60}Co) with H_2O_2	Std. solutions	200 mg/L	1.2 kGy (82.2%)	Zhang and Yu (2004)
Perfluorooctanoic acid	EB	Phosphate buffer, high alkalinity	515 mg/L	6.0 kGy (96%)	Wang et al. (2016)
Surfactants: Nonylphenol ethoxylate Octylphenol ethoxylate	EB	Wastewater	2.5 g/L 2.3 g/L	4.2 kGy (90%) 2.7 kGy (90%)	Petrovic et al. (2007)
METAL IONS					
Al and heavy metals (Cr, Fe, Zn, Co)	EB	Wastewater with formate	0.4–21 g/L	see Table 9.11	Ribeiro et al. (2004)
Cd, Cr(VI), and Pb(II)	EB	Std. solutions with formate	2 mg/L Cd 5 mg/L Pb 5 mg/L Cr	2.7 kGy (90%) 1 kGy (90%) 2.8 kGy (90%)	Pikaev et al. (1997a,b,c)
Co(II), Cu(II), Sr(II)	EB	Wastewater with humic acid and UV/TiO$_2$ treatment	0.85 mM Co 0.79 mM Cu 0.57 mM Sr	50 kGy (100%) for all	Zaki and El-Gendy (2014)

Compound	Concentration	Matrix	Method	Dose (result)	Reference
Hg (II) and Pb(II)	1 mM Hg 1 mM Pb	Std. solutions	Gamma (^{60}Co) and EB	Hg: 3 kGy (100%) γ Pb: 30 kGy (96%) EB	Schmelling et al. (1998a,b,c)
	1 mM Hg(II) 1 mM Pb(II)	Std. solutions with ethanol	EB	3 kGy (>99.9%) 40 kGy (96%)	Chaychian et al. (1998)
Organolead compounds and Pb(II)	0.2 mM each	Std. solutions for Pb(II) with formate	EB	Pb(II): 4 kGy (75%) Met$_3$Pb: 4 kGy (75%) Et$_3$Pb: 2 kGy (90%)	Unob et al. (2003)

PESTICIDES

Compound	Concentration	Matrix	Method	Dose (result)	Reference
Acetochlor	1.0 mM	20% acetonitrile	Gamma (^{60}Co)	12.2 kGy (46%)	Liu et al. (2005)
Atrazine	1.15 µM	Std. solutions with humic acid	Gamma (^{60}Co) also with ozonation	0.8 kGy (100%) in N_2 saturated solution	Leitner et al. (1999)
	20 µg/L	Ground water	EB with ozone	0.15 kGy (99%)	Gehringer and Eschweiler (1996)
	0.1 mM	Std. solutions	Gamma (^{60}Co)	6.5 kGy (90%)	Varghese et al. (2006)
	0.464 µM	Ground waters with humic acid	Gamma (^{60}Co)	0.6 kGy (70%–95%)	Mohamed et al. (2009a,b)
	40 µM	Std. solutions	Gamma (^{60}Co)	0.8 kGy (99%)	Wang et al. (2012)
	8.11 µM	Std. solutions	Gamma (^{60}Co)	3.5 kGy (100%)	Khan et al. (2015)
	6 mg/L	Std. solutions	EB	1.0 kGy (95%)	Xu et al. (2015)
Carbendazim	42 µM (1:20 diluted waste)	Std. solutions and diluted wastewater	EB and gamma (^{60}Co)	26 kGy (100%) by EB	Bojanowska-Czajka et al. (2011)
Chlorfenvinphos	50 mg/L	Std. solutions	Gamma (^{60}Co)	0.20 kGy (100%)	Bojanowska-Czajka et al. (2010)
Chlorpyrifos	13.7 µM	Acetonitrile	Gamma (^{60}Co)	10 kGy (100%)	Mori et al. (2006)
	0.5 mg/L	Std. solutions	Gamma (^{60}Co)	0.57 kGy (100%)	Ismail et al. (2013)
	5 mg/L	Std. solutions	Gamma (^{60}Co)	10 kGy (100%)	Hossain et al. (2013)

(Continued)

TABLE 9.6 (Continued)

Substrate	Ionizing Radiation Used	Matrix	Initial Concentration	Dose Required (Yield, %)	References
Clopyralid	EB also with H_2O_2	Std. solutions	100 mg/L	10 kGy (100%)	Xu et al. (2011)
2,4-D	Gamma (^{60}Co)	Std. solutions	0.5 mM	4.0 kGy (100%)	Zona et al. (2002)
	Gamma (^{60}Co)	Std. solutions	50 µM 500 µM	0.45 kGy (100%) 4.20 kGy (100%)	Zona and Solar (2003)
	Gamma (^{60}Co)	Std. solution	0.21 mM	2.12 kGy (100%)	Peller et al. (2003)
	EB with ozonation	Tap water	100 mg/L	2.66 kGy with ozonation	Drzewicz et al. (2004)
	Gamma (^{60}Co)	Std. solutions	0.2 mM	8 kGy (100%)	Homlok et al. (2010)
	Gamma (^{60}Co) also with H_2O_2	Std. solutions	3 mg/L	2 kGy (98%)	Solpan and Torun (2012)
DDT	Gamma (^{60}Co)	2-Propanol	20 mM	48 kGy (75%)	Sherman et al. (1971)
	Gamma (^{137}Cs)	Technical DDT with water	n.d.	Resistant to 284 kGy	Kimbrough and Gaines (1971)
	Gamma (^{60}Co)	On silica gel in air-saturated aqueous suspension	10 µM	16 kGy (100%)	Woods and Akhtar (1974)
	Gamma (spent reactor fuel)	Ethanol	100 mg/L	22 kGy (78%)	Mincher et al. (1991)
	Gamma (^{60}Co)	2-Propanol, cycloxane, cyclohexene	100 mg/L	10 kGy (99.2%) 10 kGy (71.8%) 10 kGy (47.8%)	Lepine et al. (1994)
Diazinon	X-ray	Std. solution	40 mg/L	0.36 kGy (78%) 120 min irradiation	Trebse and Aron (2003)
	Gamma (^{60}Co)	Ground waters	0.329 µM	1.6 kGy (70–90%)	Mohamed et al. (2009a,b)
	Gamma (^{60}Co)	Std. solutions	0.329 µM 3.29 µM	3.5 kGy (100%) 6.0 kGy (91%)	Basfar et al. (2007)

Dicamba	EB and gamma (^{60}Co) also with H_2O_2	Std. solutions and ground water	0.5 mM in std. solution	6.0 kGy (90%) and 2.3 kGy with ozonation	Drzewicz et al. (2005)
Diuron	Gamma (^{60}Co)	Std. solutions	0.1 mM	2.0 kGy (100%)	Kovacs et al. (2015)
Endrin	Gamma (^{60}Co)	Std. solutions	50 mg/L	6.0 kGy (100%)	Riaz and Butt (2010)
Fenuron	Gamma (^{60}Co)	Std. solutions	0.1 mM	10 kGy (80%)	Kovacs et al. (2014)
Fenvalerate	EB and gamma (^{60}Co) with O_3	Std. solutions	5 mg/L	EB—8 kGy (78%), γ/O_3—8 kGy (100%)	Abdel Aal et al. (2001)
Hexachlorobenzene	Gamma (^{60}Co)	Solution with detergent and N_2O	175 M	50 kGy (75%)	Zacheis et al. (2000)
Lannate	EB and gamma (^{60}Co) with O_3	Std. solutions	5 mg/L	EB—8 kGy (85%), γ/O_3—8 kGy (100%)	Abdel Aal et al. (2001)
Lindane	Gamma (spent reactor fuel)	Ethanol	250 mg/L	50 kGy (88%)	Mincher et al. (1991)
Malathion	Gamma (^{60}Co)	Std. solutions	0.2 mg/L	25 kGy (88%)	Mohamed et al. (2009a,b)
	Gamma (^{60}Co)	Std. solutions	0.2 mg/L	2.0 kGy (98.7%)	Mohamed et al. (2009a,b)
MCPA	Gamma (^{60}Co) also with H_2O_2	Std. solutions, wastewater	500 mg/L (wastewater)	5.0 kGy (100%)	Bojanowska-Czajka et al. (2006)
	Gamma (^{60}Co)	Std. solutions	0.5 mM	2.7 kGy (100%)	Zona et al. (2012)
Parathion	Gamma (^{60}Co)	Methanol	95 mg/L	30 kGy (99%)	Luchini et al. (1999)
	Gamma (^{60}Co) with ozonation	Std. solutions	15 mg/L	0.5 kGy (100%) with ozonation	Bojanowska-Czajka et al. (2012)

PHARMACEUTICALS

Alkaloids (thebaine, papaverine, noscapine)	Gamma (^{60}Co)	Methanol	5 mg/L	>50 kGy (100%)	Kantoglu and Ergun (2016)
Amoxicillin	Gamma (^{137}Cs)	Std. solutions	1.0 mM	12.8 kGy (90%)	Song et al. (2008a,b)

(Continued)

TABLE 9.6 (Continued)

Substrate	Ionizing Radiation Used	Matrix	Initial Concentration	Dose Required (Yield, %)	References
-Blockers (atenolol, metoprolol, propranolol)	Gamma (^{60}Co)	Std. solutions	1.0 mM	1.7–2.2 kGy (50%)	Song et al. (2008a,b)
Carbamazepine	Gamma (^{60}Co)	Wastewater	5.0 M	2.0 kGy (100%)	Kimura et al. (2012)
	EB	Std. solutions, surface water	50 g/L	1.0 kGy (90%) 2.0 kGy (90%)	Zheng et al. (2014)
	Gamma (^{60}Co)	Std. solution river water Hospital effluent	10 mg/L 10 g/L	0.65 kGy (90%) 0.5 kGy (62%)	Bojanowska-Czajka et al. (2015)
Cefaclor	Gamma (^{60}Co)	Std. solutions	30 mg/L	1 kGy (100%)	Yu et al. (2008)
Ceftriaxone	Gamma (^{60}Co)	Std. solutions	20 mg/L	5.0 kGy (98%)	Zhang et al. (2011)
Chloramphenicol	Gamma (^{60}Co)	Std. solutions	0.1 mM	2.5 kGy (100%)	Csay et al. (2012)
Chlortetracycline	EB	Std. solutions	30 mg/L	0.6 kGy (90%)	Kim et al. (2012)
Clofibric acid	Gamma (^{60}Co)	Wastewater	5.0 M	1.0 kGy (100%)	Kimura et al. (2012)
Cytarabine	Gamma (^{137}Cs)	Std. solutions, surface water (SW), ground water (GW), wastewater	5–20 mg/L	0.4–1.0 kGy (100%) in Std. solutions 1.0 kGy (30%) in SW and GW	Ocampo-Perez et al. (2011)
Diclofenac	Gamma (^{60}Co)	Std. solutions	0.1 mM	1 kGy (100%)	Homlok et al. (2011)
	Gamma (^{60}Co)	Std. solutions	20.5 mg/L	1.0 kGy (90%)	Liu et al. (2011)
	Gamma (^{60}Co)	Wastewater	5.0 M	1.0 kGy (100%)	Kimura et al. (2012)
	Gamma (^{60}Co)	Std. solutions	1 mM	6.5 kGy (90%)	Yu et al. (2013)
	Gamma (^{60}Co)	Std. solutions, river water, Hospital effluent	50 mg/L 10 g/L	1.3 kGy (90%) in std. solutions 0.25 kGy (100%)	Bojanowska-Czajka et al. (2015)

Compound	Radiation	Matrix	Concentration	Dose (removal)	Reference
Diatrizoate	Gamma (^{137}Cs)	Std. solutions, waters, wastewater	25 mg/L	1.0 kGy (90%)	Gala et al. (2013)
Fibrate pharmaceuticals (clofibric acid, bezafibrate, gemfibrozil)	Gamma (^{137}Cs)	Std. solutions	0.6 mM	2.1–8.0 kGy (90%)	Razavi et al. (2009)
Ibuprofen	Gamma (^{60}Co) also with H_2O_2	Std. solutions	28.3 mg/L	1.1 kGy (100%)	Zheng et al. (2011)
Ibuprofen	Gamma (^{60}Co)	Std. solutions	0.1 mM	0.7 kGy (90%)	Illes et al. (2013)
	Gamma (^{60}Co)	Std. solution river water, hospital effluent	10 mg/L, 10 g/L	0.35 kGy (90%), 0.25 kGy (100%)	Bojanowska-Czajka et al. (2015)
Iopromide (X-ray contast)	EB also with sulfite	Std. solutions	0.1 mM	19.6 kGy (90%), 0.9 kGy with sulfite	Kwon et al. (2012)
Ketoprofen	Gamma (^{60}Co)	Std. solutions	0.4 mM	2.0 kGy (100%)	Illes et al. (2012)
	Gamma (^{60}Co)	Wastewater	5.0 M	2.0 kGy (100%)	Kimura et al. (2012)
Levofloxacin lactate	Gamma (^{60}Co)	Std. solutions	10 mg/L	1 kGy (> 99%)	Cao et al. (2011)
Mefenamic acid	Gamma (^{60}Co)	Wastewater	5.0 M	2.0 kGy (100%)	Kimura et al. (2012)
Metoprolol	EB and gamma (^{60}Co)	Std. solution	1 g/L	EB: 25 kGy (90%), γ: 39 kGy (90%)	Slegers and Tilquin (2006)
Naproxen	Gamma (^{60}Co)	23.6 mg/L	23.6 mg/L	1.0 kGy (98%)	Zheng et al. (2012)
Nitroimidazoles (metronidazole—MNZ, dimetridazole, tinidazole)	Gamma (^{60}Co)	Std. solutions, surface water, wastewater	140 mM	0.45–0.75 kGy (MNZ)	Sanchez-Polo et al. (2009)
Paracetamol	Gamma (^{60}Co)	Std. solutions	0.1 mM	1.0 kGy (100%)	Szabo et al. (2012)
	Gamma (^{60}Co) also with ozone	Std. solutions	50 mg/L	3.2 kGy (90%), 2.0 kGy with ozonation	Torun et al. (2015)
Penicillin G Penicillin V	Gamma (^{137}Cs)	Std. solutions	1.0 mM	5.5 kGy (90%), 6.0 kGy (90%)	Song et al. (2008a,b)

(Continued)

TABLE 9.6 (Continued)

Substrate	Ionizing Radiation Used	Matrix	Initial Concentration	Dose Required (Yield, %)	References
Primidone	EB	Std. solutions	25–100 mg/L	0.5–2.2 kGy (90%)	Liu et al. (2015)
Sulfadiazine	Gamma (^{60}Co), with H_2O_2 and Fe(II)	Std. solutions	2–50 mg/L	0.2–4.0 kGy (90%)	Guo et al. (2012a,b)
Sulfa drugs (sulfamethazine, sulfamethiazole, sulfamethoxazole, sulfamerazine—SMZ)	Gamm (^{60}Co)	Std. solutions	0.1–1.0 mM	~3.0 kGy (50% in 0.5 mM SMZ)	Mezyk et al. (2007)
Sulfamethazine	Gamma also with H_2O_2	Std. solutions	20 mg/L	0.84 kGy (90%) 0.55 kGy with H_2O_2	Liu and Wang (2013)
Sulfamethoxazole	Gamma (^{60}Co)	Std. solutions	0.1 mM	1.2 kGy (90%)	Sagi et al. (2016)
Sulfamethoxazole	EB	Std. solutions	30 mg/L	1.1 kGy (90%)	Kim et al. (2012)
Tetracycline antibiotics (tetracycline, chlortetracycline, oxytetracycline, doxycycline)	Gamma (^{137}Cs)	Std. solutions	0.5 mM	5–10 kGy (100%)	Jeong et al. (2010)
PHENOLIC COMPOUNDS					
Catechol	EB with ozone	Std. solutions, tap water	0.5 mM (55 mg/L)	7 kGy (100%) with O_3 in tap water	Kubesch et al. (2005)
2-Chlorophenol	Gamma (^{60}Co)	Std. solutions	0.05 mM	0.55 kGy (100%)	Zona et al. (1999)
	Gamma (^{60}Co) also with ozone	Std. solutions	1 mM (130 mg/L)	10 kGy (90%)	He et al. (2002)
	Gamma (^{60}Co) also with H_2O_2	Std. solutions	50 mg/L	20 kGy (100%)	Basfar et al. (2017)
2,4-Dichlorophenol	Gamma (^{60}Co)	Std. solutions, river water	23 mg/L 10.5 mg/L	0.2 kGy (100%) 0.65 kGy (90%)	Trojanowicz et al. (1997)
	Gamma (^{60}Co)	Std. solutions	0.05 mM	0.45 kGy (100%)	Zona et al. (1999)
	Gamma (^{60}Co)	Std. solutions	20 mg/L	1.0 kGy (95%)	Trojanowicz et al. (2002)

	Radiation	Matrix	Concentration	Dose (result)	Reference
	Gamma (^{60}Co) also with ozone	Std. solutions	0.44 mM	2.8 kGy (90%)	He et al. (2002)
	Gamma (^{60}Co)	Std. solutions	0.5 mM	0.72 kGy (100%) N_2O sat.	Peller and Kamat (2005)
	Gamma (^{60}Co)	Std. solutions	0.5 mM	5 kGy (100%)	Shim et al. (2009)
Diphenolic acid	Gamma (^{137}Cs)	Std. solutions, waters, wastewater	20 mg/L	1.0 kGy (95%)	Abdel daiem et al. (2013)
Gallic acid	EB and gamma (^{60}Co)	Std. solutions	1 mM	γ: 50 kGy (90%) EB: 24 kGy (90%)	Melo et al. (2009)
Pentachlorophenol	Gamma (^{60}Co)	Std. solutions	0.1 mM	0.15 kGy (55%)	Fang et al. (1998)
Phenol	Gamma (^{60}Co)	Std. solutions	1.0 mM	7.5 kGy (90%)	Hashimoto et al. (1979)
	Gamma (^{60}Co) with	Std. solution (flow γ system)	5 mg/L	80% decomposition in 15 min with O_2 saturation	Hashimoto and Kawakami (1979)
	Gamma (^{60}Co) with ozone	Std. solutions	100 mg/L (as TOC)	1.8 kGy with ozonation	Sakumoto and Miyata (1984)
	Gamma (^{60}Co)	Std. solutions	0.01 to 1.0 mM	1 kGy reduces 100% 0.01 mM, 96% 0.1 mM, 47% 1 mM	Getoff (1986)
	EB	Std. solutions, tap water, wastewater	0.02 to 0.6 mM	0.5–7.0 kGy	Lin et al. (1995)
	EB with ozone	Std. solutions	0.5 mM (47 mg/L)	10.5 kGy	Kubesch et al. (2005)
	Gamma (^{60}Co)	Std. solutions	2.0 mM	Increase of yield at high temperature	Miyazaki et al. (2006)
	Gamma (^{60}Co)	Std. solutions	0.5 mM	20 kGy (100%)	Shim et al. (2009)
	Gamma (^{60}Co)	Std. solutions	0.1 mM	1.0 kGy (100%)	Kozmer et al. (2016)

(Continued)

TABLE 9.6 (Continued)

Substrate	Ionizing Radiation Used	Matrix	Initial Concentration	Dose Required (Yield, %)	References
2,4,6-Trichlorophenol	Gamma (^{60}Co)	Std. solutions river water	24 mg/L 12 mg/L	0.2 kGy (100%) 1.0 kGy (85%)	Trojanowicz et al. (1997)
	Gamma (^{60}Co)	Std. solutions	0.05 mM	0.45 kGy (100%)	Zona et al. (1999)
POLYCHLORINATED BIPHENYLS					
Commercial mixtures of PCBs Arochlor 1254 Delor 103 Kanechlor-300	Gamma (^{60}Co) EB Gamma (^{60}Co)	Alkaline 2-propanol	0.25% 0.5% 0.5%	G(-PCB) = 2.18 M/J 100 kGy (83%) 510 kGy (100% dechlorination)	Singh and Kremers (2002) Mucka et al. (1997) Sawai et al. (1974)
2-Chlorobiphenyl	EB	5% 2-propanol in water	50 M	12 kGy (90%)	Drenzek et al. (2004)
1,2-Dichlorobiphenyl	Gamma (^{60}Co) and EB	50% methanol	1 mM	6 kGy (90%)	Al-Sheikhly et al. (1997)
2,6-Dichloro-, 2,2',6,6'-taterachloro-, and decachlorobiphenyl (PCB 209)	Gamma (^{60}Co)	2% aqueous Triton-100 solution	100 M PCB 209	10 kGy (90%)	Schmelling et al. (1998a,b,c)
2,2',4,4',6,6'-Hexachlorobiphenyl	Gamma (^{60}Co)	Isooctane	100 mg/L	40 kGy (100%)	Mincher et al. (1996)
2,2',3,3',4,5',6,6'-Octachlorobiphenyl	Gamma (spent reactor fuel)	2-Propanol	100 mg/L 194 mg/L	80 kGy (94.5%) 103 kGy (92.3%)	Mincher et al. (1991) Mincher et al. (1994)
PCBs—10 individual congeners	Gamma (^{60}Co)	Methanol	10 mg/L	12.5 kGy (34%–60%)	Lepine and Masse (1990)
PCBs—25 individual congeners	Gamma (spent reactor fuel)	Neutral 2-propanol	200 mg/L 2,2',3,3',4,5',6,6'-octachloro-biphenyl	~90 kGy (90%)	Arbon and Mincher (1994)

VARIOUS OTHER ENVIRONMENTAL POLLUTANTS

Benzene	Gamma (^{60}Co)	Simulated wastewater	100 mg/L	7.75 kGy (99%)	Cooper et al. (1996)
Benzo(a)pyrene	EB	Std. solutions	2 M	0.25 kGy (100%)	Nickelsen et al. (1994)
Bromate	Gamma	50% methanol	5 mg/L	10 kGy (100%)	Butt et al. (2005)
Caffeine	EB	Phosphate buffer	99.8 mg/L	2.0 kGy (100%)	Wang et al. (2016)
2,3-Dihydroxynaphthalene	Gamma (^{60}Co) with H_2O_2 and O_3	Std. solutions	50 mg/L	1.2 kGy with H_2O_2 0.2 kGy with O_3	Torun et al. (2014)
17-Estradiol	Gamma (^{60}Co)	Std. solutions	25 M	1.2 kGy (80%)	Wasiewicz et al. (2006)
Methyl t-butyl ether	Gamma (^{60}Co)	Std. solutions	1.8 nM	10 kGy (98%)	Kimura et al. (2004)
Nitrite	Gamma (^{60}Co)	Std. solutions	148 g/L	120 Gy (100%)	Hsieh et al. (2004)
Ochratoxin A	Gamma (^{60}Co)	Std. solutions	0.3 mg/L	0.5 kGy (98.6%)	Guo et al. (2008a,b)
2,4,6-Trinitrotoluene	EB with H_2O_2	Std. solutions	0.1 mg/L	3 kGy (100%)	Peng et al. (2015)
	Gamma (^{60}Co)	Std. solutions	0.22 mM	8 kGy (40% with N_2, 64% with O_2)	Schmelling et al. (1998a,b,c)

EB, Electron beam; PCB, Polychlorinated biphenyl; Std. solutions, Pure aqueous solutions.

state-of-the-art in this field, to the best of the authors' knowledge, it provides probably about 60%−70% of the published papers in international journals. One can easily see that those studies are predominated by the use of gamma-irradiation from ^{60}Co sources. The EB treatment, and sporadically only gamma-rays from ^{137}Cs sources, or from a spent nuclear reactor fuel, and also in single applications—X-rays are less employed. Those studies, in fact almost fundamental, are most commonly carried out either in pure aqueous solutions or in mixed or organic solvents as in the case of polychlorinated biphenyls (PCBs). Several groups of organic pollutants were distinguished, which are especially intensively examined, and they include dyes, chlorinated hydrocarbons, phenolic compounds, and PCBs. Since 1990s, a more intense interest is focused also on pesticides, which constitute a significant environmental problem. Last but not least, a large increase in interest about pharmaceutical residues was observed in the last decade. This is due to a dramatic increase of use of pharmaceuticals, creating then new environmental problems.

The appropriate choice of the concentration level of the examined pollutant for its radiolytic decomposition, and determination of the yield of decomposition (not only in pure standard solutions, but also in real media) is particularly significant for the real utility usefulness of those investigations. A quite substantial part of those studies were carried out at relatively high concentration levels (sub-mM or mg/L), which are usually much higher, than practically occurring in natural waters. However, this approach simplifies the analytical monitoring of processes taking place during the irradiation. On the other hand, in wastewaters of a different origin this can be quite similar level. Evaluation of the yield of decomposition, even in a very simplified form—by providing the magnitude of absorbed dose need for decomposition—can be a valuable information, and hence those values are shown in Table 9.6.

The optimization of chemical conditions for the most efficient radiolytic decomposition of a given pollutant should involve several important parameters. They should comprise: (1) an initial concentration of decomposed compound, (2) a pH of irradiated solution as it affects the radiation chemical yield of radicals produced in water radiolysis (Fig. 9.1), (3) the level of dissolved oxygen, and/or (4) the presence of radical scavengers. The results of studies on gamma-decomposition of industrial pollutant bisphenol A (BPA) (Guo et al., 2012a,b) are an excellent example showing the influence of those factors on the yield of its decomposition. Fig. 9.8 shows the effect of initial concentration, pH and oxygen content in function of the absorbed dose on the yield of BPA decomposition. In another example selected from published reports, one can find the extent of the decrease of yield of gamma decomposition of X-ray contrast media diatrizoate, observed in several different natural matrices (Tables 9.6 and 9.7) (Han et al., 2002).

The vast literature on radiolysis of different compounds under various conditions contains many other examples of such matrix effects. They concern usually a decrease of the yield of decomposition of the target pollutant. In another example, the gamma-radiolysis of TCE and PCE was examined in the presence of carbonate ions (CO_3^{2-}) (Yoon et al., 2002). The complete decomposition of TCE at 10 mg/L level requires the 300 Gy dose, in which the rate-constants for TCE reactions with $^•OH$ and hydrated electrons are 1.7×10^9 and $1.3 \times 10^{10} \, M^{-1} \, s^{-1}$, respectively. In the presence of 1 mM carbonate, only 60% removal of TCE was observed for 500 Gy absorbed dose. Interestingly, upon addition of 10 mM carbonate decomposition of TCE was stopped. This observation was rationalized in terms of scavenging of $^•OH$ by CO_3^{2-} (the rate constant for reaction $CO_3^{2-} + {}^•OH$ is $3.9 \times 10^8 \, M^{-1} \, s^{-1}$).

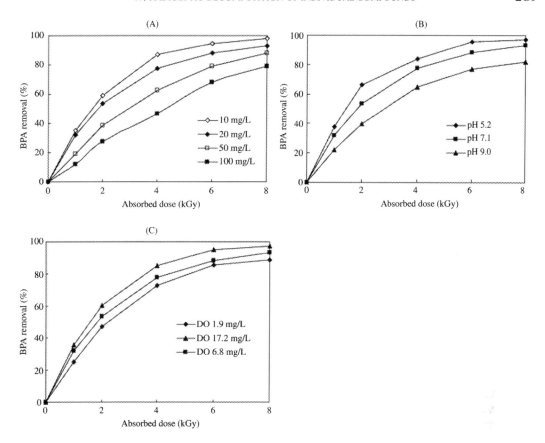

FIGURE 9.8 Effect of different experimental factors on the yield of radiolytic decomposition of industrial pollutant bisphenol A (BPA) by gamma radiation (*With permission, adapted from Guo, Z., Q. Dong, D. He, C. Zhang, 2012. Chem. Eng. J. 183:10–14*). (A) effect of initial concentration (C_0) of BPA at pH 7.1 and dissolved oxygen concentration (DO) 6.8 mg/L. (B) effect of pH of irradiated solution at $C_0 = 20$ mg/L and DO 6.8 M. (C) effect of dissolved oxygen concentration at pH 7.1, and $C_0 = 20$ mg/L.

TABLE 9.7 Inhibitory Capacity of the Water Matrices Evaluated for Gamma Decomposition of X-ray Contrast Media Diatrizoate (Han et al., 2002)

Water	% Inhibition		
	HO$^\bullet$	H$^\bullet$	e_{aq}^-
Ultrapure water (UW)	18.52	20.63	60.85
Surface water (SW)	11.03	27.50	61.47
Groundwater (GW)	8.23	0.60	91.17
Wastewater (WW)	65.72	4.86	29.42

Decomposition of PCE by gamma-irradiation was also examined in the presence of selected heavy metal ions. An inhibition of decomposition was observed in the presence of Cu(II) and Mn(II), contrary to the presence of Co(II) and Ni(II), and especially Fe(III) which enhanced the yield of PCE decomposition (Chung et al., 2002). Gamma-radiolysis of pesticide atrazine is another example of enhancement of decomposition by the components of matrices, in which the presence of humic substances and bicarbonate enhanced the yield of decomposition (Basfar et al., 2009). This effect was attributed to the scavenging of $^\bullet$OH radicals, in case when the target compound reacts faster with hydrated electrons than with $^\bullet$OH.

The fact that experimental conditions employed for complete decomposition of target compounds are practically never sufficient for a complete mineralization (controlled, e.g., by TOC measurements) is a fundamental rule in the majority of those processes. For instance, decomposition of ibuprofen (cited above) with 90% yield by applying 1 kGy dose, in all of examined natural matrices, no change of TOC values was observed (Gala et al., 2013).

The kinetics of radiolytically induced processes are determined by the order and the rate-constants for particular radical reactions, whereas the yield of the irradiation process is commonly expressed by its *G-value*, the value of *dose constant (d)*, and the *dose magnitude* required, e.g., for 50% or 90% decomposition of the solute (Mincher and Curry, 2000). The *G-value* represents a measure of radiation chemical yield, i.e., the number of molecules of reactant consumed, or product formed, per 100 eV of energy absorbed (or in SI units— number of micromoles per one Joule of absorbed energy), and can be calculated for a given absorbed dose using the following equation (Kurucz et al., 1991a,b):

$$G = C_0 - C_D \, N_A \,/ D \, K_f$$

where C_0 is the initial concentration of the analyte in the irradiated solution [M], C_D is the concentration of the analyte [M] after absorbing dose (D), expressed in Grays [Gy], N_A is the Avogadro number (6.02×10^{23} molecules/mol), and K_f is a conversion factor for *G*-value units -from 100 eV/L to Gy ($= 6.24 \times 10^{17}$). Although *G*-values are commonly reported in applied radiolysis, they cannot be considered very useful for predicting the dose required in practice to decompose a compound, since the dose usually depends on its solute concentration, and initial G-values are often calculated for the smallest applied dose at the beginning of the irradiation, i.e., from a linear kinetics approximation that is not fully representative of the actual reaction. The dose constant (*d*) determines the decomposition rate as a function of the absorbed radiation dose, and is calculated as the slope of the plot of $\ln(C_D/C_0)$ against the absorbed dose (D) expressed in (Gy). It is usually considered as a more reliable measure than *G*-value, because it uses all the data from the irradiation procedure. For instance, the *d* values can be used to determine the magnitude of the absorbed dose required for 90% decomposition ($D_{0.90}$) of a compound, as follows: $D_{0.90} = (\ln 10)/d$ (Kurucz et al., 1991a,b). Since the values of *G* and/or *d* are not commonly reported in the papers on radiolytic processes, for the sake of comparison, the values of absorbed doses for decomposition of the target pollutant with a given yield (%) are listed in Table 9.6, together with experimental conditions indicated by authors as optimal.

Another approach in description relating removal and dose are first-order or second-order models, where the changes of pollutant concentration with respect to change in dose are proportional to initial concentration or proportional to the square of initial concentration (Kurucz et al., 2002). Although the fitting of both models to experimental data are not very different, data from both gamma and EB irradiation seem to suggest that removal might be second order with respect to dose for contaminant concentrations under 100 g/L.

It is also shown in Table 9.6 that for the majority of cited cases the absorbed doses needed to 34%−100% decomposition are in the range from fraction to several kGy. The results presented for processes carried out in organic solvents, for instance decomposition of pesticide DDT, and especially PCBs, are exceptions. Studies demonstrating a comparison of radiation processes carried out with gamma and EB irradiation are especially valuable. As it was mentioned above, a key difference between them is a large disparity of dose-rates, which in practice means a very different duration of the process, even of several orders of magnitude. The evident effect of the dose rate only for gamma-irradiation was shown already in very early work on radiolytic dechlorination of DDT in 2-propanol (Sherman et al., 1971). The determined G values (molecules per 100 eV) varied from 25 to 81 for DDT dechloriantion. A large dose-rate difference between gamma and EB irradiation may affect efficiency decomposition of different compounds. However, for instance in decomposition of pharmaceutical metoprolol, only slightly more efficient decomposition was observed in EB process, where for 80% decomposition yield about 16 kGy dose was needed (for relatively large concentration 1 g/L), while 50% larger—in case of gamma-irradiation (Slegers and Tilquin, 2006) (Fig. 9.9B).

A complete removal with both types of radiation required about a 50 kGy dose. A similar observation was reported for calculated dose values for TCE decomposition at initial level of

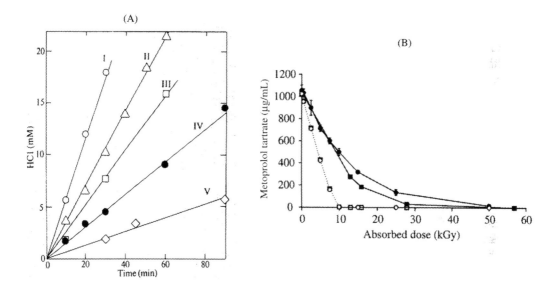

FIGURE 9.9   Dose-rate effects observed in radiolytic decomposition of pesticide DDT (A) and pharmaceutical metoprolol (B). In (A) Effect of dose rate on the dechlorination of pesticide DDT in 2-propanol by gamma-irradiation (*With permission, adapted from Sherman, W.V., R. Evans, E. Nesyto, C. Radlowski, 1971. Nature 232:118−119*): I, 1.36; II, 0.75; III, 0.46; IV, 0.21; V, 0.049 x 10^{18} eV/mL/min. In (B) The dose-rate effect in radiolytic decomposition of metroplol tartrate in function of absorbed dose in 1 mg/mL solutions saturated with nitrogen for irradiations with electron beam (■), gamma irradiation (●), EB simulation (□), and gamma simulation (○). (*With permission, adapted from Slegers, C., B. Tilquin, 2006. Radiat. Phys. Chem. 75:1006−1017*).

8 M, but with slightly faster gamma-process (Kurucz et al., 2002) (Fig. 9.10A, Plot a). The same plot shows also that the complete decomposition with gamma irradiation with the dose-rate 0.15 Gy/s takes almost 30 min, while with use of EB with the dose-rate 2.5 kGy/s − 0.1 s. The calculated concentrations of $^\bullet$OH radicals and hydrated electrons produced by gamma and EB irradiation were significantly different for both kinds of irradiations

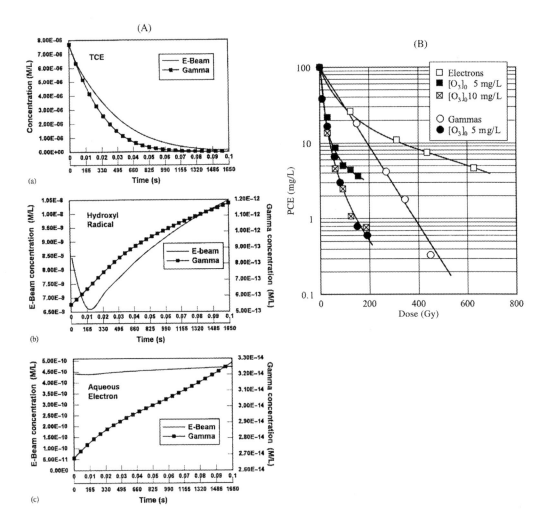

FIGURE 9.10 Comparison of irradiation of chlorinated hydrocarbons by gamma-rays and EB. In (A) Changes of calculated concentrations of trichloroethylene (TCE) (panel a), hydroxyl radicals (panel b), and hydrated electrons (e_{aq}^-) (panel c) in time as result of irradiation with a 0.25 kGy dose using a EB with dose rate 2.50 kGy/s, and by gamma-rays with dose rate 0.15 Gy/s (*With permission, adapted from Kurucz, C.N., T.D. Waite, S.E. Otano, W.J. Cooper, M.G. Nickelsen, 2002. Radiat. Phys. Chem. 65, 367–378*). In (B) Effect of dose-rate and ozone addition on radiolytic decomposition of perchloroethylene (PCE) in tap water using EB irradiation and a 0.9 Gy/s gamma irradiation from ^{60}Co source with dose-rate 0.9 G/s (*With permission, adapted from Gehringer, P., H. Eschweiler, 2002. Radiat. Phys. Chem. 65:379–386*).

(Fig. 9.10A, Plots b and c). A much smaller yield of the EB process in comparison to gamma-irradiation was reported, for instance, for decomposition of perchloroethylene (PCE) in tap water matrix at initial concentration 100 g/L (in the absence of ozone) (Gehringer and Eschweiler, 2002) (Fig. 9.10B). This can be interpreted by the increase of probability of radical recombinations and interradical reactions at large dose-rates. This effect can be significantly damped in the presence of ozone, with simultaneous enhancement of the yield of PCE decomposition as shown in Fig. 9.10B. On the other hand, one can find reports in which no effect of the dose-rate was observed in a certain range, as it was shown for gamma-irradiation of dye Acid Base 62 in dyeing wastewater (Perkowski et al., 2003).

The yield of the radiolytic decomposition may be also improved by conducting it at elevated pressures. This was demonstrated, for instance, for gamma-radiolysis of phenol and benzophenone in aqueous solutions up to supercritical conditions (Miyazaki et al., 2006) (Fig. 9.11). This effect was rationalized in terms of change of the G-value of water decomposition.

A simultaneous monitoring of toxicity of irradiated solutions is another important aspect contributing to the efficiency of radiolytic decomposition studies of pollutants aimed at optimization of the process conditions. This approach provides some additional information, essential from the environmental point of view, as well as health care of humans in case of purification of waters and wastewaters for municipal use. Its importance is based on the fact, that in many laboratory and industrial AOPs, or even occurring in the environmental processes, the 100% decomposition of a given pollutant, may be a source of formation of new products of increased toxicity (Boxall et al., 2004). For instance, in case of radiolytic decomposition of steroid hormones' residues, it was found that in decomposition of 17 -estradiol by gamma-irradiation, the primary products of decomposition also have estrogen activity, and these products should be further decomposed by addition irradiations (Kimura et al., 2004).

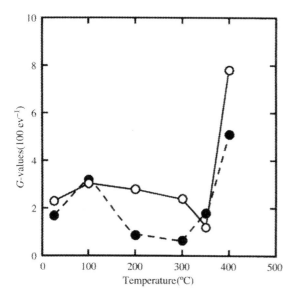

FIGURE 9.11 Temperature dependence of G values in gamma-radiolysis for phenol (●) and benzophenone (○) under pressure of 25 MPa. *With permission, adapted from Miyazaki, T., Katsumura, Y., Lin, M., Muroya, Y., Kudo, H., Taguchi, M., et al., 2006. Radiolysis of phenol in aqueous solution at elevated temperatures. Radiat. Phys. Chem. 75:408–415.*

In recent years, an increasing number of research groups working on radiation processes include toxicity monitoring to their investigations. The most commonly employed toxicity test Microtox is based on the measurement of decay of fluorescence of bacteria *Vibrio fisheri* as a result of an interaction with a given pollutant. Its application to the process of gamma-radiolysis of pharmaceutical ibuprofen showed, that at practically complete decomposition of a target compound, the toxicity of irradiated solution reaches the maximum value (Fig. 9.12A) (Illes et al., 2013). Decline of toxicity is observed with the application of the

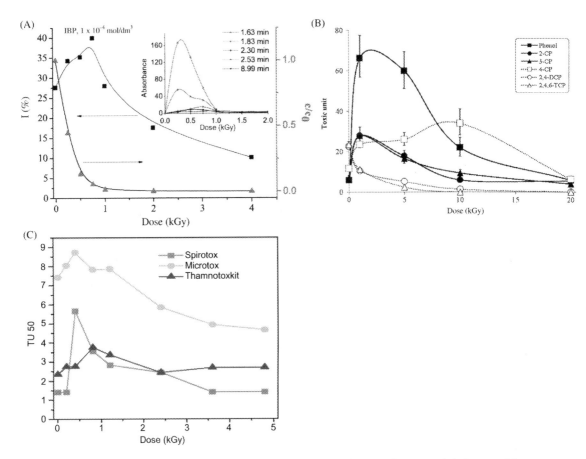

FIGURE 9.12   Toxicity changes recorded for gamma-irradiation of different pollutants. In (A) changes of fluorescence inhibition (I,%) in Microtox toxicity text, and the relative concentration of decomposed pharmaceutical ibuprofen as a function of irradiation dose observed in 0.1 mM solution saturated with air (*With permission, adapted from Illes, E., E. Takacs, A. Dombi, K. Gajda-Schrantz, G. Racz, K. Gonter, L. Wojnarovits, 2013. Sci. Total Environ. 447:286–292*); the inset shows the absorbance changes of products determined after HPLC separation. In B: toxicity changes of phenol and chlorophenols by gamma-ray treatment with initial concentration 0.5 mM (*With permission, adapted from Shim, S.B., H.J. Jo, J. Jung, 2009. J. Radioanal. Nucl. Chem. 280:41–46*). In (C) toxicity changes recorded in gamma-irradiated 50 mg/L solution of pharmaceutical diclofenac measured with different toxicity tests (*With permission, adapted from Bojanowska-Czajka, A., G. Kciuk, M. Gumiela, S. Bobrowiecka, G. Nałęcz-Jawecki, A. Koc, J.F. Garcia-Reyes, D. Solpan-Ozbay and M. Trojanowicz, 2015. Env. Sci. Poll. Res. 22:20255–20270*).

several times larger dose. Similar observations were reported for gamma-radiolysis of phenol and certain chlorophenols (Fig. 9.12B) (Shim et al., 2009). The same trends in toxicity changes in a function of the absorbed dose were reported not only for Microtox but also for two other toxicity tests in gamma-irradiation of diclofenac, the other common pharmaceutical (Fig. 9.12C). Those other tests included Spirotox, based on a very large ciliated protozoma *Sympetrum ambiguum*, and Thamnotox kit, based on using larvae of the freshwater anostracan crustacean *T. platyurus* (Bojanowska-Czajka et al., 2015).

9.5 CHEMICAL ENHANCEMENT OF RADIOLYTIC PROCESSES

The concentration of the primary reactive products derived from water radiolysis, which can potentially react with the solutes present in an irradiated solution, depends on several experimental parameters. Primarily, it is a linear function of the absorbed dose of irradiation and the G-value for particular formed species (equal to the slope of the straight line) which is directly illustrated by plots in Fig. 9.13. The predominance of particular produced reactive species, with a simultaneous enhancement of its concentration, e.g., $^{\bullet}OH$ radicals, can be achieved by irradiation in the presence of the appropriate scavenger (e.g., N_2O), which, however is not applicable for technological installations. As it was already shown, the presence or absence of oxygen in irradiated solutions plays an important role — via occurrence of various other radical reactions and protonation equilibria.

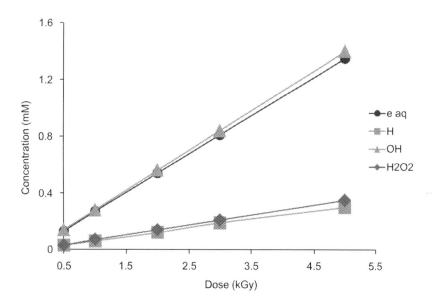

FIGURE 9.13 Estimated concentration of transient reactive species arising from high energy irradiation of water. *With permission, adapted from Mak, F.T., Zele, S.R., Cooper, W.J., Kurucz, C.N., Waite, T.D., Nickelsen, M.G., 1997. Kinetic modeling of carbon tetrachloride, chloroform and methylene chloride removal from aqueous solution using the electron beam process. Wat. Res. 31:219—228.*

The methods that are very widely employed for increasing the yield of formation of $^\bullet$OH radicals are conducting the irradiation in the presence of ozone, which was also carried out in pilot installations shown in Figs. 9.5C and 9.6. Ozone reacts with products of water radiolysis as follows (Sehested et al., 1983):

$$H + O_3 \rightarrow HO_3 \quad k = 3.6 \times 10^{10} \text{ M}^{-1}\text{s}^{-1} \tag{9.24}$$

$$e_{aq}^- + O_3 \rightarrow O_3^- \quad k = 3.7 \times 10^{10} \text{ M}^{-1}\text{s}^{-1} \tag{9.25}$$

and taking into account the protonation reaction $HO_3 \leftrightharpoons O_3^- + H^+$ ($pK_a = 6.1$) also:

$$O_3^- + H^+ \rightarrow OH + O_2 \quad k = 9 \times 10^{10} \text{ M}^{-1}\text{s}^{-1} \tag{9.26}$$

$$OH + O_3 \rightarrow HO_2 + O_2 \quad k = 1.1 \times 10^8 \text{ M}^{-1}\text{s}^{-1} \tag{9.27}$$

$$O_2^- + O_3 \rightarrow O_3^- + O_2 \quad k = 1.5 \times 10^9 \text{ M}^{-1}\text{s}^{-1} \tag{9.28}$$

This way of enhancement of the yield of radiolytic decomposition was employed, e.g., for dyes (Perkowski and Kos, 2003), halogenated hydrocarbons (Gehringer and Matschiner, 1998; Gehringer and Eschweiler, 2002; Jung et al., 2002a,b; Gehringer et al., 1992), different types of pesticides (Leitner et al., 1999; Gehringer and Eschweiler, 1996; Drzewicz et al., 2004; Abdel Aal et al., 2001; Bojanowska-Czajka et al., 2012), pharmaceuticals (Torun et al., 2015; Guo et al., 2012a,b), and phenolic compounds (Sakumoto and Miyata, 1984; Kubesch et al., 2005; He et al., 2002; Kubesh et al., 2003) as it is illustrated by data in Table 9.6. For instance, in the case of decomposition of perchloroethylene in tap water matrix by EB at its initial concentration 100 g/L, in irradiation leading to 95% decomposition, the addition of 5 g/L ozone results in a decrease of the required dose from 600 to 100 Gy (Duarte et al., 2002) (Fig. 9.10B). Moreover, the same figure shows that a smaller effect of the dose rate was observed between EB and gamma irradiation carried out with simultaneous ozonation. In the earlier work, a strong effect of the addition of ozone was observed for decomposition of TCE and PCE in pure aqueous solutions, while no effect of the addition of ozone was found for the EB decomposition of 1,1,1-trichloroethane (Gehringer and Matschiner, 1998) (Fig. 9.14A). The same effect of ozone addition was shown by the same authors for decomposition of TCE in tap water matrix, as well as a dose rate effect between gamma and EB irradiation (Gehringer et al., 1992) (Fig. 9.14B). Interestingly, gamma-irradiation of hexachlorobiphenyl in isooctane performed in the presence of ozone resulted in a drop of efficiency, compared to that in the absence of ozone (Mincher et al., 1996). This was interpreted by scavenging of hydrated electrons by ozone, as $^\bullet$OH radical produced in such a reaction do not react with PCBs.

In the work mentioned above (Gehringer and Matschiner, 1998), the effect of the addition of persulfate anions ($S_2O_8^{2-}$) on the gamma-induced decomposition of 1,4-dioxane and benzene in tap water matrix was also examined. According to authors, hydrated electrons (e_{aq}^-) formed during irradiation react with $S_2O_8^{2-}$ anions forming sulfate radical anions (SO_4^-) which react subsequently with $^-$OH leading to additional amount of $^\bullet$OH radicals:

$$e_{aq}^- + S_2O_8^{2-} \rightarrow SO_4^- + SO_4^{2-} \tag{9.29}$$

$$SO_4^- + OH^- \rightarrow SO_4^{2-} + OH \tag{9.30}$$

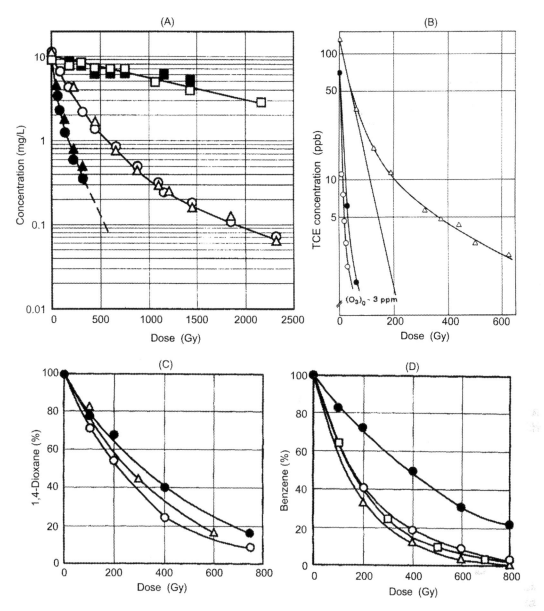

FIGURE 9.14 Effect of ozone (A,B) and persulfate (C,D) on yield of radiolytic decomposition of chlorinated solvents (A,B), 1,4-dioxine (C), and benzene (D). In (A) electron beam induced decomposition of trichloroethylene (△), perchloroethylene (○), and 1,1,1-trichloroethane (□) in deionized water without (open symbols) and with (full symbols) ozone (*With permission, adapted from Gehringer, P., H. Matschiner, 1998. Wat. Sci. Tech. 37:195–201*). In (B) gamma (△) and EB (–) decomposition of trichloroethylene in tap water matrix in the presence (○,●), and absence (△,–) of 3 mg/L ozone (*With permission, adapted from Gehringer, P., E. Proksch, H. Eschweiler, W. Szinovatz, 1992. Appl. Radiat. Isot. 43:1107–1115*). In (C) gamma radiation induced decomposition of 1,4-dioxane in tap water without (●) and in the presence of 1 mM (○) and 2.5 mM (△) persulfate (*With permission, adapted from Gehringer, P., H. Matschiner, 1998. Wat. Sci. Tech. 37:195–201*). In (D) the same as in C for benzene, also for 5 mM added persulfate (□) (*With permission, adapted from Gehringer, P., H. Matschiner, 1998. Wat. Sci. Tech. 37, 195–201*).

It was shown that such processes do not affect decomposition of 1,4-dioxane (Fig. 9.14C), but affect decomposition of benzene (Fig. 9.14D).

Among the other reagents examined for enhancement of radiolytic decomposition of organic pollutants, a large attention was focused on hydrogen peroxide (H_2O_2), which is a common, inexpensive and easy to handle reagent. The addition of H_2O_2 is also a source of an additional amount of $^{\bullet}OH$ radicals formed according to the following reactions (Christensen et al., 1994):

$$e_{aq}^- + H_2O_2 \rightarrow \ OH + OH^- \quad k = 1.5 \times 10^7 \ M^{-1}s^{-1} \tag{9.31}$$

$$H \ + H_2O_2 \rightarrow \ OH + H_2O \quad k = 2.4 \times 10^{10} \ M^{-1}s^{-1} \tag{9.32}$$

Since H_2O_2 reacts, also with $^{\bullet}OH$ radicals:

$$OH + H_2O_2 \rightarrow H_2O + HO_2 \quad k = 3.1 \times 10^{10} \ M^{-1}s^{-1} \tag{9.33}$$

$$2 \ HO_2 \rightarrow H_2O_2 + O_2 \tag{9.34}$$

therefore in certain conditions, the addition of H_2O_2 can cause an opposite effect, namely the scavenging of $^{\bullet}OH$ radicals. The literature data presented in Table 9.6 show numerous applications of H_2O_2 to enhance the yield of radiolytic process for decomposition of dyes (Solpan and Guven, 2002; Wang et al., 2006), halogenated hydrocarbons (Wu et al., 2002), different types of pesticides (Xu et al., 2011; Solpan and Torun, 2012; Drzewicz et al., 2005; Bojanowska-Czajka et al., 2006), pharmaceuticals (Zheng et al., 2011; Guo et al., 2012a,b; Liu and Wang, 2013), phenolic compounds (Basfar et al., 2017; Kubesh et al., 2003), and also caffeine (Torun et al., 2014) or ochratoxin (Schmelling et al., 1998a,b,c).

In order to illustrate more closely irradiation in the presence of H_2O_2 some selected examples are presented. For instance, gamma-induced decomposition of chloroform present at relatively high initial concentration in aqueous solutions (8 g/L) (with 60 kGy absorbed dose) leads only to ~60% decomposition. On the other hand, in the presence of about 60 mM H_2O_2, the 30 kGy absorbed dose is required for 100% decomposition. For gamma-irradiation of aqueous solutions containing polyvinyl alcohol (PVA) at concentrations ranging from 100 to 300 mg/L with different dose rates (17.2–55.7 G/min), addition of 2–3 mM H_2O_2 increased the yield of decomposition (Zhang and Yu, 2004). Depending on the dose rate and the initial PVA concentration this enhancement was in the range of 20%–30%, However, further increase of concentration of H_2O_2 resulted in the lower yield of decomposition.

This way of improvement of radiolytically-induced decomposition processes was also employed in real wastewater samples containing pesticide MCPA (2-methyl-4-chlorophenoxyacetic acid) where the presence of the equimolar amount of H_2O_2 reduced by half the dose required for a complete pesticide decomposition (Bojanowska-Czajka et al., 2006). This is accompanied by an increase of the yield of main decomposition products. Unfortunately, an unfavorable increase of the Microtox toxicity was observed, which was interpreted by formation of products, which are more resistant to radiolysis, and more toxic than the examined pollutant, and main products of decomposition.

The enhancement of radiolytic decomposition by simultaneous production of other reactive radicals can be also utilized for the EB treatment of hazardous substances. This was shown, for instance, for especially strong chemically and thermally resistant industrial pollutant perfluorooctanoic acid (PFOA) (Wang et al., 2016). During its decomposition by EB at the initial level of 0.5 mg/L a delivery of the absorbed dose 10 kGy led to the yield of defluorination in the range of 35%−54%. The addition of 20 mg/L nitrate anions (NO_3^-), or alkalization corresponding to the addition of 100 mg/L $CaCO_3$, allowed to increase that yield up to 93%−95%. In the case of NO_3^-, it was attributed by authors to the formation of oxidizing radicals, as nitrate can scavenge hydrated electrons with the formation of nitrate radical NO_3^{2-}, i.e., able to reduce organic compounds. In the case of high alkalinity, the increase of the yield of defluorination was explained by the formation and reactivity of carbonate radical anion $CO_3^{\cdot-}$.

9.6 PURIFICATION OF WASTEWATERS OF DIFFERENT ORIGIN

Since pioneering investigations on the possibility of application of ionizing radiation techniques for environmental protection and water management, the main target of those studies were, first of all, sludges and wastewaters of a different origin, and to less extent ground waters and surface waters (Dunn, 1953; Touhill et al., 1969; Brusentseva et al., 1978; Perkowski et al., 1984; Woodbridge et al., 1972; Trump et al., 1984; Craft and Eichholz, 1971; Cappadona et al., 1975). A good review of the earliest works is given by Gerrard (1971). Also, in the case of first pilot installations or large-scale treatment plants in various countries, the first works have already been devoted to a treatment of wastewaters or contaminated surface or ground waters (e.g., Cleland et al., 1984; Pikaev and Shubin, 1984; Gehringer et al., 1984).

The main driving force of these research studies was well documented in those years, inability of numerous existing and routinely used technologies, or other technologies being currently investigated, to remove certain groups of hazardous anthropogenic pollutants. So, even in the recent decade one can easily find reports on not complete removal of different groups of pollutants in wastewater treatment plants (WTP) (Petrovic et al., 2003), occurring of numerous pharmaceuticals residues and their metabolites in tap waters in many countries (Mompelat et al., 2009), estrogen hormones in effluents from WTPs (Miege et al., 2009), or pesticide residues in effluents from WTPs (Köck-Schulmeyer et al., 2013). Then, not earlier than a decade or two later, after first experiences with the treatment of real wastewaters, it was realized that there was a need for more detailed, parallel studies on optimizing experimental conditions. This required an understanding of radical processes in order to elaborate the mechanisms of radiolytic decomposition of particular pollutants, especially those most hazardous. About 20 original papers are cited in Table 9.6, in which, the matrix effects of various wastewaters on efficiency of radiolytic decomposition of particular pollutants was examined. The examples of the radiation treatment of real wastewaters reported in the literature are listed in Table 9.8.

Authors of this review do not know the completeness of above information, although they hope, that these examples are sufficiently illustrative to present a variety of experiences collected already by different research groups all over the world during the last four decades. The carried out studies involved typical laboratory investigations on gamma and EB irradiation, and also pilot systems and several large-scale installations for industrial

TABLE 9.8 Examples of Application of Ionizing Radiation for the Treatment of Real Wastewaters of a Different Origin

Origin of Wastewater	Type of Radiation (Employed Absorbed Doses)	Monitored Parameters	Changes of Monitored Parameters	Remarks	References
Municipal	EB (up to 15 kGy)	BOD, COD, sulfates, iron, ammonia, petro-chemicals, surfactants	BOD: 40–3.5 mg/L COD: 120–10 mg/L Sulfates: 256–11 mg/L Elimination of surfactants and petrochemicals	Changes for 1.3 kGy irradiation in aerosol mode	Podzorova (1995)
	EB (up to 5 kGy) with ozone	BOD, COD, color, disinfection, conc. of selected pollutants	3.2 kGy dose required for 90% reduction of COD, decreased to 1.4 kGy in 12 h after EB/O_3 process	Process more cost-effective than EB irradiation alone	Pikaev et al. (1997a,b,c)
Municipal wastewater	Gamma (^{60}Co) up to 10 kGy	Concentration of spiked pharmaceuticals	Clofibric acid, carbama-zepine, diclofenac −100% degradation at 1 kGy; mefanamic acid and keto-profen at 2 kGy	Proposed model for the predicttion of decomposition based on TOC value	Kimura et al. (2012)
Municipal and industrial wastewaters	Gamma (^{60}Co)	BOD, turbidity, dissolved oxygen, disinfection	BOD form 360 to 3 mg/L, 100% removal of E. coli and 99.9% coliform bacteria	No data on dose used	Woodbridge et al. (1972)
	EB (up to 20 kGy) with biological treatment	Concentrations of 11 organic pollutants	Dose for 99% removal varies from 5 to 20 kGy	Yield depends on step of purification process of WTP from which sample was taken	Duarte et al. (2000)
	EB also with ozone	Concentration of 6 spiked industrial pollutants		Irradiation with ozone superior to removal of TOC by other combined methods	Sakumoto and Miyata (1984)
	Gamma (^{60}Co) (up to 20 kGy)	BOD, COD, TOC, inactivation of bacteria	BOD: 8–1 (20 kGy) COD: 24–0 (14 kGy) TOC: 7–4 (20 kGy) All data in mg/L	Yield showed for samples after chlorination	Basfar and Rehim (2002)
	Gamma (^{60}Co) (up to 15 kGy) also with TiO_2	BOD, COD, TOC, color, disinfection	15 kGy reduced BOD by 85%, COD 64%, TOC 34%, and color 88%. Bacteria removed by 0.3 kGy	Secondary effluent for STP TiO_2 enhances purification	Jung et al. (2002a,b)

Waste type	Treatment	Parameters measured	Results	Notes	Reference
	EB (up to 10 kGy)	DOC concentration of selected pollutants	Absorbed dose 2 kGy reduced examined organics by about 2 orders of magnitude	Irradiation without and combined with ozonation	Gehringer et al. (2006)
Hospital wastewater	Gamma (^{60}Co) (up to 5 kGy)	Concentrations of carbamazepine, diclofenac and ibuprofen	At 500 Gy dose complete decomposition of IBP and DCF, and 63% CBZ	Gamma-irradiated waste spiked with 10 g/L of each drug	Bojanowska-Czajka et al. (2015)
Industrial effluents	EB (up to 15 kGy)	Concentrations of 10 organic pollutants	Doses required for 100% decomposition 2–3 kGy	Not fully decomposed with 15 kGy 1,1-dichloroethane, benzene, CCl_4, phenol	Sampa et al. (1998)
	Gamma (^{60}Co) up to 35 kGy after electro-Fenton	COD, color, fecal coliforms	At 35 kGy 95% reduction of COD, 90% color, and 99.9% coliforms		Barrera-Diaz et al. (2003)
	EB (up to 150 kGy) after photocatalytic step	Co, Cu, Sr concentrations	100% removal of 0.57 to 0.85 mM metals at 50 kGy dose	Combined treatment with UV/TiO_2 with humic acid	Zaki and El-Gendy (2014)
Industrial from molasses processing	EB with ozonation	BOD, COD	70% COD reduction with ozone and 2.7 kGy dose	Comparison of yield with ozone/biology system	Gehringer and Matschiner (1998)
	EB (up 20 kGy)	BOD, COD, color, suspended solids	5 kGy COID resuction form 800 to ~40 mg/L	Two cascade electron accelerators with subsequent coagulation	Pikaev et al. (2001)
	EB (up to 20 kGy) with coagulation	BOD, COD, color	3.5 kGy results COD decrease from 800–5200 mg/L to 30–60	Bubbling air irradiation followed by coagulation	Pikaev (2002)
Industrial from chemical plant	EB (20 kGy)	G values for selected organic pollutants	90% removal of the most organic pollutants at 20 kGy		Duarte et al. (2002)
Industrial from pesticide production	Gamma (^{60}Co) (1.25 kGy)	Concentration of pesticides DDT, HCCH, DDVT	Total contaminant decreased from 48 to ~0.1 mg/L	Model effluent. Radiationenhances column adsorption process of purification (removal of DDT)	Brusentseva et al. (1986)

(Continued)

TABLE 9.8 (Continued)

Origin of Wastewater	Type of Radiation (Employed Absorbed Doses)	Monitored Parameters	Changes of Monitored Parameters	Remarks	References
	EB (50 kGy)	Concentration of spiked carbendazim	90% decomposition of CBZ at 50 kGy for 1:10 diluted waste, and 25 kGy for 1:20 diluted	G values for 100 M CBZ in different chemical conditions are in the range 0.123–0.293 M/J	Bojanowska-Czajka et al. (2011)
	Gamma (^{60}Co) (up to 10 kGy)	MCPA concentration and toxicity changes	10 kGy dose needed for 0.5 g/L MCPA decomposition in waste	Examined irradiation in the presence of H_2O_2	Bojanowska-Czajka et al. (2006)
Industrial from textile dyeing	Gamma (^{60}Co) up to 50 kGy with coagulation	BOD, COD, surfactants, color	For 50 kGy: 92% color reduction, 63% COD, 99% removal of anionic surfactants	$FeSO_4$ as coagulant	Perkowski and Kos (1988)
	EB with coagulation	BOD, COD, TOC, Color index	Irradiation with 3 kGy BOD: 1620–700 mg/L TOC: 1000–305 mg/L	Aerated solutions with coagulations	Ponomarev et al. (1999)
	EB (up to 20 kGy) with biodegradation	BOD, COD, TOC	2 kGy dose prior biodegradation reduces treatment time twice for same removal degree	Pilot plant with 1000 m^3/day output	Han et al. (2002)
	EB (1 kGy)	BOD, COD	Increase of BOD/COD ratio from 0.68 to 0.79 after irradiation	Results indicate formation of easier biodegradable species for further treatment	Kim et al. (2007)
	EB (up to 20 kGy)	BOD, COD, TOC	Increase of removal efficiencies up to 30%–40%	Irradiation combined with biological treatment	Kuk et al. (2011); Han et al. (2012)
		Color removal Toxicity	55%–96% removal 33%–55% reduction	Data for three distinct effluents from textile industry	Borrely et al. (2016)
Industrial from paper mill	Gamma (^{60}Co) (0.8 kGy)	AOX, BOD, COD	0.8 Gy dose reduces AOX 75.5%, COD 14.3%, and increases BOD 2.5 times	Decrease of ratio of COD/BOD from 14 to 4 indicates improvement of biodegradability	Wang et al. (1994)

Source	Treatment (dose)	Parameter	Result	Notes	Reference
	Gamma (^{60}Co) up to 70 kGy	AOX	In biotreated effluent 96% AOX removal was in absence of oxygen for dose 10 kGy and pH 12	In aerated solution this yield was 80% for 10 kGy	Taghipour and Evans (1996)
	EB (1 kGy)	BOD, COD, TOC Color indices	Absorbed dose 1 kGy allows to decrease COD and TOC below 25 ppm	Irradiation combined with coagulation + flocculation and biological treatment	Shin et al. (2002)
Industrial from petroleum production	EB (up to 200 kGy)	Benzene, toluene, xylene, phenol and ethylbenzene	100 kGy dose removed more than 90% of all organic compounds	Concentrations of organic species from 15 to 6000 M	Duarte et al. (2004)
Industrial from biological STP	EB (up to 8 kGy)	Benzene, toluene, m-xylene, o-xylene	7.9 kGy for 1 mM benzene removed 90.6%, and for 16 mM—73.4%	Process with 3.3 mM methanol Slightly affected by wastewater matrix compared to tap water	Nickelsen et al. (1992)
	EB (5 kGy)	Concentration of selected surfactants	Decrease of residues to <0.1 ppb at 2–3 kGy absorbed dose		Petrovic et al. (2007)
	EB (up to 10 kGy)	BOD, COD, bacterial counts	1 kGy dose accelerates biodegradation process	10 kGy dose slightly decreases COD, and essentially BOD	Rawat and Sarma (2013)
Industrial from STP (inflow and effluent)	Gamma (^{60}Co) up to 80 kGy	DOC	26% decrease of DOC at 80 kGy	Enhancement to 70% by discontinuous radiolysis	Becerril et al. (2016)
	Gamma (^{60}Co) up to 37 kGy	COD	65% reduction of COD at 18 kGy	Combined with flocculation	Bao et al. (2002)
	Gamma (up to 8 kGy)	Cd and Pb concentrations	40%–60% removal at 8 kGy absorbed dose	Essentially affected by the presence of organic species	Guo et al. (2008a,b)
	EB (up to 500 kGy)	Concentration of elements (Ca, Si, P, Al, Fe, Cr, Zn, Co, As, Se, Cd, Hg)	Removal from 71% to 96% for different absorbed doses from 20 to 500 kGy	>96% removal of ppm levels of Cr, Fe, Zn, and Co at 20 kGy absorbed dose	Ribeiro et al. (2004)
Metalworking fluids	EB (up to 50 kGy)	COD	EB treatment without impact on COD removal	EB pretreatment enabled more effective biotreatment	Thill et al. (2016)

AOX, adsorbable organohalogens; BOD, biological oxygen demand; COD, chemical oxygen demand; DOC, dissolved organic carbon; EB, electron beam; STP, sewage treatment plant; TOC, total organic carbon; WTP, waterworks treatment plant.

applications. A large variety of examined wastewaters obeys municipal (domestic) wastewaters, effluents from different sewage treatment plants (STPs), water works treatment plants, and also selection wastewaters from different branches of industry. In the majority of cases, the efficiency of treatment was monitored by measurements of general water quality parameters, such as biological oxygen demand (BOD), chemical oxygen demand (COD), total organic carbon (TOC), dissolved organic carbon (DOC) or adsorbable organohalogenes (AOX), or microbiologically to determine the degree of disinfection under given conditions and applied doses. In part of the studies, individual control of selected, critical pollutants was also individually performed, usually by chromatographic methods. The example of a wide scope of such approach can be illustrated, e.g., for contaminated groundwater simulants, treated by an EB in a large-scale facility in Miami, FL, USA [73] (Kurucz et al., 1995b) (Table 9.9).

In the g/L concentration level of pollutants to be decomposed, a very efficient decomposition was observed at a dose 8 kGy, and both—compounds oxidized by $^\bullet$OH radicals and reduced by hydrated electrons were decomposed. This shows, that from this point of view, it is a unique process compared to other AOPs. In the earlier work of the same research group and with the same EB installation, it was shown that inactivation of the total bacteria flora by five orders of magnitude was achieved at the dose 6.5 kGy (Kurucz et al., 1991a,b). The effectiveness of decomposition of the same pollutants in wastewater, expressed by the dose needed, was usually several percent lower. It was also observed that solutions containing up to 10% solids (sludges) can be irradiated with the similar yield of decomposition. Experiments were carried out with the flow rate up to 610 L/min.

A detailed monitoring of the concentration of several classes of ionic and nonionic surfactants has shown that they can be very efficiently decomposed in a continuous flow installation by EB irradiation (Petrovic et al., 2007). For their initial concentrations ranging from 1.8 to 20 g/L in the effluent of biological sewage treatment plant, the concentration of all monitored species dropped down to <0.1 g/L at a dose of 3 kGy and flow rate 3 m^3/h (Table 9.10). It was concluded that upgrading of the existing treatment plant by the EB-irradiation as a tertiary treatment, may result in a significant reduction of the discharge of surfactants into the environment.

The EB-treatment of waters and wastewaters may also result in removal of ionized or complexed metal ions (see some examples in Table 9.6). As it was shown in simulated and also real wastewater samples (Ribeiro et al., 2004), at examined concentration level (ranging from 0.5 to 21 g/L in real wastewater), this requires much larger doses, than in the case of trace level of organic contaminants (Table 9.11). The yield of the removal of metal ions can be enhanced in the presence of the $^\bullet$OH radical scavengers, such as formate ions (Pikaev et al., 1997a,b,c; Unob et al., 2003) or ethanol (Chaychian et al., 1998), since the main mechanism of removal involves reduction by hydrated electrons and H atoms, and further precipitation. Further examples comprise, removal of mercury at 1 mM level by using the 3 kGy dose (Chaychian et al., 1998) in the presence of ethanol, or from 5 mg/L solution with 3 kGy dose (Pikaev et al., 1997a,b,c) in the presence of formate and Cr(VI), and also removal of triethyl- and trimethyl-lead at the initial level 0.2 mM with 4 kGy dose, with the yield of about 95% and 80%, respectively (Unob et al., 2003).

Removal of many organic contaminants from waters and wastewaters can be significantly more effective than by biodegradation. This was shown, e.g., in the case of several

TABLE 9.9 Summary of Overall Removal Efficiencies (%) for Various Organic Compounds Tested in Contaminated Water Simulants at the EB Research Facility in Miami, and Bench Scale Studies Using ^{60}Co ([a]) [73]

Compound	Removal (%)	Required Dose (kGy)
Chloroform	83–99	5.86–6.5
Bromodichloromethane	>99	0.80
Dibrochloromethane	>99	0.80
Bromoform	>99	0.80
Carbon tetrachloride	>99	0.80
Trichloroethylene (TCE)	>99	0.57–5.0
Tetrachloroethylene (PCE)	>99	2.41–5.0
Trans-1,2-dichloroethane	93	8.0
1,1-Dichloroethane	>99	8.0
1,2-Dichloroethane	60	8.0
Hexachloroethane	>99	8.0
1,1,1-Trichloroethane	89	6.5
1,1,2,2-Tetrachloroethane	88	6.5
Methylene chloride	77	8.0
Benzene	>99	0.49–6.5
Toluene	97	0.45–6.5
Chlorobenzene	97	6.5
Ethylbezene	92	6.5
1,2-Dichlorobenzene	88	6.5
1,3-Dichlorobenzene	86	6.5
1,4-Dichlorobenzene	84	6.5
1*m*-Xylene	91	6.5
o-Xylene	92	6.5
Dieldrin	>99	8.0
Total phenol	88	0.37–8.0
TNT[a]	40	8.0–17.0
DEMP[a]	90	1.5–7.8
DMMP[a]	90	2.2–9.5

Source: With permission, adapted from Kurucz, C.N., Waite, T.D., Cooper, W.J., 1995b. The Miami electron beam research facility: a large scale wastewater treatment application. Radiat. Phys. Chem. 45:299–308.

TABLE 9.10 EB-Irradiation Induced Elimination of Ionic and Nonionic Surfactants in Effluent From Selected Sewage Treatment Plant in Spain Determined by LC/MS

Compound	Before EB Treatment	After EB Treatment		
		1 kGy	2 kGy	3 kGy
$\sum$LAS	48.1	11.9	4.60	<0.1
C_{10} LAS	2.5	0.62	0.20	<0.1
C_{11} LAS	20	5.1	2.0	<0.1
C_{12} LAS	17	4.2	1.6	<0.1
C_{13} LAS	8.6	2.0	0.80	<0.1
$\sum$AS	82.9	9.81	2.95	<0.1
C_{12} AS	44	6.1	0.35	<0.1
C_{14} AS	27	2.8	0.15	<0.1
C_{16} AS	6.4	0.54	<0.1	<0.1
C_{18} AS	5.5	0.40	<0.1	<0.1
$\sum$AES	32	0.28	<0.1	<0.1
$(C_{10}-C_{18}, n_{EO}=1-10)$				
$\sum$AEO	2.5	<0.1	<0.1	<0.1
$(C_{10}-C_{18}, n_{EO}=1-10)$				
$\sum$CDEA	1.8	<0.1	<0.1	<0.1
(C_7-C_{15})				

LAS, linear alkylbenzene sulfonates; AS, alkyl sulfates; AES, alkylether sulfates; AEO, alcohol ethoxylates; CDEA, coconut diethanol amides. Concentrations are given in g/L.

Source: With permission, adapted from Petrovic, M., Gehringer, P., Eschweiler, H., Barcelo, D., 2007. Radiolytic decomposition of multi-class surfactants and their biotransformation products in sewage treatment plant effluents Chemosphere 66:114–122.

commonly used pharmaceuticals, comparing the yield of removal after 8 h lasting biodegradation with gamma-irradiation using the 2 kGy dose (Fig. 9.15A and B).

In another example the comparison of ten different methods for purification of dyehouse wastewater, including gamma-irradiation with 25 kGy dose without and with oxygenations, and gamma irradiation with 5 kGy with simultaneous ozonation, was carried out (Perkowski and Kos, 2003). The gamma-irradiation is the least efficient, as 50% dyes, only, were removed (Fig. 9.16). In comparison to the most efficient ozonation, or UV irradiation in the presence of H_2O_2 or ozone, only 5 kGy irradiation with ozonation, gives a similar degree of wastewater purification.

The effect of the matrix of the treated medium, assuming that the main reactions occur with $^{\bullet}$OH radicals, can be taken into account quantitatively in the way proposed by Kimura et al. (2012):

$$C_D = C_0 \exp -AD/OC \qquad (9.35)$$

TABLE 9.11 Removal of Selected Elements From Simulated and Actual Industrial Effluents Using EB Treatment

Removed Element	Initial Concentration	Removal (%) at Given Absorbed Dose			
		20 kGy	100 kGy	200 kGy	500 kGy
SIMULATED WASTE					
Se	2 mg/L	—	29.1	61.5	96.5
Cd	15 mg/L	—	21.0	27.0	44.0
Hg(II)	17 mg/L	—	99.0	99.0	99.0
ACTUAL WASTE					
Al	11 g/L	63.6	97.8	97.8	—
Cr	2 g/L	97.3	99.6	99.6	—
Fe	21 g/L	96.2	99.9	99.9	—
Zn	2 g/L	99.95	99.95	99.5	—
Co	0.419 g/L	96.2	96.7	99.8	—

In the case of simulated effluent, the irradiated solution contained sodium formate.
Source: *With permission, adapted from Ribeiro, M.A., Sato, I.M., Duarte, C.L., Sampa, M.H.O., Salvador, V.L.R., Scapin, M.A., 2004. Application of the electron-beam treatment for Ca, Si, P, Al, Fe, Cr, Zn, Co, As, Se, Cd and Hg removal in the simulated and actual industrial effluents. Radiat. Phys. Chem. 71:423–426.*

where C is an initial concentration of decomposed compound, CD —is the concentration after irradiation with a dose D, [OC] is the TOC value for an irradiated medium, while A is described as follows:

$$A = G_{OH} / 1001 \times 6 \times 10^{-19} N_A \quad k_s/k_{OC} \tag{9.36}$$

where G_{OH} is the G-value for $^\bullet$OH radicals formation, N_A is Avogadro's number, k_s and k_{OC} are the rate constants for the decomposed solute and many organic compounds with $^\bullet$OH radicals. The k_{OC} value is approximated as $1 \times 10^8 \, M^{-1} \, s^{-1}$. Based on the proposed relationship, the simulation of radiolytic decomposition of carbamazepine drug was made, assuming that in the real wastewater [OC] is 50 mg/L. The results of simulation are shown in Fig. 9.15C, where a good agreement with experimental data was obtained for [OC] = 50 mg/L. They also show a predicted decrease of the yield of decomposition for much larger wastewater loading at [OC] = 300 mg/L. It shows about a fivefold increase of the required dose for a complete removal of pollutant.

The effect of real matrices by comparing river waters and wastewaters from hospitals on the yield of decomposition of selected drugs and industrial pollutant bisphenol A (BPA) is illustrated in Fig. 9.17. For the initial concentration of carbamazepine 10 g/L (which is 100-fold smaller than used in the data shown in Fig. 9.15C), 90% decomposition in river waters was observed after irradiation with a dose 100 Gy, while in a hospital wastewater spiked with the same amount is 38%, only.

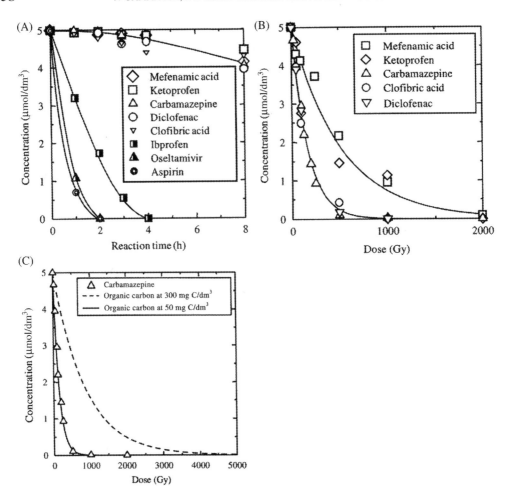

FIGURE 9.15   Comparison of efficiency of biodegradation of selected pharmaceuticals added to municipal wastewater (A), with decomposition by gamma-irradiation (B), and simulation curves for decomposition of carbamazepine in wastewater having organic carbon at 50 and 300 mg/L. *With permission, adapted from Kimura, A., Osawa, M., Taguchi, M., 2012. Decomposition of persistent pharmaceuticals in wastewater by ionizing radiation. Radiat. Phys. Chem. 81:1508–1512.*

In vast literature on this subject one can find numerous other methods of improvement of the yield of radiolytic decomposition of pollutants. A very essential factor is a design of irradiator and the mode of introducing the medium to be treated into the EB due to the limited penetration of aqueous solutions by the beam of electrons (vide supra). The importance of this factor is presented in Fig. 9.18A for the EB irradiation of the municipal wastewater.

It shows how a big difference in the yield of purification (expressed by the COD values) is observed between two irradiated disperse systems: a solution with air bubbling, and an aerosol of a wastewater (Podzorova, 1995). One can estimate that 80% reduction of COD

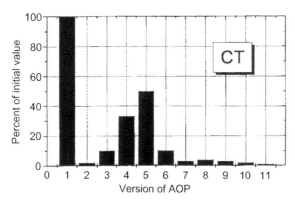

FIGURE 9.16 Comparison of efficiency of different AOP treatments for decoloration of the model dyehouse wastewater expressed by the color threshold (CT) values (*With permission, adapted from Perkowski, J., L. Kos, 2003. Fib. Text. East. Eur. 11:67–70*): 1-untreated sample, 2-ozonation 1.46 g/L/h ozone, 3-H_2O_2 treatment, 4-UV, 5-gamma irradiation with 25 kGy does, 6-gamma irradiation (25 kGy) with oxygenation, 7-UV with ozonation, 8-gamma irradiation (5 kGy) with ozonation, 9-ozonation with H_2O_2, 10-UV with H_2O_2, 11-UV with ozonation and H_2O_2.

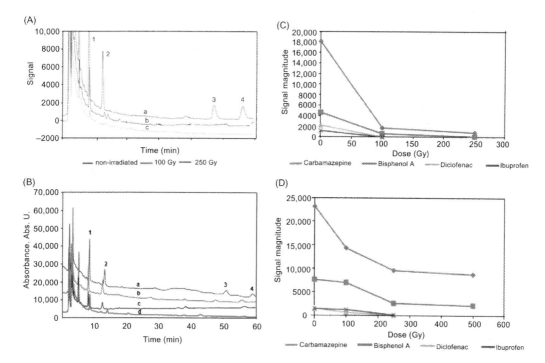

FIGURE 9.17 Effect of natural matrics on the yield of decomposition of selected environmental pollutants using gamma irradiation in river water (A,C), and in hospital wastewater (B,D) spiked with 10 g/L of each pollutant (*With permission, adapted from Bojanowska-Czajka, A., G. Kciuk, M. Gumiela, S. Bobrowiecka, G. Nałęcz-Jawecki, A. Koc, J.F. Garcia-Reyes, D. Solpan-Ozbay, M. Trojanowicz, 2015. Env. Sci. Poll. Res. 22:20255–20270*). A,B: HPLC chromatograms recorded for 250-fold preconcentrated samples for river water (A), and hospital wastewater (B), for samples irradiated prior to the irradiation (a), and after irradiation with 100 (b), 250 (C), and 500 (d) Gy doses; peak assignments: 1-carbamazepine, 2-bisphenol A, 3-dicolfenac, 4-ibuprofen. B,D: HPLC evaluated decomposition of examined pollutants in spiked samples of river water (C), and hospital waste (D) for carbamazepine (♦), bisphenol (■), diclofenac (▲), and ibuprofen (▲).

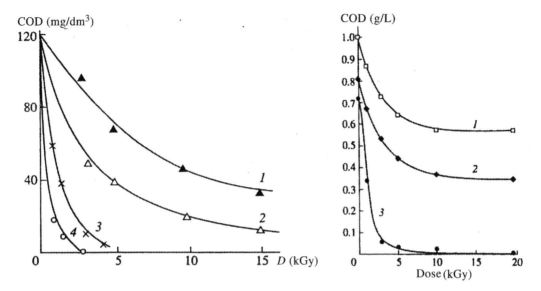

FIGURE 9.18 Illustration of different methods of enhancement of wastewater treatment efficiency by EB-irradiation. A. Effect of different forms of EB irradiation in disperse systems on reduction of COD values of municipal wastewaters (*With permission, adapted from Podzorova, E.A., 1995. High Energy Chem. 29:256–259*): 1,2—bubbling, 3,4—formation of aerosol. B. Effect of combining EB treatment with coagulation on purification of industrial wastewater expressed by COD (*With permission, adapted from Ponomarev, A.V., I.E. Makarov, A.V. Bludenko, V.N. Minin, D.K. Kim, B. Han, A. K. Pikaev, 1999. High Energy Chem. 33:145–149*). 1-deaerated wastewater after filtration, 2-deaerated wastewater after coagulation, 3-aerated wastewater after coagulation.

requires about the 13 kGy dose in the system employing bubbling, while about the 1 kGy dose only, in the case of aerosol, based on the mean values for two different irradiated samples. A significant improvement of the yield of wastewater purification can be also gained by carrying out irradiation with a suspension of TiO_2 (Jung et al., 2002a,b). In the data obtained for gamma-irradiation of a secondary effluent from a sewage treatment plant with 10 kGy, about 12% reduction of COD is observed without TiO_2, and about 60% reduction after adding TiO_2 (Fig. 9.19A). A similar but less pronounced effect was observed by the monitoring of BOD values and changes of color intensity (Fig. 9.19B and C). The observed improvement of efficiency of decomposition was interpreted by a joint effect of a higher production of $^\bullet OH$ radicals in the presence of TiO_2, partial adsorption of pollutants onto TiO_2 surface, and possible certain catalytic effects., It was also observed that the decrease of DOC value was 26% in a continuous irradiation of effluent from WTP in the presence of TiO_2 suspension using 80 kGy dose. On the other hand, using a discontinuous mode, where 3 min break was made after first 20 kGy dose, a significant improvement of the yield up to 74% was observed (Becerril et al., 2016). This was interpreted by a partial enhancement of the yield of $^\bullet OH$ production. In another coupling of photocatalysis with UV irradiation in the presence of TiO_2 (and also with humic acid) conducted prior to EB irradiation with 50 kGy dose, for removal of metal ions from wastewater, a complete removal of 0.57 to 0.85 mM Sr, Co and Cu was observed (Zaki and El-Gendy, 2014). The EB irradiation

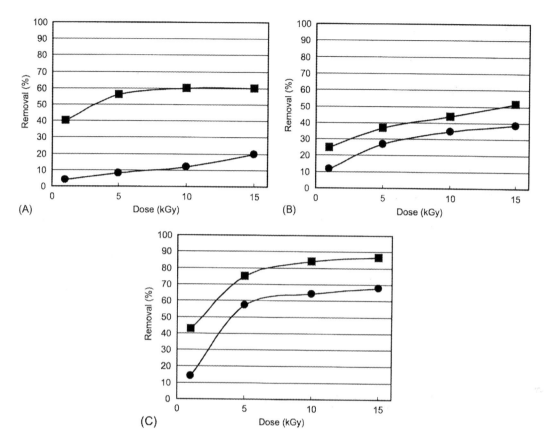

FIGURE 9.19 Effect of the presence of TiO$_2$ suspension in gamma-irradiated secondary effluent from the treatment plant on reduction of COD (A), BOD (B)values, and color removal (C) (*Adapted from Jung, J., J.H. Yoon, H.H. Chung, M.J. Lee, 2002. Radiat. Phys. Chem. 65:533–537. With permission*): (●) without TiO$_2$, (■) with TiO$_2$.

without additional treatment at 125 kGy dose resulted in the removal yield from 41% to 87%.

The improvement of the yield of radiolytic decomposition can also be obtained by hyphenating this process also with some other treatment methods like coagulation reported in several works. For treatment of wastewater for textile dyeing with the initial COD values ranging from 445 to 1450 mg/L it was shown that in the coagulation process with FeSO$_4$, the COD was reduced by 44%, while after preliminary gamma-irradiation with the 50 Gy dose, the COD was reduced by 63% (Perkowski and Kos, 1988). The effects of a similar approach were equally satisfactory for the EB treatment of industrial wastewater from the molasses distillery (Pikaev, 2002; Ponomarev et al., 1999). The irradiation after filtration, only, and deaeration gave about 40% reduction of COD with 10 kGy dose, while the same treatment with preliminary coagulation gave almost 100% COD reduction (Fig. 9.18B). Certain enhancement of the radiolytic reduction of COD in effluent from WTP can be also obtained when the gamma-irradiated solution contains a cationic copolymer, as a result of a flocculation process (Bao et al., 2009). The preliminary gamma-irradiation of a domestic sewage with 5.5 kGy dose can enhance reduction of dissolved carbon by precipitation with Ca$(OH)_2$ to 26.2%, while without irradiation that reduction was 17.1% (Campbell, 1971).

The use of adsorption on activated carbon by additional gamma-irradiation of the adsorptive column is another example of an approach used for the enhancement of wastewater purification (Brusentseva et al., 1986). For the simulated industrial wastewater containing mg/L level of three pesticides and also some petroleum products, it was shown that the irradiation of 2 m adsorptive column with 1.25 kGy dose resulted in almost twice the increase of the column lifetime, which was interpreted by partly occurring destruction of contaminants being adsorbed on a carbon surface.

Some support of gamma-radiolytic treatment of the wastewater from the sewage stream of an industrial complex can be provided by electrochemical oxidation by the use of electro-Fenton process in the presence of H_2O_2 added to solution with an iron anode (Barrera-Diaz et al., 2003). While gamma-irradiation with 17.5 kGy dose provided 71% reduction of initial COD value 3400 mg/L, with preliminary electrooxidation the COD value was reduced to 80%.

A degree of mineralization of organic pollutants in waters and wastewaters by the use of not too large radiation doses (several kGy) may be significantly increased when irradiation is followed by biodegradation. Even partial radiolytic degradation of compounds resistant to biodegradation creates better conditions for more efficient biodegradation. This was also proved for wastewaters of a different origin. For instance, for a dyeing complex wastewater in continuous EB treatment with a flow rate 1000 m^3/day, it was shown that 40% reduction of TOC by biodegradation requires about 19 h, while after irradiation with 2 kGy dose, it has shortened to 10 h (Han et al., 2002). In domestic wastewater about 20% enhancement of biodegradation yield expressed by COD value was reported after the EB irradiation with 1 kGy dose (Rawat and Sarma, 2013). Simultaneously observed, the increase of the BOD/COD ratio with the increase of the absorbed dose confirms the conversion of the non-biodegradable contaminants into biodegradable compounds. A similar effect can be reached in the case of a sequential EB-irradiation, biodegradation, zero-valent iron treatment and biodegradation to purification of heavy-loaded and toxic metalworking fluids (Thill et al., 2016). The preliminary irradiation enhances further decrease of COD by biodegradation by about 70% for a pristine fluid, and about 24% for an exhausted fluid.

The process of radiolytic treatment of wastewaters is in large-scale industrial processing plants incorporated into the integrated, multistep systems for remediation of wastewaters of a different origin. Most likely, the first system was designed with gamma-irradiation step as a commercial treatment plant near Palmdale, Florida already in the early 1970s (Mann, 1971). It consisted of aeration, biodegradation, irradiation, primary filtering, and optional sorption on activated carbon. With the capacity 38 m^3/day, this setup was used for sewage processing, in which after application of 0.5 kGy dose, a decrease of BOD from 65 to 6 mg/L was obtained, and the yield of disinfection was four orders of magnitude for bacteria population.

Such multistep purification systems for wastewater treatment are much more often designed with the use of electron accelerators. The example of such installations is the system based on three electron accelerators with a total beam power 300 kW for treatment of

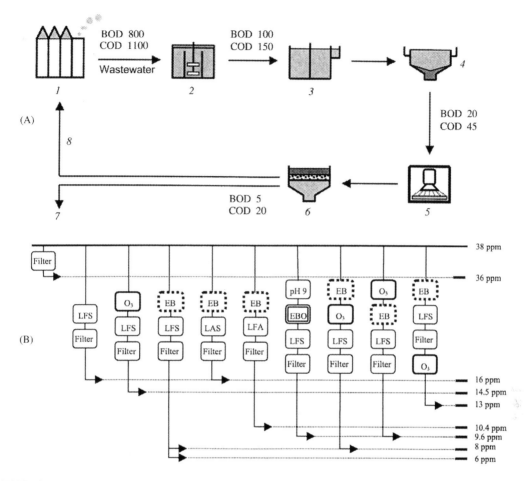

FIGURE 9.20 Process flow of the electron beam facility for paper mill wastewater treatment (A) and total organic compounds (TOC) values after treatment with different configuration of the treatment steps (B). In (A) 1-input of wastewater, 2-first coagulation, 3-biological treatment, 4-sedimentation, 5-EB treatment with subsequent coagulation, 6-filtration with sand filter, 7-effluent (20%–30%), 8-recirculated wastewater (70%–80%). In (B) EB-EB treatment, EBO-EB with ozonation, LFS-treatment with ferric sulfate coagulant and with polyacrylamide solution, LAS-treatment with aluminum sulfate coagulant and then by polyacrylamide solution, O₃-ozonation. *With permission, adapted from Shin, H.S., Kim, Y.R., Han, B., Makarov, I.E., Ponomarev, A.V., Pikaev, A.K., 2002. Application of electron beam to treatment of wastewater from papermill. Radiat. Phys. Chem. 65:539–547.*

industrial wastewater from paper mill with output 15000 m³/day (Shin et al., 2002). The whole setup consists of several conventional treatment devices for coagulation, biological treatment, and filtration. The suitable dose of radiation was evaluated as 1 kGy for the wastewater with above mentioned flow rate (Fig. 9.20A). As it is shown by the scheme in, for a treatment of wastewater with the initial values of COD and BOD 1100 and 800 mg/L, respectively, depending on combination of different processing steps, a different degree of reduction of COD (and also BOD) can be obtained (scheme in Fig. 9.20B). Under optimized conditions of the treatment, 80% of the treated wastewater can be reused for a paper production, with estimated cost of treatment 1.03 USD/m^3. Some more details on employed 1 MeV accelerator with irradiators can be found in the paper on its application for treating wastewater from the textile dyeing industry (Kuk et al., 2011).

The efficiency of purification of textile sewages and municipal wastewater was also examined in the multistep system, combining EB-irradiation with a 10 MeV accelerator (with maximum power 100 kW), coagulation, biodegradation, and absorption using activated carbon and ion-exchangers (Emami-Meibodi et al., 2016). For the textile dyeing wastewater with initial COD value 982 mg/L, irradiation with 1.1 kGy dose combined with coagulation provided 50% reduction of COD. Moreover, with a further biodegradation step this yield was increased up to 94.6%. The greatest effect of the EB-irradiation was observed in discoloration measurements for that system.

9.7 ECONOMIC ASPECTS

The fundamental conditions of industrial application of a given process or technology must always take into account a deep and competent analysis of its cost-efficiency. Because of the availability of numerous alternative methods for wastewater treatment, the application of ionizing radiation for this purpose requires a thorough consideration of all factors contributing to the final cost of application of this technology—usually expressed at the end by the cost of treatment per volume unit of treated fluids. It seems, however, that all related criteria have to be sometime reevaluated, when particular technology is not optimal in terms of the final cost, but this is the only technology that can remove/decompose certain group of pollutants of unusually strong chemical resistance.

As it has been already mentioned, the concept and pioneering investigations on application of radiolytic treatment of wastewaters were initiated in the early 1950s (Gerrard, 1971 and references herein). Already then, in the report from the Illinois Institute of Technology Research, the irradiation with the isotopic sources was one out of the several recommended for further study based on the comparison of ten sewage-processing methods (UV, chemical, thermal, and several types of ionizing radiation).

The economic analysis of radiation processes was also carried out since the 1950s, including among others, even such an opinion, that building a pipeline 5 miles long to dump sewage in the ocean, far from the shore, is cheaper than using an irradiation process and a short pipeline (Gerrard, 1971). In one of the first ^{60}Co installations designed for the wastewater treatment by the Florida Institute of Technology, and built in Palmdale, Florida in 1970, the cost of wastewater treatment was evaluated as 0.39 USD/m^3, and the effluent was pure enough to use for irrigation (Woodbridge et al., 1970). Detailed analysis of the cost-efficiency for ^{60}Co source employed for radiation treatment of industrial wastewaters was carried out with assumption that a reasonable cost of treatment should be between about 0.05 and 0.26 USD/m^3 (Mytelka, 1971). The reported considerations involved the application of ^{60}Co gamma source providing unit cost per 1 Curie radioactivity and a possible capacity in flow installations with flow-through irradiator employing ^{60}Co rods. The conclusions were based on experimental data on doses required for efficient treatment of simulated wastewater containing surfactant 2-naphthalene sulfonic acid sodium salt and a real one with Chlorantine Fast Red dye. For the cost-efficiency, the overall efficiency of oxidation reactions in those two cases must be increased by factors from 2 to 25 times, which required a better understanding of the chemistry involved. This clearly justifies a vast following of literature on radiolytic decomposition of different classes of environmental pollutants.

The evaluated costs for conducting the radiolytic gamma treatment with other isotopic source, ^{137}Cs, are substantially higher, which was given in the paper on decomposition of X-ray contrast media diatrizoate in different water matrices (Gala et al., 2013). Evaluation was carried out for decomposition of 25 mg/L concentration, in which a complete decomposition requires 1 kGy dose. Based on energy that provides the disintegration of ^{137}Cs, the cost of radioisotope needed to get required activity, and assumed flow rate 1 m^3/s, the treatment price was calculated as 16 USD/m^3. This is almost two orders of magnitude higher cost than that evaluated for different installation involving electron accelerators (see Table 9.12).

With no doubt, much more attention is focused on technological applications of electron accelerators for this purpose. Several attempts of estimating the cost efficiency of such systems were reported in the literature. In all cases they obviously point out a high capital cost requirement, which is estimated between 1700 and 5000 kUSD (Table 9.12). This includes first of all, the cost of the accelerator, but also auxiliary equipment, transport, construction, and installation. The cost of the accelerator needed for particular application depends on required dose (D [kGy]) and assumed flow-rate (throughput capacity) of the

TABLE 9.12 Examples of Evaluation of the Cost Efficiency of the Electron Beam Irradiation Facilities for the Waters and Wastewaters Treatment

Type of Installation	Application	Capital Requirement (kUSD)	Treatment Capacity	Cost of Treatment (USD/m^3)	Reference
EB (0.5 MeV, 25 kW) with ozonation	Ground water remediation	1720	1200 m^3/h	0.075	Gehringer et al. (1995)
EB (1.5 MeV, 75 kW)	Wastewaters (for kGy dose requirement)	2350	36.6 m^3/h	0.65	Kurucz et al. (1995b)
EB (1 MeV, 40 kW) with several conventional methods	Papermill wastewater	5000	15,000 m^3/day	1.03	Shin et al. (2002)
EB (0.5 MeV, 75 kW) with ozonation	Wastewater from molasses processing	2300	50 m^3/h	3.17	Gehringer and Fiedler (1998)
EB (5 MeV, 400 kW) with coagulation	Wastewater from molasses processing	2110	7000 m^3/day	0.25	Pikaev et al. (2001)
EB (1 MeV, 400 kW) with biodegradation	Textile dyeing wastewater	4000	10,000 m^3/day	0.3	Kuk et al. (2011) Han et al. (2012)
EB (1 MeV, 400 kW)	Mixed municipal and industrial	4000	575 m^3/h	0.2	Gehringer et al. (2006)

wastewater through the installation (W [kg/h]). It raises with increase of capacity and required dose, leading to the need of use of accelerator of higher beam power (P [kW]), according to the equation (Gehringer et al., 1995): $W = 3600 \, PD^{-1}\eta$, where η is the beam utilization factor. The η value is evaluated dosimetrically, and its value is usually between 0.6 and 0.7. The essential feature of EB accelerators is also the efficiency of conversion of electricity power to EB power, which for modern accelerators reaches 80%, while in the case of UV lamps is usually 20%−30%. An additional advantageous feature of those devices, mentioned already, is extremely high dose-rate reaching MGy/min.

With a given employed electron accelerator operating with a particular power of beam, the volume capacity of treatment will essentially depend on the magnitude of dose required for efficient decomposition of wastewater with particular types of contaminants. So, for instance in installation with 1.5 MeV accelerator working at power 75 kW, at required dose 5 kGy, the flow rate should be 610 L/min (36.6 m^3/h), while for a very small dose 50 Gy sufficient for disinfection—8043 L/min (483 m^3/L) (Kurucz et al., 1995b). It is obvious, that this factor significantly affects the cost efficiency. For the optimized dose 8 kGy the total treatment cost was estimated approximately 0.4−0.5 US cents/L. Another essential factor is the operation cost, including fixed costs of interest and depreciation, and variable costs of electricity, labor, maintenance etc. (Han et al., 2012).

Several examples of the evaluation of cost-efficiency for different reported installations, expressed by the cost of treatment of cubic meter of wastewater are listed in Table 9.12. In the case of installation for the treatment of textile dyeing wastewater, involving three window accelerators (1 MeV, 400 kW), with using 1 kGy dose irradiation, and followed by bio-degradation, the cost of treatment was evaluated as 0.3 USD/m^3 at treatment capacity 10,000 m^3/day (Han et al., 2012). This is comparable to the cost 0.25 USD/m^3 for a groundwater treatment in an earlier reported installation with accelerator 0.5 MeV and 40 kW beam power at 146 m^3/h (Gehringer et al., 1995). The application in the same work of simultaneous ozonation of irradiated solution at treatment capacity 1200 m^3/h, resulted in cost reduction down to 0.075 USD/m^3. The same EB/O$_3$ system but with beam power 75 kW was applied at 50 m^3/h capacity for treatment of wastewater from molasses processing, in which operation parameters included the BOD value of treated medium not exceeded 20 mg/L and 70% reduction of COD, which required the use of absorbed dose 2.7 kGy (Gehringer and Fiedler, 1998). In this case, the cost of treatment increased to 3.17 USD/m^3, and was comparable to conventional ozone/biological treatment.

The cost of EB treatment of wastewater from paper mill was compared with other existing technologies. The EB cost 1.03 USD for EB treatment was lower than three other technologies, namely the use of activated carbon filter including secondary coagulation (1.22 USD), reverse osmosis (1.67 USD), and evaporation (over 3.00 USD) (Shin et al., 2002). As the example of a more detailed specification of different costs involved in industrial application of EB for treatment of industrial wastewater from textile dyeing industry, Table 9.13 presents the construction and operating cost for the system installed in 2005 in South Korea (Kuk et al., 2011).

Another measure suggested for economic comparison of wastewater treatment methods is evaluation of the *electrical energy per order* (EE/O) figure-of-merit, which is defined as the electrical energy (kWh) required to reduce the concentration of the pollutant by an order of magnitude in 1000 L volume of liquid (Woodbridge et al., 1970). The EE/O value

TABLE 9.13 Costs of an Industrial Plant Employing the Electron Accelerator for Treating Industrial Wastewater From Textile Dyeing, Developed and Built by EB-Tech Co. Ltd. and Operated Since 2005 in South Korea (All Costs in M USD)

CONSTRUCTION COST		
	Cost	Remarks
Accelerator—1 MeV, 400 kW	2.0	
Water reactor and raw material.		Cost for land, R&D, and approval form authorities are not included
Installation cost-welding/piping/ inspection etc.	1.5	
Design		
Shield room & construction		
Transportation, tax, others	0.5	
Total	**4.0**	

OPERATING COST			
Items		Addition of electron beam	
Operating cost	Interest	0.24	6%
	Depreciation	0.20	20 years
	Electricity	0.32	800 kW
	Labor	0.10	3 shift
	Maintenance, etc.	0.08	2%
Total		**0.94**	**~1 M USD/year**

Source: *With permission, adapted from Kuk, S.H., Kim, S.M., Kang, W.G., Han, B., Kuksanov, N.K., Jeong, K.Y., 2011. High-power accelerator for environmental applications. J. Kor. Phys. Soc. 59:3485-3488.*

is specific for decomposition of the particular compound in strictly defined experimental conditions, and for given initial concentration. Their values evaluated, e.g., for 10 different organic small molecule contaminants and Methylene Blue dye, can be spanned for different conditions, for about one order of magnitude for each (Cooper et al., 2001). Among examined compounds, the smallest values 0.05–0.5 kWh/1000 L/order were found for Methylene Blue due, whereas the largest, which means the need of the use of largest doses for radiolytic decomposition—for chloroform (0.5–3.2 kWh/1000 L/order). Based on their values, one can evaluate the dose (D) required for the decrease of initial contaminant concentration C_0 to required values C_D in a given radiolytic process according to the following expression (Cooper et al., 2001):

$$D = EE/Ox \log C_0/C_D \qquad (9.37)$$

For mixture of different compounds that simultaneously decompose during irradiation (common condition in the irradiation of real wastewaters), the energy required for a treatment is not additive. It is determined by the decomposition of the most resistant component for which the largest dose is required.

Based on the bleaching of Methylene Blue and the decay of phenol, the EE/O values were compared for EB, UV/H_2O_2, and UV/TiO_2 AOPs, and values < 3 were reported for EB and UV/H_2O_2, while > 50 for UV/TiO_2 (Bolton et al., 1998). This was attributed to the efficiency of generation of $^\bullet$OH radicals, which was reported as 1.0 M/kWh, while 1.4 for UV/H_2O_2 and 0.087 for UV/TiO_2. A similar comparison was reported for the decomposition of two antibiotics sulfamethazole (SMX), and chlortetracycline (CTCN) in model aqueous solutions (Kim et al., 2012). The energy consumption EE/O parameter was compared for EB process carried out with 1 MeV, 40 kW accelerator, ozonation, and UV irradiation at 210 nm. The obtained values EE/O 0.46, 27.53 and 1.50 for SMX, and 0.19, 7.15, and 15.5 for CTCN, for EBV, ozone, and UV, respectively, show that among investigated AOPs, the EB treatment was electrically more energy efficient than ozonation or UV irradiation. It was concluded, that for technological choosing the type of AOP such factors like chemicals, capital, construction, and maintenance should be considered all together.

9.8 CONCLUSIONS

A limited efficiency and poor management of water systems is nowadays a fundamental issue for human water security and ecosystem sustainability (Grant et al., 2012). Among processes responsible for pollutant removal in water and waste purification systems, the last decades brought significant development of novel processes based on efficient reactions of free radicals, where of special importance are advanced oxidation processes based on the use of hydroxyl radicals $^\bullet$OH. One of the most efficient ways of their production is water radiolysis by the use of ionizing radiation, although surprisingly, it is often even not mentioned in extensive and comprehensive reviews of the field.

The effectiveness of radiolytic decomposition carried out by ionizing radiation was demonstrated already for numerous groups of organic compounds, to which many hazardous, anthropogenic environmental pollutants belong. Since the middle of the previous century, applications of this methodology are demonstrated by various research groups all over the world, and in various pilot installations, for the treatment of wastes of a different origin. Especially for such complex media, it is important that it is a nonselective process, which can be used for contaminants of different structures and properties. Its unique advantage to other AOP methods is the possibility of carrying out parallel the decomposition both on the oxidative and reductive ways. In addition, it does not require basically the use of any additional reagents, although in some cases, a simultaneous presence of some chemicals, supporting production of active radicals, may enhance the yield and improve the cost-efficiency of the process. Another important advantage of the EB-irradiation, compared to other AOPs is the speed of radiolytic processes (occurring in less than 1 s), incomparable to other methods, that results from EB dose-rate, evaluated, e. g., for 1 MeV, 40 kW accelerator as 40 kGy/s (Kuk et al., 2011). This technology when used for waste treatment does not exhibit any danger of inducing radioactivity in irradiated media, and also can be considered as a "clean technology," producing no organic sludge or air emission. Therefore, it can be also employed efficiently as a pretreatment step prior to the application of biodegradation.

Studies on application of this technology for treatment of waters and wastewaters were especially intensively carried out by numerous research groups in the 1980s and 1990s. In spite of the construction of many pilot installations, they still have not gotten a deserved recognition for the industrial applications, although some groups demonstrated their numerous advantages, including cost efficiency. It seems that the main barriers for such applications are high capital requirements in comparison to other instrumentally simpler AOP methods, rather limited knowledge about safety of those processes, and also often irrational fear of nuclear technology. The last couple decades have brought enlarging knowledge about the increase of danger by numerous classes of environmental pollutants (industrial, residues of pharmaceuticals, endocrine disruptors etc.), of which many are strongly resistant, and are not eliminated from waters and wastes completely by conventional methods. This creates an authentic need to search for especially efficient, nonselective methods of their decomposition. This results in a new trend of construction of hybrid installations, in which a particularly efficient additional step is the use of EB irradiation, added to conventional physico-chemical and biological treatment. In such circumstances the radiolytic processes can be accrued out at economically acceptable doses, as it was shown, e.g., for treatment of textile dyeing wastewaters (Han et al., 2012), or metalworking fluids (Thill et al., 2016). This may also encourage some new interesting applications in the future for the improvement of environmental protection and more efficient regeneration of water systems for human security. This is also associated with the progress in the accelerator technology, which is expressed not only by their increasing production for different industrial applications, but also by lowering their costs, increase of available dose rates, and reduction of size, which widens the spectrum of their applications, including building of installations for wastewater treatment of different scale.

Acknowledgments

This work was partly supported by a grant from the Polish National Center of Science (NCN); project OPUS 8, number 2014/15/B/ST4/04601. One of us (KB) would like to acknowledge the US Department of Energy Office of Science, Office of Basic Energy Science under Award Number DE-FC02-04ER15533. This is document number NDRL-5178 from the Notre Dame Radiation Laboratory.

References

Abdel Aal, S.E., Dessouki, A.M., Sokker, H.H., 2001. Degradation of some pesticides in aqueous solution by electron beam and gamma-radiation. J. Radioanal. Nucl. Chem. 250, 329–333.

Al-Sheikhly, M., Silverman, J., Neta, P., Karam, L., 1997. Mechanisms of ionizing radiation-induced destruction of 2,6-dichlorophenyl in aqueous solutions. Environ. Sci. Technol. 31, 2473–2477.

Arbon, R.E., Mincher, B.J., 1994. γ-ray destruction on individual PCB congeners in neutral 2-propanol. Environ. Sci. Technol. 28, 2191–2196.

Ballantine, D.S., 1971. Potential role of radiation in waste-water treatment. Isot. Radiat. Technol. 8, 416–420.

Ballantine, D.S., Miller, L.A., Bishop, D.F., Rohrman, F.A., 1969. Practicality of using atomic radiation for wastewater treatment. J. Wat. Poll. Cont. Fed. 41, 445–458.

Bao, H., Liu, Y., Jia, H., 2002. A study of irradiation in the treatment of wastewater. Radiat. Phys. Chem. 62, 633–636.

Bao, H., Gao, J., Liu, Y., Su, Y., 2009. A study of biodegradation/γ-irradiation on the degradation of p-chloronitrobenzene. Radiat. Phys. Chem. 78, 1137–1139.

Barrera-Diaz, C., Urena-Nunez, F., Campos, E., Palomar-Pardave, M., Romeri-Romo, M., 2003. A combined electrochemical-irradiation treatment of highly colored and polluted industrial wastewater. Radiat. Phys. Chem. 67, 657–663.

Basfar, A.A., Rehim, F.A., 2002. Disinfection of wastewater form Riyadh wastewater treatment plant with ionizing radiation. Radiat. Phys. Chem. 65, 527–532.

Basfar, A.A., Mohamed, K.A., Al-Abduly, A.J., Al-Kuraiji, T.S., Al-Shahrani, A.A., 2007. Degradation of diazinon contaminated waters by ionizing radiation. Radiat. Phys. Chem. 76, 1474–1479.

Basfar, A.A., Mohamed, K.A., Al-Abduly, A.J., Al-Shahrani, A.A., 2009. Radiolytic degradation of atrazine aqueous solutions containing humic substances. Ecotox. Environ. Safety 72, 948–953.

Basfar, A.A., Muneer, M., Alsager, O.A., 2017. Degradation and detoxification of 2-chloro-phenol aqueous solutions using ionizing gamma radiation. Nukleonika 62, 61–68.

Becerril, J.J., Lopez, A.M., Reyes, M.J., 2016. Radiocatalytic degradation of dissolved organic compounds in wastewater. Nukleonika 61, 473–476.

Bensasson, R.V., Land, E.J., Truscott, T.G., 1993. Excited States and Free Radicals in Biology and Medicine. Contributions From Flash Photolysis and Pulse Radiolysis. Oxford Univesity Press Inc., New York.

Bielski, B.H.J., Cabelli, D.E., Arudi, R.L., Ross, A.B., 1985. Reactivity of $HO2/O^{-}_{2}$ radicals in aqueous solution. J. Phys. Ref. Data 14, 1041–1100.

Bobrowski, K., 2012. Radiation induced radical reactions. In: Chatgilialoglu, C., Studer, A. (Eds.), Encyclopedia of Radicals in Chemistry, Biology and Materials, Vol. 1. John Wiley & Sons Ltd, Hoboken, NJ, pp. 395–432.

Bobrowski, K., 2017. Radiation chemistry of liquid systems. In: Sun, Y., Chmielewski, A.G. (Eds.), In Applications of Ionizing Radiation in Material Processing. Institute of Nuclear Chemistry and Technology, Warszawa, pp. 81–116.

Bojanowska-Czajka, A., Nichipor, H., Drzewicz, P., Kozyra, C., Nałęcz-Jawecki, G., Sawicki, J., et al., 2006. Radiolytic degradation of herbicide 4-chloro-2-methyl phenoxyacetic acid (MCPA) by gamma radiation for environmental protection. Ecotox. Environ. Safety 65, 265–277.

Bojanowska-Czajka, A., Drzewicz, P., Zimek, Z., Nichipor, G., Nałęcz-Jawecki, G., Sawicki, J., et al., 2007. Radiolytic degradation of pesticide 4-chloro-2-methyl-phenoxyacetic acid (MCPA) — experimental data and kinetic modelling. Radiat. Phys. Chem. 76, 1806–1814.

Bojanowska-Czajka, A., Gałęzowska, A., Marty, J.L., Trojanowicz, M., 2010. Decomposition of pesticide chlorfenvinphos in aqueous solutions by gamma-irradiation. J. Radioanal. Nucl. Chem. 285, 215–221.

Bojanowska-Czajka, A., Nichipor, H., Drzewicz, P., Szostek, B., Gałęzowska, A., Męczyńska, S., et al., 2011. Radiolytic decomposition of pesticide carbendazim in waters and waste for environmental protection. J. Radioanal. Nucl. Chem. 289, 303–314.

Bojanowska-Czajka, A., Torun, M., Kciuk, G., Wachowicz, M., Solpan-Ozbay, D., Guven, O., et al., 2012. Analytical and toxicological studies of decomposition of insecticide parathion after gamma-irradiation and ozonation. J.AOAC Intern. 85, 1378–1384.

Bojanowska-Czajka, A., Kciuk, G., Gumiela, M., Bobrowiecka, S., Nałęcz-Jawecki, G., Koc, A., et al., 2015. Analytical, toxicological and kinetic investigation of decomposition of the drug diclofenac in waters and wastes using gamma radiation. Env. Sci. Poll. Res. 22, 20255–20270.

Bolton, J.R., Valladares, J.E., Zanin, J.P., Cooper, W.J., Nickelson, M.G., Kajdi, D.C., et al., 1998. Figures-of-merit for advanced oxidation technologies: a comparison of homogeneous UV/H_2O_2, hetero-geneous UV/TiO_2 and electron beam processes. J. Adv. Oxid. Technol. 3, 174–181.

Borrely, S.I., Cruz, A.C., Mastro, N.L., Sampa, M.H.O., Somessari, E.S., 1998a. Radiation processing of sewage and sludge. A review. Prog. Nucl. Energy 33, 3–21.

Borrely, S.I., Mastro, N.L., Sampa, M.H.O., 1998b. Improvement of municipal wastewater by electron beam accelerator in Brazil. Radiat. Phys. Chem. 52, 333–337.

Borrely, S.I., Morais, A.V., Rosa, J.M., Badaro-Pedroso, C., Pereira, M.C., Higa, M.C., 2016. Decoloration and detoxification of effluents by ionizing radiation. Radiat. Phys. Chem. 124, 108–202.

Boxall, A.B.A., Sinclair, C.J., Fenner, K., Kolpin, D., Maund, S.J., 2004. When synthetic chemicals degrade in the environment. Environ. Sci. Technol. 38, 368A–354A.

Brusentseva, S.A., Makarenko, Z.N., Dolin, P.I., 1978. Radiation decoloration of solutions of humic substances. High Energy Chem. 12, 189–192.

Brusentseva, S.A., Shubin, V.N., Nikonorova, G.K., Zorin, D.M., Sosnovskaya, A.A., Petryaev, E.P., et al., 1986. Radiation-adsorption purification of effluents containing pesticides. Radiat. Phys. Chem. 28, 569–572.

Butt, S.B., Qureshi, R.N., Ahmed, S., 2005. Monitoring of radiolytic degradation of benzo(a)pyrene using γ-rays in aqueous media by HPLC. Radiat. Phys. Chem. 74, 92–95.

Buxton, G.V., 1982. Basic radiation chemistry of liquid water. In: Baxendale, J.H., Busi, F. (Eds.), In the Study of Fast Processes and Transient Species by Electron Pulse Radiolysis. D. Riedel Publishing Company, Dodrecht, pp. 241–266.

Buxton, G.V., 1987. Radiation chemistry of the liquid state: (1) Water and homogeneous aqueous solutions. In: Farhataziz, Rodgers, M.A.J. (Eds.), Radiation Chemistry. Principles and Application. VCH Publishers, Inc., New York, pp. 321–350.

Buxton, G.V., 2008. An overview of the radiation chemistry of liquids. In: Spotheim-Maurizot, M., Mostafavi, M., Douki, T., Belloni, J. (Eds.), Radiation Chemistry: From Basics to Applications in Material and Life Sciences. EDP Sciences, France, pp. 3–16.

Buxton, G.V., Greenstock, C.L., Helman, W.P., Ross, A.B., 1988. Critical review of rate constants for reactions of hydrated electrons, hydrogen atoms and hydroxyl radicals ($^\bullet$OH/$^\bullet$O$-$) in aqueous solution. J. Phys. Ref. Chem. Data 17, 513–886.

Campbell, L.A., 1971. Gamma irradiation as a pretreatment to chemical precipitation in the purification of domestic sewage. Isot. Radiat. Technol. 8, 449–450.

Cao, D.M., Zhang, X.H., Zhao, S.Y., Guan, Y., Zhang, H.Q., 2011. Appropriate dose for degradation of levofloxacin lactate: gamma radiolysis and assessment of degradation product activity and cytotoxicity. Envir. Eng. Sci. 28, 183–189.

Cappadona, C., Guariano, P., Calderaro, E., Petruso, S., Ardica, S., 1975. Possible use of high-level radiation for the degradation of some substances present in urban and industrial waters. Radiat. Clean Environ. Proc. Int. Symp. 265–284.

Chaychian, M., Al-Sheikhly, M., Silverman, J., McLaughlin, W.L., 1998. The mechanisms of removal of heavy metals from water by ionizing radiation. Radiat. Phys. Chem. 53, 145–150.

Chen, Y.P., Liu, S.Y., Yu, H.Q., Yin, H., Li, Q.R., 2008. Radiation-induced degradation of methyl orange in aqueous solutions. Chemosphere 72, 532–536.

Chmielewski, A.G., Han, B., 2016. Electron beam technology for environmental pollution control. Top. Curr. Chem. 374 (68), 32.

Christensen, H., Sehested, K., Logager, T., 1994. Temperature dependence of the rate constant for reactions of hydrated electrons with H, $^\bullet$OH and H$_2$O$_2$. Radiat. Phys. Chem. 43, 527–531.

Chung, H.H., Jung, J., Yoon, J.H., Lee, M.J., 2002. Effect of dissolved metal ions on PCE decomposition by gamma-rays. Chemosphere 47, 977–980.

Cleland, M.R., Fernald, R.A., Malcof, S.R., 1984. Electron beam process design for the treatment of wastes and economic feasibility of the process. Radiat. Phys. Chem. 24, 179–190.

Compton, D.M.J., Whittemo, W.L., Black, S.J., 1970. An evaluation of applicability of ionizing radiation to treatment of municipal waste waters and sewage sludge. Trans. Amer. Nucl. Soc. 13, 71–76.

Cooper, W.J., Nickelsen, M.G., Meacham, D.E., Cadavid, E.M., Waite, T.D., Kurucz, C.N., 1992. High energy electron beam irradiation: an innovative process for the treatment of aqueous based organic hazardous waste. J. Environ. Sci. Health 1, 219–244.

Cooper, W.J., Meacham, D.E., Nickelsen, M.G., Lin, K., Ford, D.B., Kurucz, C.N., et al., 1993. The removal of tri-(TCE) and tetrachloroethylene (PCE) from aqueous solution using high energy electrons. J. Air Waste Manage. Assoc. 43, 1358–1366.

Cooper, W.J., Dougal, R.G., Nickelsen, M.L., Waite, T.FD., Kurucz, C.N., Lin, K., et al., 1996. Benzene destruction in aqueous waste — I. Bench-scale gamma irradiation experi-ments. Radiat. Phys. Chem. 48, 81–87.

Cooper, W.J., Curry, R.D., O'Shea, K.E., 1998. Environmental Applications of Ionizing Radiation. Wiley, New York, 752 pp.

Cooper, W.J., Nickelsen, M.G., Tobien, T., Mincher, B.J., 2001. Radiation-induced Oxidation: the electron beam process for waste treatment. In: Oh, C.H. (Ed.), Hazardous and Radioactive Waste Treatment Technologies Handbook. CRC Press, Boca Raton, FL, pp. 5.5-1−5.5.-16.

Cotton, I.M., Lockingen, L.S., 1963. Inactivation of bacteriophage by chloroform and X irradiation. Biochemistry 50, 363−367.

Craft, T.F., Eichholz, G.G., 1971. Synergistic treatment of textile dye wastes by irradiation and oxidation. Intern. J. App. Radiat. Isotop. 22, 543−547.

Csay, T., Racz, G., Takacs, E., Wojnarovits, L., 2012. Radiation induced degradation of pharmaceutical residues in water: chloramphenicol. Radiat. Phys. Chem. 81, 1489−1494.

Curry, R.D., Mincher, B.J., 1999. The status of PCB radiation chemistry research: prospects for waste treatment in nonpolar solvents and soils. Radiat. Phys. Chem. 56, 493−502.

Daiem, M.M.A., Utrilla, J.R., Perez, R.O., Polo, M.S., Penalver, J.J.L., 2013. Treatment of water contaminated with diphenolic acid by gamma radiation in the presence of different compounds. Chem. Eng. J. 219, 371−379.

De Vleeschouver, F., Van Speybroeck, V., Waroquier, M., Geerlings, P., De Proft, F., 2007. Electrophilicity and nucleophicility index for radicals. Org. Lett. 9, 2721−2724.

Draganic, I.G., Draganic, Z.D., 1971. The Radiation Chemistry of Water, vol. 26. Academic Press, New York.

Drenzek, N.J., Nyman, M.C., Clesceri, N.L., Block, R.C., Stenken, J.A., 2004. Liquid chromatographic aqueous product characterization of high-energy electron beam irradiated 2-chlorobiphenyl solutions. Chemosphere 54, 387−395.

Drzewicz, P., Bojanowska-Czajka, A., Trojanowicz, M., Nałęcz-Jawecki, G., Sawicki, J., Wołkowicz, S., 2003. Application of ionizing radiation for degradation of organic pollutants in waters and wastes. Pol. J. App. Chem. 3, 127−136.

Drzewicz, P., Trojanowicz, M., Zona, R., Solar, S., Gehringer, P., 2004. Decomposition of 2,4-dichlorophenoxyacetic acid by ozonation, ionizing radiation as well as ozonation combined with ionizing radiation. Radiat. Phys. Chem. 69, 281−287.

Drzewicz, P., Gehringer, P., Bojanowska-Czajka, A., Zona, R., Solar, S., Nałęcz-Jawecki, G., et al., 2005. Radiolytic degradation of the herbicide dicamba for environmental protection. Arch. Env. Contam. Toxicol. 48, 311−322.

Duarte, C.L., Sampa, M.H.O., Rela, P.R., Oikawa, H., Cherbakian, E.H., Sena, H.C., et al., 2000. Application of electron beam irradiation combined to conventional treatment to treat industrial effluents. Radiat. Phys. Chem. 57, 513−518.

Duarte, C.L., Sampa, M.H.O., Rela, P.R., Oikawa, H., Silveira, C.G., Azevedo, A.L., 2002. Advanced oxidation process by electron-beam-irradiation-induced decomposition of pollutants in industrial effluents. Radiat. Phys. Chem. 63, 647−651.

Duarte, C.L., Geraldo, L.L., Aquino, O., Junior, P., Borrely, S.I., Sato, I.M., et al., 2004. Treatment of effluents from petroleum production by electron beam irradiation. Radiat. Phys. Chem. 71, 443−447.

Dunn, C.G., 1953. Treatment of water and sewage by ionizing radiations. Sew. Ind. Wastes 25, 1277−1281.

Emami-Meibodi, M., Parsaeian, M.R., Amraei, R., Banaei, M., Anvari, F., Tahami, S.M.R., et al., 2016. An experimental investigation of wastewater treatment using electron beam irradiation. Radiat. Phys. Chem. 125, 82−87.

Fang, X., He, Y., Liu, J., Wu, J., 1998. Oxidative decomposition of pentachlorophenol in aqueous solution. Radiat. Phys. Chem. 53, 411−415.

Farooq, S., Kurucz, C.N., Waite, T.D., Cooper, W.J., 1993. Disinfection of wastewater: high-energy electron vs gamma irradiation. Wat. Res. 27, 1177−1184.

Gala, I.V., Penalver, J.J.L., Polo, M.S., Utrilla, J.R., 2013. Degradation of X-ray contrast media diatrizoate in different water matrices by gamma irradiation. J. Chem. Technol. Biotechnol. 88, 1336−1343.

Gall, B.L., Dorfman, L.M., 1969. Pulse radiolysis studies. XV. Reactivity of the oxide radical ion and of the ozonide ion in aqueous solution. J. Am. Chem. Soc. 91, 2199−2204.

Gehringer, P., Eschweiler, H., 1996. The use of radiation-induced advanced oxidation for water reclamation. Wat. Sci. Tech. 34, 343−349.

Gehringer, P., Eschweiler, H., 2002. The dose rate effect with radiation processing of water − an interpretative approach. Radiat. Phys. Chem. 65, 379−386.

Gehringer, P., Fiedler, H., 1998. Design of a combined ozone/electron beam process for waste water and economic feasibility of the process. Radiat. Phys. Chem. 52, 345−349.

Gehringer, P., Krenmayr, P., 1973. Die radiolytische zersetzung des perfluormethylcyclohexans-I. radiolysepro-dukte. Internat. J. Radiat. Phys. Chem. 5, 113–125.

Gehringer, P., Matschiner, H., 1998. Radiation induced pollutant decomposition in water. Wat. Sci. Tech. 37, 195–201.

Gehringer, P., Proksch, E., Szinovatz, W., 1984. Radiation-induced degradation of trichloroethylene and tetrachlor-oethylene in drinking water. Int. J. Appl. Radiat. Isot. 36, 313–314.

Gehringer, P., Proksch, E., Eschweiler, H., Szinovatz, W., 1992. Remediation of groundwater polluted with chlori-nated ethylenes by ozone-electron beam irradiation treatment. Appl. Radiat. Isot. 43, 1107–1115.

Gehringer, P., Eschweiler, H., Fiedler, H., 1995. Ozone-electron beam treatment for groundwater remediation. Radiat. Phys. Chem. 46, 1075–1078.

Gehringer, P., Eschweiler, H., Leth, H., Pribil, W., Pfleger, S., Cabaj, A., et al., 2003. Bacteriophages as viral indica-tors for radiation processing of water: a chemical approach. App. Radiat. Isot. 59, 651–656.

Gehringer, P., Eschweiler, H., Weiss, S., Reemtsma, T., 2006. Decomposition of aqueous naphthalene-1,5-disulfonic acid by means of oxidation processes. Ozone Sci. Eng. 28, 437–443.

Gerrard, M., 1971. Sewage and waste-water processing with isotopic radiation: survey of the literature. Isot. Radiat. Technol. 8, 429–435.

Getoff, N., 1986. Radiation induced decomposition of biological resistant pollutants in water. App. Radiat. Isot. 37, 1103–1109.

Getoff, N., 1989. Advancements of radiation induced degradation of pollutants in drinking and waste water. Appl. Radiat. Isot. 40, 585–594.

Getoff, N., 1996. Radiation-induced degradation of water pollutants-state of the art. Radiat. Phys. Chem. 47, 581–593.

Getoff, N., 1999. Radiation chemistry and the environment. Radiat. Phys. Chem. 54, 377–384.

Getoff, N., 2002. Factors influencing the efficiency of radiation-induced degradation of water pollurtants. Radiat. Phys. Chem. 65, 437–446.

Grant, D.L., Sherwood, C.R., Mccully, K.A., 1969. Degradation and anticarboxylesterase activity of disulfoton and phorate after ^{60}Co gamma irradiation. J. Assoc. Offic. Anal. Chem. 52, 805–811.

Grant, S.B., Saphores, J.-D., Feldman, D.L., Hamilton, A.J., Fletcher, T.D., Cook, P.L.M., et al., 2012. Taking the "waste" out of "wastewater" for human water security and ecosystem sustainability. Science 337, 681–686.

Guo, Z., Tang, D., Liu, X., Zheng, Z., 2008a. Gamma irradiation-induced Cd^{2+} and Pb^{2+} removal from different kinds of water. Radiat. Phys. Chem. 77, 1021–1026.

Guo, Z., Zheng, Z., Gu, C., Zheng, Y., 2008b. Gamma irradiation-induced removal of low-concentration nitrite in aqueous solution. Radiat. Phys. Chem. 77, 702–707.

Guo, Z., Dong, Q., He, D., Zhang, C., 2012a. Gamma radiation for treatment of bisphenol A solution in presence of different additives. Chem. Eng. J. 183, 10–14.

Guo, Z., Zhou, F., Zhao, Y., Zhang, C., Liu, F., Bao, C., et al., 2012b. Gamma irradiation-induced sulfadiazine deg-radation and its removal mechanisms. Chem. Eng. J. 191, 256–262.

Haag, W.R., Yao, C.C.D., 1992. Rate constants for reaction of hydroxyl radicals with several drinking water con-taminants. Environ. Sci. Technol. 26, 1005–1013.

Han, B., Ko, J., Kim, J., Kim, Y., Chung, W., Makarov, I.E., et al., 2002. Combined electro-beam and biological treatment of dyeing complex wastewater. Pilot plant experiments. Radiat. Phys. Chem. 64, 53–59.

Han, B., Kim, J.K., Kim, Y., Choi, J.S., Jeong, K.Y., 2012. Operation of industrial-scale electron beam wastewater treatment plant. Radiat. Phys. Chem. 81, 1475–1478.

Han, B., Kim, J., Kang, W., Choi, J.S., Jeong, K.-Y., 2016. Development of mobile electron beam plant for environ-mental applications. Radiat. Phys. Chem. 124, 174–178.

Hart, E.J., Gordon, S., Thomas, J.K., 1964. Rate constant of hydrated electron reactions with organic compounds. J. Phys. Chem. 68, 1271–1274.

Hashimoto, S., Kawakami, W., 1979. Application of process control techniques to radiation treatment of wastewa-ter. Ind. Eng. Chem. Process Des. Dev. 18, 169–274.

Hashimoto, S., Miyata, T., Washino, M., Kawakami, W., 1979. A liquid chromatographic study on the radiolysis of phenol in aqueous solution. Environ. Sci. Technol. 13, 71–75.

He, Y., Liu, J., Lu, Y., Wu, J., 2002. Gamma radiation treatment of pentachlorophenol, 2-4-dichlorophenol and 2-chlorophenol in water. Radiat. Phys. Chem. 65, 565–570.

Homlok, R., Takacs, E., Wojnarovits, L., 2010. Radiolytic degradation of 2,4-dichlorophenoxyacetic acid in dilute aqueous solution: pH dependence. J. Radioanal. Nucl. Chem. 284, 415–419.

Homlok, R., Takacs, E., Wojnarovits, L., 2011. Elimination of diclofenac from water using irradiation technology. Chemosphere 85, 603–608.

Homlok, R., Takacs, E., Wojnarovits, L., 2012. Ionizing radiation induced reactions of 2,6-dichloroaniline in dilute aqueous solution. Radiat. Phys. Chem. 81, 1499–1502.

Hossain, M.S., Fakhruddin, A.N.M, Chowdhury, M.A.Z., Alam, M.K., 2013. Degradation of chlorpyrifos, an organophosphorus insecticide in aqueoul solution with gamma irradiation and natural sunlight. J. Environ. Chem. Eng. 1, 270–274.

Hsieh, L.L., Lin, Y.L., Wu, C.H., 2004. Degradation of MTBE in dilute aqueous solution by gamma radiolysis. Wat. Res. 38, 3627–3633.

Hubbell, J.H., Seltzer, S.M., 2004. X-Ray mass attenuation coefficients https://www.nist.gov/pml/x-ray-mass-attenuation-coefficients.

Husman, A.M.R., Bijkerk, P., Lodder, W., Berg, H., Pribil, W., Cabaj, A., et al., 2004. Calicivirus inactivation by nonionizing (253.7-nanometer-wavelenght [UV] and ionizing (gamma) radiation. App. Environ. Microbiol. 70, 5089–5093.

Illes, E., Takacs, E., Dombi, A., Gajda-Schranz, K., Gonter, K., Wojnarovits, L., 2012. Radiation induced degradation of ketoprofen in dilute aqueous solution. Radiat. Phys. Chem. 81, 1479–1483.

Illes, E., Takacs, E., Dombi, A., Gajda-Schrantz, K., Racz, G., Gonter, K., et al., 2013. Hydroxyl radical induced degradation of ibuprofen. Sci. Total Environ. 447, 286–292.

Ismail, M., Khan, H.M., Sayed, M., Cooper, W.J., 2013. Advanced oxidation for the treatment of chlorpyrifos in aqueous solution. Chemosphere 93, 645–651.

Janata, E., Schuler, R.H., 1982. Rate constant for scavenging $e_{aq}-$ in N_2O-saturated solutions. J. Phys. Chem. 86, 2078–2084.

Jeong, J., Song, W.H., Cooper, W.J., Jung, J., Greaves, J., 2010. Degradation of tetracycline antibiotics: mechanisms and kinetic studies for advanced oxidation/reduction processes. Chemosphere 78, 533–540.

Jung, J., Yoon, J.H., Chung, H.H., Lee, M.J., 2002a. Radiation treatment of secondary effluent from a sewage treatment plant. Radiat. Phys. Chem. 65, 533–537.

Jung, J., Yoon, J.H., Chung, H.H., Lee, M.J., 2002b. TCE and PCE decomposition by combination of gamma-rays ozone and titanium dioxide. J. Radioanal. Nucl. Chem. 252, 451–454.

Kantoglu, O., Ergun, E., 2016. Radiation induced destruction of thebaine, papaverine and noscapine in methanol. Radiat. Phys. Chem. 124, 184–190.

Kawakami, W., Hashimoto, S., Nishimura, K., Miyata, T., Suzuki, N., 1978. Electron-beam oxidation treatment of a commercial dye by use of a dual-tube bubbling column reactor. Environ. Sci. Technol. 12, 189–194.

Khan, J.A., Shah, N.S., Nawaz, S., Ismail, M., Rehman, F., Khan, H.M., 2015. Role of e_{aq}, $^\bullet OH$ and $^\bullet H$ in radiolytic degradation of atrazine: a kinetic and mechanistic approach. J. Hazard. Mat. 288, 147–157.

Kim, T.H., Lee, J.K., Lee, M.J., 2007. Biodegrability enhancement of textile wastewater by electron beam irradiation. Radiat. Phys. Chem. 76, 1037–1041.

Kim, T.H., Kim, S.D., Kim, H.Y., Lim, S.J., Lee, M., Yu, S., 2012. Degradation and toxicity assessment of sulfamethoxazole and chlortetracycline using electron beam. ozone and UV. J. Hazard. Mater. 227 (2012), 237–242.

Kimbrough, R.D., Gaines, T.B., 1971. γ-irradiation of DDT, radiation products and their toxicity. J. Agr. Food. Chem. 19, 1037–1038.

Kimura, A., Taguchi, M., Arai, H., Hiratsuka, H., Namba, H., Kojima, T., 2004. Radaition-induced decomposition of trace amounts of 17 -estradiol in water. Radiat. Phys. Chem. 69, 295–301.

Kimura, A., Osawa, M., Taguchi, M., 2012. Decomposition of persistent pharmaceuticals in wastewater by ionizing radiation. Radiat. Phys. Chem. 81, 1508–1512.

Köck-Schulmeyer, M., Villagrasa, M., Lopez de Alda, M., Cespedes-Sanchez, R., Ventura, F., Barcelo, D., 2013. Occurrence and behavior of pesticides in wastewater treatment plants and their environmental impact. Sci. Total Env. 458–460, 466–476.

Kovacs, K., Mile, V., Csay, T., Takacs, E., Wojnarovits, L., 2014. Hydroxyl radical-induced degradation of fenuron in pulse and gamma radiolysis: kinetic and product analysis. Environ. Sci. Pollut. Res. 21, 12693–12700.

Kovacs, K., He, S., Mile, V., Csay, T., Takacs, E., Wojnarovits, L., 2015. Ionizing radiation induced degradation of diuron in dilute aqueous solution. Chem. Cent. J. 9–21.

Kozmer, Z., Takacs, E., Wojnarovits, L., Alapi, T., Hernadi, K., Dombi, A., 2016. The influence of radical transfer and scavenger materials in various concentration on the gamma radiolysis of phenol. Radiat. Phys. Chem. 124, 52–57.

Krapfenbauer, K., Wolfger, H., Getoff, N., Hamblett, I., Navaratnam, S., 2000. Pulse radiolysis and chemical analysis of azo dyes in aqueous solution. I. p-phenylazoaniline. Radiat. Phys. Chem. 58, 21–27.

Kubesch, K., Zona, R., Solar, S., Gehringer, P., 2005. Degradation of catechol by ionizing radiation, ozone and the combined process ozone-electron-beam. Radiat. Phys. Chem. 72, 447–453.

Kubesh, K., Zona, R., Solar, S., Gehringer, P., 2003. Ozone, electron, beam and ozone-electron beam degradation of phenol. A comparative study. Ozone. Sci. Eng. 25, 377–382.

Kuk, S.H., Kim, S.M., Kang, W.G., Han, B., Kuksanov, N.K., Jeong, K.Y., 2011. High-power accelerator for environmental applications. J. Kor. Phys. Soc. 59, 3485–3488.

Kurucz, C.N., Waite, T.D., Cooper, W.J., Nickelsen, M.G., 1991a. High energy electron beam irradiation of water, wastewater and sludge. Chapter. In: Lewins, J., Becker, M. (Eds.), Advances in Nuclear Science and Technology, vol. 23. Plenum Press, New York, pp. 1–43.

Kurucz, C.N., Waite, T.D., Cooper, W.J., Nickelsen, M.J., 1991b. High energy electron beam irradiation of water, wastewater and sludge. Adv. Nucl. Sci. Technol. 22, 1–43.

Kurucz, C.N., Waite, T.D., Cooper, W.J., Nickelsen, M.G., 1995a. Empirical models for estimating the destruction of toxic organic compounds utilizing electron beam irradiation at full scale. Radiat. Phys. Chem. 45, 805–816.

Kurucz, C.N., Waite, T.D., Cooper, W.J., 1995b. The Miami electron beam research facility: a large scale wastewater treatment application. Radiat. Phys. Chem. 45, 299–308.

Kurucz, C.N., Waite, T.D., Otano, S.E., Cooper, W.J., Nickelsen, M.G., 2002. A comparison of large-scale electron beam and bench-scale ^{60}Co irradiation of simulated aqueous waste streams. Radiat. Phys. Chem. 65, 367–378.

Kwon, M., Yoon, Y., Cho, E., Jung, Y., Lee, B.C., Paeng, K.J., et al., 2012. Removal of iopromide and degradation characteristics in electron beam irradiation process. J. Hazard. Mater. 227, 126–134.

Leitner, N.K., Berger, P., Gehringer, P., 1999. γ-irradiation for the removal of atrazine in aqueous solution containing humic substances. Radiat. Phys. Chem. 55, 317–322.

Lepine, F., Masse, R., 1990. Degradation pathways of PCB upon gamma irradiation. Environ. Health Persp. 89, 183–187.

Lepine, F.L., Brochu, F., Milot, S., Mametr, A., Pepin, Y., 1994. γ-irradiation-induced degradation of DDT and it metabolites in organic solvents. J. Agri. Food Chem. 42, 2012–2018.

Lesse, T., Suess, A., 1984. Ten year experience in operation of a sewage sludge treatment plant using gamma irradiation. Radiat. Phys. Chem. 24, 3–16.

Lin, K., Cooper, W.J., Nickelsen, M.G., Kurucz, C.N., Waite, T.D., 1995. Decomposition of aqueous solutions of phenol using high energy electron beam irradiation- a large scale study. Appl. Radiat. Isot. 46, 1307–1316.

Liu, N., Wang, T., Zheng, M., Lei, J., Tang, L., Hu, G., et al., 2015. Radiation induced degradation of antiepileptic drug primidone in aqueous solution. Chem. Eng. J. 270, 66–72.

Liu, Q., Luo, X., Zheng, Z., Zheng, B., Zhang, J., Zhao, Y., et al., 2011. Factors that have an effect on degradation of diclofenac in aqueous solution by gamma ray irradiation. Environ. Sci. Pollut. Res. 18, 1243–1252.

Liu, S.Y., Chen, Y.P., Yu, H.Q., Zhang, S.J., 2005. Kinetics and mechanisms of radiation-induced degradation of acetochlor. Chemosphere 59, 13–19.

Liu, Y.K., Wang, J.L., 2013. Degradation of sulfamethazine by gamma irradiation in the presence of hydrogen peroxide. J. Hazard. Mat. 250, 99–105.

Lowe Jr., H.N., Lacy, W.J., Surkiewicz, B.F., Jaeger, R.F., 1956. Destruction of microorganisms in water, sewage, and sewage sludge by ionizing radiations. J. Am. Water Works Assoc. 48, 1363–1372.

Luchini, L.C., Peres, T.B., Rezende, M.O., de, O., 1999. Degradation of the insecticide parathion in methanol by gamma-irradiation. J. Radioanal. Nucl. Chem. 241, 191–194.

Mak, F.T., Zele, S.R., Cooper, W.J., Kurucz, C.N., Waite, T.D., Nickelsen, M.G., 1997. Kinetic modeling of carbon tetrachloride, chloroform and methylene chloride removal from aqueous solution using the electron beam process. Wat. Res. 31, 219–228.

Mann, L.A., 1971. Biological - gamma-radiation system for sewage processing. Isot. Radiat. Technol. 8, 439–444.

Melo, R., Leal, J.P., Takacs, E., Wojnarovits, L., 2009. Radiolytic degradation of gallic acid and its derivatives in aqueous solution. J. Hazard. Mat. 172, 1185–1192.

Mezyk, S.P., Neubauer, T.J., Cooper, W.J., Peller, J.R., 2007. Free-radical-induced oxidative and reductive degradation of sulfa drugs in water: absolute kinetics and efficiencies of hydroxyl radical and hydrated electron reactions. J. Phys. Chem. A 111, 9019–9024.

Miege, C., Karolak, S., Gabet, V., Jugan, M.-L., Oziol, L., Chevreuil, M., et al., 2009. Evaluation of estrogenic disrupting potency in aquatic environments and urban wastewaters by combining chemical and biological analysis. Trends Anal. Chem. 28, 186–195.

Mincher, B.J., Curry, R.D., 2000. Considerations for choice of a kinetic fig. of merit in process radiation chemistry for waste treatment. Appl. Radiat. Isotop. 52, 189–193.

Mincher, B.J., Meikrantz, D.H., Murphy, R.J., Gresham, G.L., Connolly, M.J., 1991. Gamma-ray induced degradation of PCBs and pesticides using spent reactor fuel. Appl. Radiat. Isot. 42, 1061–1066.

Mincher, B.J., Arbon, R.E., Knighton, W.B., Meikrantz, D.H., 1994. Gamma-ray-induced degradation of PCBs in neutral isopropanol using spent reactor fuel. App. Radiat. Isot. 45, 879–887.

Mincher, B.J., Liekhus, K., Arbon, R.E., 1996. PCB radiolysis in isooctane in the presence of ozone. Appl. Radiat. Isot. 47, 713–715.

Miyazaki, T., Katsumura, Y., Lin, M., Muroya, Y., Kudo, H., Taguchi, M., et al., 2006. Radiolysis of phenol in aqueous solution at elevated temperatures. Radiat. Phys. Chem. 75, 408–415.

Mohamed, K.A., Basfar, A.A., Al-Kahtani, H.A., Al-Hamad, K.S., 2009a. Radiolytic degradation of malathion and lindane in aqueous solution. Radiat. Phys. Chem. 78, 994–1000.

Mohamed, K.A., Basfar, A.A., Al-Shahrani, A.A., 2009b. Gamma-ray induced degradation of diazinon and atrazine in natural groundwaters. J. Hazard. Mat. 166, 810–814.

Mompelat, S., Le Bot, B., Thomas, O., 2009. Occurrence and fate of pharmaceutical products and by-products, from resource to drinking water. Environ. Int. 35, 803–814.

Mori, M.N., Oikawa, H., Sampa, M.H.O., Duarte, C.L., 2006. Degradation of chlorpyrifos by ionizing radiation. J. Radioanal. Nucl. Chem. 270, 99–102.

Mucka, V., Siber, R., Kropacek, M., Pospisil, M., Klisky, V., 1997. Electron beam degradation of polychlorinated biphenyls. Radiat. Phys. Chem. 50, 503–510.

Mucka, V., Silber, R., Pospisil, M., Klisky, V., Bartonicek, B., 1999. Electron beam degradation of PCE and a large volume flow-through equipment for radiation degradation of halogenated organic compounds. Radiat. Phys. Chem. 55, 93–97.

Mucka, V., Polakova, D., Pospisil, M., Silber, R., 2003. Dechlorination of chloroform in aqueous solutions influenced by nitrate ions and hydrocarbonate ions. Radiat. Phys. Chem. 68, 787–791.

Mytelka, A.I., 1971. Radiation treatment of industrial waste waters: an economic analysis. Isot. Radiat. Technol. 8, 444–449.

Nagai, T., 1976. The radiation-induced degradation of antraquinone dyes in aqueous solutions. Internat. J. App. Radiat. Isot. 27, 699–705.

Nakagawa, S., Shimokawa, T., 2002. Degradation of halogenated carbons in alkaline alcohol. Radiat. Phys. Chem. 63, 151–156.

Neta, P., Huie, R.E., Ross, A.B., 1988. Rate constants for reactions of inorganic radicals in aqueous solution. J. Phys. Ref. Chem. Data 17, 1027–1284.

Nickelsen, M.G., Cooper, W.J., Kurucz, C.N., Waite, T.D., 1992. Removal of benzene and selected alkyl-substituted benzenes from aqueous solution utilizing continuous high-energy electron irradiation. Environ. Sci. Technol. 26, 144–152.

Nickelsen, M.G., Cooper, W.J., Lin, K., Kurucz, C.N., Waite, T.D., 1994. High energy electron beam generation of oxidants for the treatment of benzene and toluene in the presence of radical scavengers, Wat. Res., 28. pp. 1227–1237.

Nickelsen, M.G., Kajdi, D.C., Cooper, W.J., Kurucz, C.N., Waite, T.D., Gensel, F., et al., 1998. Field application of a mobile 20 kW electron-beam treatment system on contaminated groundwater and industrial wastes. In: Cooper, W.J., Curry, R.D., O'Shea, K.E. (Eds.), Environmental Application of Ionizing Radiation. Wiley, New York, pp. 451–466.

Pastuszek, F., Vacek, K., Vondruska, V., 1993. "In situ" radiation cleaning of underground water contaminated with cyanides- six years of experience. Radiat. Phys. Chem. 42, 699–700.

Peller, J., Kamat, P.V., 2005. Radiolytic transformations of chlorinated phenols and chlorinated phenoxyacetic acids. J. Phys. Chem. 109, 9528–9535.

Peller, J., Wiest, O., Kamat, P.V., 2003. Mechanism of hydroxyl radical-induced breakdown of the herbicide 2,4-dichlorophenoxyacetic acid (2,4-D). Chem. Eur. J. 9, 5579–5587.

Peller, J.R., Cooper, W.J., Ishida, K.P., Mezyk, S.P., 2011. Evaluation of parameters influencing removal efficiencies for organic contaminant degradation in advanced oxidation processes. J. Wat. Supp. Reserch. Technol-Aqua 60.2 69–78.

Peng, C., Ding, Y., An, F., Wang, L., Li, S., Nie, Y., et al., 2015. Degradation of ochratoxin A in aqueous solutions by electron beam irradiation. J. Radioanal. Nucl. Chem. 306, 39–46.

Perez, R.O., Utrilla, J.R., Sanchez-Polo, M., Lopez-Penalver, J.J., Leyva-Ramos, R., 2011. Degradation of antineoplastic cytarabine in aqueous solution by gamma radiation. Chem. Eng. J. 174, 1–8.

Perkowski, J., 1989. Purification of textile wastewaters by radiation enhanced sorption on activated carbon (Pol.). Przegl. Włók. 113–115.

Perkowski, J., Kos, L., 1988. Purification of wastewaters from textile dyeing using radiation and coagulation (Pol.). Przegl. Włók. 444–446.

Perkowski, J., Kos, L., 2003. Decolouration of model dyehouse wastewater with advanced oxidation processes. Fib. Text. East. Eur. 11, 67–70.

Perkowski, J., Mayer, J., 1989. Gamma radiolysis of anthraquinone dye aqueous solution. J. Radioanal. Nucl. Chem. 132, 269–280.

Perkowski, J., Kos, L., Rouba, J., 1984. Irradiation in industrial −waste treatment. Effluent. Wat. Treat. J. 24, 335–342.

Perkowski, J., Kos, L., Ledakowicz, S., Żyłła, R., 2003. Decomposition of anthraquinone dye acid blue 62 by the decoloration of textile wastewater by advanced oxidation process. Fib. Text. East. Eur. 11, 88–94.

Petrovic, M., Gonzalez, S., Barcelo, D., 2003. Analysis and removal of emerging contaminants in wastewater and drinking water. Trends Anal. Chem. 22, 685–696.

Petrovic, M., Gehringer, P., Eschweiler, H., Barcelo, D., 2007. Radiolytic decomposition of multi-class surfactants and their biotransformation products in sewage treatment plant effluents. Chemosphere 66, 114–122.

Pikaev, A.K., 1999. New developments in radiation processing in Russia: a review. High Energy Chem. 33, 1–8.

Pikaev, A.K., 2000. Current status of the Application of ionizing radiation to Environmental protection. II wastewater and other liquid wastes (a review). High Energy Chem. 34, 55–73.

Pikaev, A.K., 2002. New data on electron-beam purification of wastewater. Radiat. Phys. Chem. 65, 515–526.

Pikaev, A.K., Shubin, V.N., 1984. Radiation treatment of liquid wastes. Radiat. Phys. Chem. 24, 77–97.

Pikaev, A.K., Makarov, I.E., Ponomarev, A.V., Kim, Y., Han, B., Yang, Y.W., et al., 1997a. A combined electron-beam and coagulation method of purification of water from dyes. Mend. Comm. Electr. Vers. 5, 169–212.

Pikaev, A.K., Podzorova, E.A., Bakhtin, O.M., 1997b. Combined electron-beam and ozone treatment of wastewater in the aerosol flow. Radiat. Phys. Chem. 49, 155–157.

Pikaev, A.K., Kartasheva, L.I., Zhestkova, T.P., Yurik, T.K., Chulkov, V.N., Didenko, O.A., et al., 1997c. Removal of heavy metals from water by electron-beam treatment in the presence of an hydroxyl radical scavenger. Mendel. Comm. 7 (2), 47–86.

Pikaev, A.K., Ponomarev, A.V., Bludenko, A.V., Minin, V.N., Elizar'eva, L.M., 2001. Combined electron-beam and coagulation purification of molasses distillery slops. Features of the method, technical and economic evaluation of large-scale facility. Radiat. Phys. Chem. 61, 81–87.

Podzorova, E.A., 1995. Purification of municipal waste waters by irradiation with accelerated electrons in an aerosol flow. High Energy Chem. 29, 256–259.

Podzorova, E.A., Pikaev, A.K., Belyshev, V.A., Lysenko, S.L., 1998. New data on electron-beam treatment of municipal wastewater in the aerosol flow. Radiat. Phys. Chem. 52, 361–364.

Polo, M.S., Penalver, J.L., Joya, G.P., Garcia, M.A.F., Utrilla, J.R., 2009. Gamma irradiation of pharmaceutical compounds, nitroimidazoles, as a new alternative for water treatment. Wat. Res. 43, 4028–4036.

Ponomarev, A.V., Makarov, I.E., Bludenko, A.V., Minin, V.N., Kim, D.K., Han, B., et al., 1999. Combined electron-beam and coagulation treatment of industrial wastewater with high concentrations of organic substances. High Energy Chem. 33, 145–149.

Proksch, E., Gehringer, P., Szinovatz, W., Eschweiler, H., 1987. Radiation-induced decomposition of small amounts of perchloroethylene in water. Appl. Radiat. Isot. 38, 911–919.

Rauf, M.A., Ashraf, S.S., 2009. Radiation induced degradation of dyes-an overview. J. Hazard. Mat. 166, 6–16.

Rawat, K.P., Sarma, K.S.S., 2013. Enhanced biodegradation of wastewater with electron beam pretreatment. App. Radiat. Isot. 74, 6–8.

Razavi, B., Song, W.H., Cooper, W.J., Greaves, J., Jeong, J., 2009. Free-radical-induced oxidative and reductive degradation of fibrate pharmaceuticals: kinetic studies and degradation mechanisms. J. Phys. Chem. A 113, 1287–1294.

Report 90, 2014. Journal of the International Commission on Radiation Units and Measurements. 14 (1), 1–110.

Riaz, M., Butt, S.B., 2010. Gamma radiolytic degradation of the endrin insecticide in methanol and monitoring of radiolytic degradation products by HPLC. J. Radioanal. Nucl. Chem. 285, 697–701.

Ribeiro, M.A., Sato, I.M., Duarte, C.L., Sampa, M.H.O., Salvador, V.L.R., Scapin, M.A., 2004. Application of the electron-beam treatment for Ca, Si, P, Al, Fe, Cr, Zn, Co, As, Se, Cd and Hg removal in the simulated and actual industrial effluents. Radiat. Phys. Chem. 71, 423–426.

Richter, H.W., 1998. Radiation chemistry: principles and applications. In: Wishart, J.F., Nocera, D.G. (Eds.), Photochemistry and Radiation Chemistry. Complementary Methods for the Study of Electron Transfer. American Chemical Society, Washington, DC, pp. 5–33.

Roshani, B., Leitner, N.K.V., 2011. Effect of persulfate on the oxidation of benzotriazole and humic acid by e-beam irradiation. J. Hazard. Mat. 190, 403–408.

Sagi, G., Kovacs, K., Bersenyi, A., Csay, T., Takacs, E., Wojanrovits, L., 2016. Enhancing the biological degradability of sulfamethoxazole by ionizing radiation treatment in aqueous solution. Radiat. Phys. Chem. 124, 179–183.

Sakumoto, A., Miyata, T., 1984. Treatment of waste water by a combined technique of radiation and conventional method. Radiat. Phys. Chem. 24, 99–115.

Sampa, M.H.O., Borrely, S.I., Silva, B.L., Vieira, J.M., Rela, P.R., Calvo, W.A.P., et al., 1995. The use of electron beam accelerator for the treatment of drinking water and wastewater in Brazil. Radiat. Phys. Chem. 46, 1143–1146.

Sampa, M.H.O., Duarte, C.L., Rela, P.R., Somessari, E.S.R., Silveira, C.G., Azevedo, A.I., 1998. Remotion of organic compounds of actual industrial effluents by electron beam irradiation. Radiat. Phys. Chem. 52, 365–369.

Sampa, M.H.O., Takacs, E., Gehringer, P., Rela, P.R., Ramirez, T., Amro, H., et al., 2007. Remediation of polluted waters and waste by radiation processing. Nukleonika 52, 137–144.

Sawai, T., Shimokawa, T., Shinozaki, Y., 1974. The radiolytic dechlorination of polychlorinated biphenyls in alkaline 2-propanol solutions. Bull. Chem. Soc. Jap. 47, 1889–1893.

Schmelling, D., Poster, D., Chaychian, M., Neta, P., McLaughlin, W., Silverman, J., et al., 1998a. Applications of ionizing radiation to the remediation of materials contaminated with heavy metals and polychlorinated biphenyls. Radiat. Phys. Chem. 52, 371–377.

Schmelling, D.C., Gray, K.A., Kamat, P.V., 1998b. Radiation-induced reactions of 2,4,6-trinitrotoluene in aqueous solution. Environ. Sci. Technol. 32, 971–974.

Schmelling, D.C., Poster, D.L., Chaychian, M., Neta, P., Silverman, J., Al-Sheikhly, M., 1998c. Degradation of polychlorinated biphenyls induced by ionizing radiation in aqueous micellar solutions. Environ. Sci. Technol. 32, 270–275.

Sehested, K., Holcman, J., Hart, E.J., 1983. Rate constants and products of the reactions of e_{aq}, dioxide(1-) (O_2-) and proton with ozone in aqueous solutions. J. Phys. Chem. 87, 1951–1954.

Sherman, W.V., Evans, R., Nesyto, E., Radlowski, C., 1971. Dechlorination of DDT in solution by ionizing radiation. Nature 232, 118–119.

Shim, S.B., Jo, H.J., Jung, J., 2009. Toxicity identification of gamma-ray treated phenol and chlorophenols. J. Radioanal. Nucl. Chem. 280, 41–46.

Shin, H.S., Kim, Y.R., Han, B., Makarov, I.E., Ponomarev, A.V., Pikaev, A.K., 2002. Application of electron beam to treatment of wastewater from papermill. Radiat. Phys. Chem. 65, 539–547.

Singh, A., Kremers, W., 2002. Radiolytic dechlorination pf polychlorinated biphenyls using alkaline 2-propanol solutions. Radiat. Phys. Chem. 65, 467–472.

Singh, A., Kremera, W., Smalley, P., Bennett, G.S., 1985. Radiolytic dechlorination of plychlorinated biphenyls. Radiat. Phys. Chem. 25, 11–19.

Sladkova, A.A., Sosnovskaya, A.A., Edimecheva, I.P., Shadyro, O.O., 2012. Radiation-induced destruction of hydroxyl-containing amino acids and dipeptides. Radiat. Phys. Chem. 81, 1896–1903.

Slegers, C., Tilquin, B., 2006. Final product analysis in the e-beam and gamma radiolysis of aqueous solutions of metoprolol tartrate. Radiat. Phys. Chem. 75, 1006–1017.

Solpan, D., Guven, O., 2002. Decoloration and degradation of some textile dyes by gamma irradiation. Radiat. Phys. Chem. 65, 549–558.

Solpan, D., Torun, M., 2012. The removal of chlorinated organic herbicide in water by gamma-irradiationl. J. Radioanal. Nucl. Chem. 293, 21–38.

Solpan, D., Guven, O., Takacs, E., Wojnarovits, L., Dajka, K., 2003. High-energy irradiation treatment of aqueous solutions of azo dyes: steady-state gamma radiolysis experiments. Radiat. Phys. Chem. 67, 531–534.

Song, W., Chen, W., Cooper, W.J., Greaves, J., Miller, G.E., 2008a. Free-radical destruction of -lactam antibiotics in aqueous solution. J. Phys. Chem. A 112, 7411–7417.

Song, W.H., Cooper, W.J., Mezyk, S.P., Greaves, J., Peake, B.M., 2008b. Free radical destruction of beta-blockers in aqueous solution. Environ. Sci. Technol. 42, 1256–1261.

von Sonntag, C., 1987. The Chemical Basis of Radiation Biology. Taylor and Francis, New York, p. 515.

Spinks, J.W.T., Woods, R.J., 1990. Introduction to Radiation Chemistry, third ed. Wiley, New York.

Szabo, L., Toth, T., Homlok, R., Takacs, E., Wojnarovits, L., 2012. Radiolysis of paracetamol in dilute aqueous solution. Radiat. Phys. Chem. 81, 1503–1507.

Taghipour, F., 2004. Ultraviolet and ionizing radiation for microorganism inactivation. Wat. Res. 38, 3940–3948.

Taghipour, F., Evans, G.J., 1996. Radiolytic elimination of organochlorine in pulp mill effluent. Environ. Sci. Technol. 30, 1558–1564.

Taghipour, F., Evans, G.J., 1997. Radiolytic dechlorination of chlorinated organics. Radiat. Phys. Chem. 49, 257–264.

Thill, P.G., Ager, D.K., Vojnovic, B., Tesh, S.J., Scott, T.B., Thompson, I.P., 2016. Hybrid biological, electron beam and zero-valent nano iron treatment of recalcitrant metalworking fluids. Wat. Res. 90, 214–221.

Thomas, J.K., 1969. Elementary processes and reactions in the radiolysis of water. In: Burton, M., Magee, J.L. (Eds.), In Advances in Radiation Chemistry. Wiley-Interscience, New York, pp. 103–198.

Thompson, J.E., Blatchley, E.R., 1999. Toxicity effects of γ- irradiated wastewater effluents. Wat. Res. 33, 2053–2058.

Torun, M., Abbasova, D., Solpan, D., Guven, O., 2014. Caffeine degradation in water by gamma irradiation, ozonation and ozonation/gamma irradiation. Nukleonika 59, 25–35.

Torun, M., Gultekin, O., Solpan, D., Guven, O., 2015. Mineralization of paracetamol in aqueous solution with advanced oxidation processes. Environ. Technol. 36, 970–982.

Touhill, C.J., Martin, E.C., Fujihara, M.P., Olesen, D.E., Stein, J.E., Mcdonnel, G., 1969. Effects of radiation on Chicago metropolitan sanitary district municipal and industrial wastewaters. J. Wat. Poll. Contr. Fed. 41, 44–60.

Trebse, P., Aron, I., 2003. Degradation of organophosphorous compounds by X-ray irradiation. Radiat. Phys. Chem. 67, 527–530.

Trojanowicz, M., Chudziak, A., Bryl-Sandelewska, T., 1997. Use of reversed-phase HPLC with solid-phase extraction for monitoring of radiolytic degradation of chlorophenols for environmental protection. J. Radioanal. Nucl. Chem. 224, 131–136.

Trojanowicz, M., Drzewicz, P., Panta, P., Gluszewski, W., Nałęcz-Jawecki, G., Sawicki, J., et al., 2002. Radiolytic degradation and toxicity changes in γ-irradiated solutions of 2,4-dichlorophenol. Radiat. Phys. Chem. 65, 357–366.

Trump, J.G., Mettill, E.W., Wright, K.A., 1984. Disinfection of sewage and sludge by electron treatment. Radiat. Phys. Chem. 24, 55–66.

Unob, F., Hagege, A., Lakkis, D., Leroy, M., 2003. Degradation of organolead species in aqueous solutions by electron beam irradiation. Wat. Res. 37, 2113–2117.

Unterweger, M.P., 2002. Half-life measurements at the National Institute of Standards and Technology. Appl. Radiat. Isotopes 56 (1–2), 125–130.

Vahdat, A., Bahrami, S.H., Arami, M., Motahari, A., 2010. Decomposition and decoloration of direct dye by electron beam radiation. Radiat. Phys. Chem. 79, 33–35.

Vahdat, A., Bahrami, S.H., Arami, M., Bahjat, A., Tabakh, F., Khairkhah, M., 2012. Decoloration and mineralization of reactive dyes using electron beam irradiation, Part I: effect of the dye structure, concentration and absorbed dose (single, binary and ternary systems). Radiat. Phys. Chem. 81, 851–856.

Varghese, R., Mohan, H., Manoj, P., Manoj, V.M., Aravind, U.K., Vandana, K., et al., 2006. Reactions of hydrated electrons with triazine derivatives in aqueous medium. J. Agric. Food Chem. 54, 8171–8176.

Vysotskaya, N.A., Bortum, L.N., Ogurtsov, N.A., Migdalovich, E.A., Revina, A.A., Volodko, V.V., 1986. Radiolysis of anthraquinone dyes in aqueous solutions. Radiat. Phys. Chem. 28, 469–472.

Wang, J., Chu, L., 2016. Irradiation treatment of pharmaceutical and personal care products (PPCPs) in water and wastewater: an overview. Radiat. Phys. Chem. 125, 56–64.

Wang, J.Q., Zhang, J.B., Ma, Y., Zheng, B.G., Zheng, Z., Zhao, Y.F., 2012. Degradation of atrazine in water by gamma-ray irradiation. Fresenius Environ. Bull. 21, 2778–2784.

Wang, L., Batchelor, B., Pillai, S.D., Botlaguduru, V.S.V., 2016. Electron beam treatment for potable water reuse: removal of bromate and perfluorooctanoic acid. Chem. Eng. J. 302, 58–68.

Wang, M., Yang, R., Wang, W., Shen, Z., Bian, S., Zhu, Z., 2006. Radiation-induced decomposition of reactive dyes in the presence of H_2O_2. Radiat. Phys. Chem. 75, 286–291.

Wang, T., Waite, T.D., Kurucz, C., Cooper, W.J., 1994. Oxidant reduction and biodegradability improvement of paper mill effluent by irradiation. Wat. Res. 28, 237–241.

Wardman, P., 1989. The reduction potentials of one-electron couples involving free radicals in aqueous solution. J. Phys. Chem. Ref. Data 18, 1637–1753.

Wasiewicz, M., Chmielewski, A.G., Getoff, N., 2006. Radiation-induced degradation of aqueous 2,3-dihydroxy-naphthalene. Radiat. Phys. Chem. 75, 201–209.

Weeks, J.L., Rabani, J., 1966. The pulse radiolysis of deaerated aqueous carbonate solutions: I. Transient optical spectrum and mechanism. II pK for OH radicals. J. Phys. Chem. 70, 2100–2105.

Winarno, E.K., Getoff, N., 2002. Comparative studies on the degradation of aqueous 2-chloroaniline by O_3 as well as by UV-light and γ-rays in the presence of ozone. Radiat. Phys. Chem. 65, 387–395.

Wojarovits, L., Takacs, E., 2016. Radiation induced of organic pollutants in waters and waste-waters. Top. Curr. Chem. (Z) 374 (1), 35 pp.

Wojnarovits, L., Takacs, E., 2008. Irradiation treatment of azo dye containing wastewater: an overview. Radiat. Phys. Chem. 77, 225–244.

Wojnarovits, L., Takacs, E., 2014. Rate coefficients of hydroxyl radical reactions with pesticide molecules and related compounds: a review. Radiat. Phys. Chem. 96, 120–134.

Wojnarovits, L., Takacs, E., 2017. Wastewater treatment with ionizing radiation. J. Radioanal. Nucl. Chem. 311, 973–981.

Wojnarovits, L., Palfi, T., Takacs, E., Emmi, S.S., 2005. Reactivity differences of hydroxyl radicals and hydrated electrons in destructing azo dyes. Radiat. Phys. Chem. 74, 239–246.

Woodbridge, D.D., Mann, L.A., Garrett, W.R., 1970. Applications of gamma radiation to sewage treatment. Nucl. News 13, 60–64.

Woodbridge, D.D., Mann, L.A., Garbett, W.R., 1972. Usable water from raw sewage. Bull. Environ. Cont. Toxicol. 7, 80–86.

Woods, R.J., Akhtar, S., 1974. Radiation-induced dechlorination of chloral hydrate and 1,1,1-trichloro-2,2-bis(p-chlorophenyl)ethane (DDT). J. Agr. Food Chem. 22, 1132–1133.

Wu, X.Z., Yamamoto, T., Hatashita, M., 2002. Radiolytic degradation of chlorinated hydrocarbons in water. Bull. Chem. Soc. Jpn. 75, 2527–2532.

Xu, G., Bu, T., Wu, M., Zheng, J., Liu, N., Wang, L., 2011. Electron beam induced degradation of clopyralid in aqueous solution. J. Radioanal. Nucl. Chem. 288, 759–764.

Xu, G., Yao, J., Tang, L., Yang, X., Zheng, M., Wang, H., et al., 2015. Electron beam induced degradation of atrazine in aqueous solution. Chem. Eng. J. 275, 374–380.

Yoon, J.H., Jung, J., Chung, H.H., Lee, M.J., 2002. EPR characterization of carbonate ion effect on TCE and PCE decomposition by gamma-rays. J. Radioanal. Nucl. Chem. 253, 217–219.

Yu, H., Nie, E., Xu, J., Yan, S., Cooper, W.J., Song, W., 2013. Degradation of diclofenac by Advanced Oxidation Reduction Processes: kinetic studies, degradation pathways and toxicity assessments. Wat. Res. 47, 1909–1918.

Yu, S.H., Lee, B.J., Lee, M.J., Cho, I.H., Chang, S.W., 2008. Decomposition and mineralization of cefaclor by ionizing radiation: kinetics and effects of the radical scavengers. Chemosphere 71, 2106–2112.

Zacheis, G.A., Gray, K.A., Kamat, F.V., 2000. Radiolytic reduction of hexachlorobenzene in surfactant solution: a steady-state and pulse radiolysis study. Environ. Sci. Technol. 34, 3401–3407.

Zaki, A.A., El-Gendy, N.A., 2014. Removal of metal ions from wastewater using EB irradiation in combination with HA/TiO$_2$/UV treatment. J. Hazard. Mat. 271, 275–282.

Zhang, S.J., Yu, H.Q., 2004. Radiation-induced degradation of polyvinyl alcohol in aqueous solutions. Wat. Res. 38, 309–316.

Zhang, S.J., Yu, H.Q., Zhao, Y., 2005. Kinetic modeling of the radiolytic degradation of Acid Orange 7 in aqueous solutions. Wat. Res. 39, 839–846.

Zhang, X.H., Cao, D.M., Zhao, S.Y., Gong, P., Hei, D.Q., Zhang, H.Q., 2011. Gamma radiolysis of ceftriaxone sodium for water treatment: assessment of the activity. Wat. Sci. Tech. 63, 2767–2774.

Zheng, B., Zheng, Z., Zhang, J., Liu, Q., Wang, J., Luo, X., et al., 2012. Degradation kinetics and by-products of naproxen in aqueous by gamma irradiation. Envir. Eng. Sci. 29, 386–391.

Zheng, B.G., Zheng, Z., Zhang, J.B., Luo, X.Z., Wang, J.Q., Liu, Q., et al., 2011. Degradation of the emerging contaminant ibuprofen in aqueous solution by gamma irradiation. Desalination 276, 379–385.

Zheng, M., Xu, G., Pei, J., He, X., Xu, P., Liu, N., et al., 2014. EB-radiolysis of carbamazepine: in pure-water with different ions and in surface water. J. Radioanal. Nucl. Chem. 302, 139–147.

Zimina, G.M., Tkhorzhnitskii, G.P., Krasnyi, D.V., Vannikov, A.V., 2002. Radiation-chemical degradation of azo dye solution. High Energy Chem. 36, 7–9.

Zona, R., Solar, S., 2003. Oxidation of 2,4-dichloroacetic acid by ionizing radiation: degradation, detoxification and mineralization. Radiat. Phys. Chem. 66, 137–143.

Zona, R., Schmid, S., Solar, S., 1999. Detoxification of aqueous chlorophenols solutions by ionizing radiation. Wat. Res. 33, 1314–1319.

Zona, R., Solar, S., Gehringer, P., 2002. Degradation of 2,4-dichlorophenoxyacetic acid by ionizing radiation: influence of oxygen concentration. Wat. Res. 36, 1369–1374.

Zona, R., Solar, S., Sehested, K., 2012. OH-radical induced degradation of 2,4,5-trichlorophenoxyacetic acid (2,4,5-T) and 4-chloro-2-methylphenoxyacetic acid (MCPA): a pulse radiolysis and gamma-radiolysis study. Radiat. Phys. Chem. 81, 152–159.

10

Supercritical Water Oxidation

Violeta Vadillo, Jezabel Sánchez-Oneto, Juan R. Portela
and Enrique J. Martínez de la Ossa
University of Cádiz, Puerto Real, Spain

10.1 INTRODUCTION

The increasing pollution over recent decades requires the evolution of clean technologies capable to destroy wastes with the aim to obtain non-hazardous products. Often, conventional oxidation methods do not achieve the removal of organic compounds present in the water. For this reason, it is necessary to use another kind of chemical oxidation process called advanced oxidation processes (AOPs). AOPs consist of an aqueous oxidation process that involves highly active radicals such as hydroxyl radical ($^{\bullet}OH$); it oxidizes organic matter to CO_2 and generates nontoxic reaction products. AOPs were developed over recent decades and they were divided into AOPs using strong reactive species mainly as O_3/H_2O_2, O_3/UV, H_2O_2/UV, Fenton reactive (H_2O_2 + ferrous salts), and photo-Fenton reactive (H_2O_2 + ferrous salts + UV), and oxidation using molecular oxygen at high pressure and high temperature in case of high concentrated wastewaters (hydrothermal processes: wet air oxidation and supercritical water oxidation) (Gassó and Baldasano, 1996). Fig. 10.1 shows different advanced oxidation technologies.

Supercritical water oxidation (SCWO) is a promising technology to treat a wide variety of industrial wastewaters. The main advantage of SCWO over other treatment methods such as landfill is that it is a destruction method. Destruction methods based on oxidation of organic matter include biological treatment, incineration, AOP, wet air oxidation, and supercritical water oxidation. Choosing a method to be used depends on the organic content of wastewater. For organic contents up to 1%, biological and AOP treatments are suitable. On the other hand, incineration is suitable to highly concentrated wastewaters but in the range of 1%−20% organic matter, SCWO is a better option, due to the toxic gases produced, and the high cost of incineration.

Advanced Oxidation Processes for Wastewater Treatment
DOI: https://doi.org/10.1016/B978-0-12-810499-6.00010-3

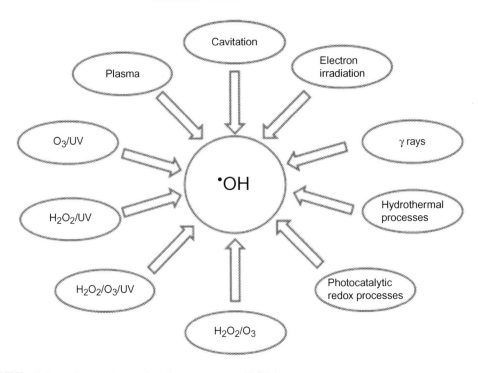

FIGURE 10.1   Different advanced oxidation processes (AOPs).

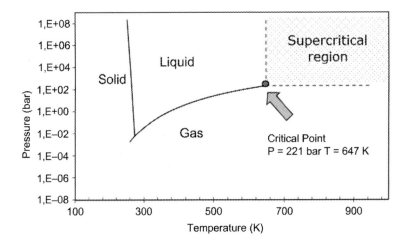

FIGURE 10.2   Phase diagram (P,T) for water.

Fig. 10.2 shows a water phase diagram. Along the vaporization curve, as increasing pressure and temperature for liquid and vapor phases in equilibrium, intermolecular interactions diminish in the liquid due to thermal expansion. Conversely, effect of compression prevails against thermal expansion for steam, resulting in increased interactions. Thus, the properties of the liquid and vapor are approaching, until, at the critical point both equate. At that point, there is only one phase (supercritical phase) with intermediate properties between those for liquids and gases (Vadillo et al., 2012).

Properties of water at ambient pressure and temperature were widely studied and its behavior is well known. However, these properties undergo major changes near the critical point and they are not well characterized. Due to these changes, the study of supercritical water as a reaction medium is very desirable. Main properties of SCW are described as follows.

10.1.1 Density

Density is a function extremely dependent on pressure near critical point as the fluid is highly compressible. Thus, compressibility tends to infinite in the critical point. For this reason, properties that are dependent on density such as solubility parameter, dielectric constant, and partial molar volume of solute change due to small changes in pressure and temperature. Fig. 10.3 shows a density-temperature diagram for water at 250 bar of pressure.

As can be seen in the figure, water density can be controlled from liquid to gas values just only changing temperature of the system. In the critical point, density is known as critical density and the value is 325 kg/m^3. Above the critical point, water density value is around 100 kg/m^3 (Weingartner and Franck, 2005).

10.1.2 Dielectric Constant

Dielectric constant value for water at 25 C and 1 atm is 80 because molecules are joined by hydrogen bonds. However, its value decreases to 6 at the critical point (Weingartner and Franck, 2005) and 1−2 at 450 C. For this reason, behavior of water is similar to nonpolar solvent under these conditions. Thus, the low value of dielectric constant produces salt precipitation. For example, NaCl solubility is under 100 ppm and 10 ppm for $CaCl_2$ (Tester et al., 1993).

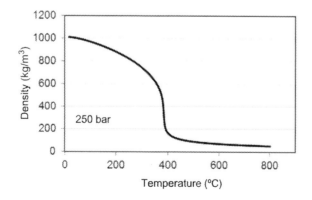

FIGURE 10.3 Density-temperature diagram for water at 250 bar of pressure.

Under these conditions, dipole forces predominate and hence, supercritical water is a successful solvent for organic compounds and it is completely miscible with many gases such as oxygen. For this reason, reaction medium is homogeneous and it consists of organic compounds, supercritical water, and oxygen. Thus, no mass transfer limitations and oxidation reactions take place at a high rate.

10.1.3 Ionic Product

Ionic product for water (K_w) is up to three orders of magnitude higher in liquid region near critical point than at ambient temperature. For this reason, there is a high concentration of H^+ and OH^- ions that favors organic reactions catalyzed by acids and bases. However, once the critical point is achieved, K_w drastically diminishes; so water is a suitable medium for free radical reactions. Thus, supercritical water oxidation takes place by a free radical mechanism (Li et al., 1991). Fig. 10.4 shows a plot of dielectric constant and water ionic product vs. temperature at 250 bar.

10.1.4 Viscosity

Supercritical water viscosity is one order of magnitude lower than liquid water. In this way, diffusion coefficients and ion mobility are one order of magnitude greater. At

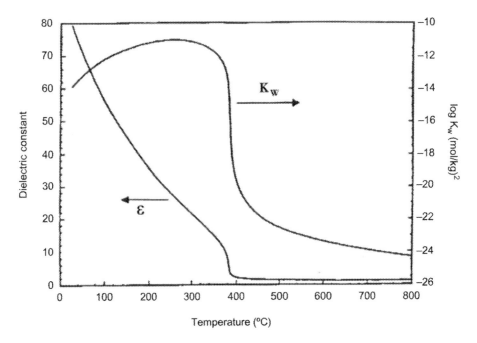

FIGURE 10.4   Dielectric constant ( ) and ionic product of water (K_w) as a function of temperature at 250 bar. *Adapted from Tester, J.W., Holgate, H.R., Armellini, F.J., Webley, P.A., Killilea, W.R., Hong, G.T., Barner HT, ACS Symp Ser, 518, 35–75, 1993. With permission.*

supercritical conditions, water viscosity is low so solute molecules diffuse easily through supercritical water and therefore, it is a reaction medium, where reactions take place at high velocity (Fig. 10.5).

10.1.5 Heat Capacity

Its value varies with temperature and pressure. At 250 bar of pressure, heat capacity value tends to infinity around the critical temperature. It is due to the fact that water suffers a strong thermal expansion that demands a high amount of energy around the critical point (Shaw et al., 1991). Fig. 10.6 shows the variation of the water heat capacity with the temperature at 250 bar.

The increase of the specific heat (C_p) around the critical point makes water heating difficult from 350 C to 380 C at 250 bar, so the heat needed is very high. This aspect is a technical obstacle that hinders the SCWO process scale up (Vadillo et al., 2013).

10.1.6 Thermal Conductivity

At constant pressure, water thermal conductivity increases with temperature up to a maximum value around 200 C. Under supercritical conditions, thermal conductivity decreases due to the rupture of the hydrogen bonds. For this reason, heat transfer coefficients are higher around the critical point and diminish above 400 C (Fig. 10.7). Thus, heat losses are higher around the critical point than in supercritical conditions, where losses decrease as a result of the decrease of thermal conductivity.

SCWO is an oxidation process that takes place above the critical point of water, that is, 374 C and 22.1 MPa. Water polarity is a function of temperature and pressure. At supercritical conditions, water is a nonpolar solvent and it is completely miscible with organics and gases like oxygen (Tester and Cline, 1999). Under these conditions, there is a homogeneous reaction medium, where there are no mass transfer limitations. Furthermore, as the reaction takes place at high velocity due to high temperature used, the residence time necessary to achieve high destruction levels (>99%) are lower than 1 min (Svanström et al.,

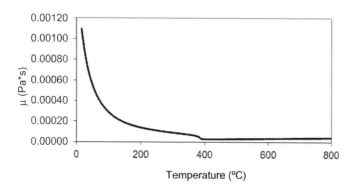

FIGURE 10.5   Viscosity ( ) as a function of temperature at 250 bar.

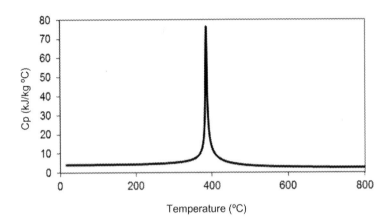

FIGURE 10.6 Heat capacity (C_p) as a function of temperature at 250 bar.

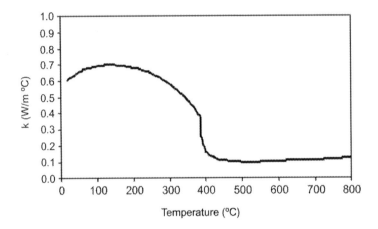

FIGURE 10.7 Thermal conductivity (k) as a function of temperature at 250 bar.

2004a). Furthermore, the reaction products are not toxic. Production of NO_x, SO_x, and dioxins are negligible because the temperature is too low for these compounds to be produced (Kritzer and Dinjus, 2001).

10.2 DEVELOPMENT OF SCWO

During recent decades, numerous researchers focused their studies on obtaining SCWO kinetic parameters of model compounds at laboratory scale to design SCWO reactors. Several authors reviewed studies of model compounds (Bermejo and Cocero, 2006b; Brunner, 2009; Vadillo et al., 2014). Table 10.1 shows an actualization of the compilation of main model compounds studied by different workers (based on Vadillo et al., 2014).

TABLE 10.1 Compilation of Main Model Compounds Studied by Different Workers

Model Compounds	References
Acetonitrile, Acetic acid, Ammonia, Ammonium, Aniline	Merchant (1992), Levec and Smith (1976), Mateos et al. (2005), Imamura et al. (1985), Helling and Tester (1988), Webley and Tester (1991), Oshima (2001), Goto et al. (1999), Segond et al. (2002), Ploeger et al. (2006), Dinaro et al. (2000), Ding et al. (2001), Shimoda et al. (2016)
Benzene, Biphenyl, PCB's, Butyric acid	Chen et al. (2001), Anitescu et al. (2004), Fang et al. (2004), Anitescu and Tavlarides (2005), Fang et al. (2005), OBrien et al. (2005), Williams et al. (1973), Sánchez-Oneto et al. (2006)
Caprylic acid	Sánchez-Oneto et al. (2004)
Dichloroethane, Diethanolamine, Dimethil methilphosphate, Dodecane, Dyes	Limousin et al. (1999), Martín (1998), Chen et al. (1999), Baillod et al. (1982), Mishra et al. (1995), Veriansyah et al. (2006)
EDTA, Ethanol, Ethylenglycol	Helling and Tester (1988), Lee et al. (2004), Rice and Croiset (2001), Xiang et al. (2002), Schanzenbächer et al. (2002), Hayashi et al. (2007), Kim et al. (2003b)
Formic acid	Foussard et al. (1989), Shende and Levec (1999), Imamura (1988)
Glucose, glyoxalic acid	Shende and Mahajani (1994), Shishido et al. (2001)
Ion exchange resins, Isopropanol	Leybros et al. (2010), Abelleira et al. (2013), Queiroz et al. (2013)
Lactic acid	Bianchetta et al. (1999)
Methylphosphonic acid, Methane, Methanol, Methylamine	Tester et al. (1993), Mateos et al. (2005), Webley and Tester (1991), Hayashi et al. (2007), Sullivan (2003), Sullivan and Tester (2004), Rofer and Streit (1988, 1989), Savage et al. (1998), Broell et al. (2002), Webley et al. (1991), Brock et al. (1996), Dagaut et al. (1996), Phenix (1997), Anitescu et al. (1999), Kruse et al. (2000), Jian and Wang (2004), Vogel et al. (2005), Fujii et al. (2011), Benjamin and Savage (2005), Benjamin et al. (2009), Yao et al. (2000)
Nitro-alkanes, Nitrobenzene	Ding et al. (2001), Anikeev et al. (2004), Zhang and Hua (2003), Svishchev and Plugatyr (2006)
Oleic acid, Oxalic acid, o-Dichlorobenzene	Sánchez-Oneto et al. (2004), Shende and Mahajani (1994), Imamura (1982)
3-Hydroxypropionic acid, Propionic acid, Phenol, 2-Chlorophenol, Phenol derivatives, p-aminophenol, p-nitrophenol, Pyridine, Polyethylene glycol, Propene	Merchant (1992), Pruden and Le (1976), Shende and Levec (1999), Mateos et al. (2005), Ding et al. (2001), O'Brien, et al. (2005), Yang and Eckert (1988), Joglekar et al. (1991), Li et al. (1992), Martino and Savage (1997), Shibaeva et al. (1969), Helling et al. (1981), Baillod and Faith (1983), Harris et al. (1983), Jaulin and Chornet (1987), Willms et al. (1987), Thornton et al. (1991), Thornton and Savage (1990, 1992b), Thornton and Savage (1992a), Chang et al.

(Continued)

TABLE 10.1   (Continued)

Model Compounds	References
	(1995), Kolaczkowski et al. (1997), Gopalan and Savage (1995), Krajnc and Levec (1996), Koo et al. (1997), Portela et al. (1997), Oshima et al. (1998), Ju and Feng (2000), Portela et al. (2001a), Henrikson and Savage (2004), Fourcault et al. (2009), García-Jarana et al. (2010), Ge et al. (2003), Crain et al. (1993); Aki and Abraham (1999a, 1999b), Otal et al. (1997), Broll et al. (2002), Li et al. (1993); Lee et al. (2002), Bruce et al. (2003), Liu et al. (2009),Wang et al. (2003), Dong et al. (2015)
Quinazoline	Gong et al. (2016)
Sodium 3,5,6-trichloropyridin-2-ol, Sulfides	Lachance et al. (1999), Zu et al. (2004)
Thiodiglycol, Toluene, 2,4,6-Trinitrotoluene (TNT)	Kim et al. (2002), Kim et al. (2003a), Sánchez-Oneto et al. (2007)
Xylene	Thornton et al. (1991), Aki and Abraham (1999b), Portela et al. (2001a), Shende and Levec (2000)

In relation to reaction pathways, first, Li et al. (1991) stated that SCWO reactions take place by free radical mechanisms. This issue was claimed by Chang and Liu (2007) who proposed a degradation mechanism for 2,4,6-trinitrotoluene based on a free radical pathway identifying the intermediate compounds involved in this reaction. Later, Leybros et al. (2010) studied the free radical mechanism of ion exchange resins SCWO and recently, Gong et al. (2016) studied the reaction mechanism of quinazoline and checked the effect of process variables on the yield of intermediate products.

On the other hand, in case of real wastewaters, which are usually a mixture of compounds, the kinetic study is necessary to design and achieve an optimal operation of the SCWO process. Since the identification of all intermediate compounds involved and the establishment of their complex reaction mechanisms require more effort and high investment, the development of generalized kinetic models is very useful. In this way, several authors developed generalized kinetic models. First, Li et al. (1991) proposed a generalized kinetic model considering acetic acid, the most representative intermediate compound. Later, this reaction mechanism was modified to represent wastewaters containing nitrogen and chlorinated compounds being ammonia (Webley et al., 1991; Li et al., 1993) and methyl chloride, the corresponding representative intermediate compounds, respectively.

Later, Portela et al. (2001b) proposed a new generalized kinetic model based on the carbon monoxide formation as a refractory intermediate compound for cases, in which the formation of acetic acid was not relevant and therefore, the application of the model proposed by Li et al. (1991) would not be applicable. This model was satisfactorily used in the case of cutting oil used in metal working industry. On the other hand, Shende and Mahajani (1994) developed a two step kinetic model and used it to fit the results of hydrothermal oxidation of glyoxalic acid with and without catalyst. Later this model was used in other works, Shende and Levec (2000) used it to model the hydrothermal oxidation of maleic and fumaric acids while Lei et al. (2000) modeled the hydrothermal oxidation of textile industrial wastewaters. Sánchez-Oneto et al. (2007) used this model because it was the most adequate to represent the hydrothermal oxidation of cutting oils with excess of oxidant.

A further step in the development of SCWO was the study of the process on real wastewaters. First, it was satisfactorily applied to a high amount of organic wastewaters at laboratory and pilot plant scale achieving removal efficiencies up to 99.9% with residence times in the order of seconds. A compilation of more representative industrial wastewaters studied for last two decades is shown in Table 10.2. As it is clear from the table, most of the studies were conducted at laboratory scale, because pilot plant and industrial scale studies are scarce.

Due to the harsh operational conditions of the SCWO process, there are far fewer facilities in supercritical water oxidation than in wet air oxidation. However, numerous pilot plant facilities have been built up to date (Table 10.3).

On the other hand, Table 10.4 shows SCWO industrial facilities commercialized to date, highlighting the first commercial SCWO plant built by Eco Waste Technologies for the company Hunstman Chemical Corporation in Austin (Texas). It has been operating since 1994 and shut down in 1999.

10.3 DETECTED PROBLEMS

SCWO commercial development has been hindered by some technical drawbacks, high investment, and high operational costs (Table 10.4).

Corrosion and salt precipitation are the main technical problems of SCWO. Both are caused by the combination of a high temperature and harsh chemical environment inherent to this application (Marrone and Hong, 2009).

Corrosion depends on several factors such as the dissociation of acids, salts and bases, solubility of gases, solubility of corrosion products, and the stability of the protecting oxide layer. However, it is difficult to find a material that can stand all the conditions so it is necessary to choose the material as function of the operation conditions (Brunner, 2009). Since corrosion problem have limited the SCWO development, it is necessary to put effort into solving it. For this reason, during the last decade, several researchers have focused their work on the detailed study of this phenomenon (Kritzer and Dinjus, 2001; Brunner, 2009; Bermejo et al., 2006a, 2006b). Corrosion can be managed by several ways (Marrone and Hong, 2009) including a cooling strategy to avoid the conditions of high temperature and density, (which are the conditions of high corrosion rates) (Kritzer and Dinjus, 2001), and new reactor concepts as transpiring wall or film-cooled reactors (Vadillo et al., 2013).

As a consequence of the low solubility of inorganic compounds in supercritical water, precipitation of salts occurs in SCWO reactors and can lead to equipment plugging (Vadillo et al., 2013). Solid salts form conglomerations that cover the walls of the equipment, reduce the heat transfer in heat exchangers, and produce plugging in pipes and reactor. Besides, there is a dead place between the salt layer and the wall of the reactor, where heavy corrosion can occur (Kritzer and Dinjus, 2001).

TABLE 10.2 Industrial Wastewaters Treated by SCWO

Waste	Reactor Type	Operation Mode	Flow Rate	Reactor Volume (L)	References
Sewage sludge	Tubular	Batch		0.004	Goto et al. (1998)
		Batch		0.572	Qian et al. (2015)
Alcohol distillery wastewater	Tubular	Continuous	0.06 L/h	0.013	Goto et al. (1998)
Olive mill wastewater	Tubular	Continuous	0.084 kg/h	0.0035	Erkonak et al. (2008)
	Tubular	Continuous	3.56 L/h	0.08	Rivas et al. (2001)
	Tubular	Continuous	2.8 kg/h	(n/a)	Chkoundali et al. (2008)
Sulfonated lignin waste	Tubular	Semi-continuous	0.03 kg/h	0.016	Drews et al. (2000)
Sewage and industrial sludge	Tubular	Batch		0.020	Shanableh (2005)
Automobile manufacturer painting effluent	Tubular	Continuous	20 kg/h	0.75	Abeln et al. (2007)
Wastewaters from LCD (liquid crystal display) manufacturing	Tubular	Continuous	0.63 kg/h	0.020	Veriansyah et al. (2005)
Wastewater from acrylonitrile manufacturing plant	Tubular	Continuous	1.15 L/h	1/8 i.d.	Shin et al. (2009)
Oily sludge	Stirred tank	Batch		0.650	Cui et al. (2009)
Cutting oil wastes	Tubular	Continuous	2 L/h	0.08	Sánchez-Oneto et al. (2007)
PCB-contaminated mineral transformer oil	Tubular	Continuous	0.138 L/h	0.057	Marulanda and Bolaños (2010)
	Tubular	Continuous	30 kg/h	(n/a)	Kim et al. (2010)
Ion exchange resins	Double shell stirred	Continuous	0.5 kg/h	0.44	Leybros et al. (2010)
Strength coking wastewater	Tubular	Continuous	1.2 L/h	0.075	Du et al. (2013)
Flammable industrial wastewaters	Tubular	Continuous	10 L/h	1.23	Vadillo et al. (2011)
Coal	Tubular	Continuous	2 L/h	0.202	Wang et al. (2011)
Landfill leachate	Transpiring wall	Continuous	2 L/h	0.032	Weijin and Xuejun (2010)
P-nitrophenol in wastewater	Tubular	Continuous	50 g/min	12.8 mm i.d.	Dong et al. (2015)
Pesticide wastewater	Tubular	Batch		0.250	Xu et al. (2015)
		Continuous	1 L/h	0.808	Xu et al. (2015)

n/a: No available data.

TABLE 10.3   SCWO Pilot Plants, (P   25 MPa and T   550 C)

Process	Reactor Type	Oxidant Type	Wastewater Treated	Status
Komatsu & Kurita Group (General Atomic) (www. kurita.co.jp)	Tank Double shell	Air	Wastewater and sewage sludge	400 kg/h 1000 h tests (Japan)
Organo Corporation (General Atomics) (Rossignol, 2001; Cansell and Farhi, 2002)	Tank Double tank	Air	PCB[a], dioxins, sewage sludge, and radioactive wastes	100 kg/day, 2 t/day 1995, Tokyo (Japan)
EBARA (Cansell and Farhi, 2002)	Tank with flame generation	Oxygen	Wastewater from incineration plants	60 kg/h Operating before 2002, Tsukuba (Japan)
US Navy (General Atomics) (Beslin et al., 1997)	Tank Double tank	Air	Propellant	250 kg/h 1997, UTAH (USA)
US Navy (General Atomics) (Beslin et al., 1997)	Tank Double tank	Air	Dangerous wastewaters from military	250 kg/h 1998, mobile plant (USA)
US Navy (Foster Wheeler) (Crooker et al., 2000)	Tubular with porous wall	Air	Toxics wastewaters	250 kg/h 1999, mobile plant, (USA) Nowadays inactive
US Department of Defense (McFarland et al., 1993)	Coated tank	(n/a)	Radioactive wastes	0.4 kg/h 1997, Los Alamos (USA)
Aquacritox (Super Critical Fluids International) (Gidner and Stenmark, 2000)	Tubular	Oxygen	Industrial wastewaters	250 kg/h 1999, Karlskoga (Sweden) 2008, Cork (Ireland)
PIOS HOO (Cansell, 2002)	Tubular and tank combined	Oxygen	Toxic wastewaters	100 kg/h 2004, SME-St Médard in Jalles (France)

(*Continued*)

TABLE 10.3 (Continued)

Process	Reactor Type	Oxidant Type	Wastewater Treated	Status
Universidad de Valladolid (Cocero et al., 2002)	Transpiring wall	Air	Cutting oils, polyethylene terephthalate (PET), production wastewater, industrial wastewaters	40 kg/h, Valladolid, (Spain)
Universidad de Valladolid (Cocero, et al., 2003)	Transpiring wall and cool wall	Oxygen	Industrial wastewater	200 kg/h, Valladolid, (Spain)
Universidad de Cádiz (García-Jarana et al., 2010)	Tubular	Air	Industrial wastewater	25 kg/h, Cádiz, (Spain)
School of Energy and Power Engineering, China (Xu et al., 2012)	Transpiring wall combined with reverse flow tank	Oxygen	Sewage sludge	125 kg/h 2011, (China)
University of British Columbia (Canada) (Asselin et al., 2008)	Tubular	Oxygen	Ammoniacal sulfate solution	120 kg/h
Boreskov Institute of Catalysis, Russia (Anikeev and Yermakova, 2011)	Tubular	Air	Explosive manufacture waste	40−60 kg/h
Super Water Solutions, City of Orlando	Tubular	Oxygen	Sewage sludge	5 dried matter t/day
Super Critical Fluids International (LO2X Project)	Tubular	Oxygen	Sewage sludge, agri-food waste, industrial wet waste, and landfill leachates	6 t/day

*a*PCB: Polychlorinated biphenyls; n/a: No available data.

Fig. 10.8 shows a summary reported by Vadillo et al. (2013) including the main problems that can occur in different steps of the conventional SCWO process.

The solution to these problems is necessary to advance in the scale up of the SCWO process. Vadillo et al. (2013) summarized main solutions to these problems such as: use of high corrosion resistance materials (Inconel 625 and Hastelloy 600), use of liners, use of coating, design SCWO systems including transpiring wall/film-cooled wall reactors (Marrone and Hong, 2009), assisted hydrothermal oxidation (Tateishi et al., 2000), use of a base to pre-neutralize the feed stream, cold (ambient temperature) feed injection, addition of quench water, optimization of process operating conditions (Kritzer and Dinjus, 2001), to avoid corrosive feeds or the pretreatment of the feed to remove corrosive species (Hong et al., 1996) to avoid corrosion. In case of avoiding salt precipitation, Marrone et al. (2004) studied salt precipitation and scale control in supercritical water oxidation focused on commercial scale applications. The methods are specific reactor designs (such as reverse flow, tank reactor with brine pool, transpiring wall reactor, adsorption/reaction on fluidized solid phase, reversible flow in tubular reactor, and centrifuge reactor) and specific techniques (such as high velocity flow, mechanical brushing, rotating scraper, reactor

TABLE 10.4 Commercially Designed Full Scale SCWO Plants

Company Responsible for the Project	Customer	Wastewater	Kind of Reactor	Capacity	Comments
ACTIVES					
Joint Venture of Bechtel National Inc., and Parsons Government Services Inc.	US Army (Richmond, Kentucky)	Blue Grass VX and GB (sarin) nerve agent wastes	TWR	(n/a)	About to be started up
Innoveox	Private company	Hazardous industrial waste	Tubular	100 kg/h	Operating since June 2011
Mitsubishi heavy industries	Japan Environmental Safety Corp. (Tokyo Bay)	PCB's[a]	Fluidized bed tank reactor followed by tubular reactor	306 t/day	Operating in late 2005
INACTIVES					
General Atomics	U.S. Army (McAlester)	Pink water	Tank	6.5 t/day	Operating since 2001
General Atomics	U.S. Army (Richmond)	Chemical Weapons	Tank	36 t/day	6000 hours of tests in 2011
General Atomics	Toele Army (Toele)	Hydrolysate of conventional explosive devices	Tank	18 t/day (3 gpm)	Operating since 2008
General Atomics	Bluegrass Army Depot	Explosives and propellants	Tank	50 t/day	(n/a)
General Atomics	Pacific Environmental Corporation Alaska	Industrial wastewaters	(n/a)	10 t/day	(n/a)
Chematur Engineering	Johnson Matthey, Brimsdown (UK)	Spent catalyst (recovery of precious metals)	Tubular	80 t/day	Operating since 2002, from October 2004 in charge of Johnson Matthey
Hanwha Chemical	Namhae Chemical Corp. (Korea)	Wastewater from DNT production	Tank	53 t/day	Commissioned in 2002
Hanwha Chemical	Samah Petrochemical Corp. (Korea)	Wastewater from TPA production	Tubular	145 t/day	Operating since end of 2006
Organo (MODAR)	National University of Japan	Laboratory wastewater	Tank	(n/a)	Operating since 2002

(*Continued*)

TABLE 10.4 (Continued)

Company Responsible for the Project	Customer	Wastewater	Kind of Reactor	Capacity	Comments
EcoWaste Technologies	Hunstman Chemical, Austin (Texas).	Alcohols, glycols, and amines	Tubular	29 t/day	Operating between 1994–1999
Foster Wheeler Development Corp.	U.S. Army, Pinebluff Arsenal	Obsolete weapons	Tank	3.8 t/day	Built in 1998, stopped in 2001 due to mechanical problems and financial restriction
Organo (MODAR)	Nittetsu semiconductor, Tateyama, Japan	Semiconductor industry waste	Tank	2 t/day	Built in 1998, nowadays out of operation
Hydroprocessing	Harlingen wastewater treatment plant, Texas	Sewage sludge	Tubular	150 t/day	Built in 2001, stopped in 2002 due to corrosion in heat exchanger
Hydrothermal oxidation option	SYMPESA, (France)	Food industry wastewater	Tubular	2.7 t/day	Operating between 2004–2008

[a]PCB: Polychlorinated biphenyl.
n/a: No available data.

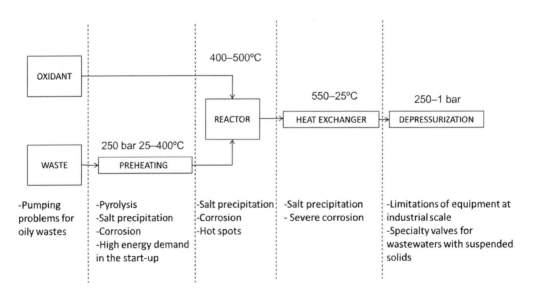

FIGURE 10.8 Main problems that can occur in different steps of the conventional SCWO process. *Adapted from Vadillo, V., Sánchez-Oneto, J., Portela, J.R., Martínez de la Ossa, E.J., 2013. Ind. Eng. Chem. Res., 52, 7617–7629. With permission.*

flushing, additives, low turbulence, homogeneous precipitation, crossflow filtration, density separation, and extreme pressure operation).

In this way, to avoid corrosion and salt precipitation problems, industrial wastewaters to be treated by means of SCWO in tubular reactors, need to satisfy certain requirements as described by Vadillo et al. (2011).

In addition to corrosion and salt precipitation phenomena Vadillo et al. (2013) summarized several technical drawbacks that caused the delay of the SCWO process scale up and these are:

- Treatment of insoluble waste such as oily wastes or organic solvents. The pumping of oily wastes to be treated by SCWO is a challenge due to the non-homogeneous mixture of oil and water. Vadillo et al. (2012) reported the disadvantages of pumping a heterogeneous wastewater. Thus, if the feed is a mixture of oil and wáter, there will be moments, in which the pump will only introduce pure oil in the reactor producing a dangerous increase in the reactor temperature. That is the reason why, in SCWO reactors, it is advisable to use a secondary feed pump to introduce insoluble wastewaters in the reactor. Also, due to the variation in concentration of wastewater feedstock during the operation, peaks of temperature along the reactor can occur, jeopardizing the safe operation of SCWO.

- Another important limitation at industrial scale is the energy required in the start-up of the process. For this reason, the only way to achieve economic feasibility at industrial scale is to reduce the interruptions of the process that implies continuous process start-up. Vadillo et al. (2013) reported that it was necessary to achieve 95% of availability. It is necessary to design a heater system to provide enough energy to the high pressure stream to achieve supercritical temperature. For instance, in case of plants of capacity 250 t/day the thermal energy required is around 5 MW. In order to optimize the operation costs during the start-up, Benjumea et al. (2015) proposed a new strategy to minimize the energy required in this step. It consisted of a slow and static heating up of the whole plant (at near critical pressure, containing a mixture of water and isopropanol) before starting the circulation of the wastewater and feeding the oxidant. Then, the wastewater and the oxidant were fed at a low flow rate and the oxidation reactions started immediately releasing the heat required for autothermal operation.

- In order to retain safe conditions and carry out an optimal energy recovery, a SCWO plant with tubular reactor treating highly concentrated wastewaters must have stringent thermal control. This can be conducted by means of cooling water injections and multioxidant injections in different points along the reactor. Several authors have developed and studied new configurations of reactor to improve the thermal control of the process (Gidner and Stenmark, 2000; Vielcazals et al., 2006; Chkoundali et al., 2008; Benjumea et al., 2014). Recently, Benjumea et al. (2017) improved a 25 kg/h SCWO pilot plant to control the amount of heat released to avoid any safety hazards installing both cooling water and oxidant split injectors. Thus, experimental tests were carried out to compare the operation of the pilot plant before and after those two improvements. These improvements increased the range of wastewater concentration that could be treated in a SCWO process. Besides, it increased the energy recovery rate and improved the safety control of the pilot plant, while the same COD removal efficiency level was maintained.

- Depressurization of effluent process is a key aspect in SCWO plants because a correct operation implies an adequate control of system pressure (Vadillo et al., 2013). Wastewaters that contain suspended solid particles can produce problems during the effluent depressurization because they can erode the internal parts of the "back pressure regulator" valves. Besides, depressurization step can produce problems at industrial scale, because at high flow rate, the use of one valve is not recommended to get the overall pressure drop (250 bar) in a single step. Therefore, several studies were made on the use of capillary devices to achieve the pressure drop. Results obtained by Marias et al. (2007) showed that a depressurization system for a flow rate of 20 kg/h should consist of 1/16″ and 21.5 m long pipe. O'Regan et al. (2008) proposed a capillary system, where pressure drop is achieved distributing the total flow in several capillary devices with a high length. They claimed that the use of only one valve produced extreme velocities and severe erosion problems.
- Later, Benjumea et al. (2011) satisfactorily conducted the high pressure reduction (up to 240 bar) in a SCWO pilot plant using a system of variable length based on three series of coiled pipes. The system proposed was tested under different flow rates of water and air to check its suitability in the control of the pressure. Furthermore, the experimental results were compared with those obtained by the modelling carried out using different friction factor correlations from the literature, for both: single-phase and two-phase systems. None of those correlations were used before for pressure drops higher than 50 bar. Therefore, in the case of industrial plants, the use of capillary devices and a combination of valves is desirable.
- Some wastewaters can suffer pyrolysis and hydrolysis due to the preheating step (in the absence of oxygen) that is necessary to reach 400 C at the tubular reactor entrance. As a consequence of these undesirable reactions, plugging in the preheating system can occur in addition to the presence of gas compounds in the effluent such as CH_4 and CO. The solution patented by Portela et al. (2005) avoids this phenomenon due to the absence of organic material in the preheating line of the system.

10.4 ENERGY RECOVERY IN SCWO PLANTS

Recent SCWO studies were focused on energy production. Queiroz et al. (2015) considered supercritical water oxidation as a potential clean energy generation process. SCWO process has to be considered not only as a process, in which a residue is completely removed, but as a possible source of energy production due to the possibility of exploiting the exothermicity of the oxidation reactions. Thus, it is possible to decrease the operating costs related to extreme operating conditions (T > 374 C, P > 22.1 Mpa), that hindered the industrial scale up and commercialization of SCWO technology. For this reason, various researchers have addressed the issue through theoretical studies and simulation. Cocero et al. (2002) developed an energy study for a 2 m³/h SCWO plant using the software Aspen Plus®. Later, Bermejo et al. (2004) studied SCWO of coal including energy recovery as an alternative to carbon power plants. However, Marias et al. (2008) claimed that energy recovery proposed by Bermejo et al. (2004) was not feasible because it was based on particle separation at 650 C and 30 Mpa and separator and turbine can not been nowadays built using current technology. As an alternative, they proposed energy recovery in SCWO using an auxiliary fluid that receives the energy produced in SCWO process. On the other hand, Svanström et al. (2004a), simulated energy recovery from 10% weight sewage sludge SCWO using Aspen Plus® software.

Later, Jiménez-Espadafor et al. (2011) proposed that it is possible to decrease SCWO treatment costs by recovering energy at low temperature and high fluid pressure, such as water heating and steam generation. For a supercritical flow of 1000 kg/h (water and air), the recovered energy goes from 118 kW (1700 m^3/h of hot water at 65 C) to 75 kW (100 kg/h of steam flow at 1.1 bar and 170 C). Besides, considering a rate of 15.9 kg/h of cutting fluid with a conversion of 95%, the waste heat energy recovered would be between 71% and 45.6% of the energy content of the organic compound, a level that can be considered as high thermal efficiency.

If wastewaters have a low reaction heat, the use of auxiliary fuels is desirable to increase the temperature profile along the reactor and to achieve the autothermal operation (Bermejo and Cocero, 2006b). Thus, hydrothermal flames are generated in SCWO reactors, including devices specially designed for it. Several authors studied hydrothermal flames for more than twenty years. A hydrothermal flame is a combustion flame produced in aqueous medium under conditions of temperature and pressure above the critical point of water. They are characterized by high temperatures, typically above 1000 C, higher reaction rates, and residence times around 10–100 ms in which complete oxidation of the reactants is achieved (Vadillo et al., 2013).

The preheating of the fuel and the oxidant until a high temperature is required to produce the flame. If temperature is high enough for the waste concentration, the autoignition of the mixture is produced (Augustine and Tester, 2009). Deeper studies on hydrothermal flames formation can be found in the literature (Wellig et al., 2005, 2009).

From an engineering point of view, the design of SCWO reactors with flame generation was studied by Bermejo et al. (2011). Specifically, they studied the scale up process in a transpiring wall reactor with hydrothermal flame as an internal heat source. That work showed the remarkable effect of the fluid velocity over the injection temperature to get a stable flame generation. At low velocity in the injector extinction, temperature is low as well as it being possible to inject at 170 C and 0.04 m/s. They designed a transpiring wall reactor (L = 0.4 m Di = 0.33 m) to treat 3000 kg/h of waste injected at 25 C using modeling. Refractory compounds such as acetic acid and ammonia can be oxidized due to the high temperature in SCWO reactors with hydrothermal flames.

Thus, Cabeza et al. (2011) used a tubular reactor with a residence time of 0.7 s to oxidize those compounds using IPA (isopropyl alcohol) as cofuel. In the case of acetic acid, temperature was increased to 750 C achieving 99% of removal. However, in the case of ammonia, removal was under 94% even if temperature was higher than 710 C. Besides, it was found that concentration under 2% weight did not produce ignition of ammonia solutions. Recently, Queiroz et al. (2015) studied deoxidation under a hydrothermal flame of biomass. They claimed that it can intensify the SCWO process in order to develop microcombustors to produce high pressure and temperature steam from wastes. However, this execution needs the development of technical solutions for injecting biomass in a hydrothermal flame combustor, and equipment for energy recovery must be developed and engineered.

10.5 ECONOMIC ASPECTS

Due to the nature and characteristics of SCWO process, high investment costs are required. On one hand, equipment must be able to work at high pressure and temperature and, on the other hand, corrosion problems lead to use high corrosion resistance alloys to build reactors and heat exchangers (preferably alloys with high nickel content). In addition to high material costs, the maintenance and repair costs of equipment that works under extreme conditions are very high too. Therefore, the initial investment required is high and the only way to make the SCWO process economically feasible is reaching the auto-thermal regime in the reactor. During the last years, several authors have studied the process' economics. Abeln et al. (2007) estimated that tubular reactor costs represent 10% of the overall equipment costs in a SCWO plant able to treat 100 kg/h of wastewater.

Later, Marulanda and Bolaños (2010) estimated that reactor costs represent 7% of the overall equipment costs in a mobile SCWO plant able to treat 1 L/h of organic wastewater. Besides, the choice of the oxidant is a key point from the point of view of operational cost, Bermejo and Cocero (2006a) claimed that it is cheaper to use pure oxygen instead of air because at industrial scale, the compression cost is very high. On the other hand, Savage et al. (1999) suggested that catalytic SCWO process is a more competitive alternative because using a catalyst, the temperature necessary to reach removal efficiencies higher than 99% is reduced significantly decreasing the energy demand.

Gidner and Stenmark (2000) estimated operational costs of a sewage sludge SCWO plant based on a flow rate 7 m^3/h of sewage sludge being 137 €/t dried sludge. Svanström et al. (2004b) estimated total cost for a 1 t/day plant being 243 $/t dried sludge. O'Regan et al. (2008) claimed that treatment cost of sewage sludge SCWO is in the range $36.6 - 73.15$ €/t. Abeln et al. (2007) first estimated treatment cost of an ideal wastewater made of a mixture of ethanol 10% weight and water using air as oxidant in a plant of 100 kg/h with two different reactors: tubular and transpiring wall reactor, and the treatment costs 406 €/t and 660 €/t, respectively. Later, they estimated the treatment cost for a 1 t/h plant being 330–430 €/t for the transpiring wall reactor plant and 203–264 €/t for the tubular reactor plant. Marulanda and Bolaños (2010) also estimated that treatment cost for a mobile SCWO unit able to treat 1 L/h of oils contaminated with PCBs amounts to 75 $/L.

Based on the results obtained at pilot plant scale, Vadillo et al. (2011) elaborated an economic analysis for an industrial SCWO plant capable of treating 1 m^3/h of flammable industrial wastewater with a concentration of 90 g O_2/L. Thus, the main components considered in the SCWO plant were: a tubular reactor made of stainless steel AISI 316, a counter current heat exchanger, a high pressure pump, a cooler, and two back-pressure regulators. The oxidant was pure oxygen supplied by a cryogenic oxygen storage plant bearing in mind the use of 10% of excess. All in all, the estimated overall cost for the SCWO treatment of wastewater in a 1 m^3/h plant with tubular reactor technology was 109 €/t wastewater, bearing in mind depreciation time over 10 years and an operating period of 330 days/year.

Economic results showed that although SCWO technology was initially shown as a technology suitable for all kind of wastes, research conducted over the last three decades showed that this technology can only be applied at industrial scale using tubular reactor and to treat wastewaters that satisfy the requirements described by Vadillo et al. (2011).

10.6 CONCLUSIONS

SCWO has to be considered as a feasible technology that respects the environment and achieves the removal of those toxic wastes that can not be destroyed by any conventional technology. SCWO has been developed for decades from laboratory scale to commercial scale. Although SCWO technology was initially shown as a technology suitable for a wide range of organic wastewaters, this technology has some technical limitations and can only be applied at industrial scale to treat wastewaters that satisfy a list of requirements. In order to advance in the commercial development of SCWO, it is crucial to select an appropriate wastewater and to choose the most suitable reactor concept (as transpiring wall reactor, multioxidant reactor, reactor including hydrothermal flames, etc.), the correct operational method (between the conventional feed system or a direct feed injection), and the appropriate technical solution (to handle solids, to avoid pyrolysis/hydrolysis, etc.). Besides, from an economical point of view, the design of SCWO plants has to include an energy recovery system in order to achieve the economic feasibility of the process.

Acknowledgements

This work was supported by Project P11-RNM-7048 promoted by Junta de Andalucía (Spain).

References

Abelleira, J., Sánchez-Oneto, J., Portela, J.R., Martínez de la Ossa, E.J., 2013. Kinetics of supercritical water oxidation of isopropanol as an auxiliary fuel and co-fuel. Fuel 111, 574–583.

Abeln, J., Kluth, M., Pagel, M., 2007. Results and rough cost estimation for SCWO of painting effluents using a transpiring wall and a pipe reactor. J. Adv. Oxid. Technol. 10 (1), 169–176.

Aki, S., Abraham, M.A., 1999a. Catalytic supercritical water oxidation of pyridine: comparison of catalysts. Ind. Eng. Chem. Res. 38, 358–367.

Aki, S.N., Abraham, M.A., 1999b. Catalytic supercritical water oxidation of pyridine: kinetic and mass transfer. Chem. Eng. Sci. 54, 3533–3542.

Anikeev, V., Yermakova, A., Goto, M., 2004. Decomposition and oxidation of aliphatic nitro compounds in supercritical water. Ind. Eng. Chem. Res. 43 (26), 8141–8147.

Anikeev, V.I., Yermakova, A., 2011. Technique for complete oxidation of organic compounds in supercritical water. Russ. J. Appl. Chem. 84, 88–94.

Anitescu, G., Tavlarides, L.L., 2005. Oxidation of biphenyl in supercritical water: Reaction kinetics, key pathways, and main products. Ind. Eng. Chem. Res. 44, 1226–1232.

Anitescu, G., Tavlarides, L.L., Munteanu, V., 2004. Decomposition of monochlorobiphenil isomers in supercritical water in the presence of metanol. AIChE J. 50 (7), 1536–1544.

Anitescu, G., Zhang, Z., Tavlarides, L.L., 1999. A kinetic study of methanol oxidation in supercritical water. Ind. Eng. Chem. Res. 38 (6), 2231–2237.

Asselin, E., Alfantazi, A., Rogak, S., 2008. Thermodynamics of the corrosion of alloy 625 supercritical water oxidation reactor tubing in ammoniacal sulfate solution. Corrosion 64, 301–314.

Augustine, C., Tester, J.W., 2009. Hydrothermal flames: from phenomenological experimental demonstrations to quantitative understanding. J. Supercrit. Fluid. 47 (3), 415–430.

Baillod, C.R., Faith, B.M., 1983. Wet oxidation and ozonation of specific organic pollutants; EPA-600/2-83-060.

Baillod, C.R., Faith, B.M., Masi, O., 1982. Fate of specific pollutants during wet oxidation and ozonation. Environ. Progress 1 (3), 217–227.

Benjamin, K.M., Meyer, J., Sefa, F., Lane, S., 2009. Quantum chemical and detailed chemical kinetic modeling of methylamine oxidation: applications to atmospheric and supercritical water chemistries. Abstracts of papers, 237th ACS National Meeting, Salt Lake City, UT, United States, March 22–26, 2009.

Benjamin, K.M., Savage, P.E., 2005. Detailed chemical kinetic modeling of methylamine in supercritical water. Ind. Eng. Chem. Res. 44 (26), 9785–9793.

Benjumea, J.M., Sánchez-Oneto, J., Portela, J.R., Martínez de la Ossa, E.J., 2011. Use of helical coil pipes for depressurization in a supercritical water oxidation (SCWO) pilot plant: Experimental results & simulation. Chem. Eng. Trans. 24, 283–288.

Benjumea, J.M., Portela, J.R., Sánchez-Oneto, J. Martínez de la Ossa, E.J., 2014. Simulation of a counter current refrigeration system for a SCWO reactor. 14th European Meeting on Supercritical Fluids EMSF'14. 18–21 May 2014. Marsella, Francia.

Benjumea, J.M. Portela, J.R., Sánchez-Oneto, J. Martínez de la Ossa, E.J., 2015. New strategy for supercritical water oxidation plant start-up with low energy comsuming. 10th European Congress of Chemical Engineering, September 27th–October 1st, 2015, Niza, Francia.

Benjumea, J.M., Portela, J.R., Sánchez-Oneto, J., Martínez de la Ossa, E.J., 2017. Temperature control in a supercritical water oxidation reactor: assessing strategies for highly concentrated wastewaters. J. Supercrit. Fluid 119, 72–80.

Bermejo, M.D., Cocero, M.J., 2006a. Destruction of an industrial wastewater by supercritical water oxidation in a transpiring wall reactor. J. Hazard. Mater. 137 (2), 965–971.

Bermejo, M.D., Cocero, M.J., 2006b. Supercritical water oxidation: a technical review. AIChE J. 52 (11), 3933–3951.

Bermejo, M.D., Cocero, M.J., Fernández-Polanco, F., 2004. A process for generating power from the oxidation of coal in supercritical water. Fuel 83, 195–204.

Bermejo, M.D., Cocero, M.J., Fernández-Polanco, F., 2006a. Experimental study of the operational parameters of a transpiring wall reactor for supercritical water oxidation. J. Supercrit. Fluid 39 (1), 70–79.

Bermejo, M.D., Cocero, M.J., Fernández-Polanco, F., 2006b. Effect of the transpiring wall on the behavior of a supercritical water oxidation reactor: modelling and experimental results. Ind. Eng. Chem. Res. 45, 3438–3446.

Bermejo, M.D., Cabeza, P., Queiroz, J.P.S., Jimenez, C., Cocero, M.J., 2011. Analysis of the scale up of a transpiring wall reactor with a hydrothermal flame as a heat source for the supercritical water oxidation. J. Supercrit. Fluid 56, 21–32.

Beslin, P., Cansell, F., Garrabos, Y., Demazeau, G., Berdeu, B., Sentagnes, D., 1997. Le traitement hydrothermal des déchets: Une solution innovante. Déchets Sci. Tech. 5, 17–21.

Bianchetta, S., Li, L., Gloyna, E.F., 1999. Supercritical water oxidation of methylphosphonic acid. Ind. Eng. Chem. Res. 38 (8), 2902–2910.

Brock, E.E., Oshima, Y., Savage, P.E., Barker, J.R., 1996. Kinetics and mechanism of methanol oxidation in supercritical water. J. Phys. Chem. 100, 15834–15842.

Broell, D., Kraemer, A., Vogel, H., 2002. Heterogeneously catalyzed partial oxidation of methane in supercritical water. Chem. Ingenieur Tech. 74 (6), 795–800.

Broll, D., Kramer, A., Vogel, H., 2002. Partial oxidation of propene in sub- and supercritical water. Chem. Ing. Tech. 74 (1-2), 81–85.

Bruce, D.A., Thies, M.C., O'Brien, C., 2003. Supercritical water oxidation of the PCB congener 2-chlorobiphenil: a kinetic analysis. Am. Chem. Soc. Div. Environ. Chem. 43 (2), 142–147.

Brunner, G., 2009. Near and supercritical water. Part II: oxidative processes. J. Supercrit. Fluid. 47, 382–390.

Cabeza, P., Bermejo, M.D., Jimenez, C., Cocero, M.J., 2011. Experimental study of the supercritical water oxidation of recalcitrant compounds under hydrothermal flames using tubular reactors. Water Res. 45 (8), 2485–2495.

Cansell, F., 2002. Method for treating waste by hydrothermal oxidation. International Patent WO 0220414. 14-03-2002.

Cansell, F., Farhi, R., 2002. Rapport de la mission d'évaluation du développement industriel du procedé d'oxydation hydrothermale au Japon organisée par le Ministère des Affaires Etrangères: période du 04 au 08 février 2002. Pessac: ICMCB, 6 p.

Chang, C.J., Li, S.S., Ko, C.M., 1995. Catalytic wet oxidation of phenol and p-chlorophenol contaminated waters. J. Chem. Technol. Biotechnol. 64, 245–252.

Chang, S.J., Liu, Y.C., 2007. Degradation mechanism of 2,4,6-trinitrotoluene in supercritical water oxidation. J. Environ. Sci. 19, 1430–1435.

Chen, Gm, Lei, L., Yue, P.L., 1999. Wet oxidation of high-concentration reactive dyes. Ind. Eng. Chem. Res. 38, 1837–1843.

Chen, F., Wu, S., Chen, J., Rong, S., 2001. COD removal efficiencies of some aromatic compounds in supercritical water oxidation. Chin. J. Chem. Eng. 9 (2), 137–140.

Chkoundali, S., Alaya, S., Launay, J.C., Gabsi, S., Cansell, F., 2008. Hydrothermal oxidation of olive oil mill wastewater with multi-injection of oxygen simulation and experimental data. Environ. Eng. Sci. 25 (2), 173–180.

Cocero, M.J., Alonso, E., Sanz, M.T., Fernández-Polanco, F., 2002. Supercritical water oxidation process under energetically self-sufficient operation. J. Supercrit. Fluid. 24, 37–46.

Cocero, M.J., Martín, A., Bermejo, M.D., Santos, M., Rincón, D., Alonso, E., et al., 2003. Supercritical water oxidation of industrial waste from pilot to demonstration scale. Proceedings of the 6th International Symposium on Supercritical Fluids, Versailles, France, April 28-30.

Crain, N., Tebbal, S., Li, L., Gloyna, E.F., 1993. Kinetics and reaction pathways of pyridine oxidation in supercritical water. Ind. Eng. Chem. Res. 32, 2259–2268.

Crooker, P.J., Ahluwalia, K.S., Fan, Z., 2000. Operating results from supercritical water oxidation plants. Ind. Eng. Chem. Res. 39, 4865–4870.

Cui, B., Cui, F., Jing, G., Xu, S., Huo, W., Liu, S., 2009. Oxidation of oily sludge in supercritical water. J. Hazard. Mater. 165 (1-3), 511–517.

Dagaut, P, Cathonnet, M., Boettner, J.C., 1996. Chemical kinetic modelling of the supercritical water oxidation of methanol. J. Supercrit. Fluids 98, 33–42.

Dinaro, J.L., Howard, J.B., Green, W.H., Tester, J.W., Bozzelli, J.W., 2000. Analysis of an elementary reaction mechanism for benzene oxidation in supercritical water. Proc. Combust. Inst. 28 (2), 1529–1536.

Ding, J.-W., Chen, F.-G., Wu, S.-F., Rong, S.-X., 2001. Kinetics of aniline oxidation in supercritical water. Gaoxiao Huaxue Gongcheng Xuebao 15 (1), 66–70.

Dong, X., Gan, Z., Lu, X., Jin, W., Yu, Y., Zhang, M., 2015. Study on catalytic and non-catalytic supercritical water oxidation of p-nitrophenol wastewater. Chem. Eng. J. 277, 30–39.

Drews, M.J., Barr, M., Williams, M., 2000. A kinetic study of the SCWO of a sulfonated lignin waste stream. Ind. Eng. Chem. Res. 39, 4784–4793.

Du, X., Zhang, R., Gan, Z., Bi, J., 2013. Treatment of high strength coking wastewater by supercritical water oxidation. Fuel 104, 77–82.

Erkonak, H., Sogut, O.O., Akgun, M., 2008. Treatment of olive mill wastewater by supercritical water oxidation. J. Supercrit. Fluid. 46 (2), 142–148.

Fang, Z., Xu, S., Butler, I.S., Smith, R.L., Kozinski, J.A., 2004. Destruction of decachlorobiphenyl using supercritical water oxidation. Energy Fuels 18, 1257–1265.

Fang, Z., Xu, S.K., Smith Jr., R.L., 2005. Destruction of deca-chlorobiphenyl in supercritical water under oxidizing conditions with and without Na_2CO_3. J. Supercrit. Fluids 33, 247–258.

Fourcault, A., García-Jarana, B., Sánchez-Oneto, J., Marias, F., Portela, J.R., 2009. Supercritical water oxidation of phenol with air: experimental results and modelling. Chem. Eng. J. 152, 227–233.

Foussard, J.N., Debellefontaine, H., Besombes-Vailhé, J., 1989. Efficient elimination of organic liquid wastes. Wet air oxidation. J. Environ. Eng. 115 (2), 367–385.

Fujii, T., Hayashi, R., Kawasaki, S., Suzuki, A., Oshima, Y., 2011. Water density effects on methanol oxidation in supercritical water at high pressure up to 100 MPa. J. Supercrit. Fluids 58 (1), 142–149.

García-Jarana, M.B., Sánchez-Oneto, J., Portela, J.R., Nebot, E., Martínez de la Ossa, E.J., 2010. Simulation of supercritical water oxidation with air at pilot plant scale. Int. J. Chem. React. Eng. 8, 58. Available from: https://doi.org/10.2202/1542-6580.2259.

Gassó, S., Baldasano, J.M., 1996. Tratamiento de aguas residuales industriales mediante procesos de oxidación avanzada. Residuos 29, 37–42.

Ge, H.G., Chen, K.X., Zhang, Z.J., Zheng, L., 2003. Supercritical water oxidation of p-aminophenol with H_2O_2. Guocheng Gongcheng Xuebao 3 (4), 381–384.

Gidner, A., Stenmark, L., 2000. Supercritical water oxidation of sewage sludge-state of art. Chematur Engineering AB, Box 430, 69127, Karlskoga, Sweden.

Gong, Y., Guo, Y., Wang, S., Song, W., 2016. Supercritical water oxidation of quinazoline: Effects of conversion parameters and reaction mechanism. Water Res. 100 (1), 116–125.

Goto, M., Nada, T., Ogata, A., Kodama, A., Hirose, T., 1998. Supercritical water oxidation for the destruction of municipal excess sludge and alcohol distillery wastewater of molasses. J. Supercrit. Fluids 13, 277–282.

Goto, M., Shiramizu, D., Kodama, A., Hirose, T., 1999. Kinetic analysis for ammonia decomposition in supercritical water oxidation of sewage sludge. Ind. Eng. Chem. Res. 38 (11), 4500–4503.

Gopalan, S., Savage, P.E., 1995. A reaction network model for phenol oxidation in supercritical water. AIChE J. 41 (8), 1864–1873.

Harris, M.T., Jolley, R.L., Oswald, G.E., Rose, J.C., 1983. Wet oxidation of phenol and naphthalene (as a surrogate PAH) in aqueous and sludge solution: application to coal-conversion waste-water and sludge treatment. Report ORNL/TM-8576, 1-46. Oak Ridge. National Laboratory.

Hayashi, R., Onishi, M., Sugiyama, M., Koda, S., Oshima, Y., 2007. Kinetic analysis on alcohol concentration and mixture effect in supercritical water oxidation of methanol and ethanol by elementary reaction model. J. Supercrit. Fluids 40, 74–83.

Helling, R.K., Strobel, M.K., Torres, R.J., 1981. Kinetics of wet oxidation of biological sludges from coal conversion wastewater treatment. ORNL/MIT-332.

Helling, R.K., Tester, J.W., 1988. Oxidation of simple compound and mixtures in supercritical water: carbon monoxide, ammonia and ethanol. Environ. Sci. Technol. 22, 1319–1324.

Henrikson, J.T., Savage, P.E., 2004. Potencial explanations for the inhibition and acceleration of phenol SCWO by water. Ind. Eng. Chem. Res. 43, 4841–4847.

Hong, G.T., Killilea, W.R., Bourhis, A.L., 1996. Method for treating halogenated hydrocarbons prior to hydrothermal treatment, U. S. Patent No. 5, 492, 634.

Imamura, S., 1982. Wet oxidation of acetic acid catalyzed by Co-Bi complex oxides. Ind. Eng. Chem. Res. 21, 570–575.

Imamura, S., 1988. Wet oxidation of organic compounds catalyzed by ruthenium supported on cerium (IV) oxides. Ind. Eng. Chem. Res. 27, 718–721.

Imamura, S., Dol, A., Ishida, S., 1985. Wet oxidation of ammonia catalyzed by cerium-based composite oxides. Ind. Eng. Chem. Prod. Res. Dev. 24 (1), 75–80.

Jaulin, L., Chornet, E., 1987. High shear jet-mixers as two phase reactors: An application to the oxidation of phenol in aqueous media. J. Can. Chem. Eng. 65 (2), 64–70.

Jian, C., Wang, Y., 2004. Kinetic modeling for the oxidation of methanol in supercritical water. Sichuan Daxue Xuebao, Gongcheng Kexueban 36 (5), 35–39.

Jiménez-Espadafor, F., Portela, J.R., Vadillo, V., Sánchez-Oneto, J., Becerra, J.A., Torres, M., et al., 2011. Supercritical water oxidation of oily wastes at pilot plant: simulation for energy recovery. Ind. Eng. Chem. Res. 50, 775.

Joglekar, H., Samant, S.D., Joshi, J.B., 1991. Kinetics of wet air oxidation of phenol and substituted phenols. Water Res. 25 (2), 135–145.

Ju, M.-T., Feng, C.-W., 2000. Kinetics of phenol oxidation in supercritical water. Shuichuli Jishu 26 (2), 105–109.

Kim, Y.-L., Kim, J.-D., Lim, J.S., Lee, Y.-W., Yi, S.-C., 2002. Reaction pathway and kinetics for uncatalyzed partial oxidation of p-xylene in sub- and supercritical water. Ind. Eng. Chem. Res. 41 (23), 5576–5583.

Kim, Y.-L., Kim, J.-D., Lim, J.S., Lee, Y.-W., Yi, S.-C., 2003a. Effects of reaction conditions on selectivity of terephthalic acid in uncatalyzed partial oxidation of p-xylene under subcritical and supercritical water. Hwahak Konghak 41 (1), 26–32.

Kim, B.-J., Won, Y.-S., Lee, J.-H., 2003b. Kinetics of ethylene glycol oxidation in supercritical water. Kongop Hwahak. 14 (2), 182–188.

Kim, K., Son, S.H., Kim, K., Kim, K., Kim, Y.C., 2010. Environmental effects of supercritical water oxidation (SCWO) process for treating transformer oil contaminated with polychlorinated biphenyls (PCBs). Chem. Eng. J. 165, 170.

Koo, M., Lee, W.K., Lee, C.H., 1997. New reactor system for supercritical water oxidation and its application on phenol destruction. Chem. Eng. Sci. 52 (7), 1201–1213.

Kolaczkowski, S.T., Beltrán, F.J., McLurgh, D.B., 1997. Wet air oxidation of phenol: Factors that may influence global kinetics. Trans IChemE 75 (B), 257–265.

Krajnc, M., Levec, J., 1996. On the kinetics of phenol oxidation in supercritical wáter. AIChE J. 42 (7), 1977–1984.

Kritzer, P, Dinjus, E., 2001. An assessment of supercritical water oxidation (SCWO). Existing problems, possible solutions and new reactor concepts. Chem. Eng. J. 83, 207–214.

Kruse, A., Ederer, H., Mas, C., Schmieder, H., 2000. Kinetic studies of methanol oxidation in supercritical water and carbon dioxide. Supercrit. Fluids NATO Sci. Ser Ser E Appl. Sci. 366, 439–450.

Lachance, R., Paschkewitz, J., DiNaro, J., Tester, J.W., 1999. Thiodiglycol hydrolysis and oxidation in sub- and supercritical water. J. Supercrit. Fluids 16 (2), 133–147.

Lee, G.-H., Nunoura, T., Matsumura, Y., Yamamoto, K.J., 2002. Global kinetics of 2-chlorophenol disappearance with NaOH in supercritical water. Chem. Eng. Jpn. 35 (12), 1252–1256.

Lee, H.C., In, J.H., Hwang, K.Y., Lee, C.H., 2004. Decomposition of ehylenediaminetetraacetic acid by supercritical water oxidation. Ind. Eng. Chem. Res. 43, 3223–3227.

Lei, L., Hu, X., Chen, G., Porter, J.F., Lock Yue, P., 2000. Wet air oxidation of desizing wastewater from the textile industry. Ind. Eng. Chem. Res. 39, 2896–2901.

Levec, J., Smith, J.M., 1976. Oxidation of acetic acid solutions in a trickle-bed reactor. AIChE J. 22 (1), 159–168.

Leybros, A., Roubad, A., Guichardon, P., Boutin, O., 2010. Ion exchange resins destruction in a stirred supercritical water oxidation reactor. J. Supercrit. Fluids 51, 369.

Li, L., Chen, P., Gloyna, E.F., 1991. Generalized kinetic model for wet oxidation of organic compounds. AIChE J. 37 (11), 1687–1697.

Li, L., Chen, P., Gloyna, E.F., 1993. Kinetic model for wet oxidation of organic compounds in subcritical and supercritical water. Supercrit. Fluid Eng. Sci. 24, 305–313.

Li, R., Thornton, T.D., Savage, P.E., 1992. Kinetics of CO_2 formation of phenols in supercritical water. Environ. Sci. Technol. 26 (12), 2388–2395.

Limousin, G., Joussot-Dubien, C., Papet, S., Garrabos, Y., Sarrade, S., 1999. Hydrothermal oxidation of organic compounds oxidation kinetics of acetic acid and dodecane. Recents Progres en Genie des Procédés 13 (71), 165–172.

Liu, N., Cui, H.Y., Yao, D., 2009. Decomposition and oxidation of sodium 3,5,6-trichloropyridin-2-ol in sub- and supercritical water. Process Saf. Environ. Prot. 87, 387–394.

Marias, F., Vielcazals, S., Cezac, P., Mercadier, J., Cansell, F., 2007. Theoretical study of the expansion of supercritical water in a capillary device at the output of a hydrothermal oxidation process. J. Supercrit. Fluids 40, 208–217.

Marias, F., Mancini, F., Cansell, F., Mercadier, J., 2008. Energy recovery in supercritical water oxidation process. Environ. Eng. Sci. 25 (1), 123–130.

Marrone, P.A., Hodes, M., Smith, K.A., Tester, J.W., 2004. Salt precipitation and scale control in supercritical water oxidation-Part B: commercial/full-scale applications. J. Supercrit. Fluid. 29, 289–312.

Marrone, P.A., Hong, G.T., 2009. Corrosion control methods in supercritical water oxidation and gasification processes. J. Supercrit. Fluid. 51 (2), 83–103.

Martín, M.I., 1998. Oxidación húmeda y en agua supercrítica de disoluciones acuosas de compuestos modelo y de aguas residuales industriales. Tesis Doctoral. Universidad Complutense de Madrid.

Martino, C.J., Savage, P.E., 1997. Supercritical water oxidation kinetics, products, and pathways for CH_3- and CHO- substituted phenols. Ind. Eng. Chem. Res. 36, 1391–1400.

Marulanda, V., Bolaños, G., 2010. Supercritical water oxidation of a heavily PCB-contaminated mineral transformer oil: Laboratory-scale data and economic assessment. J. Supercrit. Fluids 54 (2), 258–265.

Mateos, D., Portela, J.R., Mercadier, J., Marias, F., Marraud, C., Cansell, F., 2005. New approach for kinetic parameters determination for hydrothermal oxidation reaction. J. Supercrit. Fluids 34, 63–70.

McFarland, R.D., McGuinness, T.G., Moore, S.W, 1993. Preliminary design of a hydrothermal processing unit for hanford tank waste simulant-300 GPD Pilot Unit, Los Alamos National Laboratory Internal Report LA-UR-93-1846.

Merchant, K.P., 1992. Studies in Heterogeneous Reactions, Ph. D. Thesis. University of Bombay, Mumbai, India.

Mishra, V.S., Mahajani, V.V., Joshi, J.B., 1995. Wet air oxidation. Ind. Eng. Chem. Res. 34, 2–48.

O'Brien, C.P., Thies, M.C., Bruce, D.A., 2005. Supercritical water oxidation of the PCB congener 2-chlorobiphenyl in methanol solutions: a kinetic analysis. Environ. Sci. Technol. 39 (17), 6839–6844.

O'Regan, J., Preston, S., Dunne, A., 2008. Supercritical water oxidation of sewage sludge- An update. 13th European Biosolids & Organic Resources Conference and Work shop.

Oshima, Y., 2001. Decomposition of hardly decomposable substances by supercritical water oxidation process using solid catalyst and its reaction engineering analysis. Chorinkai Saishin Gijutsu 5, 44−48.

Oshima, Y., Hori, K., Toda, M., Chommanad, T., Koda, S., 1998. Phenol oxidation kinetics in supercritical water. J. Supercrit. Fluids 13, 241−246.

Otal, E., Mantzavinos, D., Delgado, M.V., Hellenbrand, R., Lebrato, J., Metcalfe, I.S., et al., 1997. Integrated wet air oxidation and biological treatment of polyethylene glycol-containing wastewaters. J. Chem. Technol. Biotechnol. 70, 147−156.

Phenix, B., 1997. Hydrothermal Oxidation of Simple Organic Compounds, Ph. D. thesis. Massachusetts Institute of Technology, Boston, USA.

Ploeger, J.M., Madlinger, A.C., Tester, J.W., 2006. Revised global kinetic measurements of ammonia oxidation in supercritical water. Ind. Eng. Chem. Res. 45 (20), 6842−6845.

Portela, J.R., López, J., Nebot, E., Martínez de la Ossa, E.J., 1997. Kinetics of wet air oxidation of phenol. Chem. Eng. J. 67, 115−121.

Portela, J.R., Nebot, E., Martínez de la Ossa, E.J., 2001a. Kinetic comparison between subcritical and supercritical water oxidation of phenol. Chem. Eng. J. 81 (1-3), 287−299.

Portela, J.R., Nebot, E., Martínez de la Ossa, E.J., 2001b. Generalized kinetic models for supercritical water oxidation of cutting oil wastes. J. Supercrit. Fluids 21, 135−145.

Portela Miguélez, J.R., Sánchez-Oneto, J., Nebot Sanz, E., Martínez de la Ossa, E.J., 2005. System and method for the hydrothermal oxidation of water-insoluble organic residues. Patent No. US20090266772 (EP1834928B1).

Pruden, B.B., Le, H., 1976. Wet air oxidation of soluble components in waste water Can. J. Chem. Eng. 54 (4), 319−325.

Qian, L., Wang, S., Xu, D., Guo, Y., Tang, X., Wang, L., 2015. Treatment of sewage sludge in supercritical water and evaluation of the combined process of supercritical water gasification and oxidation. Biores. Technol. 176, 218−224.

Queiroz, J.P.S., Bermejo, M.D., Cocero, M.J., 2013. Kinetic model for isopropanol oxidation in supercritical water in hydrothermal flame regime and analysis. J. Supercrit. Fluids 76, 41−47.

Queiroz, J.P.S., Bermejo, M.D., Mato, F., Cocero, M.J., 2015. Supercritical water oxidation with hydrothermal flame as internal heat source: Efficient and clean energy production from waste. J. Supercrit. Fluids 96, 103−113.

Rice, S.F., Croiset, E., 2001. Oxidation of simple alcohols in supercritical water III. Formation of intermediates from ethanol. Ind. Eng. Chem. Res. 40 (1), 86−93.

Rivas, F.J., Gimeno, O., Portela, J.R., Martínez de la Ossa, E.J., Beltrán, F.J., 2001. Supercritical water oxidation of olive mill wastewater. Ind. Eng. Chem. Res. 40, 3670−3674.

Rofer, C.K., Streit, G.E., 1988. Kinetics and mechanisms of methane oxidation in supercritical water. Los Álamos National Laboratory Report, LA 11439-MS DOE/HWP-64.

Rofer, C.K., Streit, G.E., 1989. Oxidation of hydrocarbons and oxygenates in supercritical water. Los Álamos National Laboratory Report, LA 11700-MS DOE/HWP-90.

Rossignol, A., 2001. Utilisation des fluides supercritiques: Dépêche de l' Ambassade de France au Japon. Service pour la Science et la Technologie. Ambassade de France au Japon, Tokyo, 23 p.

Sánchez-Oneto, J., Portela, J.R., Nebot, E., Martínez-de-la-Ossa, E.J., 2004. Wet air oxidation of long-chain carboxylic acids. Chem. Eng. J. 100 (1-3), 43−50.

Sánchez-Oneto, J., Portela, J.R., Nebot, E., Martínez de la Ossa, E.J., 2006. Kinetics and mechanism of wet air oxidation of butyric acid. Ind. Eng. Chem. Res. 45 (12), 4117−4122.

Sánchez-Oneto, J., Portela, J.R., Nebot, E., Martínez de la Ossa, E.J., 2007. Hydrothermal oxidation: application to the treatment of different cutting fluid wastes. J. Hazard. Mater. 144, 639−644.

Savage, P.E., Yu, J., Stylski, N., Brock, E., 1998. Kinetics and mechanism of methane oxidation in supercritical water. J. Supercrit. Fluids 12, 141−153.

Savage, P.E. Yu, J., Zhang, X., 1999. Catalytic oxidation in supercritical water. Proceedings of the 6th meeting on supercritical fluids, chemistry and materials. April, 1999. Nottingham, U.K., 421.

Schanzenbächer, J., Taylor, J.D., Tester, J.W., 2002. Ethanol oxidation and hydrolysis rates in supercritical water. J. Supercrit. Fluids 22 (2), 139−147.

Segond, N., Matsumura, Y., Yamamoto, K., 2002. Determination of ammonia oxidation rate in sub- and supercritical water. Ind. Eng. Chem. Res. 41 (24), 6020–6027.

Shanableh, A., 2005. Generalized first-order kinetic model for biosolids decomposition and oxidation during hydrothermal treatment. Environ. Sci. Technol. 39, 355–362.

Shaw, R.W., Brill, T.B., Clifford, A.A., Echert, C.A., Franck, E.U., 1991. Supercritical water. A medium for chemistry. Chem. Eng. News. 23, 26–39.

Shende, R.V., Levec, J., 1999. Wet oxidation kinetics of refractory low molecular mass carboxylic acids. Ind. Eng. Chem. Res. 38, 3830–3837.

Shende, R.V., Levec, J., 2000. Subcritical aqueous-phase oxidation kinetics of acrylic, maleic, fumaric, and muconic acids. Ind. Eng. Chem. Res. 39, 40–47.

Shende, R.V., Mahajani, V.V., 1994. Kinetics of wet air oxidation of glyoxalic and oxalic acid. Ind. Eng. Chem. Res. 33, 3125–3130.

Shibaeva, L.V., Metelitsa, D.I., Denisov, E.T., 1969. Oxidation of phenol with molecular oxygen in aqueous solutions. Kinet. Catal. 10 (5), 832–836.

Shimoda, E., Fujii, T., Hayashi, R., Oshima, Y., 2016. Kinetic analysis of the mixture effect in supercritical water oxidation of ammonia/methanol. J. Supercrit. Fluids 116, 232–238.

Shin, Y.H., Shin, N.C., Veriansyah, B., Kim, J., Lee, Y.W., 2009. Supercritical water oxidation of wastewater from acrylonitrile manufacturing plant. J. Hazard. Mater. 163 (2-3), 1142.

Shishido, M., Okubo, K., Saisu, M., 2001. Effect of oxygen concentration on supercritical water oxidation of glucose. Kagaku Kogaku Ronbunshu 27 (6), 806–811.

Sullivan, P.A., 2003. Ph. D. Thesis. Cambridge, MA, USA: Massachusett Institute of Technology.

Sullivan, P.A., Tester, J.W., 2004. Methylphosphonic acid oxidation kinetics in supercritical water. AIChE J. 50 (3), 673–683.

Svanström, M., Modell, M., Tester, J., 2004a. Direct energy recovery from primary and secondary sludges by supercritical water oxidation. Water Sci. Technol. 49 (10), 201–208.

Svanström, M., Fröling, M., Modell, M., Peters, W.A., Tester, J., 2004b. J. Environmental assessment of supercritical water oxidation of sewage sludge. Resour. Conserv. Recy. 41, 321–338.

Svishchev, I.M., Plugatyr, A., 2006. Supercritical water oxidation of o-dichlorobenzene: degradation studies and simulation insights. J. Supercrit. Fluids. 37, 94–101.

Tateishi, M., Tsuchiyama, Y., Yamauchi, Y., Fukuzumi, T., Hatano, T., 2000. PCB decomposition process, U. S. Patent n 6. 162. 958 2000.

Tester, J.W., Cline, J.A., 1999. Hydrolysis and oxidation in subcritical and supercritical water: connecting process engineering science to molecular interactions. Corrosion 51, 1088–1100.

Tester, J.W, Holgate, H.R, Armellini, F.J, Webley, P.A, Killilea, W.R, Hong, G.T, et al., 1993. Emerging technologies in hazardous waste management III. Acs. Symp. Ser. 518, 35–75.

Thornton, T.D., LaDue, D.E., Savage, P.E., 1991. Phenol oxidation in supercritical water: formation of dibenzofuran, dibenzo-p-dioxin, and related compounds. Environ. Sci. Technol. 25, 1507–1510.

Thornton, T.D., Savage, P.E., 1990. Phenol oxidation in supercritical water. J. Supercrit. Fluids 3, 240–248.

Thornton, T.D., Savage, P.E., 1992a. Kinetics of phenol oxidation in supercritical water. AIChE J. 32 (3), 321–327.

Thornton, T.D., Savage, P.E., 1992b. Phenol oxidation pathways in supercritical water. Ind. Eng. Chem. Res. 31 (11), 2451–2456.

Vadillo, V., García-Jarana, M.B., Sánchez-Oneto, J., Portela Miguélez, J.R., Martínez de la Ossa, E.J., 2011. Supercritical water oxidation of flammable industrial wastewaters: Economic perspectives of an industrial plant. J. Chem. Technol. Biotechnol. 86, 1049.

Vadillo, V., García-Jarana, M.B., Sánchez-Oneto, J., Portela, J.R., Martínez de la Ossa, E.J., 2012. New feed system for water-insoluble organic and/or highly concentrated wastewater in the supercritical water oxidation process. J. Supercit. Fluid 72, 263–269.

Vadillo, V., Sánchez-Oneto, J., Portela, J.R., Martínez de la Ossa, E.J., 2013. Problems in supercritical water oxidation process and proposed solutions. Ind. Eng. Chem. Res. 52, 7617–7629.

Vadillo, V., Sánchez-Oneto, J., Portela, J.R., Martínez de la Ossa, E.J., 2014. Supercritical water oxidation for wastewater destruction with energy recovery. In: Anikeev, V, Fan, M (Eds.), Supercritical Fluid Technology for Energy and Environmental Applications. Elsevier, pp. 181–190In: Anikeev, V, Fan, M (Eds.), Supercritical Fluid Technology for Energy and Environmental Applications. Elsevier, pp. 181–190.

Veriansyah, B., Park, T.-J., Lim, J.-S., Lee, Y.-W., 2005. Supercritical water oxidation of wastewater from LCD manufacturing process: kinetic and formation of chromium oxide nanoparticles. J. Supercrit. Fluid. 34, 51–61.

Veriansyah, B., Kim, J.D., Lee, Y.W., 2006. Decomposition kinetics of dimethyl methylphospate (chemical agent simulant) by supercritical water oxidation. J. Environ. Sci. 18 (1), 13–16.

Vielcazals, S., Mercadier, J., Marias, F., Mateos, D., Bottreau, M., Cansell, F., et al., 2006. Modeling and simulation of hydrothermal oxidation of organic compounds. AIChE J. 52 (2), 818–825.

Vogel, F., Blanchard, J.L.D., Marrone, P.A., Rice, S.F., Webley, P.A., Peters, W.A., et al., 2005. Critical review of kinetic data for the oxidation of methanol in supercritical water. J. Supercrit. Fluids 34 (3), 249–286.

Wang, T., Xiang, B., Liu, J., Shen, Z., 2003. Supercritical water oxidation of sulfide. Environ. Sci. Technol. 37 (9), 1955–1961.

Wang, S., Guo, Y., Wang, L., Wang, Y., Xu, D., Ma, H., 2011. Supercritical water oxidation of coal: Investigation of operating parameters' effects, reaction kinetics and mechanism. Fuel Process. Technol. 92 (3), 291.

Webley, P.A., Tester, J.W., 1991. Fundamental kinetics of methane oxidation in supercritical wáter. Energy Fuels 5, 411–419.

Webley, P.A., Tester, J.W., Holgate, R.H., 1991. Oxidation kinetics in ammonia and ammonia-methanol mixtures in supercritical water in the temperature range 530-700$_\mathrm{o}$C at 246 bar. Ind. Eng. Chem. Res. 30, 1745–1754.

Weijin, G., Xuejun, D., 2010. Degradation of landfill leachate using transpiring-wall supercritical water oxidation (SCWO) reactor. Waste Manage. 30 (11), 2103–2107.

Weingartner, H., Franck, E.U., 2005. Supercritical water as a solvent. Angew. Chem. Int. Ed. 44 (18), 2672–2692.

Wellig, B., Lieball, K., Rudolf von Rohr, Ph., 2005. Operating characteristics of a transpiring-wall SCWO reactor with a hydrothermal flame as internal heat source. J. Supercrit. Fluid. 34 (1), 35–50.

Wellig, B., Weber, M., Lieball, K., Prikopsky, K., Rudolf von Rohr, Ph., 2009. Hydrothermal methanol diffusion flame as internal heat source in a SCWO reactor. J. Supercrit. Fluid. 49 (1), 59–70.

Williams, P.E.L., Day, D.C., Hudgins, R.R., Silveston, P.L., 1973. Wet air oxidation of low molecular weight organic acids. Water Pollut. Res. Canada 8, 224–237.

Willms, R.S., Reible, D.D., Wetzel, D.M., Harrison, D.P., 1987. Aqueous-phase oxidation: rate enhancement studies. Ind. Eng. Chem. Res. 26, 606–612.

Xiang, B., Wang, T., Shen, Z., 2002. Study on the reaction pathway and kinetics of ethanol wastewater disposal by supercritical water oxidation. Huanjing Kexue Xuebao 22 (1), 21–23.

Xu, D., Wang, S., Zhang, J., Tang, X., Guo, Y., Huang, C., 2015. Supercritical water oxidation of a pesticide wastewater. Chem. Eng. Res. Des. 94, 396–406.

Xu, D., Wang, S., Tang, X., Gong, Y., Guo, Y., Wang, Y., et al., 2012. Design of the first pilot scale plant of China for supercritical water oxidation of sewage sludge. Chem. Eng. Res. Des. 90 (2), 288–297.

Yang, H.H., Eckert, C.A., 1988. Homogeneus catalysis in the oxidation of p-clorophenol in supercritical water. Ind. Eng. Chem. Res. 27, 2009–2014.

Yao, H., Wu, S.-F., Chen, F.-Q., Ding, J.-W., 2000. Study on supercritical water oxidation in aromatic organic waste water treatment. Chem. React. Eng. Technol. 16 (3), 301–304.

Zhang, G, Hua, I., 2003. Supercritical water oxidation of nitrobenzene. Ind. Eng. Chem. Res. 42 (2), 285–289.

Zu, X., Wang, Q., Jiang, C., 2004. Toluene oxidation to benzaldehyde in supercritical (subcritical) fluids. Huagong Xuebao 55 (12), 2001–2007.

Further Reading

Li, L, Portela, JR, Vallejo, D, Gloyna, E.F, 1999. Oxidation and hydrolisis of lactic acid in near-critical water. Ind. Eng. Chem. Res. 38 (7), 2599–2606.

CHAPTER

11

Electrochemical Oxidation Processes

Kuravappullam V. Radha and Karunamoorthy Sirisha

Anna University, Chennai, India

11.1 INTRODUCTION

Conventional methods for water treatment are inefficient to remove toxic and bio-recalcitrant organic micropollutants, when compared to electrochemical processes, which implement clean and effective techniques for the direct or indirect generation of hydroxyl radicals.

Organic micropollutants like persistent organic pollutants (POPs), herbicides, and pesticides are compounds, which have been found at relatively high contents in the aquatic environment, and have increased the toxicity levels in water where the conventional treatments have not found a solution for this problem successfully. The electrochemical methods have been a key to resolve such major drawbacks.

These procedures called advanced oxidation processes (AOPs) use a very strong oxidizing agent such as hydroxyl radical ($^{\bullet}OH$) with E ($^{\bullet}OH/H_2O$) = 2.8 V/NHE, generated in situ in the reaction medium. These radicals react rapidly with organics mainly either by dehydrogenation or hydroxylation.

Electrochemical advanced oxidation process was described earlier, which focused on electrochemistry in general and environmental electrochemistry in particular. Electro-Fenton was the first technology to be considered as an EAOP because of the production and the active role of hydroxyl radical on the oxidation of organics.

This technology is based on the processes:

- the electrochemical regeneration of iron(II) from iron(III) species on the cathodic surface, and
- the cathodic formation of hydrogen peroxide from the reduction of oxygen.

These processes lead to the catalytic decomposition of hydrogen peroxide into hydroxyl radicals.

A brief difference between AOPs and EAPs are given in Table 11.1.

Advanced Oxidation Processes for Wastewater Treatment
DOI: https://doi.org/10.1016/B978-0-12-810499-6.00011-5

TABLE 11.1   Different Types of AOPs and EAPs

Advanced Oxidation Processes	Electrochemical Advanced Processes
<ul><li>Fenton, Photo-Fenton, Solar Photo-Fenton</li><li>O_3 based processes</li><li>Heterogenous Photocatalysis: Nanomaterials, Nanofibers, UVA-LED</li></ul>	<ul><li>Electrocoagulation (Electroflotation)</li><li>Hydrogen recovery</li><li>New cell/Reactors design</li><li>Electro-Fenton, Solar photoelectro-Fenton<ul><li>Electrooxidation</li></ul></li></ul>

11.2 ELECTROCHEMICAL OXIDATION PROCESSES

An electron donor (reductant) transfers one or more electrons to an electron acceptor (oxidant), which has a higher affinity. A chemical transformation takes place during the electron transfer in both the oxidant and the reductant, giving rise to chemical species with an odd number of valence electrons. These species, known as radicals, are highly unstable and reactive due to the presence of this unpaired electron.

One of the key focuses is to clarify the high efficiencies by EAOPs in the evacuation of natural toxins by intervened oxidation forms in the general oxidation completed amid the treatment. Intervened oxidation in EAOPs can be caught on as the oxidation of toxins contained in wastewater by the substance response between these mixes and oxidants created already on the anode surfaces. In this way, AO does not just lead to the immediate oxidation of natural contaminants on the anode surface; however, it additionally advances the development of tremendous measures of oxidants, which can act on the surface of the cathodes as well as amplify the oxidation procedure to the arrangement majority of the treated waste (Panizza and Cerisola, 2009). The sort and augmentation of the generation of oxidants rely upon many information sources, the anode material being the most important and the event of reasonable crude matter for the creation of oxidants in the wastewater. Their impact on the proficiency of EAOPs is imperative in light of the fact that the oxidation of poisons is stretched out from the region of the anode surface to the main part of the electrolyte. Notwithstanding, it ought to be taken into account that these oxidants generally influence the instruments of the oxidation of contaminants and, periodically, they can prompt the arrangement of undesirable intermediates or last stable items. At times, the species that advance the development of oxidants are not contained in the wastewater, be that as it may included as reagents, which brings about understanding of exceptionally successful procedures. A standout amongst the most fascinating cases is the treatment of squanders with Ag (II), whose arrangement was shown to be extremely viable with conductive cathodes (Panizza et al., 2000) (Fig. 11.1).

However, the most referenced example of mediated electrochemical oxidation arises from the effect of chlorides on the oxidation of organics. Chlorides are commonly contained in most of the wastewater flow streams and they are known to be easily oxidized to chlorine by many types of anode materials. This gaseous oxidant diffuses into the wastewater and forms hypochlorite and chloride in the reaction medium by disproportionation. Deprotonation of hypochlorous acid produces hypochlorite. Since hypochlorite is the primary final product, in literature, it is common to find the direct transformation of chloride

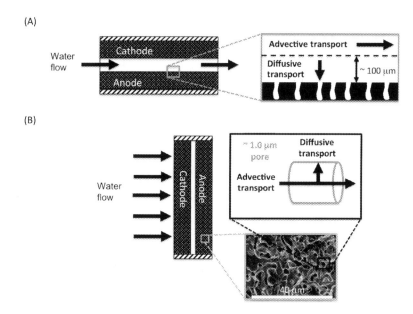

FIGURE 11.1 Development of stable electrode materials that efficiently generate high yields of hydroxyl radicals ($^{\bullet}$OH) (e.g., boron-doped diamond (BDD), doped-SnO$_2$, PbO$_2$, sub-stoichiometric- and doped-TiO$_2$). (A) Flow-by Electrode Operation. (B) Flow-through Electrode Operation. *Adapted from Chaplin, B.P. 2014. Critical review of electrochemical advanced oxidation processes for water treatment applications. Environ. Sci.: Proc. Impacts, 16, 1182–1203 Chaplin (2014). With permission.*

into hypochlorite instead of the complete set of reactions. However, the oxidation in that media is carried out by a mixture of reagents and the particular concentration of each species depends on the concentration and pH.

$$2Cl^- \rightarrow Cl_2 + 2e^- \tag{11.1}$$

$$Cl_2 + H_2O \rightarrow HClO + H^+ + Cl^- \quad \text{Acidic} \tag{11.2}$$

$$Cl_2 + 2OH^- \leftrightarrow ClO^- + Cl^- + H_2O \quad \text{Alkaline} \tag{11.3}$$

$$HClO + OH^- \leftrightarrow ClO^- + H_2O \tag{11.4}$$

$$Cl^- + H_2O \rightarrow ClO^- + 2H^+ + 2e^- \tag{11.5}$$

The subsequent blend (chlorine, hypochlorite, and hypochloric corrosive) is very responsive with numerous organics, being effective for their mineralization (Comninellis and Nerini, 1995; Panizza and Cerisola, 2003). Notwithstanding, it is likewise known to frame numerous organo-chlorinated species as intermediates and lasts that can be considerably more unsafe than the crude poison. These species are regularly very difficult and even the arrangement of low sub-atomic weight items, for example, (a chemical that causes unconsciousness, if you breathe its vapors) turns into a very huge issue. This will be expanding (almost completely/basically) and the costs of the remediation are huge (Cañizares et al., 2003).

Regardless of the way that chlorine intervened oxidation is extremely outstanding, it is by all account not the only instance of interceded oxidation forms and, obviously, it is not the most critical one. Subsequently, when the goal is centered on the advancement of interceded oxidation, two vital angles ought to be considered:

- direct electrochemical generation of oxidants on the anode surface from non-oxidant species contained in the waste, and
- transport of these species towards the mass (wastewater). The crude matter for the generation of oxidants ought to be contained in the wastewater or dosed and, normally, it can be a particle (i.e., chloride, sulfate, and so forth), a natural poison (acidic corrosive), broke down gases (oxygen) or even water.

11.2.1 Photoelectrochemical Processes

There is an expanding enthusiasm for the utilization of photoelectrochemical procedures for water and wastewater remediation. These photo-assisted medications depend on the illumination of a debased arrangement or a photoactive anode with UV or a sun oriented light source; UV-A (= 315–400 nm), UV-B (= 285–315 nm), and UV-C (< 285 nm) lights provided by UV lights as vitality sources are usually utilized. The force and wavelength of such radiations have noteworthy impact on the pulverization rate of natural contaminants. Even when the cost is above normal, the use of UV-assisted procedures is very useful. This is settled in sun power-assisted forms, in which daylight (> 300 nm) is utilized as a free, economical and sustainable power source, although every now and again, the impact of this radiation has been evaluated utilizing a sun powered test system gadget. In the future, the qualities and fundamental uses of the most intriguing photoelectrochemical forms including photoelectro-Fenton (PEF), sun powered photoelectro-Fenton (SPEF), photoelectrocatalysis (PEC), and cross breed frameworks, are depicted.

11.2.2 Photoelectro-Fenton (PEF) and Solar Photoelectro-Fenton (SPEF) Processes

PEF with simulated UV-A light and SPEF(its determined daylight assisted technique), have been conceived and broadly created by the Brillas' gathering (Brillas et al., 2009; Sirés and Brillas, 2012). These procedures include the treatment of the sullied arrangement under EF conditions alongside the concurrent illumination with UV-A or sun based light to quicken the mineralization rate of organics. Oxidizing hydroxyl radicals are delivered from Fenton's response, while the undesired amassing of unmanageable Fe(III) particles that decelerate the treatment are maintained at a strategic distance from the reductive photolysis of [Fe (OH)]$^{2+}$. The transcendent Fe(III) species are in arrangement at pH 2.8–3.5, as indicated by the photograph (Fig. 11.2), in the manner of recovering Fe^{2+} (i.e., the impetus in Fenton's response) and delivering more radicals (Brillas et al., 2009; Sirés and Brillas, 2012) (Fig. 11.3).

The radiation can also promote the photolysis of some oxidation intermediates or their Fe(III) complexes that allows the regeneration of Fe^{2+} as similar to photodecarboxylation reactions.

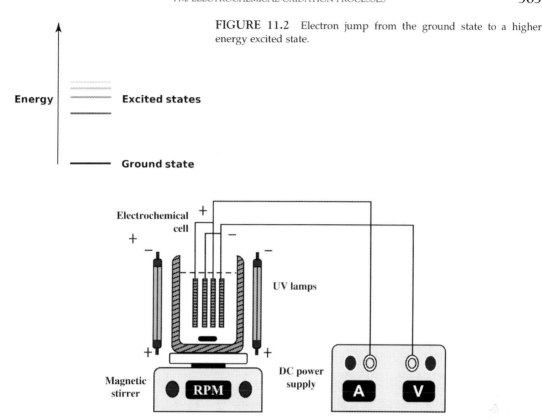

FIGURE 11.2 Electron jump from the ground state to a higher energy excited state.

FIGURE 11.3 Schematic representation of photoelectro-Fenton process.

The investigation of the SPEF procedure has been stretched out to a 10 L pre-pilot plant (Isarain-Chavez et al., 2011). This plant has an indistinguishable part with a reactor of 90.3 cm² cathode zone coupled to a 1.57 L solar compound parabolic collector (CPCs) as the photoreactor. The examination was centered on the advancement of the SPEF treatment of the paracetamol utilizing a Pt/adenine cell by reaction surface philosophy (Almeida et al., 2011). The ideal factors were observed to be 5 A, 0.4 mM Fe^{2+}, and pH 3.0 for 157 mg/L paracetamol with 0.05 M Na_2SO_4, yielding 75% TOC decrease 93 kWh (kg TOC)$^{-1}$ vitality utilization and 71% current productivity at 120 min.

High Performance Liquid Chromotagraphy (HPLC) investigation of electrolyzed arrangements permitted recognizing hydroquinone and different benzoquinones as sweet-smelling intermediates, which were expelled by $^\bullet$OH, though maleic, fumaric, succinic, lactic, oxalic, formic, and oxamic acids were distinguished as carboxylic acids. As of late, the SPEF treatment of 297 mg/L of the azo color sunset yellow FCF in 0.05 M Na_2SO_4 and 0.5 mM Fe^{2+} of pH 3.0 utilizing a similar framework at 7.0. It was exhibited that aggregate decolorization was practical at 120 min. About 94% mineralization with 197 kWh (kg TOC)$^{-1}$ vitality utilization was accomplished in 150 min.

The SPEF corruption of 100 mg/L TOC of arrangements with the beta-blockers atenolol, metoprolol tartrate, and propranolol hydrochloride in 0.10 M Na_2SO_4 with 0.5 mM Fe^{2+} at pH 3.0 was tried utilizing single Pt/adenine and BDD/adenine cells and furthermore, their blend with a Pt/CF cell to upgrade Fe^{2+} recovery from Fe^{3+} decrease (Isarain-Chavez et al., 2011). Some processes demonstrate the outline of the consolidated BDD/Adenine-Pt/CF cell. For instance, some processes highlight for a 0.246 mM metoprolol tartrate arrangement, the prevalence of joined cells over single ones, BDD over Pt and SPEF over EF in regards to the TOC decrease. This can be clarified by the more noteworthy creation of •OH from Fenton's response in the consolidated cells, the higher oxidizing energy of BDD (•OH) and the photolytic activity of daylight in SPEF, effortlessly derived from the pseudo-arrangement of cells.

In any case, the Pt/ADE-Pt/CF cell gave the most minimal vitality utilization of 80 kWh (kg TOC)$^{-1}$ for 88%−93% mineralization, being the most practical framework for mechanical application (Fig. 11.4).

11.2.3 Photoelectrocatalysis (PEC)

The PEC strategy depends on the collaboration amongst electrochemistry and photocatalysis to give significantly more noteworthy productivity for wastewater remediation. Conventional photocatalysis has been broadly created utilizing the nanocrystalline anatase type of TiO_2 for light-initiated oxidation of natural toxins in waters (Brillas, et al., 2009; Daghrir et al., 2012; Georgieva et al., 2012; Martínez-Huitle and Ferro, 2006).

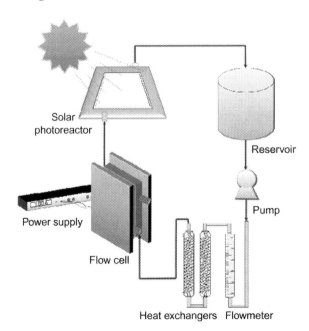

FIGURE 11.4 Solar photoelectro-Fenton system. *Adapted from Trellu, C., Ganzenko, O., Papirio, S., Oturan, M.A., 2016. Chem. Eng. J. 306, 588−596. With permission.*

This semiconductor material has extremely alluring properties for example, minimal effort, low poisonous quality, and a wide band crevice of 3.2 eV, which brings about a decent strength and anticipates photocorrosion. Exposure of anatase TiO_2 nanoparticles, either in colloidal suspension or, on the other hand, kept as a thin film on Ti, by UV photons of adequate vitality (<380 nm) promotes an electron from the valence band to the conduction band (e_{CB}^-) producing an emphatically charged opening or, a hole (h_{VB}^+) as follows:

$$TiO + hv \rightarrow e_{CB}^- + h_{VB}^+ \qquad (11.6)$$

Organics can then be specifically oxidized by the hole or by heterogeneous hydroxyl radical generated from the reaction between the photogenerated opportunity and adsorbed water.

$$h_{VB}^+ + H_2O \rightarrow {}^\bullet OH + H^+ \qquad (11.7)$$

In addition, other weaker receptive oxygen species (ROS), (like superoxide radical particles $O_2^{\bullet -}$, HO_2, and H_2O_2), and more ${}^\bullet OH$ can be created from the photoinduced electron as per the accompanying responses.

$$e_{CB}^- + O_2 \rightarrow O_2^{\bullet -} \qquad (11.8)$$

$$O_2^{\bullet -} + H^+ \rightarrow HO_2 \qquad (11.9)$$

$$2HO_2 \rightarrow H_2O_2 + O_2 \qquad (11.10)$$

$$H_2O_2 + O_2^{\bullet -} \rightarrow {}^\bullet OH + OH^- + O_2 \qquad (11.11)$$

The major drop in efficiency, results from the recombination of photoinduced electrons with either unreacted holes or adsorbed ${}^\bullet OH$.

$$e_{CB}^- + h_{VB}^+ \rightarrow TiO_2 + heat \qquad (11.12)$$

$$e_{CB}^- + {}^\bullet OH \rightarrow OH^- \qquad (11.13)$$

Then again, the PEC technique comprises the use of either a steady present or a consistent inclination anodic potential (E_{anod}) to a semiconductor-based thin film anode subjected to UV exposure for the ceaseless extraction of photoinduced electrons by an outside electrical circuit (Daghrir et al., 2012; Georgieva et al., 2012.).

The electrolytic cells utilized as a part of PEC are mixed tanks or stream reactors that allow the section of UV light specifically to the arrangement or through a quartz window to achieve the uncovered surface of the photoanode with the base loss of occurrence illumination (Fig. 11.5).

11.2.4 Hybrid Combinations of PEF and PEC

The decolorization of a few dyes has been upgraded by consolidating PEF and TiO_2 photocatalysis (TiO_2/UV) by Khataee et al. (2010, 2012), who improved the exploratory conditions utilizing reaction surface strategy. These creators used open, unified, and tube-shaped mixed tank reactors of 1 or 3 L limit outfitted with a Pt anode, a cathode made of

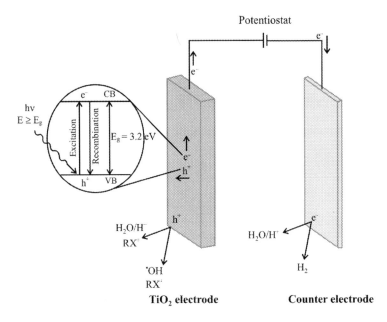

FIGURE 11.5   Enhancement of photoelectrocatalysis efficiency using nanostructured electrodes.

carbon nanotubes immobilized onto a graphite surface (CNT/graphite) with air supply; 6 W UV-C light presented in a quartz tube and TiO_2 nanoparticles immobilized on paper or glass plates covering the internal surface of the phone. The decolorization rate of azo color acid yellow 36 diminished in the following order:

$$PEF\text{-}TiO_2/UV > PEF > EF > TiO_2/UV > UV \text{ photolysis}$$

The ideal factors for the previous joined process were 25 mg/L of dye, 0.15 mM Fe^{3+}, 127 min of treatment, and 115.6 mA, yielding a most extreme shading expulsion of 83%. The operation costs identified with the utilization of the UV light were as high as 16.5 kWh/m^3, though the electrical vitality utilization was just 0.88 kWh/m^3. Then again, a few likewise endeavors have been made to improve the oxidation capacity of PEC by coupling its with EF, albeit more work is expected to affirm the utilization of such joined forms.

Xie and Li (2006) utilized quartz reactor for the removal of azo dye orange G for 5 h by various strategies. The cell was loaded with 30 mL of a 0.1 mM arrangement of the dye containing 0.01 M Na_2SO_4 at pH 6.2. 8 W, and UV-A light was utilized as a light source. Orange G was not specifically photolyzed under UV-A light, being marginally crushed (ca. 3%) by AO with a TiO_2/Pt cell at $E_{anod} = +0.71$ V/SCE and achieving 8% by TiO_2 photocatalysis; of course, if little amounts of oxidizing species ($^\bullet OH$ and additionally holes) are shaped at the TiO_2 surface. The PEC procedure utilizing the Pt/TiO_2 cell improved the dye removal to 25% because of the photogeneration of more oxidizing holes. The oxidation capacity of this procedure expanded up to yield half dye devastation, when the Pt cathode was supplanted by a reticulated vitreous carbon (RVC) anode at $E_{cat} = -0.54$ V/SCE, making the oxidation of dye conceivable with H_2O_2 created from O_2 diminishment. Strangely, orange G vanished totally, while applying PEF with a Fe/RVC cell utilizing ca. 17 mg/L Fe^{2+}, pH 3.0, and $E_{cat} = -0.71$ V/SCE. Moreover, the decolorization rate somewhat expanded in the event on coupling EF under similar conditions to PEC utilizing a TiO_2/RVC cell; thus the extra development of a lot of homogeneous $^{\bullet}OH$ radicals from Fenton's, and additionally the stress caused by Fenton response.

Similarly, PEC combined with EF gave the most noteworthy mineralization of 74% in 5 h (Peralta-Hernández et al., 2006). It portrayed a solid change of the decolorization proficiency and TOC evacuation by PEC coupled to EF in contrast with EF alone utilizing a concentric annular unified TiO_2/graphite material stream cell with a focal 75 mW/cm^2 UV-A light in bunch operation mode.

11.2.5 Sonoelectrochemical Processes

Ultrasound (US) can deliver very receptive radicals from water and also the pyrolysis of natural poisons contained in such lattices. Be that as it may, its oxidation capacity is fairly low and, subsequently, it is generally consolidated with other oxidants like H_2O_2, O_2, UV, and Fenton's reagent for water remediation (Garbellini et al., 2008; González-García et al., 2010; Oturan et al., 2008) (Fig. 11.6).

The investigation of the blend of US and terminal procedures, and the use of sonoelectrochemical innovation to the burning of organic compounds and flow dynamic are current active research fields (González-García et al., 2010). This segment depicts the qualities of sonoelectrolysis and sonoelectro-Fenton, which are the most critical sonoelectrochemical forms used to sterilize wastewater.

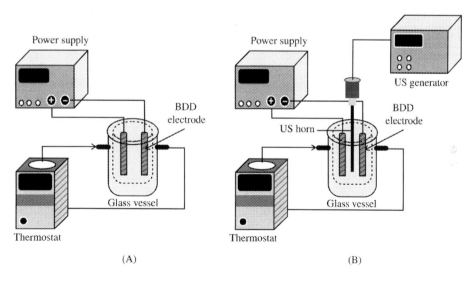

FIGURE 11.6   Schematic representation of (A) Electrochemical cell and (B) Sonoelectrochemical cell.

11.2.6 Sonoelectrolysis

The compound instrument is found at high recurrence and includes the homolytic fracture of H_2O, and broke up O_2 to yield distinctive ROS ($^\bullet OH$, HO_2, and $^\bullet O$). The physical system is alleged sonication and comprises in the creation of cavitation microbubbles, which develop and fall, starting extraordinary breaking powers with a great degree of high temperatures (up to 6000K) and weights (of the request of 104 kPa). Under these conditions, organics can be specifically pyrolyzed and the sonolysis of water can occur, where ($\bullet$) means the US, while created $\bullet OH$ in this way quickens the organics oxidation.

$$H_2O + \quad \rightarrow OH + H \tag{11.14}$$

Additionally, the solid cavitational crumple close to the anode surface improves the mass transport of the electroactive species and the cleaning of the anodes' surfaces by dissolving or setting the repressing layers (Garbellini et al., 2008; Garbellini, 2012). Every one of these marvels to a great extent enhances the decimation rate of organics, in the long run supporting the mineralization procedure in sonoelectrolysis. A few test setups have been imagined for sonoelectrochemical tests and these are (González-García et al., 2010):

- inundation of the electrochemical cell inside a US shower,
- coupling of the electrolytic cell with the ultrasonic tip through a glass divider or filled fluid chambers,
- utilization of a sonoelectrochemical cell, where the terminals and the ultrasound tip are straightforwardly dunked into the working arrangement, which turns out to be the most utilized course of action, and
- concurrent utilization of the ultrasound tip as ultrasound producer and as cathode.

By far, most of research in sonoelectrolysis has been completed at lab scale with exclusively planned frameworks in light of effective US horns dunked into customary glass electrochemical vessels. This strategy is exceptionally costly and has proved a few downsides identified with reproducibility, scale-up, and plan perspectives, which have backed off its further improvement.

11.2.7 Sonoelectro-Fenton (SEF)

SEF is a process applied under EF conditions to an acidic solution. This novel strategy has been utilized for the debasement of herbicides 2,4-D and 4,6-dinitro-o-cresol utilizing a unified Pt/CF tank reactor with a clay piezoelectric transducer put on its base to supply energies of 20, 60, and 80 W at a low recurrence of 28 kHz.

The predominance of SEF over EF and substance Fenton was shown for the treatment of the cationic dye red X-GRL (Li et al., 2010) and azure B (Martínez and Uribe, 2012). An air-soaked arrangement of 1 L of 37.5 mg/L of the previous dye with 0.05 M Na_2SO_4 and 5 mM Fe^{2+} of pH 3.0 in a unified tank reactor with a Ti/RuO$_2$ anode and an initiate carbon fiber cathode, each of 100 mm $\times$ 90 mm regions, was corrupted by EF at 8.9 mA/cm^2. Relative SEF trials were made utilizing a US arrangement of 20 kHz and energies of 80, 120, and 160 W with the horn plunged into the arrangement at 20 mm of its surface, put between the terminals. The dye was dependably totally decolorized in 180 min, in spite of the fact that shading evacuation was upgraded by SEF, where it expanded as US vitality rose. A comparable pattern was found for TOC rot, which ascended from 75% in EF to 83% in SEF at 160 W. For 0.5 mM sky blue B, it was found that the steady rate for its rot in SEF by applying $E_{cat} = -0.7$ V/SCE to a RVC cathode under US of 91 W and 24 kHz was 10-overlap that of direct sonolysis and two-fold increase than the one acquired by compound Fenton under noiseless conditions. As need be, COD was decreased 68% in 60 min by Fenton, while a much vast estimation of up to 85% was acquired by SEF.

11.3 ADVANTAGES

- prevention and remediation of pollution problems as electron is a clean reagent,
- high energy efficiency,
- amenability to automation,
- easy handling as the equipment is simple to handle, safe, and versatile, and
- applied to effluents with COD in the range from 0.1 g/L to 100 g/L.

11.4 DISADVANTAGES

- The cost is high, when related to electrical supply.
- There is low conductance of many wastewaters that require the addition of electrolytes.
- Another disadvantage is the decline of activity and shortening of the electrode lifetime by fouling due to the deposition of organic material on their surface.

11.5 APPLICATIONS

- Applications in wastewater treatment: Electrochemical processes have been a major solution for environmental remediation for the past few decades. The most significant one is by treating liquid coming from industries and electro-kinetic soil remediation processes (Barrera-Díaz et al., 2014).

  Electrolysis and electro-coagulation are the pretreatment processes, which are one of the most promising technologies in recent trends. Electro-coagulation process is capable of removing turbidity, decolorization of dyes, and breakup of wastes consisting of emulsions (Canizares et al., 2009).

  In the electrolysis process, coarse removal of pollution in the industrial waste cannot be implemented as the amount of energy required for the removal of pollutants on the concentration of pollution is large. Electrolysis treats effluents polluted with anthropogenic organic species, within a concentration range of 1000–20,000 mg chemical oxygen demand (COD) dm^{-3} for direct anodic oxidations.

  Electrolysis with conductive-diamond electrochemical oxidation exhibits properties, which are exceptionally good, when compared to other advanced oxidation processes like:
  - robustness to attain complete mineralization of almost any type of organic without producing refractory final products,

- efficiency, when it is operated under no diffusion control, current efficiencies are close to 100%, and
- integration capability, it is easily coupled with other treatment technologies and fed with green energy sources and photovoltaic panels.
- Applications on surface of electrodes: EAOPs are commonly divided into two groups, based on the mechanisms of the overall oxidation development: surface of the electrons/ bulk of the electrochemical cell (Barrera-Díaz et al., 2014).

  Surface controlled processes, also known as anodic oxidation processes, which occur near the surface of the electrode, has three categories that are distinguished based on the oxidation mechanisms:
  - direct electron-transfer processes,
  - hydroxyl radical processes, and
  - heterogeneous photocatalytic processes.
- Electrochemical oxidation process can be used for the treatment of landfill leachates and evaluation of toxicity in *allium cepa* (Klauck et al., 2013). In this study, the effectiveness of the advanced oxidation process of electrochemical oxidation in leachate treatment was evaluated, and toxicity assessment of the sample before and after treatment was also observed. During the treatment process, the removal of dye was appreciable whereas, the removal of chlorine odor was not up to the mark because chlorine ions present in abundance in the leachate can often act as oxidants in indirect electrochemical oxidation of organic matter. It was finally concluded that the treatment was effective in removing the physico-chemical parameters. However, the eco-toxicological test indicated that the effluent became as toxic as the initial sample.

  Further optimization of the parameters like pH, time, and current should be carried out, along with application of integrated treatment systems, ozone, electro- and photooxidation processes, characterization of the leachate after biological treatment, a test with other biomarkers, and characterization of organic compounds and by-products generation. All this optimization should be carried out.

11.6 CURRENT SCENARIO

Electrocatalytic activity and stability of electrode materials, optimized reactor geometry, and reactor hydrodynamics is the field, are under intense investigation due to their expanding field.

Pilot scale for EAOPs has reached an advanced stage, which has been used for commercialization for disinfectants and purification of wastewater pollutants with organic compounds.

One of the major successful large scale application of EAOPs is the automated disinfection of swimming pool water using boron doped diamond (BDD) anode products such as Oxineo® and Sysneo®. They have been developed for both private and public pools. Major advantages in implementing this method are:

- absence of chlorine smell,
- no accumulation of chemicals,

- no need of antialgae, and
- residual action to resist jagged disinfections.

Companies including CONDIAS and Advanced Diamond Technologies Inc. do research and development to supply equipments for AO with BDD anode. Major applications of these cells are water disinfection and industrial wastewater treatment.

EAOPs in EctoSys®, provide a reliable and sustainable disinfection of the ballast water in an economical and ecological way.

A BDD electro-oxidation pilot plant was installed in Marelo (Cantabria, Spain) for the treatment of landfill leachates using AOPs followed by EAOPs.

11.7 FUTURE PROSPECTS

- Points of view of improving interceded oxidation for future EAOP advancements appear to be positive. Interceded oxidation forms turn into the key indicate to achieve an improvement in the proficiency of EAOPs. It is not the generation of oxidants, but rather their initiation in the response media is all the more intriguing points of research.
- Expenses of innovation are high and changes in proficiency are not generally in the same class as to propose their coupling with EAOPs. This is an immediate outcome of the immense measures of scattered vitality with this innovation. Any curiosity in this theme needs to originate from a more proficient utilization of vitality with a specific end goal to advance the arrangement of numerous problem areas, where radical responses could be initiated.
- Interestingly, light illumination as of now is by all accounts an extremely encouraging option with great points of view to be utilized as a part of the not so distant future. The synergistic impact of the actuation of oxidants has been plainly showed and the vitality illuminated is much lower than that connected.
- The utilization of the eminent BDD anode essentially improved the oxidation power and mineralization proficiency due to the creation of supplementary hydroxyl radicals at the anode surface. Hence, this innovation turns out to be presently developed enough to pass to pilot-scale reactor plan and application to treatment of huge volumes of wastewaters. The initial step will be the origination and plan of a pilot-scale reactor. This origination can incorporate consolidated procedures to build the adequacy of the treatment.
- Group reactor and persistent (stream) reactor ought to be considered. The displaying of the procedure can be helpful to upgrade the working parameters and foresee the conduct of contaminations and can help for prudent and handy application to the genuine wastewater treatment.
- The coupling with an organic procedure as pre- or post-treatment unit is another promising way for a financially know-how treatment. Some current reviews have demonstrated the attainability of such a coupling.
- In truth, the EF procedure can change harmful as well as bio-refractory particles to biodegradable species amid a short treatment time. The entire mineralization of

arrangements in this manner acquired at that point can be accomplished by natural treatment.

- To create fiscal capacity, the utilization of a green and cheap vitality source in light of daylight driven electrical power frameworks, for example, an EF reactor specifically controlled by photovoltaic boards can likewise be considered.
- Consideration regarding propelled terminals and reactors, alongside process demonstrating can prompt an exhaustive extension of procedures into all zones of ecological conservation, including remediation of sullied waters, vaporous streams, and soils. There exists an idealistic foundation in light of the fact that a critical advance has been proven from the improvement of novel cathodes and layers, and the enhancement of the reactor setup. For instance, the advancement of more proficient numerous stage oxidation and three-dimensional terminal reactors.
- A basic examination of the prompt difficulties of the photoelectrochemical forms for the treatment of natural poisons in water needs to concentrate on the particularities of PEC and SPEF, given their prevalence.
- Expected advancements considering the uncovered outcomes on PEC proposals with the elective mix of SPEC with SPEF, a half and half framework that has not been tried yet, since it would be considerably more conservative than the PEC mixes.

References

Almeida, L.C., Garcia-Segura, S., Bocchi, N., Brillas, E., 2011. Solar photoelectro-Fenton degradation of paracetamol using a flow plant with a Pt/air-diffusion cell coupled with a compound parabolic collector: process optimization by response surface methodology. Appl. Catal. B-Environ. 103, 21−30.

Barrera-Díaz, C., Cañizares, P., Fernández, F.J., Natividad, R., Rodrigo, M.A., 2014. Electrochemical advanced oxidation processes: an overview of the current applications to actual industrial effluents. J. Mex. Chem. Soc. 58 (3), 256−275.

Brillas, E., Sirés, I., Oturan, M.A., 2009. Electro-Fenton process and related electrochemical technologies based on Fenton's reaction chemistry. Chem. Rev. 109, 6570−6631.

Cañizares, P., García-Gómez, J., Sáez, C., Rodrigo, M., 2003. Electrochemical oxidation of several chlorophenols on diamond electrodes Part I. Reaction mechanism. J. Appl. Electrochem. 33, 917−927.

Canizares, P., Martinez, F., Saez, C., Andres Rodrigo, M., 2009. The electrocoagulation, an alternative to the conventional coagulation process of wastewater. Afinidad 66 (539), 27−37.

Chaplin, B.P., 2014. Critical review of electrochemical advanced oxidation processes for water treatment applications. Environ. Sci. Proc. Impacts 16, 1182−1203.

Comninellis, C., Nerini, A., 1995. Anodic oxidation of phenol in the presence of NaCl for wastewater treatment. J. Appl. Electrochem. 25, 23−28.

Daghrir, R., Drogui, P., Robert, D., 2012. Photoelectrocatalytic technologies for environmental applications. J. Photochem. Photobiol. A Chem. 238, 41−52.

Garbellini, G.S., Salazar-Banda, G.R., Avaca, L.A., 2008. Ultrasound applications in electrochemical systems: theoretical and experimental aspects. Química Nova 31, 123−133.

Georgieva, J., Valova, E., Armyanov, S., Philippidis, N., Poulios, I., Sotiropoulos, S., 2012. Bicomponent semiconductor oxide photoanodes for the photoelectrocatalytic oxidation of organic solutes and vapours: a short review with emphasis to TiO_2-WO_3 photoanodes. J. Hazard. Mater. 211−212, 30−46.

González-García, J., Esclapez, M.D., Bonete, P., Hernández, Y.V., Garretón, L.G., Sáez, V., 2010. Current topics on sonoelectrochemistry. Ultrasonics 50, 318−322.

Isarain-Chavez, E., Rodriguez, R.M., Cabot, P.L., Centellas, F., Arias, C., Garrido, J.A., et al., 2011. Degradation of pharmaceutical beta-blockers by electrochemical advanced oxidation processes using a flow plant with a solar compound parabolic collector. Water Res. 45, 4119−4130.

Khataee, A., Safarpour, M., Zarei, M., Aber, S., 2012. Combined heterogeneous and homogeneous photodegradation of a dye using immobilized TiO_2 nanophotocatalyst and modified graphite electrode with carbon nanotubes. J. Mol. Catal. A Chem. 363, 58−68.

Khataee, A.R., Zarei, M., Asl, S.K., 2010. Photocatalytic treatment of a dye solution using immobilized TiO_2 nanoparticles combined with photoelectro-Fenton process: Optimization of operational parameters. J. Electroanal. Chem. 648, 143−150.

Klauck, C.R., Benvenuti, T., da Silva, L.B., Rodrigues, M.A.S., 2013. Electrochemical oxidation process can be used for the treatment of landfill leachate and evaluation of toxicity in *allium cepa*, 4th International Workshop - Advances in Cleaner Production, Sao-Paulo, Brazil.

Li, H., Lei, H., Yu, Q., Li, Z., Feng, X., Yang, B., 2010. Effect of low frequency ultrasonic irradiation on the sonoelectro-Fenton degradation of cationic red X-GRL. Chem. Eng. J. 160, 417−422.

Martínez, S.S., Uribe, E.V., 2012. Enhanced sonochemical degradation of azure B dye by the electroFenton process. Ultrasonics Sonochem. 19, 174−178.

Martínez-Huitle, C.A., Ferro, S., 2006. Electrochemical oxidation of organic pollutants for the wastewater treatment: direct and indirect processes. Chem. Soc. Rev. 12, 1324−1340.

Oturan, M.A., Sirés, I., Oturan, N., Pérocheau, S., Laborde, J.-L., Trévin, S., 2008. Sonoelectro Fenton process: A novel hybrid technique for the destruction of organic pollutants in water. J. Electroanal. Chem. 624, 329−332.

Panizza, M., Cerisola, G., 2003. Electrochemical oxidation of 2-naphthol with in situ electrogenerated active chlorine. Electrochim. Acta 48, 1515−1519.

Panizza, M., Cerisola, G., 2009. Direct and mediated anodic oxidation of organic pollutants. Chem. Rev. 109, 6541−6569.

Panizza, M., Duo, I., Michaud, P., Cerisola, G., Comnellis, C., 2000. Electrochemical generation of silver (II) at boron-doped diamond electrodes. Electrochem. Solid State Lett. 3, 550−551.

Peralta-Hernández, J., Meas-Vong, Y., Rodríguez, F.J., Chapman, T.W., Maldonado, M.I., Godínez, L.A., 2006. In situ electrochemical and photo-electrochemical generation of the Fenton reagent: a potentially important new water treatment technology. Water Res. 40, 1754−1762.

Sirés, I., Brillas, E., 2012. Remediation of water pollution caused by pharmaceutical residues based on electrochemical separation and degradation technologies: a review. Environ. Int. 40, 212−229.

Xie, Y.-B., Li, X., 2006. Interactive oxidation of photoelectrocatalysis and electro-Fenton for azo dye degradation using TiO_2−Ti mesh and reticulated vitreous carbon electrodes. Meter. Chem. Phys. 95, 39−50.

CHAPTER

12

Catalytic Wet Peroxide Oxidation

Ali R. Tehrani-Bagha and Tarek Balchi

American University of Beirut (AUB), Beirut, Lebanon

12.1 INTRODUCTION

The aim of wastewater treatment is to reduce the load of pollutants from water and to meet the discharge regulations. Wastewaters containing high concentrations of persistent, toxic, and nonbiodegradable organic pollutants (e.g., aromatics, pesticides, etc.) are hard to treat with conventional physical-chemical or biological methods (Munoz et al., 2015). Many techniques such as adsorption, membrane filtration, and floatation can only collect or transfer the pollutants from one phase to another. In contrast, advanced oxidation processes (AOPs) can: (1) degrade and mineralize various persistent organic pollutants effectively without addition of hazardous substances (Bali et al., 2003; Shu et al., 2016), (2) oxidize the pollutants at a relatively low concentration of the reagents (Buthiyappan et al., 2016; Meric et al., 2005), and (3) improve the biodegradability of the treated wastewater (Vogelpohl and Kim, 2004).

One of the most famous AOPs in wastewater treatment is the homogeneous Fenton process. Briefly, Fe^{2+} is oxidized by hydrogen peroxide to Fe^{3+}, forming a hydroxyl radical ($^{\bullet}OH$) and a hydroxide ion in the process. Fe^{3+} is then reduced back to Fe^{2+} by another molecule of hydrogen peroxide, forming a hydroperoxyl radical (HO_2) and a proton. A very rapid decomposition of the produced $^{\bullet}OH$ occurs and the rate of oxidation is reduced due to the slow regeneration of Fe^{2+} to Fe^{3+}. The homogeneous Fenton process is very efficient and fast in degradation of organic pollutants; but it has some shortcomings as follows (Diya'uddeen et al., 2012):

- The Fenton process is efficient in acidic pH (2—4) and inefficient at higher pH due to precipitation of ferric oxyhydroxide in solution. This means large amounts of acid and base are needed to adjust the pH for the optimum condition and for the final neutralization, respectively. Therefore, the salinity of the final wastewater increases.

Advanced Oxidation Processes for Wastewater Treatment
DOI: https://doi.org/10.1016/B978-0-12-810499-6.00012-7

- The water-soluble iron salt used for the process remains in solution after treatment and its recovery and reuse is difficult.
- Complete mineralization of organic compounds needs a large excess of the iron salt and hydrogen peroxide, which makes the process relatively cost-ineffective.

Ross and Chowdhury (1977) had one of the first papers on catalytic oxidation of wastewaters using heterogeneous Fenton process using solid catalyst. However, the term "catalytic wet peroxide oxidation (CWPO)" was coined for the first time for the heterogeneous oxidation of phenol using (Al-Cu) pillared clays as catalyst (Barrault et al., 1998). The bibliometric analysis of these keywords in Scopus database has been provided in Fig. 12.1 showing that the number of publications on heterogeneous fenton and CWPO processes is increasing exponentially. Textile dyes and phenolic compounds are among the organic pollutants that have been targeted in these studies as they are normally nonbiodegradable and resistant towards oxidation (Emami et al., 2010b; Stuber et al., 2005).

The CWPO/heterogeneous Fenton process using water-insoluble solid catalysts started to emerge as a promising approach to overcome the aforementioned shortcomings of the homogeneous Fenton process (Fig. 12.2). The main goal in the CWPO process for wastewater treatment is to develop an AOP with superior oxidation efficiency in a wide range of pH without leaching iron/other metals and producing sludge (Wang et al., 2016). Therefore, researchers are working hard to synthesize water insoluble catalysts that: (1)

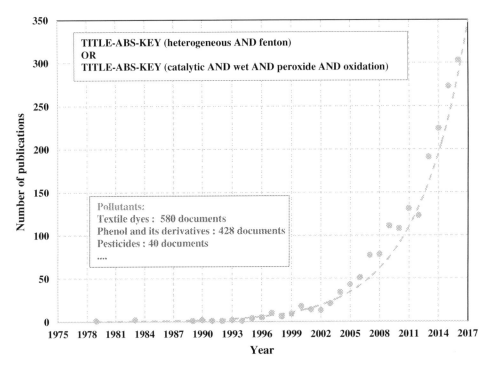

FIGURE 12.1   Evolution of the number of scientific papers published on catalytic wet peroxide oxidation or heterogeneous Fenton. *Data from Scopus.com in May 2017.*

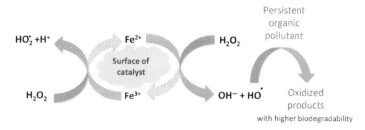

FIGURE 12.2 Scheme showing the catalytic wet peroxide oxidation (CWPO) or heterogeneous Fenton process for oxidation of organic pollutants.

are as efficient as Fe^{2+} salt in the homogeneous Fenton process, (2) are very stable with no or minimum metal leaching, (3) have large surface area and abundant catalytic sites, and (4) are active at a wide range of pH and low temperatures (Galeano et al., 2014; Munoz et al., 2015; Ribeiro et al., 2016b).

The CWPO process has already been used for treatment of nonbiodegradable pollutants found in various industrial wastewater with promising results. Many different solid catalysts have been synthesized to replace the soluble iron salt in the conventional homogeneous Fenton system (Catrinescu et al., 2012; Garrido-Ramirez et al., 2010). Using an iron-bearing solid catalyst can limit the continuous leaching of catalyst that occurs in a homogeneous system. The recovery and reuse of the iron-containing solid catalyst is much easier than soluble iron salt at the end of the reaction (Centi and Perathoner, 2008; Caudo et al., 2006).

As for the pH regulation, the heterogeneous solid catalyst can facilitate the AOP over a broad range of pH (Caudo et al., 2007; Cheng et al., 2008). This advantage can be attributed to the immobilization of the Fe^{2+}/Fe^{3+} ions inside the structure and within the pores of the solid catalyst that prevents the precipitation of iron hydroxide (Catrinescu et al., 2003; Chen and Zhu, 2006). Moreover, some of the catalysts have an adsorption capacity that can contribute to the contaminant removal (Shahbazi et al., 2014).

Compared to simple chemical or biological process, the CWPO process has a high operating cost. In order to reduce the operating cost of the treatment, the utilization of CWPO as a pre-treatment step for achieving enhanced biodegradability of wastewater that contains recalcitrant compounds is an appealing method, if further biological processes can easily degrade the produced intermediates.

12.2 CATALYSTS FOR CWPO

The main reason for using catalysts in CWPO is to accelerate the rate of hydroxyl radical generation and increase the rate of pollutant degradation. Currently, research works are focused on developing catalysts with high stability and activity. A large number of catalysts with various structures (supported, nonsupported, multimetallic, etc.) have already been synthesized and tested (He et al., 2016; Herney-Ramirez et al., 2010; Nidheesh, 2015). The oxidation reaction of organic pollutants in the presence of water-insoluble iron based

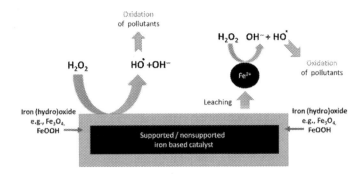

FIGURE 12.3   Scheme showing the water-insoluble iron based catalyst for CWPO process.

catalysts are shown in Fig. 12.3. The Fe^{2+}/Fe^{3+} ions may leach into the solution under acidic condition, which contributes to oxidation via homogeneous Fenton process. This is of course an unfavorable side reaction in the heterogeneous CWPO process, which reduces the activity of the catalyst.

Majority of the catalysts synthesized for the Fenton and CWPO processes are based on iron salts. However, other transition metal cations such as Cu^{2+}, Mn^{2+}, and Co^{2+} can also be used for this purpose (Gokulakrishnan et al., 2009). Various supporting materials have also been used for preparation of these catalysts. (Table 12.1).

12.2.1 Nonsupported Metal Based Catalysts

12.2.1.1 Zero Valent Iron (Fe⁰)

Fe^0 is commonly used as a reducing agent and exhibits a dual surface property in which the core is covered with iron and the shell with iron oxides (Fig. 12.3). The reaction of Fe^0 with H_2O_2 in an acidic condition generates Fe^{2+} ions (i.e., leaching). The released ferrous ions undergo further Fenton reaction and produce hydroxyl radicals in the system. A layer of iron oxide (mainly γ-FeOOH and Fe_3O_4) is generally formed at the surface of Fe^0 with relatively lower catalytic activity. Although Fe^0 is considered to be an accessible, efficient, and cheap catalyst for CWPO, the high rate of leaching Fe^{2+} ions into the solution and low surface activity are certainly two drawbacks for this type of catalyst (He et al., 2016; Nidheesh, 2015).

12.2.1.2 Iron Minerals

Iron minerals (e.g., FeO, -Fe_2O_3, -Fe_2O_3, Fe_3O_4, -FeOOH, -FeOOH, etc.) can also be used as catalyst without any support for the CWPO process. The crystalline structure and composition of Fe^0, Fe^{2+}, Fe^{3+} at the surface of the iron oxides determine their catalytic activity (Hou et al., 2014). Fe^{2+} at the surface of catalysts is crucial for hydroxyl radical generation (Fig. 12.2). Therefore, the smaller particles with larger surface area will enhance the catalytic oxidation due to the presence of more available Fe^{2+} sites (Xu and Wang, 2012). As an example, the crystalline structure of Fe_3O_4 has been provided in Fig. 12.4 (Friak et al., 2007). Fe_3O_4 is one of the iron oxides that has both Fe^{2+} and Fe^{3+} in octahedral

TABLE 12.1 Classification of Iron/Multimetallic Based Catalysts for the CWPO Process

	Category	Some Examples	References
Nonsupported catalysts	Zero valent iron (Fe^0)		Epolito et al. (2008), Litter and Slodowicz (2017), Messele et al. (2016), Danish et al. (2017)
	Iron minerals	Ferrihydrite, Ferrite, Geothite Magnetite, Schorl, Hematite, Pyrite	Munoz et al. (2017), Garrido-Ramirez et al. (2010)
	Mixed metal oxides/ Multimetallic	Fe$_{3-x}$M$_x$O$_4$ (M = Co and Mn) Mixed iron oxide with Cu, Ce, Co, Mn, or V Fe$_3$O$_4$/CeO$_2$	Tehrani-Bagha et al. (2016), He et al. (2016), Zhou et al. (2011), Costa et al. (2003)
Supported catalysts		Mixed Al/M-pillared clays (M = Fe, Cu, Mn)	Galeano et al. (2014), Kim et al. (2002), Barrault et al. (1998), Gan et al. (2017)
	Clay-based material as support	Bentonite, Kaolinite, Laponite, Vermiculite, Sepiolite, Saponite, Montmorlite, Pillared interlayered clays, Alumina, Silica, Zeolites	Djeffal et al. (2014), Azmi et al. (2014), Platon et al. (2013), Lee et al. (2013), Herney-Ramirez et al. (2010), Garrido-Ramirez et al. (2010), Centi and Perathoner (2008), Galeano et al. (2014)
	Carbon-based material as support	Activated carbon Multiwalled CNTs Graphene-based materials	Ribeiro et al. (2016b), Martinez et al. (2014), Dominguez et al. (2014a), Rey et al. (2016), Rey et al. (2009), Arshadi et al. (2016), Ribeiro et al. (2017), Wan and Wang (2017)
	Organic-based material as support	MnO$_2$@SiO$_2$ nanofibrous membrane PAN nanofiber Fe Complex	Zhao et al. (2013), Wang et al. (2017)

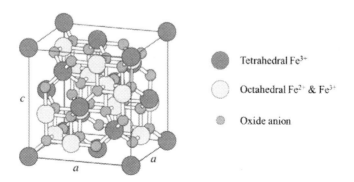

FIGURE 12.4 Face-centred cubic spinel structure of magnetite (Fe$_3$O$_4$). *Adapted from Friak, M., Schindlmayr, A., Schaffler, M., 2007. New J. Phys., 9. doi:10.1088/1367-2630/9/1/005. With permission.*

sites. The leaching and catalytic activity of Fe_3O_4 is lower than those of Fe^0 for the CWPO (Costa et al., 2008; He et al., 2016; Hou et al., 2014).

12.2.1.3 Supported or Nonsupported Mixed Metal Oxides

The incorporation of a second metal can increase the stability and catalytic activity of the catalyst (Chen et al., 2011; Lien and Zhang, 2001; Su et al., 2011). As an example, the CWPO of methylene blue (0.1 g/L) in the presence of two mixed metal oxides (i.e., $Fe_{2.47}Mn_{0.53}O_4$ and $Fe_{2.25}Co_{0.75}O_4$) is extremely faster than its degradation in the presence of Fe_3O_4 (Costa et al., 2003). The discoloration of the solution within 10 min was achieved in the presence of two mixed metal oxides, while the discoloration in the presence of Fe_3O_4 was less than 10% after 50 min (Costa et al., 2006). Mixed metal oxides of Cu, Co, Ni, and Fe, and Pt, Pd, Rh, Ru, and Ir have also been reported as effective catalysts for wastewater purification (Levec and Pintar, 1995).

The catalytic degradation of organic pollutants in aqueous streams by mixed Al/M-pillared clays (M = Fe, Cu, Mn) has been reviewed (Galeano et al., 2014). The incorporation of bimetallic iron and nickel nanoparticles on chitosan as a stabilizer was effectively used for the removal of the mixed amoxicillin and cadmium from the aqueous solution (Weng et al., 2013). Similarly, polyethyleneimine-decorated graphene oxide supported Fe-Ni demonstrated an excellent dehydrogenation (Zhou et al., 2012).

12.2.2 Supported Metal Based Catalysts

12.2.2.1 Clay-Based Material as Support

12.2.2.1.1 PILLARED INTERLAYERED CLAYS

Pillared interlayered clays (PILC) are inexpensive, highly porous solid materials exhibiting remarkable characteristics and structures (e.g., high stability). PILC are prepared by intercalating metal polycations and heated at elevated temperatures to form clusters of the conforming metal oxide by dehydration and dehydroxylation. By upholding the silicate layers apart, the formed metal oxides act as pillars, generating interlayer pores (Pan et al., 2008). Therefore, hydrogen peroxide can easily access the interlayer catalytic sites as a result of the pillaring process (Galeano et al., 2014; Sanabria et al., 2008). Various pillared interlayered clays that contain Fe, Cu, Zn, etc. have been extensively utilized as heterogeneous catalysts in the CWPO process (Basoglu and Balci, 2016; Cardona and Ocampo, 2016; Li et al., 2016; Mnasri-Ghnimi and Frini-Srasra, 2016; Tomul et al., 2016; Tomul, 2016; Ye et al., 2016a, 2016b; Zhou et al., 2014).

The pillaring process generally increases the stability of clay and the resulting catalyst shows significantly reduced metal leaching to the treated water (Caudo et al., 2008; Ramirez et al., 2007). Fe-PILC and Cu-PILC are among the most effective clay-supported catalysts and their metal leaching to the solution is almost zero (Table 12.2). Thus, these catalysts can be recovered and reused several times without noticeable loss of activity. Fe-PILCs are effective over a wider range of pH with acceptable catalytic activity at near neutral pH with a short substantial activity loss (Galeano et al., 2014). This advantage is due to the significant immobilization of Fe^{3+} species in the interlayer space of the pillared clay;

TABLE 12.2 The CWPO of p-Coumaric Acid Using Different Supported Catalysts

Sample	Preparation Method	Specific Surface Area (m²/g)	Metal Content (wt%)	Leaching After 4 h (%)	TOC Removal (%)
Fe-PILC	Pillaring	108	2	0	75
Cu-PILC	Pillaring	107	2	0	82
Cu-Al$_2$O$_3$	Impregnation	185	1	8	91
Cu-Bentonite	Ion exchange	106	0.9	3	70
Cu-ZrO$_2$	Impregnation	50	1	30	73
Cu-ZSM-5	Ion exchange	370	1	5	78

Reaction conditions: 70 C; 0.5 g catalyst; rate of 35% H$_2$O$_2$ addition: 0.5 mL/h
Source: *Adapted from Caudo, S., Senti, G., Genovese, C., Perathoner, S., 2007. Appl. Catal. B: Environ., 70, 437–446. With permission.*

thus, significantly reducing the effect of alteration in solution pH as a result of stabilizing the iron inside the PILC and limiting leaching.

12.2.2.1.2 ALUMINA

Alumina (Al$_2$O$_3$) possesses a strong alkalinity, broad range of surface area, and notable ion exchange abilities (Soon and Hameed, 2011). Alumina has been used as a support for both iron and iron oxides for the degradation of numerous organic compounds. The degradation of 2,4-dinitrophenol with CWPO processes using alumina-supported Fe^{3+} and nonsupported ferric sulfate at the same temperature and pH has been studied by Ghosh et al. (2012). The CWPO in the presence of the alumina-supported ferric ions seems to be slightly improved (Fig. 12.5). The supported catalyst was recovered and reused four times with about 20% of loss in activity due to iron leaching. It was reported that the amount of the alumina-supported ferric catalyst needed for the CWPO process is eight times higher than that of the nonsupported Fe^{3+} (Ghosh et al., 2012). Alumina based catalysts normally have a high level of metal leaching compared with pillared-based catalysts (Caudo et al., 2007).

12.2.2.1.3 ZEOLITE

Zeolite belongs to the aluminosilicate family of microporous crystalline materials that can accommodate a wide range of cations such as Na$^+$, K$^+$, Ca^{2+}, Mg^{2+}. SiO$_4$ and AlO$_4$, the two main constituents of zeolites, share an oxygen ion that forms a link between them (Petranovskii et al., 2016). The integration of Al into a silica framework (AlO$_4$) makes the framework negatively charged, which can be balanced by extra inorganic or organic cations. Ion exchange and impregnation are two methods that are generally used for introducing metals into zeolites (Misaelides, 2011). Being highly porous materials, zeolites have been widely used as support material for several catalysts. These catalysts are considered cost effective, vastly available, and recyclable. They can function at nearly neutral pH, and remove pollutants by both adsorption and oxidation (Franus et al., 2015; Hedstrom, 2001; Zhao, 2016). The porous structure of zeolites and alterations on the surface impose

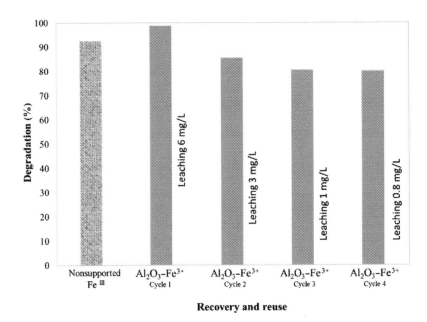

FIGURE 12.5   The CWPO of 2,4-dinitro phenol using 1 g/L of 10 wt% supported alumina-Fe^{III} and 0.125 g/L nonsupported Fe^{III} at 30 C, pH 3, $[H_2O_2]$ = 250 mg/L, time 35 min. *Based on the data of Ghosh, P., Kumar, C., Samanta, A.N., Ray, S., 2012. J. Chem. Technol. Biotechnol. 87, 914–923.*

complex sorption mechanisms. ZSM-5 as a typical microporous zeolite has a 10-ring interconnected channel system with a high Si/Al ratio (Yaripour et al., 2015). The specific surface area of ZSM-5 can be as high as 370 m^2/g (Table 12.2), which may influence the mass transfer efficiency. MCM-41, another member of mesoporous zeolites, possesses a hexagonally arranged uniformed pore structure (Wu et al., 2001). High specific surface area (more than 800 m^2/g), large pore volume, and uniform pore size makes MCM-41 a good candidate for catalyst support, or adsorbent in the CWPO process (Kumar et al., 2001; Timofeeva et al., 2007).

12.2.2.2 Carbon-Based Materials as Support

There are many scientific publications on carbon-based catalysts in liquid or in gas phase reactions (Figueiredo, 2013). Porous carbon materials can be used as supports for immobilization of metal complexes at their surface and the resulting catalysts have many interesting characteristics as follows: (1) They are stable at high temperatures and in a wider range of pH, (2) they have normally high specific surface area and porosity, (3) their surface can be modified chemically in order to enhance the diffusion of species across the surface and control the dispersion of metals, (4) they can be separated from the solution and burned to recover the costly metal phase, (5) they are relatively cheap, and (6) they can be used as adsorbents (Figueiredo, 2013; Herney-Ramirez and Madeira, 2010; Philippe et al., 2009; Rodriguez-Reinoso, 1998; Serp and Figueiredo, 2009).

The most widely used technique for the preparation of carbon-supported metal catalysts is impregnation (Figueiredo, 2013). Nonetheless, numerous additional methods have been used to prepare carbonaceous supports, for instance, precipitation or co-precipitation, liquid-phase reduction, chemical vapor deposition and physical vapor deposition (Serp and Figueiredo, 2009). Due to their remarkable characteristics and a wide range of preparation methods, the utilization of the carbon-supported catalysts is an appealing solution to overcome the shortcomings stated earlier on the utilization of homogeneous Fenton process.

12.2.2.2.1 ACTIVATED CARBON (AC)

AC is a porous material exhibiting amphoteric characteristics, and is usually used for adsorption of organic and inorganic compounds. There are some remarkable advantages of activated carbon such as, reduced operation cost, high surface area, significant stability, and tunability of the surface and structure (Britto et al., 2008; Martinez et al., 2014). AC supported catalysts can be found in two different forms (i.e., powdered and granular). The adsorption capacity of these supports is substantially affected by the nature of the carbon sorbent and the preparation methods. The CWPO process using AC supported catalysts is quite favorable compared with the conventional homogeneous Fenton process (Buthiyappan et al., 2016).

AC supported catalysts can be simply prepared by incipient impregnation of various types of AC in iron salt solution, followed by drying and heat treatment at 200 C. Complete phenol conversion and 80% mineralization can be achieved in less than 2 h using the CWPO process in the presence of these supported catalysts (Rey et al., 2009). However, the iron leaching can be very high (between 20% and 60%), which affects the reusability of the catalysts. However, there are some papers showing that the iron leaching of the AC supported catalysts is relatively low without providing any proof for their reusability (Messele et al., 2015).

The high conversion and TOC removal (%) of organic pollutants in many of the CWPO studies are directly related to the high adsorption capacity of AC (Ribeiro et al., 2012; Ribeiro et al., 2016a). Therefore, the loss of activity after each cycle of recovery and reuse is observed.

12.2.2.2.2 MULTIWALLED CARBON NANOTUBES (MWCNTS)

Carbon nanotubes (CNTs) have remarkable characteristics such as enhanced mass-transfer efficiency, increased ordering and mechanical resistance, excellent electronic characteristics, and increased thermal stability in oxidizing conditions. For preparation of the supported catalyst, CNTs are mainly synthesized by chemical vapor deposition of a hydrocarbon gas on the surface of catalysts (Kumar and Ando, 2010). A synthesized hybrid (Fe_3O_4/MWCNT) was used as a heterogeneous catalyst for the CWPO process. The supported catalyst showed superior performance than that of the nonsupported magnetite (Fe_3O_4) for the degradation of methyltestosterone. Fe_3O_4/MWCNT showed an enhanced removal rate of pollutant (86%) compared to that of Fe_3O_4 (62%) at 20 C and pH 5. This is attributed to the adsorption of the pollutant at the surface of catalyst and near the active catalytic sites. The decomposition of hydrogen peroxide was also reduced in the presence of this catalyst, indicating an enhanced hydrogen peroxide consumption efficiency (Hu et al., 2011). Ferrocene groups immobilized on aluminum-silicate and MWCNTs (Si/Al@Fe/MWCNT) were utilized as a heterogeneous catalyst for the CWPO process. The synthesized material showed a remarkable efficiency in the removal of methyl orange dye

(100% removal after 6 min) by both adsorption and CWPO processes. The catalyst was successfully recovered and reused six times without any noticeable loss of activity in the CWPO process (Arshadi et al., 2016) while the material shows a huge loss of activity as an adsorbent in the absence of hydrogen peroxide.

12.2.2.2.3 GRAPHENE-BASED MATERIALS

Graphene oxide (GO) has also been used as a support for Fe_3O_4 magnetic nanoparticles by the co-precipitation synthesis method followed by reduction with hydrazine to produce the catalyst Fe_3O_4/rGO (reduced graphene oxide). Under the same conditions, the CWPO of 4-chlorophenol in the presence of Fe_3O_4/rGO was much more efficient (98.6% removal) than those in the presence of Fe_3O_4 (17.7% removal) or GO (24.5% removal). These magnetic nanocomposites can be separated by a magnetic field, washed, and reused after each process (Liu et al., 2013). The reduced graphene oxide was impregnated by $Fe_3O_4–Mn_3O_4$ nanoparticles to produce the supported catalyst. The CWPO efficiency for sulfamethazine degradation using $Fe_3O_4–Mn_3O_4-rGO$ was remarkably higher than those of $Fe_3O_4–Mn_3O_4$ and Mn_3O_4 under the same optimal conditions: [Catalyst] = 0.5 g/L, [H_2O_2] = 6 mM at 35 C, pH 3 (Wan and Wang, 2017).

12.2.2.3 Organic-Based Materials as Support

In one of the few studies using polymers as the support of catalyst, Cu^{2+} species were adsorbed on the surface of poly(4-vinylpyridine) and the resulting composite was used as a catalyst for the degradation of 1 g/L of phenol in water. The conversion rate was relatively low (60%) after 120 min at 50 C and pH 6. The catalyst preserved its integrity during the reaction, and the leaching of Cu was above 7 mg/L for this process (Castro et al., 2013). The drawback of using a polymeric support is its in stability in the CWPO process. Hydroxyl radicals attack the polymeric support and degrade its structure gradually.

12.3 EFFICIENCY OF CWPO OF PHENOL

To compare the efficiency and performance of various heterogeneous catalysts in the CWPO process, the degradation of phenol (as an organic pollutant) in water has been summarized in Table 12.3.

- The high availability and reduced cost of magnetic natural minerals make them appealing for the CPWO application. Exceptional H_2O_2 consumption efficiency was reported for the CWPO process using naturally occurring iron minerals (i.e., magnetite, hematite, and ilmenite) compared with that of homogeneous Fenton oxidation under similar operating conditions (Zazo et al., 2011). However, their reusability is limited due to their high iron leaching and generation of iron sludge, which imposes a treatment after their use; thus, increasing the cost of operation (Munoz et al., 2017).
- The catalytic performance of magnetic nanoparticles is not noticeably better than that of natural magnetic minerals. A high concentration of nanoparticles and H_2O_2 is normally needed for the CWPO process. Much better catalysts with reduced manufacturing cost,

TABLE 12.3 The Efficiency of CWPO of Phenol Using Various Solid Catalysts

Category	Catalyst	Removal (%)	$[P]_0$ mg/L	Optimized Conditions	Stability	References
Fe^0	ZVI/AC1900 C	85% (Phenol) %76 (TOC) After 3 h	150	[Cat] = 2.5 g/L, $[H_2O_2]$ = 0.75 g/L, T = 30 C, pH = 3	Fe leaching = 0.04%	Messele et al. (2015)
Iron minerals	Ilmenite	100%(Phenol) 50 % (TOC) After 8 h	100	[Cat] = 0.45 g/L, $[H_2O_2]$ = 0.5 g/L, T = 25 C, pH = 3, UV = 30 W/m^2	Fe leaching = 2.3 mg/L	Garcia-Munoz et al. (2016)
	Magnetite	100% (Phenol) ~70% (TOC) After 4 h	100	[Cat] = 2 g/L, $[H_2O_2]$ = 0.5 g/L, T = 75 C, pH = 3	Fe leaching = 5.5 mg/L After 4 h	Munoz et al. (2017)
Mixed metal oxides /Multimetallic	Cu−Ni−Al hydrotalcite	100% (Phenol) ~80% (TOC) After 2 h	250	[Cat] = 0.8 g/L, $[H_2O_2]$ ~ 40 mM, T = 40 C, pH = 6.5	Leaching (mg/L) Cu ~ 38.8 Ni ~ 6.6 Al ~ 0.74	Zhou et al. (2011)
	Modified diatomites	100% (Phenol) 53% (TOC) After 5 h	1000	[Cat] = 2.85 g/L, $[H_2O_2]$ = 1.3 stoichiometric, T = 70 C, Neutral pH	Fe leaching = 4.3 mg/L	Inchaurrondo et al. (2017)

(Continued)

TABLE 12.3 (Continued)

Category		Catalyst	Removal (%)	$[P]_0$ mg/L	Optimized Conditions	Stability	References
Clay based materials as support	Pillared clays	Fe-Al-PILC	100% (Phenol) 88%(COD) After 3 h	250	$[Cat] = 1$ g/L, $[H_2O_2] = 3.5$ g/L, $T = 50$ C, pH = 5	Fe leaching ~0%	Catrinescu et al. (2003)
		Fe-Al-PILC	100% (Phenol) 80%(TOC) At 70 C 65%(TOC) At 25 C	470	$[Cat] = 5$ g/L, $[H_2O_2] = 3.5$ g/L, $T = 25$ C–70 C, pH 3.7	Fe leaching <0.3 mg/L At 25 C	Guelou et al. (2003)
	Zeolites	Fe-ZSM-5 zeolite membrane (25% Fe loading)	95% (Phenol) 45% (TOC) Continuous run for 7 h	1000	FFR = 2 mL/min, CBH = 2.0 cm, $[H_2O_2] = 5.1$ g/L, $T = 80$ C, pH = 3	Fe leaching <7 mg/L	Yan et al. (2015)
		Fe$_2$O$_3$/MCM-41 (25% Fe loading)	99% (phenol) 72.5% (TOC) Continuous run for 7 h	1000	FFR = 2 mL/min, CBH = 4.0 cm, $[H_2O_2] = 5.1$ g/L, $T = 80$ C, pH = 3	Fe leaching <1 mg/L	Yan et al. (2016)
		Fe$_2$O$_3$/ SBA-15(DS)	100% (Phenol) 64% (TOC) After 8 h	1000	FFR = 1 mL/min, CBH = 1.2 cm, $[Cat] = 2.9$ g, $[H_2O_2] = 5.1$ g/L, $T = 80$ C, pH = 4.3	Fe leaching <14 mg/ L	Botas et al. (2010)
	Alumina	Fe$_2$O$_3$-γ-Al$_2$O$_3$	100% (phenol)	5000	FFR = 5.4 mL/min, CBH = 11.5 cm,	Fe leaching <25%	Di Luca et al. (2014)

Category	Abbr.	Support / Catalyst	Concentration (mg/L)	Removal efficiency	Conditions	Notes	Reference
		Heat treated at 900 C		63% (TOC) After 4 h	[cat] = 20 g, [H$_2$O$_2$] = 395 mM, T = 70 C, pH = 3	–	Inchaurrondo et al. (2012a)
		BASF Cu-0226 S	1000	100% (phenol) 91% (TOC) After 4 h	[Cat] = 25 g/L, [H$_2$O$_2$] = 3.9 × stoichiometric, T = 70 C, pH = 2.4		
Carbon-based material as support	AC	Fe precursors supported on activated carbon (AC)	1000	100% (Phenol) 85% (TOC) After 2 h	[Cat] = 5 g/L, [H$_2$O$_2$] = 5 g/L, T = 50 C, pH = 3	Fe leaching = 32.5 mg/L	Rey et al. (2009)
		Au/AC	5000	100% (Phenol) 70%(TOC) After 24 h	[Cat] = 2.5 g/L, [H$_2$O$_2$] = 25 g/L, T = 80 C, pH = 3.5	–	Dominguez et al. (2014b)
	MWCNTs	Multiwalled carbon nanotubes (SA2)	4500	100% (Phenol) 67%(TOC) After 24 h	[Cat] = 2.5 g/L, [H$_2$O$_2$] = 25 g/L, T = 79.85 C, pH = 3.5	Fe leaching = 22.6%	Pinho et al. (2015)
Organic-based material as support		PVP$_2$-Cu(II)	1000	64% (Phenol) After 2 h	[Cat] = 0.5 g/L, [H$_2$O$_2$] = 1:14 Molar ratio, T = 40 C, pH = 6	Cu leaching = 7 mg/L	Castro et al. (2013)

Abbreviations: [Cat] = Catalyst dosage, FFR = Feed flow rate, CBH = Catalyst bed height.

should be established to permit their extensive application in the CWPO process (Munoz et al., 2015).

- The supported materials having higher surface area, larger pore volume, and uniform pore size normally enhance the performance of the CWPO process. The catalytic performance of Fe_2O_3/ZSM-5 (Yan et al., 2015) and Fe_2O_3/MCM-41 (Yan et al., 2016) under the same conditions can be compared.
- A high level of iron leaching from the supported/nonsupported catalysts contributes to the oxidation of the pollutants via the Fenton process. This leads to the deactivation of the catalyst due to complexation and reduces its recyclability (Inchaurrondo et al., 2017; Andas et al., 2014; Pinho et al., 2015). The minimum level of iron leaching has been reported for the pillared clay supported catalyst as of 0% (Catrinescu et al., 2003) and many of the catalysts based on MWCNT, alumina, and silica as supports have a very high level of iron leaching indeed (Andas et al., 2014; Di Luca et al., 2014; Pinho et al., 2015).
- Metal leaching from catalysts normally increases at higher temperatures (Zhou et al., 2011). High-temperature calcination of the catalysts after impregnation with various salts and addition of some additives (e.g., oxalic acid) improves the stability of the clay based catalysts (Di Luca et al., 2014).
- A remarkable performance in the mineralization of phenol in water ($\sim$75%) at a wide range of pH and using moderate amounts of oxidant has been reported for Fe_2O_3/SBA-15 nanocomposites. 75% of the stoichiometric amount required for the complete mineralization of the initial phenol concentration was used (3.8 g/L), which is fairly low. Another advantage was that the reaction medium was not acidified at the end of the process, which can reduce the operation costs since the neutralization process is not needed anymore. However, this CWPO process has been performed at a high temperature of 100 C, which is quite energy intensive (Melero et al., 2007).
- Some of the supporting materials for catalysts are very good adsorbents (e.g., AC, MWCNTs). Thus, a very high level of pollutant removal can be achieved due to adsorption and the CWPO (Messele et al., 2015; Ribeiro et al., 2015; Variava et al., 2012).
- Both Fe-ZSM-5 and Cu-ZSM-5 particle catalysts were used for the CWPO of phenol in a fixed bed reactor. 95% phenol removal and 80% H_2O_2 conversion were achieved in the CWPO. The Cu-ZSM-5 membrane catalyst showed better catalytic activity because no quinone compounds were detected in the treated wastewater. As for the stability, an increased Cu leaching concentration was reported 0.20 g/L (15% leached Cu) (Jiang et al., 2015).

12.4 EFFECT OF THE MAIN PARAMETERS

12.4.1 Effect of Initial pH

The pH of solution is a very important parameter and it has a direct effect on both the catalytic performance and the degree of iron leaching from the support. The latter determines the overall stability of the utilized catalyst, which is an important factor in the recyclability of the catalyst. The CWPO process showed its best performance at pH 2–4, which is a representative pH value for the conventional homogeneous Fenton process (Molina et al.,

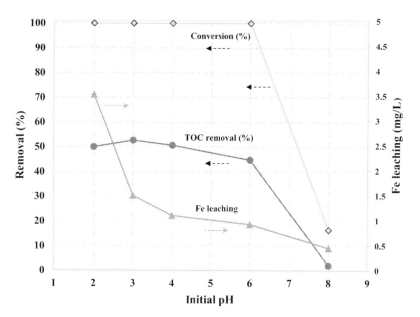

FIGURE 12.6 Effect of initial pH on m-cresol conversion, TOC removal, and Fe leaching in the CWPO process using Fe/γ-Al₂O₃ catalyst. Reaction conditions: 60 C, 120 min. *Adapted from Liu, P.J., He, S.B., Wei, H.Z., Wang, J. H., Sun, C.L., 2015. Ind. Eng. Chem. Res. 54, 130–136. With permission.*

2006). However, the CWPO process using solid catalyst (e.g., magnetic copper ferrite nanoparticles) has a wider range of working pH and the catalyst is still active at pH 8 and even 10 (Tehrani-Bagha et al., 2016). The effect of initial pH on the performance of iron oxide supported catalyst for the m-cresol conversion, TOC removal, and iron leaching has been shown in Fig. 12.6. The degradation and TOC removal (%) were more or less constant, when the initial pH was varied from 2 to 6. The m-cresol conversion (%) and TOC removal (%) noticeably decreased at pH above 6 (Liu et al., 2015). At pH 3, the highest concentration of active Fe^{2+} species is achieved accompanied by the maximum scavenging effect of the hydroxyl radicals. While at a pH value above 3, the insoluble FeOOH precipitates and decomposes significantly (Pignatello, 1992), resulting in lower catalyst activity and lower metal leaching.

12.4.2 Effect of Temperature

The CWPO and mineralization of organic pollutants normally increases noticeably by increasing the temperature from 25 to around 80 C (Fig. 12.7). This can be attributed to the higher H_2O_2 conversion to hydroxyl radicals (Di Luca et al., 2014; Martinez et al., 2014; Yan et al., 2016). The CWPO of pollutant at a temperature as high as 100 C has also been reported in one case (Melero et al., 2007). However, the CWPO process at elevated temperatures (>40 C) will increase the total cost of the treatment and the thermal decomposition of H_2O_2 into oxygen and water (i.e., inactive species). The latter will reduce the concentration

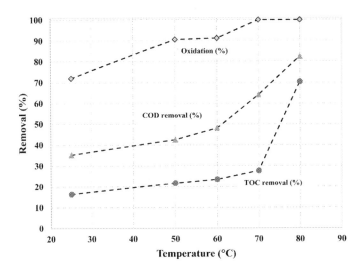

FIGURE 12.7 CWPO of 2,4-dichlorophenol (0.25 g/L) using $Ce_{0.5}Fe_{0.5}O_2$ (0.5 g/L) nanocatalyst and H_2O_2 (0.06 mol/L) at pH 4 and various temperatures after 120 min. *Adapted from Kurian, M., Kunjachan, Sreevalsan, A., 2017. Chem. Eng. J. 308, 67–77. With permission.*

of hydroxyl radicals in the solution and hence, more H_2O_2 is needed for the efficient oxidation (Herney-Ramirez et al., 2010).

12.4.3 Effect of H_2O_2 Dosage

In the CWPO process, the concentration of oxidant (H_2O_2) is a vital factor that considerably affects the removal of organic pollutants. The amount of generated hydroxyl radicals is related directly to the concentration of H_2O_2. The TOC removal efficiency of phenol in water using a commercial catalyst (BASF Cu-0226S) has been presented in Table 12.4 (Inchaurrondo et al., 2012a). The final TOC removal (%) increases by increasing the amount of H_2O_2 in the solution. However, the most efficient CWPO process (i.e., high rate of TOC removal at minimum concentration of H_2O_2) can be determined by comparing the efficiency of H_2O_2 consumption provided in this table. Excess hydrogen peroxide in the solution will increase the chance for recombination of the radicals (i.e., scavenging effect) and production of less active radicals such as HO_2 in the solution. H_2O_2 can also react with the active hydroxyl radicals in the solution to generate inactive species (e.g., H_2O and O_2). Therefore, the efficiency factor (TOC consumption over H_2O_2 consumption) should be calculated for different CWPO processes to determine the most efficient conditions (Inchaurrondo et al., 2012b; Inchaurrondo et al., 2012a).

12.4.4 Effect of the Catalyst Load

The number of accessible active sites on the surface of catalysts is directly proportional to the concentration of the catalyst in the solution. Hence, an increased

TABLE 12.4 The Effect of H_2O_2 Dosage on TOC Removal and H_2O_2 Consumption Efficiencies for Phenol in Water (1 g/L) at Different Experimental Conditions at 70 C After 180 min

Experimental Conditions		Efficiency Parameters			
				Efficiency of H_2O_2 Consumption	
Catalyst Load (mg/L)	H_2O_2 Dosage[a,b]	Final TOC Conversion (%)	Initial TOC Conversion Rate (min^{-1})	TOC Consumption mg $\overline{H_2O_2 \text{ Fed g}}$	TOC Consumption mg $\overline{H_2O_2 \text{ Consumption g}}$
1	1 dose of 3.3 mL	64	0.6	75	101
1	1 dose of 6.6 mL	70	0.64	42	60
1	1 dose of 9.9 mL	51	0.67	20	31
25	1 dose of 3.3 mL	76	2.53	90	90
25	1 dose of 6.6 mL	89	4.18	53	53
25	1 dose of 9.9 mL	94	5.12	37	37
25	3 doses of 1.1 mL	90	—	107	102
25	3 doses of 3.3 mL	91	—	36	34

[a]H_2O_2 30% was used in all the cases.
[b]1.3 times the stoichiometric ratio requires 3.3 mL addition.
Source: Adapted from Inchaurrondo, N., Ceehivi, J., Font, J., Haure, P., 2012. Appl. Catal. B Environ. 111, 641–648. With permission.

catalyst load normally enhances the catalytic performance and accelerates the reaction rate (Inchaurrondo et al., 2012a; Inchaurrondo et al., 2017; Kurian et al., 2017). When a high concentration of catalyst (25 g/L) was used, the mineralization of the pollutant increased between 10 and 40% under the same condition (Table 12.4). For reducing the cost of the CWPO process and reducing the iron leaching, the concentration of the catalyst in the solution should be optimized as it was described earlier. The minimum concentration of the catalyst that is needed for the degradation of organic pollutants with stoichiometric amount of H_2O_2 in a reasonable time is normally reported as the optimized concentration of the catalyst (Liu et al., 2015; Wang et al., 2015).

The results also show that by increasing the catalyst concentration beyond a certain value, the performance of the CWPO process is also adversely affected (Zhou et al., 2011) (Fig. 12.8). This can be explained by the higher rate of H_2O_2 decomposition and the subsequent formation of less active radicals or inactive species (water, O_2), which are less effective in oxidation of the pollutant.

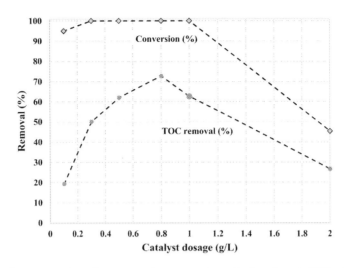

FIGURE 12.8 Effect of catalyst dosage on the conversion of phenol and TOC, Conditions: Catalyst = Cu−Ni−Al hydrotalcite; [Phenol]$_{initial}$ = 2.66 mM; [H$_2$O$_2$] = 39.85 mM; temperature = 30 C; time = 2 h; pH = 6.5. *Adapted from Zhou, S.W., Qian, Z.Y., Sun, T., Xu, J.G., Xia, C.H., 2011. Appl. Clay Sci., 53, 627−633. With permission.*

12.5 CWPO PERFORMANCE

To compare the efficiency and performance of the CWPO process with other AOPs, the degradations of two persistent organic pollutants (i.e., phenol and C.I. reactive red 120) in water have been summarized in Tables 12.5 and 12.6, respectively. There are a number of effective parameters on degradation of the organic pollutants by AOPs that have already been explained and discussed earlier. The reported conditions in these tables may not necessarily represent the optimum conditions of the degradation by the specified AOP due to incomplete optimization used in those references.

The results of these tables can be summarized as follows:

- AOPs are generally quite fast and efficient in the degradation of organic pollutants in water. Complete phenol removal in 0.5 h and complete mineralization (i.e., 100% TOC removal) in about 2 h has already been reported for UV + H$_2$O$_2$ process under a very strong UV radiation (Olmez-Hanci and Arslan-Alaton, 2013).
- The efficiency of AOPs in the degradation of organic pollutants increases by addition of catalyst dosage, H$_2$O$_2$ concentration, UV irradiation power, ozonation dosage, and sonication power to a large extent. There is generally a strong synergy between the aforementioned parameters. The combination of various oxidants (UV, O$_3$, H$_2$O$_2$, etc.) and addition of catalysts normally enhance the efficiency of the AOPs.
- Fenton and CWPO processes are both temperature dependent and the degradation rate of organic pollutants normally increases by increasing the temperature (Emami et al., 2010a, 2010b; Tehrani-Bagha et al., 2016).
- The CWPO process using various water insoluble catalysts is considered a fast and efficient AOP. Depending on the efficiency of the synthesized catalysts, the process can generate similar results as that of homogeneous Fenton process using ferrous sulfate as

TABLE 12.5 CWPO Efficiency in Degradation of Phenol in Comparison to Other AOPs

AOPs	Removal (%)	$[P]_{initial}$ (mg/L)	Conditions	References
CWPO	100% (Phenol) in 15 min	100 and 500	Catalyst:Copper hydroxyl sulfates, [Cat] = 0.33 g/L	Huang et al. (2015a)
	>85% (COD)		$[H_2O_2]$ = 1.68 g/L, 60 C, pH = 7, 1 h	
CWPO	100% (Phenol)	470	Catalyst: Fe-Al-PILC, [Cat] = 5 g/L, $[H_2O_2]$ = 3.5 g/L, T = 25 C, pH = 3.7	Guelou et al. (2003)
	65% (TOC)			
Fenton	100% (Phenol)	500	Catalyst: $FeSO_4$, $[Fe^{2+}]$ = 0.5 g/L, $[H_2O_2]$ = 0.2 g/L, 50 C, pH = 3, 1 h	Rubalcaba et al. (2007)
	91% (COD)			
Photo-Fenton	100% (Phenol) After 0.5 h,	500	Catalyst: Fe_2SO_4, [Cat] = 7.5 mg/L, $[H_2O_2]$ = 0.5 g/L, 20 C, pH = 3, 6 h	Bali et al. (2003)
	97% (TOC)		Irradiation: Low pressure,	
	After 6 h		Mercury vapor UVC lamp (16 W)	
Electro-Fenton	>80% (phenol)	200	Catalyst: $FeSO_4$, [Cat] = 4 mg/L, $[H_2O_2]$ = 0.5 g/L, current density = 12 mA/cm^2, initial pH = 3, 1h	Babuponnusami and Muthukumar (2012)
	>50% (COD)			
Sono-electro-Fenton	100% (Phenol)	200	Catalyst: $FeSO_4$, [Cat] = 4 mg/L, $[H_2O_2]$ = 0.5 g/L, current density = 12 mA/cm^2, initial pH = 3, 1h	Babuponnusami and Muthukumar (2012)
	~80% (COD)		Sonication: 34 kHz, 120 W	
Photo-electro-Fenton	100% (Phenol)	200	Catalyst: $FeSO_4$, [Cat] = 4 mg/L, $[H_2O_2]$ = 0.5 g/L, current density = 12 mA/cm^2, initial pH = 3, 1h	Babuponnusami and Muthukumar (2012)
	~80% (COD)		Irradiation: Low pressure, mercury lamp (254 nm, 8 W)	
O_3	100% (Phenol)	~100	Gas flow = 100 L/h, $[O_3]$ = ~0.25 g/h, 20 C, pH = 9.4, 1.5 h	Esplugas et al. (2002)
$O_3 + H_2O_2$	90% (Phenol)	~100	Gas flow = 100 L/h, $[O_3]$ = ~0.25 g/h, $[H_2O_2]$ = 0.62 mM, 20 C, pH = 6.8 buffered, 1.5 h	Esplugas et al. (2002)
	37% (TOC)			
$O_3 + UV$	92% (Phenol)	~100	Gas flow = 100 L/h, $[O_3]$ = ~0.25 g/h, 20 C, pH = 6.9 buffered, 1 h	Esplugas et al. (2002)
	65% (TOC)		Irradiation : Mercury vapor lamp (254 nm), flux = 28.7 einstein/s	
$O_3 + H_2O_2 + UV$	99% (Phenol)	~100	Gas flow = 100 L/h, $[O_3]$ = ~0.25 g/h, $[H_2O_2]$ = 0.07 mM, 20 C, pH = 5→3, 2 h	Esplugas et al. (2002)
	65% (TOC)		Irradiation: Mercury vapor lamp (254 nm), flux = 28.7 einstein/s	

(Continued)

TABLE 12.5   (Continued)

AOPs	Removal (%)	$[P]_{initial}$ (mg/L)	Conditions	References
UV + H_2O_2	90% (Phenol)	~100	$[H_2O_2] = 7.41$ mM, 20 C, pH = $3.2 \rightarrow 2.3$, 0.5 h Irradiation : Mercury vapor lamp (254 nm), flux = 28.7 einstein/s	Esplugas et al. (2002)
UV + H_2O_2	100% (Phenol) and 100% (TOC) after 0.5 h	~50	$[H_2O_2] = 30$ mM (1.88 g/L), 20 C, pH = 3, 0.5 h Irradiation: Low pressure, mercury vapor UVC lamp (40 W), flux = 1.6×10^{-5} einstein/L. s	Olmez-Hanci and Arslan-Alaton (2013)
UV + H_2O_2	97% (Phenol) 9% (TOC)	100	$[H_2O_2] = 0.5$ g/L, 20 C, pH = 3, 6 h Irradiation: Low pressure, mercury vapor UVC lamp (16 W)	Bali et al. (2003)
UV + Persulfate	100% (phenol) and 100% (TOC) after 0.5 h	~50	[Persulfate] = 20 mM, 20 C, pH 3, 0.5 h Irradiation: Low-pressure, mercury vapor UVC lamp (40 W), flux = 1.6×10^{-5} einstein/L. s	Olmez-Hanci and Arslan-Alaton (2013)

catalyst. However, the CWPO process normally needs a slightly higher concentration of H_2O_2 and higher temperature to show their best performance (Huang et al., 2015a; Huang et al., 2015b; Rubalcaba et al., 2007). The catalytic activity of both of these AOPs decreases noticeably at pH > 5 due to lower rate of radical formation at the surface of catalysts. Nevertheless, very efficient pollutant degradation can be still observed at pH 5–8 for the CWPO process, which is not the case for the Fenton process (Emami et al., 2010a; Emami et al., 2010b; Tehrani-Bagha et al., 2016).

- Unlike many of noncatalytic AOPs (O_3 + H_2O_2, UV + H_2O_2, etc.), the CWPO method needs a noticeable amount of catalyst in the solution for wastewater treatment. The recovery and reuse of the remaining catalyst is still a big challenge. A conventional filtration and regeneration of the catalyst can be quite difficult and expensive in a large scale treatment. For solving this issue, a catalytic fixed bed reactor (Ferentz et al., 2015; Nadejde et al., 2016; Ribeiro et al., 2016a, 2017; Yan et al., 2016) or magnetic catalyst particles (Tehrani-Bagha et al., 2016) can be used.
- In comparison with many conventional wastewater treatments (e.g., biological, coagulation/flocculation, etc.), AOPs are considered to be energy intensive and expensive. AOPs based on UV irradiation, sonication, and ozonation need special equipment and consume electricity (Chatzisymeon et al., 2013; Esplugas et al., 2002). In terms of degradation efficiency and energy requirements, the Fenton and CWPO processes are superior to UV based AOPs. However, the price of catalyst and H_2O_2 are relatively high, which affects the total cost of the treatment. Besides, the catalysts

TABLE 12.6 CWPO Efficiency in Degradation of C. I. Reactive Red 120 in Comparison to Other AOPs

AOPs	Removal (%)	$[dye]_{initial}$ (mg/L)	Conditions	References
CWPO	99% (Dye) 100% (COD)	50	Catalyst: Copper ferrite nanoparticles, [Cat] = 0.2 g/L, [H_2O_2] = 10 mM (0.034 g/L), pH = 3, 75 C, 0.5 h	Tehrani-Bagha et al. (2016)
Fenton	100% (Dye) 70% (COD)	100	Catalyst: $FeSO_4$, [Cat] = 0.046 g/L, [H_2O_2] = 3 mM, pH = 3, 25 C, 2 h	Emami et al. (2010a)
Fenton	98% (Dye) 100% (COD)	100	Catalyst: $FeSO_4$, [Cat] = 0.3 g/L, [H_2O_2] = 0.5 g/L, pH = 3, 40 C, 0.5 h	Meric et al. (2005)
O_3	100% (Dye) >50% (COD) >20% (TOC)	200	Gas flow 20 L/min, [O_3] = 12.8 mg/L, 25 C, 2.5 h	Zhang et al. (2002)
O_3 + NaOH	100% (Dye) 77% (COD)	200	[NaOH] = 1 g/L, gas flow 0.08 L/min, [O_3] = 55 g/m^3, 25 C, 0.5 h	Tehrani-Bagha (2011)
O_3 + UV	100% (Dye) 72% (COD)	200	Gas flow 0.08 L/min, [O_3] = 55 g/m^3, 25 C, 0.5 h Irradiation: UVC (9 W)	Tehrani-Bagha and Amini (2010)
O_3 + NaOH + UV	100% (Dye) 88% (COD)	200	Gas flow 0.08 L/min, [O_3] = 55 g/m^3, 25 C, 0.5 h Irradiation: UVC (9 W)	Tehrani-Bagha (2011)
UV + Fenton	98% (Dye) 85% (COD) 73% (TOC)	100	Catalyst: Fe_2SO_4, [Cat] = 0.25 mM, [H_2O_2] = 5 mM, pH = 3, 25 C, 0.25 h Irradiation: UV immersed lamp TN 15/35 (15 W)	Neamtu et al. (2003)
UV + H_2O_2	100% (Dye) 41% (COD)	50	[H_2O_2] = 20 mL/L, 25 C, 0.25 h Irradiation: Low pressure mercury lamp, 380 nm (9 W)	Ananthashankar and Ghaly (2013)

should be synthesized and prepared for the CWPO method in multiple synthetic steps from inorganic salts, which makes the catalyst much more expensive than ferrous sulfate and other inorganic salts. Many of the synthesized catalysts for the CWPO process also need a temperature as high as 75 C to show their best performance in wastewater treatment (Tehrani-Bagha et al., 2016; Huang et al., 2015b). Therefore, the operational cost of these methods should be roughly in this order: UV based AOPs > O_3 based AOPs > CWPO > Fenton (Riano et al., 2014; Chatzisymeon et al., 2013).

It should be noted that ozonation in alkaline conditions has been considered to be more cost effective than the Fenton process by some other researchers (Louhichi and Bensalah, 2014; Esplugas et al., 2002). The cost calculations on these references were only approximate and a more thorough economic analysis should be performed by considering the initial investment requirement, the price of chemicals in a large scale, maintenance, and labor costs.

Compared with homogeneous Fenton process, the CWPO process using solid catalysts shows numerous advantages (e.g., reusability of catalyst, enhanced catalytic activity in a wider range of pH, low rate of metal leaching and sludge production, etc.). However, some of the solid catalysts still have a high rate of metal leaching that will affect their activity and reusability. An additional issue for the CWPO process is the deactivation of the solid catalysts due to poisoning, fouling, thermal degradation, mechanical damage, and leaching by the reaction mixture (Bartholomew, 2001). Some of the used catalysts, especially the supported ones, have high adsorption capabilities that contribute to the final removal of pollutant from the solution. The adsorption of impurities at the surface of these catalyst can block their pores and eventually deactivate them (Pinho et al., 2015; Bartholomew, 2001).

Complete mineralization of organic pollutants in aqueous solution is achievable by the CWPO process. However, the preparation of a solid catalyst with high surface area and a large number of active sites will add to the total cost of wastewater treatment. Therefore, the utilization of this process depends on the development of solid catalysts with high efficiency, stability, reusability, and low price in a large scale treatment plant.

References

Ananthashankar, R., Ghaly, A., 2013. Effectivness of photocatalytic decolourization of reactive red 120 dye in textile effluent using UV/H_2O_2. Am. J. Environ. Sci. 9, 322–333.

Andas, J., Adam, F., Ab Rahman, I., 2014. Sol-gel derived mesoporous cobalt silica catalyst: Synthesis, characterization and its activity in the oxidation of phenol. Appl. Surf. Sci. 315, 154–162.

Arshadi, M., Abdolmaleki, M.K., Mousavinia, F., Khalafi-Nezhad, A., Firouzabadi, H., Gil, A., 2016. Degradation of methyl orange by heterogeneous Fenton-like oxidation on a nano-organometallic compound in the presence of multi-walled carbon nanotubes. Chem. Eng. Res. Design 112, 113–121.

Azmi, N.H.M., Ayodele, O.B., Vadivelu, V.M., Asif, M., Hameed, B.H., 2014. Fe-modified local clay as effective and reusable heterogeneous photo-Fenton catalyst for the decolorization of acid green 25. J. Taiwan Inst. Chem. Eng. 45, 1459–1467.

Babuponnusami, A., Muthukumar, K., 2012. Advanced oxidation of phenol: A comparison between Fenton, electro-Fenton, sono-electro-Fenton and photo-electro-Fenton processes. Chem. Eng. J. 183, 1–9.

Bali, U., Catalkaya, E.C., Sengul, F., 2003. Photochemical degradation and mineralization of phenol: a comparative study. J. Environ. Sci. Health Part A 38, 2259–2275.

Barrault, J., Bouchoule, C., Echachoui, K., Frini-Srasra, N., Trabelsi, M., Bergaya, F., 1998. Catalytic wet peroxide oxidation (CWPO) of phenol over mixed (Al-Cu)-pillared clays. Appl. Catal. B Environ. 15, 269–274.

Bartholomew, C.H., 2001. Mechanisms of catalyst deactivation. Appl. Catal. A Gen. 212, 17–60.

Basoglu, F.T., Balci, S., 2016. Catalytic properties and activity of copper and silver containing Al-pillared bentonite for CO oxidation. J. Mol. Struct. 1106, 382–389.

Botas, J.A., Melero, J.A., Martinez, F., Pariente, M.I., 2010. Assessment of Fe_2O_3/SiO_2 catalysts for the continuous treatment of phenol aqueous solutions in a fixed bed reactor. Catal. Today 149, 334–340.

Britto, J.M., De Oliveira, S.B., Rabelo, D., Rangel, M.D., 2008. Catalytic wet peroxide oxidation of phenol from industrial wastewater on activated carbon. Catal. Today 133, 582–587.

Buthiyappan, A., Aziz, A.R.A., Daud, W.M.A.W., 2016. Recent advances and prospects of catalytic advanced oxidation process in treating textile effluents. Rev. Chem. Eng. 32, 1–47.

Cardona, J.A., Ocampo, G.T., 2016. Degradation study of phenol on pillared clay catalyst. Rev. Cientifica 2, 265–279.

Castro, I.U., Fortuny, A., Stuber, F., Fabregat, A., Font, J., Bengoa, C., 2013. Heterogenization of copper catalyst for the oxidation of phenol, a common contaminant in industrial wastewater. Environ. Prog. Sustain. Energy 32, 269–278.

Catrinescu, C., Teodosiu, C., Macoveanu, M., Miehe-Brendle, J., Le Dred, R., 2003. Catalytic wet peroxide oxidation of phenol over Fe-exchanged pillared beidellite. Water Res. 37, 1154–1160.

Catrinescu, C., Arsene, D., Apopei, P., Teodosiu, C., 2012. Degradation of 4-chlorophenol from wastewater through heterogeneous Fenton and photo-Fenton process, catalyzed by Al-Fe PILC. Appl. Clay Sci. 58, 96–101.

Caudo, S., Centi, G., Genovese, C., Perathoner, S., 2006. Homogeneous versus heterogeneous catalytic reactions to eliminate organics from waste water using H_2O_2. Topics Catal. 40, 207–219.

Caudo, S., Centi, G., Genovese, C., Perathoner, S., 2007. Copper- and iron-pillared clay catalysts for the WHPCO of model and real wastewater streams from olive oil milling production. Appl. Catal. B Environ. 70, 437–446.

Caudo, S., Genovese, C., Perathoner, S., Centi, G., 2008. Copper-pillared clays (Cu-PILC) for agro-food wastewater purification with H_2O_2. Microp. Mesop. Mater. 107, 46–57.

Centi, G., Perathoner, S., 2008. Catal. by layered materials: a review. Microp. Mesop. Mater. 107, 3–15.

Chatzisymeon, E., Foteinis, S., Mantzavinos, D., Tsoutsos, T., 2013. Life cycle assessment of advanced oxidation processes for olive mill wastewater treatment. J. Clean. Prod. 54, 229–234.

Chen, J.X., Zhu, L.Z., 2006. Catalytic degradation of orange II by UV-Fenton with hydroxyl-Fe-pillared bentonite in water. Chemosphere 65, 1249–1255.

Chen, Z.X., Jin, X.Y., Chen, Z.L., Megharaj, M., Naidu, R., 2011. Removal of methyl orange from aqueous solution using bentonite-supported nanoscale zero-valent iron. J. Colloid Interface Sci. 363, 601–607.

Cheng, M.M., Song, W.J., Ma, W.H., Chen, C.C., Zhao, J.C., Lin, J., et al., 2008. Catalytic activity of iron species in layered clays for photodegradation of organic dyes under visible irradiation. Appl. Catal. B Environ. 77, 355–363.

Costa, R.C.C., De Fatima, M., Lelis, F., Oliveira, L.C.A., Fabris, J.D., Ardisson, J.D., et al., 2003. Remarkable effect of Co and Mn on the activity of $Fe_{3-x}MxO_4$ promoted oxidation of organic contaminants in aqueous medium with H_2O_2. Catal. Commun. 4, 525–529.

Costa, R.C.C., Lelis, M.F.F., Oliveira, L.C.A., Fabris, J.D., Ardisson, J.D., Rios, R.R.V.A., et al., 2006. Novel active heterogeneous Fenton system based on $Fe_{3-x}MxO_4$ (Fe, Co, Mn, Ni): The role of M^{2+} species on the reactivity towards H_2O_2 reactions. J. Hazard. Mater. 129, 171–178.

Costa, R.C.C., Moura, F.C.C., Ardisson, J.D., Fabris, J.D., Lago, R.M., 2008. Highly active heterogeneous Fenton-like systems based on $Fe-0/Fe_3O_4$ composites prepared by controlled reduction of iron oxides. Appl. Catal. B Environ. 83, 131–139.

Danish, M., Gu, X.G., Lu, S.G., Brusseau, M.L., Ahmad, A., Naqvi, M., et al., 2017. An efficient catalytic degradation of trichloroethene in a percarbonate system catalyzed by ultra-fine heterogeneous zeolite supported zero valent iron-nickel bimetallic composite. Appl. Catal. A Gen. 531, 177–186.

Di Luca, C., Massa, P., Fenoglio, R., Cabello, F.M., 2014. Improved Fe_2O_3/Al_2O_3 as heterogeneous Fenton catalysts for the oxidation of phenol solutions in a continuous reactor. J. Chem. Technol. Biotechnol. 89, 1121–1128.

Diya'uddeen, B.H., Aziz, A.R.A., Daud, W.M.A.W., 2012. On the limitation of fenton oxidation operational parameters: a review. Int. J. Chem. React. Eng. 10.

Djeffal, L., Abderrahmane, S., Benzina, M., Fourmentin, M., Siffert, S., Fourmentin, S., 2014. Efficient degradation of phenol using natural clay as heterogeneous Fenton-like catalyst. Environ. Sci. Pollut. Res. 21, 3331–3338.

Dominguez, C.M., Ocon, P., Quintanilla, A., Casas, J.A., Rodriguez, J.J., 2014a. Graphite and carbon black materials as catalysts for wet peroxide oxidation. Appl. Catal. B Environ. 144, 599–606.

Dominguez, C.M., Quintanilla, A., Casas, J.A., Rodriguez, J.J., 2014b. Kinetics of wet peroxide oxidation of phenol with a gold/activated carbon catalyst. Chem. Eng. J. 253, 486–492.

Emami, F., Tehrani-Bagha, A.R., Gharanjig, K., 2010a. Influence of operational parameters on the decolorization of an azo reactive dye (C.I. Reactive Red 120) by Fenton process. J. Color Sci. Technol. 4, 105–114.

Emami, F., Tehrani-Bagha, A.R., Gharanjig, K., Menger, F.M., 2010b. Kinetic study of the factors controlling Fenton-promoted destruction of a non-biodegradable dye. Desalination 257, 124–128.

Epolito, W.J., Yang, H., Bottomley, L.A., Pavlostathis, S.G., 2008. Kinetics of zero-valent iron reductive transformation of the anthraquinone dye Reactive Blue 4. J. Hazard. Mater. 160, 594–600.

Esplugas, S., Gimenez, J., Contreras, S., Pascual, E., Rodriguez, M., 2002. Comparison of different advanced oxidation processes for phenol degradation. Water Res. 36, 1034–1042.

Ferentz, M., Landau, M.V., Vidruk, R., Herskowitz, M., 2015. Fixed-bed catalytic wet peroxide oxidation of phenol with titania and Au/titania catalysts in dark. Catal. Today 241, 63–72.

Figueiredo, J.L., 2013. Functionalization of porous carbons for catalytic applications. J. Mater. Chem. A 1, 9351–9364.

Franus, M., Wdowin, M., Bandura, L., Franus, W., 2015. Removal of environ. pollutions using zeolites from fly ash: a review. Fresen. Environ. Bull. 24, 854–866.

Friak, M., Schindlmayr, A., Scheffler, M., 2007. Ab initio study of the half-metal to metal transition in strained magnetite. New J. Phys. 9. Available from: http://dx.doi.org/10.1088/1367-2630/9/1/005.

Galeano, L.A., Vicente, M.A., Gil, A., 2014. Catalytic degradation of organic pollutants in aqueous streams by mixed Al/M-pillared clays (M = Fe, Cu, Mn). Catal. Rev. Sci. Eng. 56, 239–287.

Gan, G.Q., Liu, J., Zhu, Z.X., Yang, Z.R., Zhang, C.L., Hou, X.H., 2017. A novel magnetic nanoscaled Fe_3O_4/CeO_2 composite prepared by oxidation-precipitation process and its application for degradation of orange G in aqueous solution as Fenton-like heterogeneous catalyst. Chemosphere 168, 254–263.

Garcia-Munoz, P., Pliego, G., Zazo, J.A., Bahamonde, A., Casas, J.A., 2016. Ilmenite ($FeTiO_3$) as low cost catalyst for advanced oxidation processes. J. Environ. Chem. Eng. 4, 542–548.

Garrido-Ramirez, E.G., Theng, B.K.G., Mora, M.L., 2010. Clays and oxide minerals as catalysts and nanocatalysts in Fenton-like reactions - a review. Appl. Clay Sci. 47, 182–192.

Ghosh, P., Kumar, C., Samanta, A.N., Ray, S., 2012. Comparison of a new immobilized Fe^{3+} catalyst with homogeneous Fe^{3+}-H_2O_2 system for degradation of 2,4-dinitrophenol. J. Chem. Technol. Biotechnol. 87, 914–923.

Gokulakrishnan, N., Pandurangan, A., Sinha, P.K., 2009. Catalytic wet peroxide oxidation technique for the removal of decontaminating agents ethylenediaminetetraacetic acid and oxalic acid from aqueous solution using efficient fenton type Fe-MCM-41 mesoporous materials. Ind. Eng. Chem. Res. 48, 1556–1561.

Guelou, E., Barrault, J., Fournier, J., Tatibouet, J.M., 2003. Active iron species in the catalytic wet peroxide oxidation of phenol over pillared clays containing iron. Appl. Catal. B Environ. 44, 1–8.

He, J., Yang, X.F., Men, B., Wang, D.S., 2016. Interfacial mechanisms of heterogeneous Fenton reactions catalyzed by iron-based materials: a review. J. Environ. Sci. 39, 97–109.

Hedstrom, A., 2001. Ion exchange of ammonium in zeolites: a literature review. J. Environ. Eng. ASCE 127, 673–681.

Herney-Ramirez, J., Madeira, L.M., 2010. Use of pillared clay-based catalysts for wastewater treatment through fenton-like processes. Pillared Clays Rel. Catal. 129–165.

Herney-Ramirez, J., Vicente, M.A., Madeira, L.M., 2010. Heterogeneous photo-Fenton oxidation with pillared clay-based catalysts for wastewater treatment: a review. Appl. Catal. B Environ. 98, 10–26.

Hou, L.W., Zhang, Q.H., Jerome, F., Duprez, D., Zhang, H., Royer, S., 2014. Shape-controlled nanostructured magnetite-type materials as highly efficient Fenton catalysts. Appl. Catal. B Environ. 144, 739–749.

Hu, X.B., Liu, B.Z., Deng, Y.H., Chen, H.Z., Luo, S., Sun, C., et al., 2011. Adsorption and heterogeneous Fenton degradation of 17 alpha-methyltestosterone on nano Fe_3O_4/MWCNTs in aqueous solution. Appl. Catal. B Environ. 107, 274–283.

Huang, K., Wang, J.J., Wu, D.F., Lin, S., 2015a. Copper hydroxyl sulfate as a heterogeneous catalyst for the catalytic wet peroxide oxidation of phenol. RSC Adv. 5, 8455–8462.

Huang, K., Xu, Y., Wang, L.G., Wu, D.F., 2015b. Heterogeneous catalytic wet peroxide oxidation of simulated phenol wastewater by copper metal-organic frameworks. RSC Adv. 5, 32795–32803.

Inchaurrondo, N., Cechini, J., Font, J., Haure, P., 2012a. Strategies for enhanced CWPO of phenol solutions. Appl. Catal. B Environ. 111, 641–648.

Inchaurrondo, N.S., Massa, P., Fenoglio, R., Font, J., Haure, P., 2012b. Efficient catalytic wet peroxide oxidation of phenol at moderate temperature using a high-load supported copper catalyst. Chem. Eng. J. 198, 426–434.

Inchaurrondo, N., Ramos, C.P., Zerjav, G., Font, J., Pintar, A., Haure, P., 2017. Modified diatomites for Fenton-like oxidation of phenol. Microp. Mesop. Mater. 239, 396–408.

Jiang, S.S., Zhang, H.P., Yan, Y., 2015. Catalytic wet peroxide oxidation of phenol wastewater over a novel Cu-ZSM-5 membrane catalyst. Catal. Commun. 71, 28–31.

Kim, S.C., Kim, D.S., Lee, G.S., Kang, J.K., Lee, D.K., Yang, Y.K., 2002. Catalytic wet oxidation of reactive dyes with H_2O_2 over mixed (Al-Cu) pillared clays. Impact of zeolites and other porous materials on the new technologies at the beginning of the new millennium, Pt. A and B. Stud. Surf. Sci. Catal. 142, 683–690.

Kumar, M., Ando, Y., 2010. Chemical vapor deposition of carbon nanotubes: a review on growth mechanism and mass production. J. NanoSci. Nanotechnol. 10, 3739–3758.

Kumar, D., Schumacher, K., Von Hohenesche, C.D.F., Grun, M., Unger, K.K., 2001. MCM-41, MCM-48 and related mesoporous adsorbents: their synthesis and characterisation. Colloids Surf. A 187, 109–116.

Kurian, M., Kunjachan, C., Sreevalsan, A., 2017. Catalytic degradation of chlorinated organic pollutants over CexFe1-xO2 (x: 0, 0.25, 0.5, 0.75, 1) nanocomposites at mild conditions. Chem. Eng. J. 308, 67–77.

Lee, Y.C., Chang, S.J., Choi, M.H., Jeon, T.J., Ryu, T., Huh, Y.S., 2013. Self-assembled graphene oxide with organo-building blocks of Fe-aminoclay for heterogeneous Fenton-like reaction at near-neutral pH: a batch experiment. Appl. Catal. B Environ. 142, 494–503.

Levec, J., Pintar, A., 1995. Catalytic-oxidation of aqueous-solutions of organics - an effective method for removal of toxic pollutants from waste-waters. Catal. Today 24, 51–58.

Li, W.B., Wan, D., Wang, G.H., Chen, K., Hu, Q., Lu, L.L., 2016. Heterogeneous Fenton degradation of Orange II by immobilization of Fe_3O_4 nanoparticles onto Al-Fe pillared bentonite. Korean J. Chem. Eng. 33, 1557–1564.

Lien, H.L., Zhang, W.X., 2001. Nanoscale iron particles for complete reduction of chlorinated ethenes. Colloids Surf. A 191, 97–105.

Litter, M.I., Slodowicz, M., 2017. An overview on heterogeneous Fenton and photoFenton reactions using zerovalent iron materials. J. Adv. Oxid. Technol. 20. https://doi.org/10.1515/jaots-2016-0164.

Liu, W., Qian, J., Wang, K., Xu, H., Jiang, D., Liu, Q., et al., 2013. Magnetically separable Fe_3O_4 nanoparticles-decorated reduced graphene oxide nanocomposite for catalytic wet hydrogen peroxide oxidation. J. Inorg. Organometallic Polym. Mater. 23, 907–916.

Liu, P.J., He, S.B., Wei, H.Z., Wang, J.H., Sun, C.L., 2015. Characterization of alpha-Fe_2O_3/gamma-Al_2O_3 catalysts for catalytic wet peroxide oxidation of m-cresol. Ind. Eng. Chem. Res. 54, 130–136.

Louhichi, B., Bensalah, N., 2014. Comparative study of the treatment of printing ink wastewater by conductive-diamond electrochemical oxidation, Fenton process, and ozonation. Sustain. Environ. Res. 24, 49–58.

Martinez, F., Pariente, I., Brebou, C., Molina, R., Melero, J.A., Bremner, D., et al., 2014. Chemical surface modified-activated carbon cloth for catalytic wet peroxide oxidation of phenol. J. Chem. Technol. Biotechnol. 89, 1182–1188.

Melero, J.A., Calleja, G., Martinez, F., Molina, R., Pariente, M.I., 2007. Nanocomposite Fe_2O_3/SBA-15: An efficient and stable catalyst for the catalytic wet peroxidation of phenolic aqueous solutions. Chem. Eng. J. 131, 245–256.

Meric, S., Selcuk, H., Gallo, M., Belgiorno, V., 2005. Decolourisation and detoxifying of Remazol Red dye and its mixture using Fenton's reagent. Desalination 173, 239–248.

Messele, S.A., Soares, O.S.G.P., Orfao, J.J.M., Bengoa, C., Stuber, F., Fortuny, A., et al., 2015. Effect of activated carbon surface chemistry on the activity of ZVI/AC catalysts for Fenton-like oxidation of phenol. Catal. Today 240, 73–79.

Messele, S.A., Bengoa, C., Stuber, F., Fortuny, A., Fabregat, A., Font, J., 2016. Catalytic wet peroxide oxidation of phenol using nanoscale zero-valent iron supported on activated carbon. Desalin. Water Treat 57, 5155–5164.

Misaelides, P., 2011. Application of natural zeolites in Environ. remediation: a short review. Microp. Mesop. Mater. 144, 15–18.

Mnasri-Ghnimi, S., Frini-Srasra, N., 2016. Effect of Al and Ce on Zr-pillared bentonite and their performance in catalytic oxidation of phenol. Russian J. Phys. Chem. A 90, 1766–1773.

Molina, C.B., Casas, J.A., Zazo, J.A., Rodriguez, J.J., 2006. A comparison of Al-Fe and Zr-Fe pillared clays for catalytic wet peroxide oxidation. Chem. Eng. J. 118, 29–35.

Munoz, M., De Pedro, Z.M., Casas, J.A., Rodriguez, J.J., 2015. Preparation of magnetite-based catalysts and their application in heterogeneous Fenton oxidation - a review. Appl. Catal. B Environ. 176, 249–265.

Munoz, M., Dominguez, P., De Pedro, Z.M., Casas, J.A., Rodriguez, J.J., 2017. Naturally-occurring iron minerals as inexpensive catalysts for CWPO. Appl. Catal. B Environ. 203, 166–173.

Nadejde, C., Neamtu, M., Hodoroaba, V.D., Schneider, R.J., Ababei, G., Panne, U., 2016. Hybrid iron-based core-shell magnetic catalysts for fast degradation of bisphenol A in aqueous systems. Chem. Eng. J. 302, 587–594.

Neamtu, M., Yediler, A., Siminiceanu, I., Kettrup, A., 2003. Oxidation of commercial reactive azo dye aqueous solutions by the photo-Fenton and Fenton-like processes. J. Photochem. Photobiol. A: Chem. 161, 87–93.

Nidheesh, P.V., 2015. Heterogeneous Fenton catalysts for the abatement of organic pollutants from aqueous solution: a review. RSC Adv. 5, 40552–40577.

Olmez-Hanci, T., Arslan-Alaton, I., 2013. Comparison of sulfate and hydroxyl radical based advanced oxidation of phenol. Chem. Eng. J. 224, 10–16.

Pan, J.X., Wang, C., Guo, S.P., Li, J.H., Yang, Z.Y., 2008. Cu supported over Al-pillared interlayer clays catalysts for direct hydroxylation of benzene to phenol. Catal. Commun. 9, 176–181.

Petranovskii, V., Chaves-Rivas, F., Espinoza, M.A.H., Pestryakov, A., Kolobova, E., 2016. Potential uses of natural zeolites for the development of new materials: Short review. Chem. Chem. Technol. Xxi Century 85. CCt 2016.

Philippe, R., Caussat, B., Falqui, A., Kihn, Y., Kalck, P., Bordere, S., et al., 2009. An original growth mode of MWCNTs on alumina supported iron catalysts. J. Catal. 263, 345–358.

Pignatello, J.J., 1992. Dark and photoassisted Fe^{3+}-catalyzed degradation of chlorophenoxy herbicides by hydrogen-peroxide. Environ. Sci. Technol. 26, 944–951.

Pinho, M.T., Gomes, H.T., Ribeiro, R.S., Faria, J.L., Silva, A.M.T., 2015. Carbon nanotubes as catalysts for catalytic wet peroxide oxidation of highly concentrated phenol solutions: towards process intensification. Appl. Catal. B Environ. 165, 706–714.

Platon, N., Siminiceanu, I., Nistor, I.D., Silion, M., Jinescu, C., Harrouna, M., et al., 2013. Catalytic wet oxidation of phenol with hydrogen peroxide over modified clay minerals. Rev. De Chim. 64, 1459–1464.

Ramirez, J.H., Costa, C.A., Madeira, L.M., Mata, G., Vicente, M.A., Rojas-Cervantes, M.L., et al., 2007. Fenton-like oxidation of Orange II solutions using heterogeneous catalysts based on saponite clay. Appl. Catal. B Environ. 71, 44–56.

Rey, A., Faraldos, M., Casas, J.A., Zazo, J.A., Bahamonde, A., Rodriguez, J.J., 2009. Catalytic wet peroxide oxidation of phenol over Fe/AC catalysts: Influence of iron precursor and activated carbon surface. Appl. Catal. B Environ. 86, 69–77.

Rey, A., Hungria, A.B., Duran-Valle, C.J., Faraldos, M., Bahamonde, A., Casas, J.A., et al., 2016. On the optimization of activated carbon-supported iron catalysts in catalytic wet peroxide oxidation process. Appl. Catal. B Environ. 181, 249–259.

Riano, B., Coca, M., Garcia-Gonzalez, M.C., 2014. Evaluation of Fenton method and ozone-based processes for colour and organic matter removal from biologically pre-treated swine manure. Chemosphere 117, 193–199.

Ribeiro, R.S., Fathy, N.A., Attia, A.A., Silva, A.M.T., Faria, J.L., Gomes, H.T., 2012. Activated carbon xerogels for the removal of the anionic azo dyes Orange II and Chromotrope 2R by adsorption and catalytic wet peroxide oxidation. Chem. Eng. J. 195, 112–121.

Ribeiro, R.S., Silva, A.M.T., Pastrana-Martinez, L.M., Figueiredo, J.L., Faria, J.L., Gomes, H.T., 2015. Graphene-based materials for the catalytic wet peroxide oxidation of highly concentrated 4-nitrophenol solutions. Catal. Today 249, 204–212.

Ribeiro, R.S., Frontistis, Z., Mantzavinos, D., Venieri, D., Antonopoulou, M., Konstantinou, I., et al., 2016a. Magnetic carbon xerogels for the catalytic wet peroxide oxidation of sulfamethoxazole in Environmentally relevant water matrices. Appl. Catal. B Environ. 199, 170–186.

Ribeiro, R.S., Silva, A.M.T., Figueiredo, J.L., Faria, J.L., Gomes, H.T., 2016b. Catalytic wet peroxide oxidation: a route towards the application of hybrid magnetic carbon nanocomposites for the degradation of organic pollutants. A review. Appl. Catal. B Environ. 187, 428–460.

Ribeiro, R.S., Silva, A.M.T., Tavares, P.B., Figueiredo, J.L., Faria, J.L., Gomes, H.T., 2017. Hybrid magnetic graphitic nanocomposites for catalytic wet peroxide oxidation applications. Catal. Today 280, 184–191.

Rodriguez-Reinoso, F., 1998. The role of carbon materials in heterogeneous Catal. Carbon 36, 159–175.

Ross, L.W., Chowdhury, A.K., 1977. Catalytic-oxidation of strong waste-waters. Mech. Eng. 99, 108.

Rubalcaba, A., Suarez-Ojeda, M.E., Stuber, F., Fortuny, A., Bengoa, C., Metcalfe, I., et al., 2007. Phenol wastewater remediation: advanced oxidation processes coupled to a biological treatment. Water Sci. Technol. 55, 221–227.

Sanabria, N., Alvarez, A., Molina, R., Moreno, S., 2008. Synthesis of pillared bentonite starting from the Al-Fe polymeric precursor in solid state, and its catalytic evaluation in the phenol oxidation reaction. Catal. Today 133, 530–533.

Serp, P., Figueiredo, J.L., 2009. Carbon Materials for Catal. John Wiley & Sons, Inc, Hoboken, New Jersey.

Shahbazi, A., Gonzalez-Olmos, R., Kopinke, F.D., Zarabadi-Poor, P., Georgi, A., 2014. Natural and synthetic zeolites in adsorption/oxidation processes to remove surfactant molecules from water. Sep. Purif. Technol. 127, 1–9.

Shu, H.Y., Chang, M.C., Huang, S.W., 2016. Decolorization and mineralization of azo dye acid blue 113 by the UV/oxone process and optimization of operating parameters. Desalin. Water Treat. 57, 7951–7962.

Soon, A.N., Hameed, B.H., 2011. Heterogeneous catalytic treatment of synthetic dyes in aqueous media using Fenton and photo-assisted Fenton process. Desalination 269, 1–16.

Stuber, F., Font, J., Eftaxias, A., Paradowska, M., Suarez-Ojeda, M.E., Fortuny, A., et al., 2005. Chemical wet oxidation for the abatement of refractory non-biodegradable organic wastewater pollutants. Process Saf. Environ. Prot. 83, 371–380.

Su, J., Lin, S., Chen, Z.L., Megharaj, M., Naidu, R., 2011. Dechlorination of p-chlorophenol from aqueous solution using bentonite supported Fe/Pd nanoparticles: synthesis, characterization and kinetics. Desalination 280, 167–173.

Tehrani-Bagha, A.R., 2011. Unpublished results.

Tehrani-Bagha, A.R., Amini, F.L., 2010. Decolorization of wastewater containing C. I. reactive red 120 by UV-enhanced ozonation. J. Color Sci. Technol. 4, 151–160.

Tehrani-Bagha, A.R., Gharagozlou, M., Emami, F., 2016. Catalytic wet peroxide oxidation of a reactive dye by magnetic copper ferrite nanoparticles. J. Environ. Chem. Eng. 4, 1530–1536.

Timofeeva, M.N., Mel'gunov, M.S., Kholdeeva, O.A., Malyshev, M.E., Shmakov, A.N., Fenelonov, V.B., 2007. Full phenol peroxide oxidation over Fe-MMM-2 catalysts with enhanced hydrothermal stability. Appl. Catal. B Environ. 75, 290–297.

Tomul, F., 2016. The effect of ultrasonic treatment on iron-chromium pillared bentonite synthesis and catalytic wet peroxide oxidation of phenol. Appl. Clay Sci. 120, 121–134.

Tomul, F., Basoglu, F.T., Canbay, H., 2016. Determination of adsorptive and catalytic properties of copper, silver and iron contain titanium-pillared bentonite for the removal bisphenol A from aqueous solution. Appl. Surf. Sci. 360, 579–593.

Variava, M.F., Church, T.L., Harris, A.T., 2012. Magnetically recoverable FexOy-MWNT Fenton's catalysts that show enhanced activity at neutral pH. Appl. Catal. B Environ. 123, 200–207.

Vogelpohl, A., Kim, S.M., 2004. Advanced oxidation processes (AOPs) in wastewater treatment. J. Ind. Eng. Chem. 10, 33–40.

Wan, Z., Wang, J.L., 2017. Degradation of sulfamethazine using Fe_3O_4-Mn_3O_4/reduced graphene oxide hybrid as Fenton-like catalyst. J. Hazard. Mater. 324, 653–664.

Wang, Y.M., Wei, H.Z., Liu, P.J., Yu, Y., Zhao, Y., Li, X.N., et al., 2015. Effect of structural defects on activated carbon catalysts in catalytic wet peroxide oxidation of m-cresol. Catal. Today 258, 120–131.

Wang, N.N., Zheng, T., Zhang, G.S., Wang, P., 2016. A review on Fenton-like processes for organic wastewater treatment. J. Environ. Chem. Eng. 4, 762–787.

Wang, X.Q., Dou, L.Y., Yang, L., Yu, J.Y., Ding, B., 2017. Hierarchical structured MnO_2@SiO_2 nanofibrous membranes with superb flexibility and enhanced catalytic performance. J. Hazard. Mater. 324, 203–212.

Weng, X.L., Lin, S., Zhong, Y.H., Chen, Z.L., 2013. Chitosan stabilized bimetallic Fe/Ni nanoparticles used to remove mixed contaminants-amoxicillin and Cd (II) from aqueous solutions. Chem. Eng. J. 229, 27–34.

Wu, Q., Hu, X., Yue, P.L., Zhao, X.S., Lu, G.Q., 2001. Copper/MCM-41 as catalyst for the wet oxidation of phenol. Appl. Catal. B Environ. 32, 151–156.

Xu, L.J., Wang, J.L., 2012. Fenton-like degradation of 2,4-dichlorophenol using Fe_3O_4 magnetic nanoparticles. Appl. Catal. B Environ. 123, 117–126.

Yan, Y., Jiang, S.S., Zhang, H.P., Zhang, X.Y., 2015. Preparation of novel Fe-ZSM-5 zeolite membrane catalysts for catalytic wet peroxide oxidation of phenol in a membrane reactor. Chem. Eng. J. 259, 243–251.

Yan, Y., Wu, X.W., Zhang, H.P., 2016. Catalytic wet peroxide oxidation of phenol over Fe_2O_3/MCM-41 in a fixed bed reactor. Sep. Purif. Technol. 171, 52–61.

Yaripour, F., Shariatinia, Z., Sahebdelfar, S., Irandoukht, A., 2015. Conventional hydrothermal synthesis of nanostructured H-ZSM-5 catalysts using various templates for light olefins production from methanol. J. Nat. Gas Sci. Eng. 22, 260–269.

Ye, W., Zhao, B.X., Ai, X.L., Wang, J., Liu, X.S., Guo, D., et al., 2016a. Catalytic wet peroxide oxidation of coking wastewater with Fe-Al pillared montmorillonite catalysts. Water Resour. Environ. 227–231.

Ye, W., Zhao, B.X., Gao, H., Huang, J.J., Zhang, X.L., 2016b. Preparation of highly efficient and stable Fe, Zn, Al-pillared montmorillonite as heterogeneous catalyst for catalytic wet peroxide oxidation of Orange II. J. Porous Mater. 23, 301–310.

Zazo, J.A., Pliego, G., Blasco, S., Casas, J.A., Rodriguez, J.J., 2011. Intensification of the Fenton process by increasing the temperature. Ind. Eng. Chem. Res. 50, 866–870.

Zhang, F.F., Yediler, A., Liang, X.M., Kettrup, A., 2002. Ozonation of the purified hydrolyzed azo dye Reactive Red 120 (CI). J. Environ. Sci. Health Part A 37, 707–713.

Zhao, Y.N., 2016. Review of the natural, modified, and synthetic zeolites for heavy metals removal from wastewater. Environ. Eng. Sci. 33, 443–454.

Zhao, X.T., Dong, Y.C., Cheng, B.W., Kang, W.M., 2013. Removal of textile dyes from aqueous solution by heterogeneous photo-fenton reaction using modified PAN nanofiber fe complex as catalyst. Int. J. Photoenergy. Available from: http://dx.doi.org/10.1155/2013/820165.

Zhou, S.W., Qian, Z.Y., Sun, T., Xu, J.G., Xia, C.H., 2011. Catalytic wet peroxide oxidation of phenol over Cu-Ni-Al hydrotalcite. Appl. Clay Sci. 53, 627–633.

Zhou, X.H., Chen, Z.X., Yan, D.H., Lu, H.B., 2012. Deposition of Fe-Ni nanoparticles on polyethyleneimine-decorated graphene oxide and application in catalytic dehydrogenation of ammonia borane. J. Mater. Chem. 22, 13506–13516.

Zhou, S.W., Zhang, C.B., Hu, X.F., Wang, Y.H., Xu, R., Xia, C.H., et al., 2014. Catalytic wet peroxide oxidation of 4-chlorophenol over Al-Fe-, Al-Cu-, and Al-Fe-Cu-pillared clays: Sensitivity, kinetics and mechanism. Appl. Clay Sci. 95, 275–283.

Index

Printed in the United States
By Bookmasters